Springer Series in
MATERIALS SCIENCE 32

Springer

Berlin
Heidelberg
New York
Barcelona
Hong Kong
London
Milan
Paris
Singapore
Tokyo

Hadis Morkoç

Nitride Semiconductors and Devices

With 271 Figures and 23 Tables

 Springer

Professor Dr. Hadis Morkoç
Department of Electrical Engineering
Virginia Commonwealth University
P.O. Box 843072
Richmond, VA 23284-3072, USA

Series Editors:

Prof. Alex Zunger
NREL
National Renewable Energy Laboratory
1617 Cole Boulevard
Golden Colorado 80401-3393, USA

Prof. R. M. Osgood, Jr.
Microelectronics Science Laboratory
Department of Electrical Engineering
Columbia University
Seeley W. Mudd Building
New York, NY 10027, USA

Prof. Robert Hull
University of Virginia
Dept. of Materials Science and Engineering
Thornton Hall
Charlottesville, VA 22903-2442, USA

Prof. H. Sakaki
Institute of Industrial Science
University of Tokyo
7-22-1 Roppongi, Minato-ku
Tokyo 106, Japan

ISSN 0933-033X

ISBN 3-540-64038-x Springer-Verlag Berlin Heidelberg New York

Library of Congress Cataloging-in-Publication Data

Morkoç, Hadis. Nitride semiconductors and devices / Hadis Morkoç. p. cm. – (Springer series in materials science; v. 32) Includes bibliographical references. ISBN 3-540-64038-X 1. Semiconductors–Materials, 2. Nitrides, 3. Gallium nitride, 4. Semiconductor lasers, 5. Light emitting diodes. I. Title, II. Series. TK7871.85.M594 1999 621.3815'2–dc21 99-14665

The use of general descriptive names, registered names, trademarks, etc. in this publication does not imply, even in the absence of a specific statement, that such names are exempt from the relevant protective laws and regulations and therefore free for general use.

Typesetting: PS™ Technical Word Processor
Cover concept: eStudio Calamar Steinen
Cover production: *design & production* GmbH, Heidelberg

SPIN: 10663800   57/3144/di - 5 4 3 2 1 0 – Printed on acid-free paper

Acknowledgement

A project of this magnitude can not realistically be born by the efforts of one individual only. The present case is no exception in that it is a testament to members of the nitride community working collectively in unison to make this product possible. The community was more than forthcoming in providing unpublished and to be published manuscripts and data. In a similar vein, I can not, in good conscience, pass up the opportunity to affectionately mention my early mentors at my elementary school in a small town in Eastern Turkey who graciously tolerated my unusual ways in good spirit and followed me through as I went up the grades, my mentors at Istanbul Technical University, the late Prof. M. Santur, and Profs. D. Leblebici, K. Sarioglu and T. Saya, my advisors at the graduate school (Profs. L.F. Castma, J. Ballentyne and G.J. Fry), my colleagues at Varian Associates, and my mentors later in life, Profs. A. Yariv and T. Tombrello of CalTech. I learned more than just technical knowledge from them all which proved pivotal as I tackled many of the secrets of nitride semiconductors.

I would like to thank Max Yoder of ONR who, in no uncertain terms and with unbounded enthusiasm, encouraged me to begin researching semiconductor nitrides and promoted the fledgling field. I can not overstate the role of Dr. G. Witt of AFOSR who commissioned several of us to visit the laboratories active in semiconductor nitrides for a report on wide bandgap semiconductors which caused to peak my interest. Persistent encouragement of the publishing editor, Dr. H. Lotsch and my colleagues Drs. C. Litton and S.N. Mohammad were very pivotal. The financial support by AFOSR (Dr. G.L. Witt) and ONR (Drs. C.E.C. Wood, R. Brandt, Y.S. Park, and Mr. M. Yoder) in the form of research grants and AFOSR in the form of an appointment at Wright Laboratory (Drs. A. Garscadden and P. Hemenger) where the requisite serenity and peace was provided.

I would like to thank my colleagues and past students, Profs. B. Segall, W.R.L. Lambrecht, E. Haller, S. N. Mohammad, H. X. Jiang, J.Y. Lin, J. J. Song, B. Gil, R. Cingolani, R. Rinaldi, D.J. Smith, J.M. Gibson, P. Ruterana, G. Nouet, R. Merlin, K.T. Tsen, A. Rockett and L. Allen; and Drs. S. Strite, M.E. Lin, A. Botchkarev, A. Salvador, G.A. Martin, F. Hamdani, H. Tang, M. Yeadon, G. Popovici, Z. Liliental-Weber, W. Shan, B. Goldenberg, S. Krishnankutty, W. Yang, B.N. Sverdlov, P. Chow, D.C. Reynolds, D.C. Look. C.W. Litton, S.C.Y. Tsen, S. Ruvimov, and Wook Kim, and Mr. Z. Fan.

Professor J. Pankove served as a source of inspiration throughout my relatively short foray into the world of semiconductor nitrides and wrote the preface. Profs. I. Akasaki and A. Amano provided a good deal of material regarding MOCVD growth and materials characterization. Dr. S. Nakamura provided enormous amount of data on LEDs and lasers, photographs, and devices. Dr. F. Ponce provided electron microscope images and preprints. Prof. P. Perlin provided many of his papers in progress on LEDs as well as precious photographs. Dr. C. Kisielowski provided the image of In distribution in InGaN wells, Drs. S. Ruvimov and Z. Liliental-Weber provided electron microscope images of ohmic contacts. Prof. J. Bernholc and Dr. C. van de Walle provided preprints of their papers dealing with defects and doping issues as well as critiquing the thoughts that are contained in this manuscript. Moreover, Dr. C. van de Walle read the chapter on defects and doping. Dr. A. Bykhovski and Prof. M.S. Shur provided their preprints on piezoelectric effect and critical thickness with Alexie Bykhovski working with me several days to iron out the issues related to piezoelectric effects. Prof. S. Porowski provided the photo of a bulk GaN. Prof. M. Osinski provided preprints of his work on reliability as well as figures, Profs. P. Zory, R. Cingolani, A. Nurmikko provided preprints of their papers as well as critiquing the laser chapter and offering suggestions, Prof. V. Fiorentini provided preprints of many of his papers on defects, band structure, piezoelectric effect, band discontinuities and dopant incorporation. Prof. A. Hangleitter provided preprints of many of his papers and offered suggestions regarding the nature of lasing in nitride semiconductors. Prof. J. Schneider provided his papers dealing with the rainbow color LEDs pumped by blue nitride LEDs. Dr. S. Lester provided figures regarding the commercial applications of LEDs. Dr. H. Jörgensen provided the Nichia white LED spectrum. Prof. S. Hersee provided his preprints on growth by MOCVD. Prof. B. Monemar, Prof. T. Tansley, and Dr. M. Suzuki graciously provided me with the preprints of their chapters which I found very useful. Profs. B.K. Meyer, A. Hoffmann, A.E. Yunovich, and J. Schetzina graciously provided many of their preprints and shared their ideas regarding optical processes in nitrides and LEDs. Dr. W. Shan and Prof. J.J. Song provided many of their papers and read the Optical Processes in Nitride Semiconductors. Dr. Y. Jogai read the Electronic Band Structure of Bulk and QW Nitrides. Prof. P. Eliseev read the Laser chapter with many comments. Dr. Ron Kaspi read the chapter on growth. My friend across the Atlantic, Prof. B. Gil, was most helpful and provided many of his papers and figures in progress, and read the Optical Processes in Nitride Semiconductors, and Electronic Band Structure of Bulk and QW Nitrides chapters. Prof. H.X. Jiang read the Optical Processes in Nitride Semiconductors chapter. Mr. G.Y. Xu was very instrumental in crunching numbers for LED and Laser chapters. Mr. L. Zhou crunched some numbers for the current-voltage characteristics. Drs. D.C. Look and D.C. Reynolds shared with me their data as they evolved on transport and optical processes in nitrides. Dr. C.W. Litton served as a valuable consultant, throughout the laborious period of putting this manuscript together, for bouncing ideas and reading LED and Laser chapters. Drs. D.C. Look, D.C. Reynolds and C.W. Litton also provided me with many opportunities around the lunch table to discuss thorny, but pertinent, GaN problems and offered their suggestions very willingly and freely. Dr. D.C. Look read the Transport, and Defects and Doping chapters. Dr. H. Tang was very helpful in the preparation of the transport chapter. Despite ample warnings to the contrary, Dr. J. Turner allowed himself to be recalled from the relative safety of retirement to read the entire book with laser precision. I really can not express my gratitude to him in words.

I am obliged to acknowledge the tremendous support and offer of assistance by my adopted brother Prof. R.J. Mattauch of the Virginia Commonwealth University. Last not least, I am truly indebted to my parents, Mustafa and Saadet Morkoç, who stretched beyond their means to send me away to school for my education, and instilled discipline and work ethics in me which is imperative in endeavors of this kind. Finally, I would like to thank my wife Dr. Amy Carol Morkoç and our son Erol Taner Morkoç who understood and accepted my absence from their lives during the preparation of this manuscript, not to mention the assistance of Erol in typing the expressions into the wee hours of many nights.

Richmond, Virginia *Hadis Morkoç*
November 1998

Contents

Foreword
A View of the Past, and a Look into the Future by a Pioneer
By Jacques I. Pankove

This forword will be a brief review of important developments in the early and recent history of gallium nitride, and also a perspective on the current and future evolution of this exciting field. Gallium nitride (GaN) was synthesized more than 50 years ago by *Johnson* et al. [1] in 1932, and also by *Juza* and *Hahn* [2] in 1938, who passed ammonia over hot gallium. This method produced small needles and platelets. The purpose of *Juza* and *Hahn* was to investiagte the crystal structure and lattice constant of GaN as part of a systematic study of many compounds. Two decades later, *Grimmeiss* et al. [3] in 1959 employed the same technique to produce small crystals of GaN for the purpose of measuring their photoluminescence spectra. Another decade later *Maruska* and *Tietjen* [4] in 1969 used a chloride transport vapor technique to make a large-area layer of GaN on sapphire. All of the GaN made at that time was very conducting n-type even *when* not deliberately doped. The donors were believed to be nitrogen vacancies. Later this model was questioned by *Seifert* et al. [5] in 1983, and oxygen was proposed as the donor. Oxygen with its 6 valence electrons on a N site (N has 5 valence electrons) would be a single donor.

The accomplishment of *Maruska* and *Tietjen* led to a flurry of activities in many laboratories, especially when Zn-doping produced the first blue LED in 1972 [6]. This was an M-i-n type of device (M: metal) (Fig. 1) that could emit either blue, green, yellow or red light depending on the Zn concentration in the light-emitting region (Fig. 2). Note that light is emitted only from the cathode. If the Zn concentration is different at the two edges of the Zn-compensated region, reversing the polarity of the bias (to make the opposite interface of the i-layer the cathode) could cause a change in color, i.e., the device could switch from blue to green or to yellow. *Maruska* et al. [8] in 1973 were the first to utilize Mg as a luminescent center in a M-i-n diode emitting violet light. Other discoveries made with the new single crystal were: anti-Stokes LEDs (2.8 eV photons emitted with only 1.5 V applied) in 1975 [9], negative electron affinity in 1974 [10], surface acoustic-wave generation in 1973 [11], and solar-blind UV photovoltaic detectors in 1971 [12]. But conducting p-type GaN was still too elusive to lauch a massive effort on devices. It was the perseverance of I. Akasaki that eventually paid off in the pursuit of conducting p-type GaN. Actually, this was an accidental discovery; *Akasaki* et al. [13] and *Amano* et al. [14]

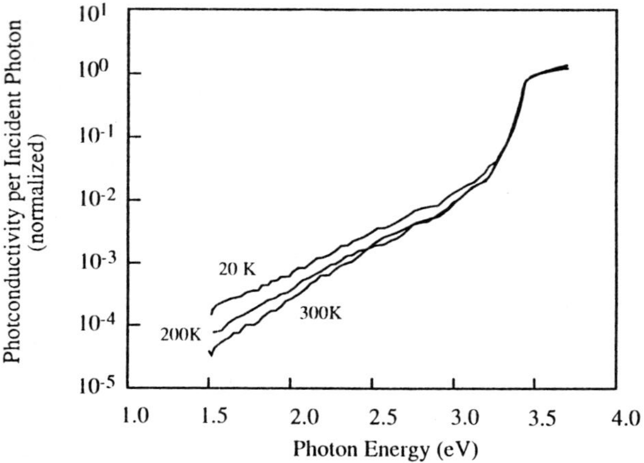

Fig. 1. Structure of GaN M-i-n LED

were observing the cathodoluminescence of GaN:Mg in an SEM (Scanning Electron Microscope) and noticed that the brightness increased with further raster scanning. A photoluminescence study of the sample before and after the Low-Energy Electron Beam Interaction (LEEBI) treatment showed that by the time luminescence was saturated, the lumiscence efficieny had increased by two orders of magnitude [14]. This surprising phenomenon of a beam-induced-type conversion was explained by *van Vechten* et al. [15] who proposed that the shallow acceptor level of Mg was compensated by a hydrogen atom complexing with the Mg acceptor (just as H complexes with acceptors in Si [16]). The energy of the electron beam releases the hydrogen atom from this complex that then becomes a shallow acceptor about 0.16 eV above the valence band [17]. The follow-up investigation of *Nakamura* et al. [18] revealed that annealing GaN:Mg above 750°C in N_2 or

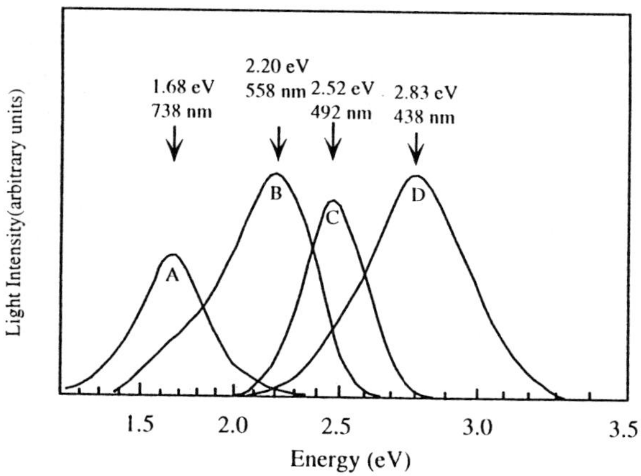

Fig. 2. Emission Spectra from GaN M-i-n LEDs, using Zn-doped i-layers

vacuum also converted the material to conducting p-type. However, annealing in NH_3 reintroduced atomic hydrogen and made GaN:Mg insulating again [18]. All this recent work lead to the brightest visible LEDs available today, especially in the blue part of the spectrum [19].

The GaN UV injection laser is the much hoped one for a solution to the high-information-density compact-disc vision since the areal packing is inversely proporsional to the square of the wavelength. A 360 nm laser would allow a factor of five increase in information storage in compact discs. Although optically-pumped stimulated-emission in GaN had been demonstrated by *Dingle* et al. [20] in 1971, the electrically pumped version had remained elusive until recently when *Nakamura* et al. [19] demonstrated the first successful injection laser in 1995. This was followed by many improvements [20]. Making a p-n junction is necessary but not sufficient. The material must be of exceptionally high quality. The Nichia bright LEDs have an enormous concentration ($10^{11}\,cm^{-2}$) of defects [21]. Furthermore, these LEDs have an extremely large concentration of impurities: the p-region has one hundred times more Mg than holes, the luminescence region is an alloy of InN and GaN with undoubtedly locally varying composition. The active region is loaded with Zn that is the luminescent center; and the donors in the undoped (and the doped) regions are either N vacancies or O atoms. Evidence for the high defect concentration appears in electron-microscopy studies of *Lester* et al. [21] and in the photoconductivity spectra of *Qui* et al. [22]. The photoconductivity spectrum reveals the presence of a high density of states in the bandgap of GaN. Unlike GaAs that has an abrupt Urbach edge, GaN exhibits an extensive absorption tail (Fig.3). In order to obtain a low-threshold injection laser with GaN, one must eliminate the losses due to absorption at the laser wavelength. Another factor affecting the threshold current is the width of the emitted spectrum because the thresold current is proportional to the spectral width. The narrowest emission spectrum is that due to exciton recombination. Hence, the most efficient lasing should be due to the stimulation of exciton recombination. However, excitons are destroyed by local fields in heavily-perturbed semiconductors. Quantum Wells

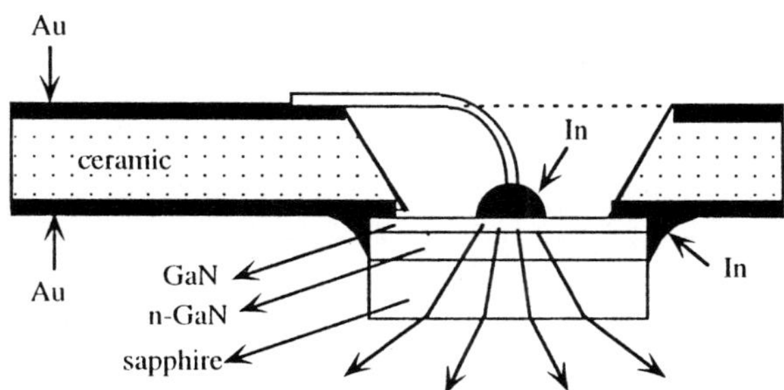

Fig. 3. Photoconductivity spectra of undoped GaN at various temperatures

(QWs) of InN between GaN barriers can be suitable for a wide range of lasing wavelengths tunable by the width of the wells. Thin strained-lattice quantum wells are desirable to overcome the lattice mismatch between InN and GaN. Doping the wells is undesirable because doping leads to level broadening and increased bandgap absorption.

Another point should be made about impurities in GaN. Column-II acceptors can either substitute for Ga to form single acceptors or they can substitute for N to form deeper triple acceptors. Since *a priori* the acceptor can occupy both sites, a column-II element should be a quadrupole acceptor with one shallow and 3 deep levels. Thus, Zn forms levels 9.57 eV, 0.88 eV, 1.2 eV, and 1.72 eV above the valence band. Mg forms the levels 0.16 eV, 0.25 eV, 0.36 eV, and 0.49 eV above the valence band. Note that also the 0.16 eV "shallow" level is many kT above the valence band, thus contributing few thermally activated carriers, the conductance of GaN:Mg is measurable because of the extremely low concentration of intrinsic carriers thermally generated across the bandgap of GaN. For a more recent historical review of GaN, see [23].

The Future of GaN

Even all the problems mentioned in the above section have not completely been solved, the understanding and the technology of GaN are progressing so rapidly that many practical applications will emerge soon.

Electronics

The large bandgap of GaN makes this material suitable for high-temperature applications. Field-Effect Transistors (FETs) with potential applications at high frequencies and/or high temperatures have already been demonstrated [24]. But these are mostly two-dimensional devices that control a sheet of current. Three-dimensional devices are not on the horizon. For example, the "Static Induction Transistor" (SIT) clusters many FETs transverse to the surface, thus increasing tremendously the power handling capability of FETs. Although the work of *Nishizawa* et al. [25] was in Si FETs, the same approach is bound to be applied to GaN FETs, too.

Bipolar Transistors

A breakthrough in transistor technology is the Heterojunction Bipolar Transistor (HBT) currently under development at Astrolux [26]. In this case the current flows transversely to the control electrode, unlike the smaller cross section parallel flow of the FET. Hence, larger powers can be controlled. Actually, HBT was originally conceived for high-temperature operation. The emitter is made of n-type GaN while the base consists of p-type SiC. When the emitter is forward-biased for maximum injection the conduction-band edges of GaN and SiC are nearly flattened, thus leaving a step in the valence-band edge. This step is equal to the difference between the bandgaps of the two materials. In the case of 6H-SiC that difference is 0.4 eV. A three times larger difference would be obtained if the base were made of 3C-SiC. The importance of this valence-band edge step is that it forms a barrier to the escape of holes from the base to the emitter, thus greatly increasing the electron injection efficiency. At high temperatures the injection efficiency decreases because the tail of the Boltzmann distribution of holes allows some holes to go over the step barrier into the valence band. But this happens only at very high temperatures. with the 6H-SiC a current gain of 100 has been achieved at 535°C (a gain comparable to that of commercial Si transistors at room temperature). At room temperature, a current gain dI_c/dI_b of 10^7 has been obtained, with I_c and I_b being the collector and base currents, respectively. Note that the use of SiC has two benefits: (i) SiC has an indirect bandgap greatly reducing the loss of electrons in the base by radiative recombination (as compared to using InGaN for the base), and (ii) SiC is a good heat conductor helping the dissipation of waste heat in the transistor. Evidently, a device that can operate at elevated temperatures should be capable of operating at high power levels. The first HBTs had operated at a current density of 2 kA/cm^2. Since, in principle, a SiC p-n junction (e.g., a collector) could operate at a voltage of 4 KV, as shown by *Kordina* et al. [27], a power-handling capability of 8 MW/cm^2 should be possible. At present the size of a HBT is limited by the micropipe concentration in the SiC (a limitation that will decrease with further development).

Photo-Transistors

Optical hole injection into the base of a HBT allows the remote control of transistors. Then, the device could be in a harsh environment and be controlled by room-temperature electronics via a UV laser beam or via a quartz optical fiber connected to a UV LED.

Thyristors

By adding another layer to the HBT, for example, a p-type layer to either the emitter or the collector, a p-n-p-n thyristor would be obtained. Thyristors are latching electronic switches that have two low-power consuming states, ON or OFF. They can be switched by a small forward bias across either of the outer junctions or by exceeding the brakdown voltage of the central reverse-biased junction. They can be switched optically, in which case they become photo-thyristors. If the inner two layers are made of a direct-bandgap III-nitride or are doped with luminescent centers, a thyristor can emit light when it is in the ON state.

Memory Devices

In computers the information is stored in various forms of memories such as magnetic polarization in tapes or as charges in semiconductors. The importance of semiconductor memories is growing rapidly because they can be integrated with low-power FETs to drive them or to interrogate them. Memory devices are based on charge storage. Long-term storage leads to nonvolatile or archival memories. In all semiconductors, the charge-carrier lifetime depends on the thermal activation of the charge out of its trapping level or the thermal activation of the opposite charge with which the stored charge will recombine. Hence, the wider the bandgap of the semiconductor, the less probable is the loss of stored charge by recombination. This is why GaN is a prime candidate for the application to memory devices. It is the most promising material for memory devices that can store for more than a century at elevated tempertures. Adding GaN bipolar transistors, these memory devices could feature the highest read-out efficiency.

Optoelectronics: Light–Emitting Diodes (LEDs)

As reported by *Nakamura* et al. [28] in 1994, ultra-bright LEDs made of GaN have already reached the commercial stage, appearing in outdoor displays and traffic lights. With the realization that LEDs are more efficient than incandescent tungsten-filament lamps, it is inevitable that LEDs are heading for the lamp replacement market. Such a replacement will lead to enormous fuel savings and a reduction of the green-house gases that are responsible for global warming. Furthermore, being solid-state devices, LEDs should provide a very long life of service. New types of lamps will emerge, for example, lamps emitting the three primary colors with individually adjusting intensities, thus permitting the choice of hue to fit one's mood. Already bright emitters of white light are available. UV LEDs will find applications in portable fluorescence instrumentation for field use in medicine and biology.

Injection Laser Diodes (LDs)

With GaAlN LDs it will become possible to extend UV emission to still shorter wavelengths and thus further increase the capacity of compact discs. With multiple-wavelength lasers focused at different depths, further increase in the information density will become possible. By now, a number of laboratories have indicated to produce edge-emitting LDs. Only Nichia reported CW operations for more that 10,000 hours at 20°C [29]. In the years to come longer CW operation at higher power levels wiil be achieved. Also surface-emitting LDs will resulte in a more collimated short-wavelength beam.

UV Detectors

As shown by *Xu* et al. [30] in 1997, a p-n junction in GaN makes a sensitive photodetector that does not respond to visible light; in other words, it is solar-blind. This property makes GaN of interest for use in space. Commercial applications will include flame detectors in furnaces.

X-Ray Detectors

GaN p-n junctions are sensitive to X rays because X rays are energetic enough to generate electron-hole pairs that are then separated by the internal field of a p-n junction to produce a photo-voltage, as shown by *Qiu* et al. [31]. The sensitivity is increased by adding a heavier element, namely indium. Hence, an i-region of InGaN inserted between n- and p-type GaN is recommended. Note that the number of electron-hole pairs generated by the X-ray photon increases with the photon energy. Hence, for a given photon flux, the signal scales with the photon energy. Such a detector has a potential for high-energy spectrometry applications.

Conducting Windows for Solar Cells

For space applications radiation-hard materials are needed. In the case of solar cells, a conducting layer of wide-bandgap n-type GaN can be utilized as a radiation-hard window material on top of a solar cell.

Optics

GaN having a polar bond between Ga and N has no inversion symmetry. Therefore, when GaN is subject to an alternating electric field, the induced polarization is not symmetric. Thus, the sine wave of an electromagnetic field is distorted, generating harmonics. This is how GaN is endowed with nonlinear optical properties. These nonlinear properteis will be used for second-harmonic generation. For example, converting GaAs 1.4 eV radiation into 2.8 eV blue light. Another potential application is a mixer that multiplies two RF, microwave or optical signals having different frequencies to generate sum and difference frequencies. These phenomena were discussed by *Miragliotta* and *Wickenden* [32].

Piezoelectronics

The same lack of inversion symmetry endows GaN with piezoelectric properties, namely, strain causes a displacement of charges and the appearance of an electric field. Conversely, an applied electric field causes atomic displacement such as a change in length. These changes occur at sonic speeds.

Surface Acoustic Wave Generation

Duffy et al. [11] showed in 1973 that interdigital electrodes on the surface of GaN can be used to lauch a Surface Acoustic Wave (SAW) that propagates along the surface to the edge of the sample where it is reflected, returning to the interdigitated electrodes that can detect the induced electric field. Such SAWs can be employed in sensors of chemical adsorbed on the surface.

Acousto-Optic Modulator

A propagating SAW periodically changes the density of GaN at the surface, alternating compressed and dilated regions. A change in density is accompanied by a change in refractive index. Therefore, a SAW can form an optical grating that will detect a transverse light beam. Changing the frequency of the SAW will change the periodicity of the grating and hence the angle of deflection. Acousto-Optic (AO) modulators are employed for raster scanning in projection TV. It is conceivable that some days the output of tricolor lasers with individual color modulation [33] will traverse a GaN AO modulator to form a full color image on a large screen.

Pyroelectricity

A piezo-related phenomenon is pyroelectricity: a change in electric field due to a temperature gradient. This is a large effect. The piezoelectric constant of GaN being 10^4 V/mK, close to the value for PZT and $BaTiO_3$ according to *Bykhovski* et al. [34].

Negative Electron Affinity

GaN doped with acceptors can have its Fermi level located near midgap, i.e., about 1.7 eV below the conduction-band edge. If the surface is coated by a low work function (f) element such as cesium (f = 1.5eV), the surface work function becomes −0.2 eV. This is called Negative Electron Affinity (NEA), a condition that allows electrons in the conduction band to escape into the vacuum, as demonstrated by *Pankove* and *Schade* [10] in 1974. This is an opportunity to make other useful devices.

References

1   W.C. Johnson, J.B. Parsons, M.C. Crew: J. Phys. Chem. **234**, 2651 (193)

2   R. Juza, E. Hahn: Z. Anorg. Allgem. Chem. **234**, 282 (1938)

3   H. Grimmeiss, H. Koelmans: Z. Naturfg. **14a**, 264 (1959)

4   H.P. Maruska, J.J. Tietjen: Appl. Phys. Lett. **15**, 367 (1969)

5   W. Seifert, R. Franzheld, E. Buttler, H. Sobotta, V. Riede: Crystal Res. Technol. **18**, 383 (1983)

6   J.I. Pankove, E.A. Miller, J.E. Berkeyheiser: J. Luminescence **5**, 84 (1972)

7   J.I. Pankove: J. Luminescence **7**, 114 (1973)

8   H.P. Maraska, D.A. Stevenson, J.I. Pankove: Appl. Phys. Lett. **22**, 303 (1973)

9   J.I. Pankove: Phys. Rev. Lett. **34**, 809 (1975); and IEEE Trans. EDUC-**22**, 721 (1975)

10   J.I. Pankove, H.E.P. Schade: Appl. Phys. Lett. **25**, 53 (1974)

11   M.T. Duffy, C.C. Wang, G.D. O'Clock, S.H. McFarlane III, P.J. Zanzucchi: J. Elect. Mater. **2**, 359 (1973)

12   J.I. Pankove, R. McIntyre: Unpubl. results (1971)

13   I. Akasaki, T. Kozowa, K. Hiramatsu, N. Sawak, K. Ikeda, Y. Ishii: J. Luminescence **40, 41**, 121 (1988)

14   H. Amano, M. Kito, K. Hiramatsu, I. Akasaki: Jpn. J. Appl. Phys. **28**, L2112 (1989)

15   J.A. van Vechten, J.D. Zook, R.D. Horning: Jpn. J. Appl. Phys. **31**, 3662 (1962)

16   J.I. Pankove, P.J. Zansucchi, C.W. Magee: Appl. Phys. Lett. **46**, 421 (1985)

17   I. Akasaki, H. Amano, M. Kito, K. Hiramatsu: J. Luminescence **48**, 666 (1990)

18   S. Nakamura, T. Mukai, M. Senoh: Jpn. J. Appl. Phys. **30**, L1998 (1991)

19   S. Nakamura, M. Senoh, N. Iwasa, S. Nagahama: Jpn. J. Appl. Phys. **34**, 1797 (1995)

20   S. Nakamura, N. Iwasa, M. Senoh, T. Mukai: Jpn. J. Appl. Phys. **31**, 1258 (1997)

21   S.D. Lester, F.A. Ponce, M.G. Crawford, D.A. Steigenvald: Appl. Phys. Lett. **66**, 1249 (1995)

22   C.H. Giu, C. Hoggatt, W. Melton, M.W. Leksono, J.I. Pankove: Appl. Phys. Lett. **66**, 2712 (1995)

23   S. Strite, H. Morkoç: J. Vac. Sci. Technol. B **10**, 1237 (1992)

24   A. Özgür, W. Kim, Z. Fan, S.N. Mohammad, A. Botchkarev, A. Salvador, B. Sverdlov, H. Morkoç: Electron. Lett. **31**, 1389 (1995)

25   J. Nishizawa, T. Terasaki, J. Shibata: IEEE Trans. ED-**22**, 185 (1975)

26   J.I. Pankove, M. Leksono, S.S. Chang, C. Walker, B. Van Zeghbroeck: MRS Internet J. NSR-1, Article 39 (1997)

27   O. Kordina, J.P. Bergman, K. Bergman: Appl. Phys. Lett. **67**, 1561 (1995)

28   S. Nakamura, T. Mukai, M. Senoh: Appl. Phys. Lett. **64**, 1687-1689 (1994)

29   S. Nakamura: In *Optical Properties of GaN and Related Materials*, ed. by S.J. Pearton (Gordon & Breach, Amsterdam 1998)

30   G.Y. Xu, A. Salvador, W. Kim, Z. Fan, C. Lu, H. Tang, H. Morkoç, G. Smith, M. Estes, B. Goldberg, W. Yank, S. Krishnankutty: Appl. Phys. Lett. **71**, 2154 (1997)

31   C.H. Qiu, J.I. Pakove, C. Rossington: In *Nitrides Based X-Ray Detectors*, ed. by J.I. Pankove, T.D. Moustakas. Semiconductors and Semimetals, Vol.50B (Academic, San Siedo 1998) pp. 1-24

32   J. Miragliotta, D.K. Wickenden: Nonlinear Optical Properties of GaN, Vol.50B, ed. by J.I. Pankove, T.D. Moustakas (Academic, San Siego 1998) p.25

33   H.K.V. Lotsch, F. Schröter: Das Laser-Farbfernsehen. Laser **2**, 37-39 (December 1970)

34   A. Bykhovski, W.W. Kaminski, M.S. Shur, Q.C. Chen, M.A. Khan: Appl. Phys. Lett. **69**, 3254 (1996)

1. Introduction

For the last three decades or so, the III-V semiconductor nitride system has been viewed as highly promising for semiconductor device applications for blue and ultraviolet wavelengths in much the same manner that its highly successful As-based and P-based counterparts have been exploited for infrared, red, and yellow wavelengths. The wurtzite polytypes of GaN, AlN and InN form a continuous alloy system whose direct bandgaps range from 1.9 eV for InN, to 3.4 eV for GaN, and to 6.2 eV for AlN. For all practical purposes, the III-V nitrides could potentially be fabricated into optical devices which are active at wavelengths ranging from the green well into the ultraviolet. By using nitride emitters as pumps, all primary and mixed colors can be obtained, too.

Nitride-based green and blue **Light Emitting Diodes** (LEDs) with efficiency, brightness, and longevity that are well in excess of those required for outdoor applications are already commercially available. In addition, blue LEDs are employed to pump integrated inorganic and organic media to produce colors reaching red, on the one hand, and white light on the other. In addition to the traditional displays, these LEDs have applications in traffic lights, moving signs, indicator lights, spot lights, and possibly light sources for accelerated photosynthesis, and medicine for diagnosis and treatment. Potentially, further improvement in LEDs would expand the applications to lighting with large energy savings as LEDs are more efficient than incandescent bulbs.

Injection lasers operating at short wavelengths have been coveted for years for digital data reading and storage applications. Semiconductor nitrides have the bandgaps required to reach these short wavelengths. Unlike display and lighting applications, digital information storage and reading requires coherent light sources, namely lasers. The output of these coherent light sources can be focused into a diffraction-limited spot, paving the way for an optical system in which bits of information can be recorded and read with ease and uncommon accuracy. As the wavelength of the light gets shorter, the focal diameter becomes smaller. Using a two-layer scheme in what has been named as the **Digital Versatile Disk** (DVD), the storage density is predicted to go up from today's 1 Gb to about 40 Gb per compact disk when blue lasers are used. For consumer applications, Continuous Wave (CW) operating lifetimes on the order of 10,000 hours at 60°C are required. Currently, the extrapolated room-temperature lifetimes of InGaN/ GaN/AlGaN injection lasers exceed that at low power levels.

Semiconductor nitrides are also prime candidates for UltraViolet (UV) photodetectors which have many potential applications in such areas as solar astronomy, missile-plume detection and combustion-process monitoring. If the bandgap of the absorbing layer is chosen such that the detector is sensitive to wavelengths of only below about 280 nm, very-low-level UV radiation can be detected since the much-reduced solar-background radiation noise is low [1.1-3]. As precursory devices, AlGaN/ GaN and GaN/ GaN p-n-junction photovoltaic detectors with very promising results have recently been reported. The AlGaN/GaN photodiodes had a maximum zero-bias responsivity of 0.12 A/W at 364 nm, which decreases by more than three orders of magnitude for wavelengths longer than 390 nm. A reverse bias of of -10 V raises the responsivity to 0.15 A/W corresponding to an internal quantum efficiency of about 60%, without any significant increase in noise. The root-mean-square (rms) noise current at 400 nm is less than 1.0 pA, limited by the measurement set-up. It corresponds to a noise-equivalent power of lees than 8.3 pW. The response time is as low as 9 ns [1.4].

Large carrier velocities, large total carrier concentrations available in two-dimensional systems, large band discontinuities in AlGaN/GaN hetero-structure systems, and a tolerance to high junction temperatures form the basis for nitride high-power electronic devices with applications requiring low cost, small space and power efficiency [1.5]. Though in an embryonic state, Modulation Doped Field Effect Transistors (MODFETs) built in a GaN environment have shown remarkable performances. Extrinsic trans-conductances of about 280 mS/mm and current levels of 1.1 A/mm have been obtained in devices with 1.5 μm gate lengths, rivaling even their august GaAs counterparts with the added advantage of much larger break-down fields and a large thermal conductivity. Output microwave power levels of about 1.5 W/mm in inverted modulation-doped field-effect transistors with 2 μm gate lengths have been obtained at 4 GHz; which is remarkable. More remarkable even is the CW total power level of about 4 W with a 0.45 μm gate length at 10 GHz.

Contrary to popular perception, GaN has been around for quite a while. Actually, *Johnson* et al. [1.6] described the synthesis of GaN in 1932 by converting metallic Ga in a NH_3 stream into GaN via the reaction $2Ga + 2NH_3 \rightarrow 2GaN + 3H_2$. What launched GaN to the realm of semiconductors worthy of investigation with device applications in mind was its synthesis in the epitaxial form on sapphire substrates by hydride vapor-phase epitaxy by *Maruska* and *Tienjen* [1.7]. It is appropriate to put some personality to these names with a story told by J. Pankove and P. Maruska. With no prior work in the area, P. Maruska, then a young Ph.D. at the RCA Laboratories, developed a hydride vapor-phase depition method for GaN while J. Pankove was on sabbatical from RCA at UC Berkeley.

The wider-bandgap cousin of GaN, AlN was first synthesized by *Tiede* et al. [1.8] from metallic Al through the conversion of metallic Al in an NH_3 stream via the reaction $2Al + 2NH_3 \rightarrow 2AlN + 3H_2$. It is not just the pure AlN that is of importance for devices, it is the ternary AlGaN used together with GaN and/or the smaller-bandgap ternary, InGaN. AlN exhibits many useful mechanical and electronic properties. For example, hardness, high thermal conductivity, resistance to high temperatures and caustic chemicals are among its attractions. It is likely that AlN maybe the isomorphic substrate that is available for the semiconductor nitride activity. Its wide bandgap is also the reason for AlN to be touted as an insulating material in semiconductor-device applications. The piezoelectric properties make AlN suitable for surface-acoustic-wave device applications [1.9]. However, the majority of interest in this semiconductor stems from its ability to form alloys with GaN producing AlGaN and allowing the fabrication of AlGaN/GaN-based electronic and optical devices, the latter of which could be active from the green wavelengths well into the ultraviolet. The smaller-bandgap cousin of GaN, InN, was first synthesized by *Juza* et al. [1.10] from $InF_6(NH_4)_3$. InN has not received the experimental attention given to GaN and AlN, in part due to difficulties associated with its deposition caused by the large nitrogen partial pressure over In. Although the impetus for good InGaN thin layers embedded between GaN or AlGaN layers can not be overstated for optical devices, the growth of high-quality InN and the enumeration of its fundamental physical properties remains for the present a purely scientific enterprise.

The evolution of semiconductor nitrides has been interesting and followed a bumpy road. Following the epitaxial synthesis of GaN on sapphire substrates, a flurry of activity sprung up to exploit this semiconductor for light emitters. In these early stages, achievement of p-type GaN, however, proved insurmountable. In retrospect, this failure resulted from compensating-defect generation during attempts to dope it p-type which is endemic to all large bandgap semiconductors, and form high background electron concentrations that are attributed to nitrogen vacancies. Nonetheless, *Pankove* and his colleagues at then RCA Laboratories managed to produce LEDs in an MIS structure in the early 1970's [1.11]. Electrons injected from the metal in a forward bias into the i region formed by Zn doping, drop to Zn centers and give off blue light. Zn centers are not only efficient radiative recombination centers, but also instrumental in decreasing the photon energy from the UV to blue. Combinations of the large forward bias needed (9 V), and the yellow emission centers in GaN which spoiled the color saturation and make the emission appear greenish, eventually proved insurmountable even though Matsushita marketed these device for a while.

While interest in the USA was waning I. Akasaki in Japan became fascinated with the topic. In fact, J. Pankove often mentions of the momentous train ride from Kyoto to Tokyo, around the time of MIS LED development, that he took with I. Akasaki who was then affiliated with Matsushita. While the overall effort in the US and elsewhere in Japan diminished, the interest of I. Akasaki remained if not intensified. He moved to Nagoya University to devote more time to nitrides. There, with A. Amano and K. Hiramatsu, he went on to develop OrganoMetallic Vapor-Phase Epitaxy (OMVPE) for the nitride growth and low-temperature AlN buffer layers in 1985. Introduction of OMVPE paved the way to deposit AlGaN and InGaN as well. Though the low-temperature buffer layer may appear to be a laboratory practice, it lowered the large background electron concentrations from the previous $10^{19} \div 10^{20}$ cm^{-3} levels to about 10^{17} cm^{-3}, which not only improved the crystal quality but set the stage for p-type doping [1.12]. Finally, the group of I. Akasaki, now at Meijo University because of the mandatory retirement in public institutions, achieved p-type GaN doped with Mg in 1989 [1.13]. This was no ordinary doping in that the sample had to be subjected to **Low-Energy Electron Beam Irradiation** (LEEBI) to activate the Mg acceptors. Hydrogen present in the deposition system which passivates the dopant, thereby preventing self-compensating defect formation, is driven out of the sample during ion-beam treatment. Several years later, S. Nakamura and colleagues at Nichia Chemicals were able to increase the p-doping using the LEEBI treatment. Soon afterwards, they found out that thermal annealing, which is much simpler, can activate Mg as well [1.14]. They also went on to improve the quality of InGaN which formed the basis for the LEDs their company commercialized in November of 1993 [1.15]. In January 1996, *Nakamura* and colleagues reported pulsed room-temperature operation of an AlGaN/GaN/InGaN injection laser [1.16]. At the time of this writing, the room-temperature CW lifetime of impoved lasers is extrapolated to be about 10,000 hours.

The evolution of the developments in the arena of light emitters was complied by I. Akasaki in a set of rather descriptive figures. Figure 1.1 illustrates the evolution of the efficiency of LEDs where the inception of the low-temperature buffer layer and attainment of p-type doping are indicated. Figure 1.2 displays the evolution of stimulated emission and later laser operation. Data points prior to the 1996 result are all by optical excitation with earlier samples having been cooled to low temperatures. As time went on and improvements ensued, stimulated emission became possible at room temperature. With the introduction of waveguides, the power densities needed to attain stimulated emission eventually dropped to as low as 20 kW/cm^2.

In this monograph, we begin our odyssey through the wonders of **semiconductor nitrides** with a review of their general properties, i.e., mechani-

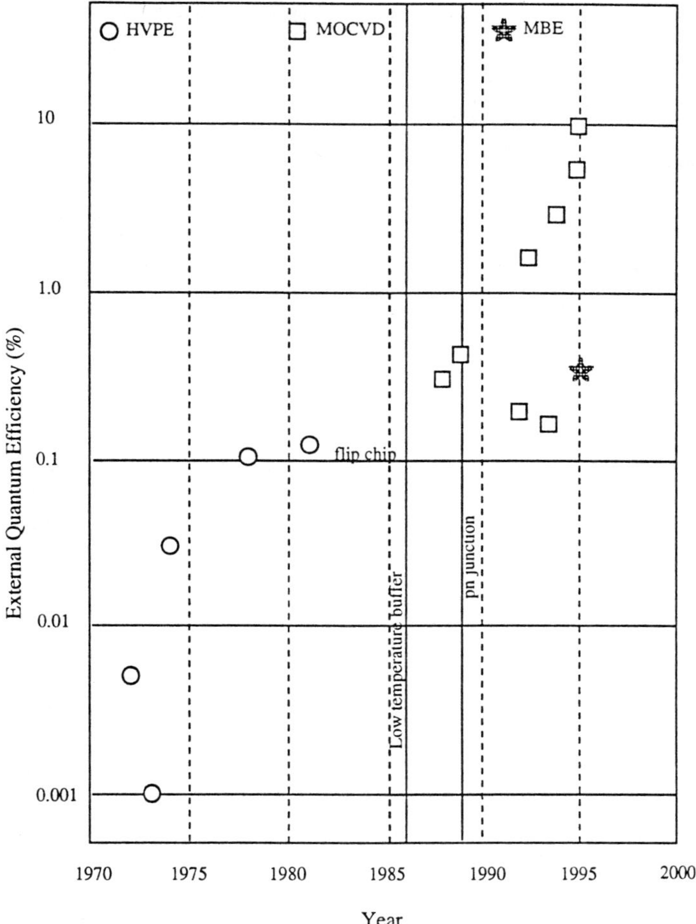

Fig. 1.1. Evolution of the LED efficiency with a low-temperature AlN buffer layer and the attainment of p-type GaN. Courtesy of I. Akasaki, Meijo University

cal, electrical and optical (Chap. 2). This is followed by a treatise of the electronic band structure of both bulk and quantum wells (Chap. 3). Then, the technology that made the recent commercial success a reality possible, the deposition of semiconductor nitrides, is discussed in Chap. 4. What follows the deposition is the treatment of, to a large extent, the point defects and doping since p-type doping is intertwined with defects (Chap. 5). To ease into the fabrication issues, a treatment of ohmic contacts, which are exacerbated by a large bandgap and large carrier effective masses, particularly for p-type GaN, commences in Chap. 6. With the help of ohmic contacts, the treatment is shifted to the determination of impurity and carrier concentrations (Chap. 7). This is followed by discussions of carrier transport in nitrides including alloys (Chap. 8). However, the treatment is limited

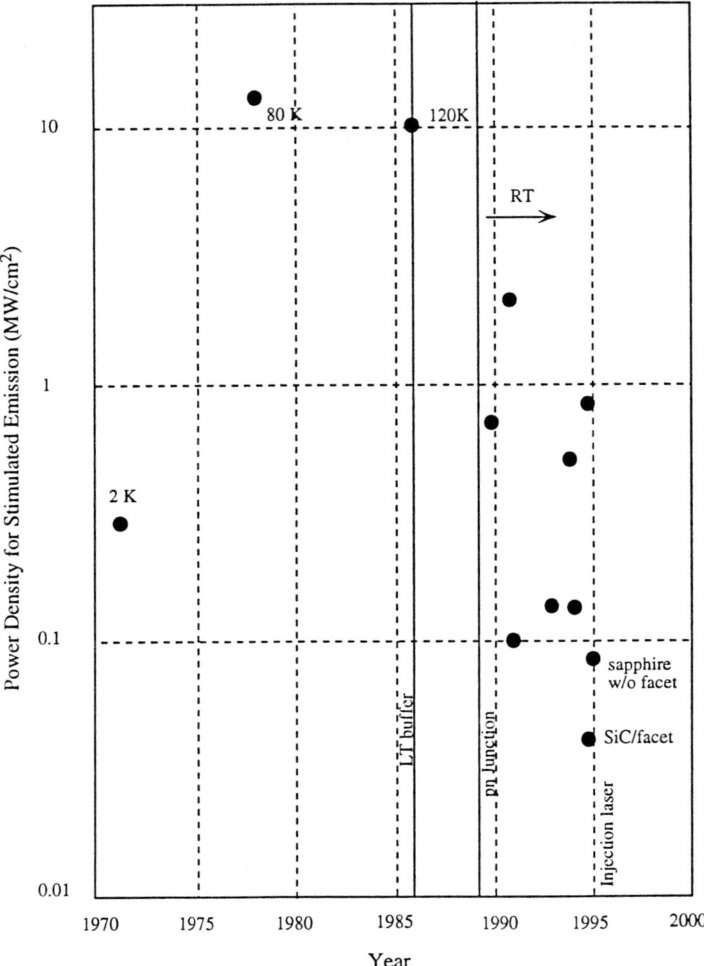

Fig. 1.2. Evolution of the threshold power for stimulated emission in GaN for optically pumped GaN bulk, and in later years, GaN waveguides. Courtesy of I. Akasaki, Meijo University

to n-type semiconductors since what is known on p-type is not yet at a level to be included here. As a preamble to optical emitters, the current conduction mechanism in nearly ideal p-n junctions is treated from a fundamental point of view (Chap. 9). This also includes issues related to band discontinuities in heterojunctions giving way to experimentally observed current-voltage characteristics.

As a gateway to optical emitters, optical processes in semiconductors are discussed in some detail with the identification of intrinsic and extrinsic transitions (Chap. 10). In Chap. 11, light emitting diodes are treated, with a good deal of attention paid to fundamentals, experimental results, and unique features of the current-voltage characteristics coupled with the nature of radiative carrier recombination which gives rise to light emission. Last but not least, an in-depth treatment of injection lasers is given in Chap. 12. Beginning with a succinct description of the driving forces for short-wavelength injection lasers, a concise treatment of double-heterojunction laser operation is provided. This is followed by the treatment of an essential part of lasers, the waveguide, with useful curves to ameliorate the design of nitride-based injection lasers. The present understanding of the lasing mechanism, together with the experimentally observed GaN-based injection-laser properties, is then treated in a comprehensive manner.

2. General Properties of Nitrides

Semiconductor nitrides have excellent material, optical and electrical properties. The respective parameters are informative in determining the utility and applicability of these materials to devices, as will be evident below and throught out the present monograph.

2.1 Crystal Structure of Nitrides

There are three common crystal structures shared by the group-III nitrides: the wurtzite, zincblende, and rocksalt structures. Under ambient conditions, the thermodynamically stable structure is wurtzite for bulk AlN, GaN, and InN. The zincblende structure for GaN and InN has been stabilized by epitaxial growth of thin films on $\{011\}$ crystal planes of cubic substrates such as Si, MgO, and GaAs. In these cases, the intrinsic tendency to form the **wurtzite** (Wz) structure is overcome by topological compatibility. The rocksalt, or NaCl, structure can be induced in AlN, GaN, and InN under very high pressures. The wurtzite structure has a hexagonal unit cell and thus two lattice constants, c and a. It contains 6 atoms of each type. The space grouping for the wurtzite structure is $P6_3mc$ (C_{6v}^4). The wurtzite structure consists of two interpenetrating **Hexagonal Close Packed** (HCP) sublattices, each with one type of atoms, offset along the c axis by 5/8 of the cell height (5c/8).

The zincblende structure has a cubic unit cell, containing four group-III elements and four nitrogen elements. The space grouping for the **zincblende** (ZB) structure is $F\bar{4}3m$ (T_d^2). The position of the atoms within the unit cell is identical to the diamond crystal structure. Both structures consist of two interpenetrating face-centered cubic sublattices, offset by one quarter of the distance along a body diagonal. Each atom in the structure may be viewed as positioned at the center of a tetrahedron, with its four nearest neighbors defining the four corners of the tetrahedron.

The zincblende and wurtzite structures are similar. In both cases, each group-III atom is coordinated by four nitrogen atoms. Conversely, each nitrogen atom is coordinated by four group-III atoms. The main difference be-

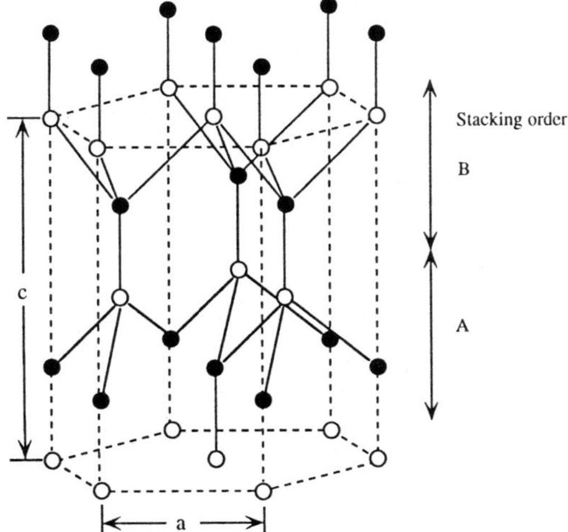

Fig. 2.1. A stick and ball diagram of a hexagonal structure

tween these two structures lies in the stacking sequence of closest packed diatomic planes. For the Wz structure, the **stacking sequence** of the (0001) plane is ABABAB in the ⟨0001⟩ direction. For the zincblende structure, the stacking sequence of the (111) plane is ABCABC in the ⟨111⟩ direction.

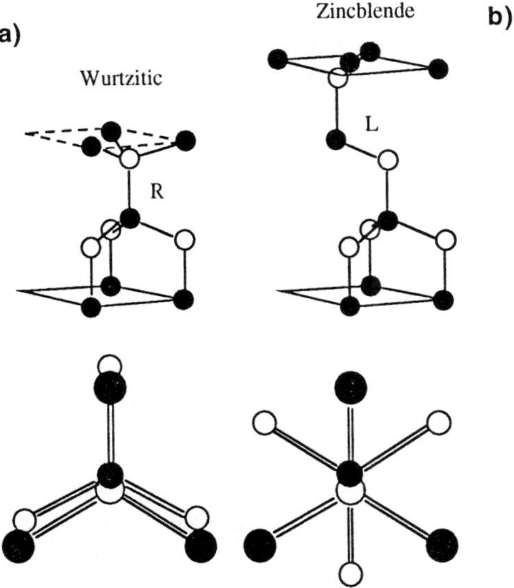

Fig. 2.2. Stick and ball stacking model of crystals with wurtzite (**a**) and zincblende (**b**) orientations. Note the rotation in the zincblende case

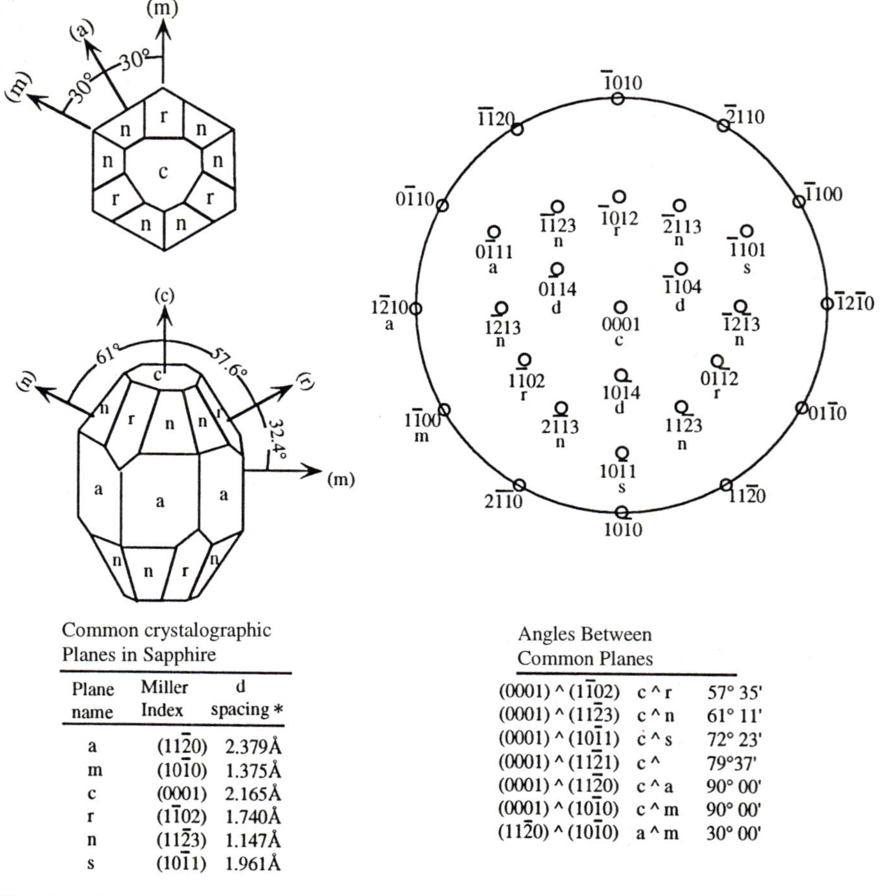

Common crystalographic Planes in Sapphire	Angles Between Common Planes

Plane name	Miller Index	d spacing *
a	(11$\bar{2}$0)	2.379Å
m	(10$\bar{1}$0)	1.375Å
c	(0001)	2.165Å
r	(1$\bar{1}$02)	1.740Å
n	(11$\bar{2}$3)	1.147Å
s	(10$\bar{1}$1)	1.961Å

Plane 1 ^ Plane 2	Label	Angle
(0001) ^ (1$\bar{1}$02)	c ^ r	57° 35'
(0001) ^ (11$\bar{2}$3)	c ^ n	61° 11'
(0001) ^ (10$\bar{1}$1)	c ^ s	72° 23'
(0001) ^ (11$\bar{2}$1)	c ^	79°37'
(0001) ^ (11$\bar{2}$0)	c ^ a	90° 00'
(0001) ^ (10$\bar{1}$0)	c ^ m	90° 00'
(11$\bar{2}$0) ^ (10$\bar{1}$0)	a ^ m	30° 00'

Fig. 2.3. Labeling of planes in hexagonal symmetry (* for sapphire)

A stick and ball representation of a wurtzite structure is depicted in Fig.2.1. The wurtzite and zincblende structures differ only in the bond angle of the second-nearest neighbor (Fig.2.2). As clearly shown, the stacking order of the wurtzite along the [0001] c direction is ABAB, meaning a mirror image but no in-plane rotation with the bond angles. In the zincblende structure along the [111] direction there is a 60° rotation which causes a stacking order of ABCABC. The point with regard to rotation is very well illustrated in Fig.2.2b. The nomenclature for various commonly used planes of hexagonal semiconductors in two- and three-dimensional versions is presented in Figs.2.3 and 4.

The wurtzite polytypes of GaN, AlN and InN form a continuous alloy system whose direct bandgaps range from 1.9 eV for InN, to 3.4 eV for GaN, to 6.2 eV for AlN. Thus, the III-V nitrides could potentially be fabricated into optical devices which are active at wavelengths ranging from the

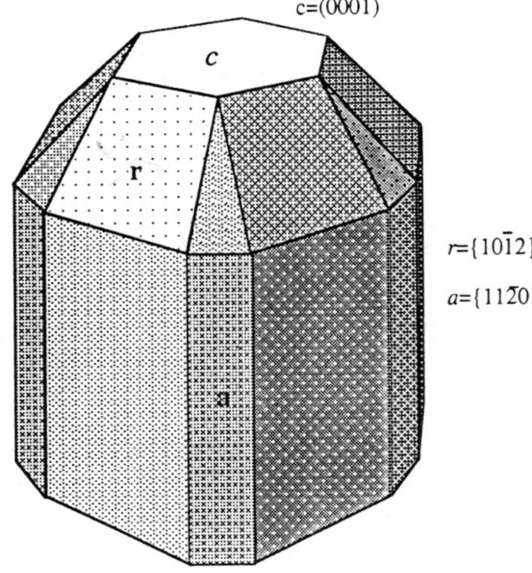

Fig. 2.4. Three-dimensional schematic view of commonly used planes and directions in a crystal with hexagonal symmetry

red well into the ultraviolet. The bandgaps of nitrides, substrates commonly used for nitrides, and other conventional semiconductors are shown in Fig. 2.5 with respect to their lattice constants.

2.2 Gallium Nitride

Although GaN has been studied far more extensively than other group-III nitrides, there is still much need for further investigations to even approach the level of understanding of technologically important materials such as Si and GaAs. GaN growth often suffers from large-background n-type carrier concentrations due to native defects and possibly impurities. The lack of commercially available native substrates exacerbates the situation. These,

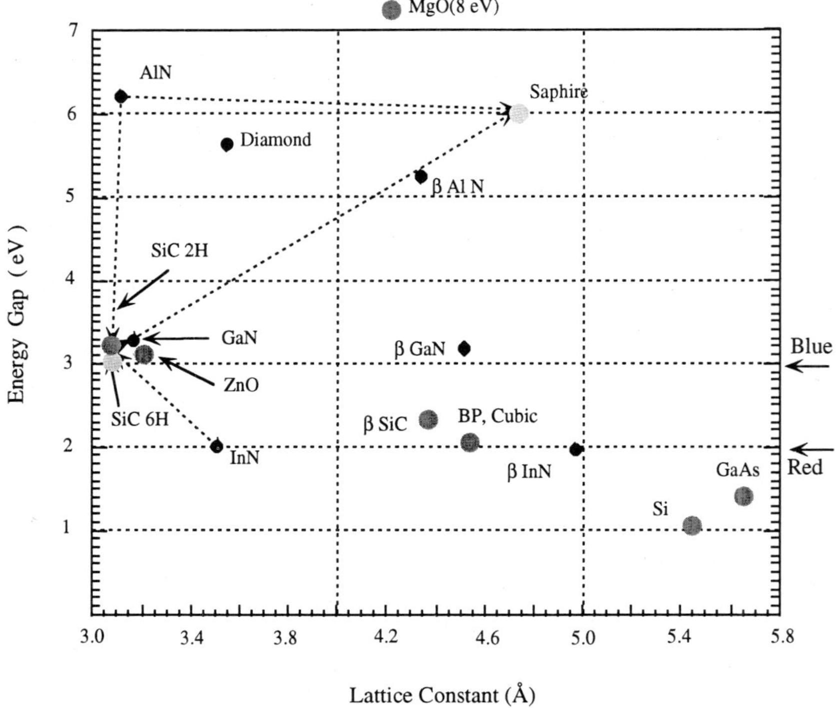

Fig. 2.5. The bandgaps of nitrides, substrates commonly used for nitrides and other conventional semiconductors vs. their lattice constants

together with the difficulties in obtaining p-type doping, and the arcane fabrication processes, catalyzed the early bottlenecks stymieing progress. Information available in the literature regarding many of the physical properties of GaN is still in the process of evolution, and naturally controversial. This is in part a consequence of measurements being made on samples of widely varying quality. When possible, the spurious determination have been disregarded. In others, measurements are too few to yield a consensus, in which case the available data are simply reported.

Recent burgeoning interest has led to an improvement in the crystal growth and processing technology, and allowed many difficulties encountered earlier to be overcome. Consequently, a number of laboratories consistently obtained high-quality GaN with room-temperature background electron concentrations as low as $5 \cdot 10^{16}$ cm^{-3}. The successful development of schemes leading to p-type GaN has led to the demonstration of excellent p-n-junction LEDs in the violet, blue, green and yellow bands of the visible spectrum with brightness suitable for outdoor displays, CW lasers, power Modulation Doped Field Effect Transistor (MODFETs), and UV detectors. However, much work remains to be done in the determination of the fundamental physical properties of GaN. What follows reports on the state of knowledge regarding the physical properties of GaN.

2.2.1 Chemical Properties of GaN

Since *Johnson* et al. [2.1] first synthesized GaN in 1932, a large body of information has repeatedly indicated that GaN is an exceedingly stable compound and exhibits significant hardness. It is this chemical stability at elevated temperatures combined with its hardness that has made GaN an attractive material for protective coatings. Moreover, owing to its wide energy bandgap, it is also an excellent candidate for device operation in high-temperature and caustic environments. As a matter of fact, the majority of GaN researchers are currently interested in semiconductor-device applications. While the thermal stability of GaN allows freedom of high-temperature processing, the chemical stability of GaN presents a technological challenge. Conventional wet-etching techniques used in semiconductor processing has not been very successful for GaN device fabrication. For example, *Maruska* and *Tietjen* [2.2] reported that GaN is insoluble in H_2O, acids, or bases at room temperature, but does dissolve in hot alkali solutions at a very slow rate. *Pankove* [2.3] noted that GaN reacts with NaOH forming a GaOH layer on the surface and prohibiting wet etching of GaN. To circumvent this difficulty, he developed an electrolytic etching technique for GaN. Low-quality GaN has been etched at reasonably high rates in NaOH [2.4, 5], H_2SO_4 [2.6], and H_3PO_4 [2.7-9]. Although these etches are useful for identifying defects and estimating their densities in GaN films, they are not very successful for the fabrication of devices. Well established chemical etching processes are required for the device-technology development. Promising possibilities are the various dry-etching processes under development, and reviewed by *Mohammad* et al. [2.10].

Various spectroscopic techniques, such as Auger electron spectroscopy, X-ray photoemission spectroscopy, and electron energy loss spectroscopy have been very useful for the study of the surface chemistry of GaN. Employing these techniques, the thermal stability and dissociation of GaN have also been examined. As indicated earlier, the materials characteristics depend, to a large extent, on the growth conditions. Because of this, the materials studied in various laboratories were obtained from various sources and had different characteristics. This led to inconsistent results from different laboratories. While some experimental studies of the stability of GaN conducted at high temperatures suggested that significant weight losses occur at temperatures as low as 750°C, others contradicted this proposal, and suggested that no significant weight loss should occur even at a temperature of 1000°C. *Morimoto* [2.11], and *Furtado* and *Jacob* [2.12] observed that

GaN is less stable in an HCl or H_2 atmosphere than in N_2. Some controversy exists regarding the process steps that dominate the decomposition of GaN. Using mass spectroscopy, *Gordienko* et al. [2.13] noted that $(GaN)_2$ dimers are the primary components of decomposition. Others [2.14, 15] found only N_2^+ and Ga^+ to be the primary components in the vapor over GaN. Based on measurements of the apparent vapor pressure, *Munir* and *Searcy* [2.16] calculated the **heat of sublimation** of GaN to be 72.4 ± 0.5 kcal/mol. *Logan* and *Thurmond* [2.17, 18] determined the equilibrium N_2 pressure of GaN as a function of temperature. In view of the fact that a number of laboratories are now producing high-quality material, this is an area which requires a revisit. The thermal stability of GaN, particularly in conjunction with metal contacts, is critical in applications which necessitate high-power or high-temperature operation and high-temperature processing. Detailed references to the aforementioned investigation can be found in the monograph by *Mohammad* and *Morkoç* [2.19].

2.2.2 Thermal and Mechanical Properties of GaN

In the hexagonal wurtzite structure GaN has a **molecular weight** of 83.728 gm/mol. At room temperature, the lattice parameters of this semiconductor are $a_0 = 3.1892 \pm 0.0009$ Å and $c_0 = 5.1850 \pm 0.0005$ Å. However, for the zincblende polytype the calculated lattice constant based on the measured Ga-N bond distance in wurtzite GaN is $a = 4.503$ Å. The measured value for this polytype varies between 4.49 and 4.55 Å, indicating that the calculated result lies within acceptable limits [2.20]. A high-pressure phase transition from the wurtzite to the rocksalt structure has been predicted and observed experimentally. The transition point is 50 GPa and the experimental lattice constant in the rocksalt phase is $a_0 = 4.22$ Å. This is slightly different from the theoretical result of $a_0 = 4.09$ Å obtained from first-principles non-local pseudopotential calculations. Table 2A.1 compiles the known properties of wurtzite GaN.

The lattice constant of early versions of GaN is a function of growth conditions, impurity concentrations, and film stoichiometry [2.21]. This can be attributed to a high concentration of interstitial and bulk extended defects. A case in point is that the lattice constants of GaN grown with higher growth rates was found to be larger. When doped heavily with Zn [2.22] and Mg [2.23] a lattice expansion occurs because, at high concentrations, the group-II element begins to occupy the lattice sites of the much smaller nitrogen atom.

Measurements made over the temperature range[1] of $300 \div 900$ K indicates the mean coefficient of thermal expansion of GaN in the c plane to be

---

[1] The symbol $\div$ is used throughout the text as a shorthand for "from – to" or "between".

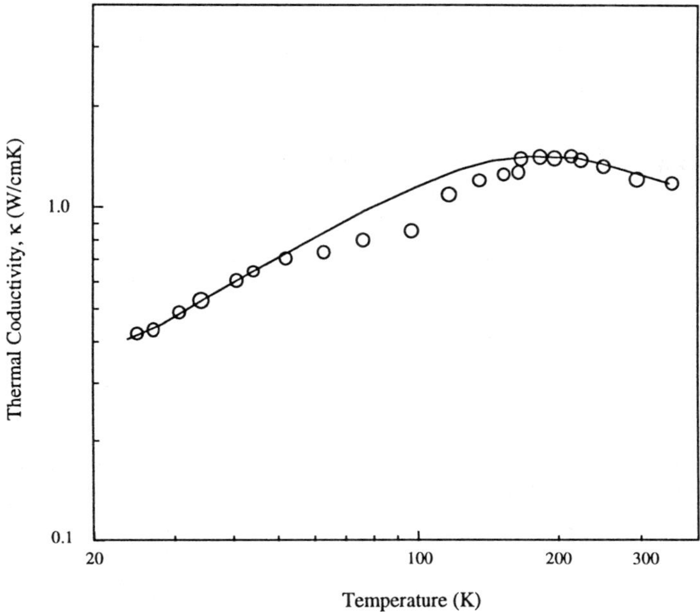

Fig. 2.6. Thermal conductivity along the c axis of GaN as a function of temperature. After [2.25]

$\Delta a/a = 5.59 \cdot 10^{-6}$ K^{-1}. Similarly, measurements over the temperature ranges of $300 \div 700$ K and $700 \div 900$ K, respectively, indicate the mean coefficient of thermal expansion in the c direction to be $\Delta c/c = 3.17 \cdot 10^{-6}$ K^{-1} and $7.75 \cdot 10^{-6}$ K^{-1}, respectively [2.2]. *Sheleg* and *Savastenko* [2.24] reported a **thermal expansion coefficient** near 600 K, perpendicular and parallel to the c-axis, of $(4.52 \pm 0.5) \cdot 10^{-6}$ K^{-1} and $(5.25 \pm 0.05) \cdot 10^{-6}$ K^{-1}, respectively.

Sichel and *Pankove* [2.25] measured the thermal conductivity of GaN for the temperature range of $25 \div 360$ K (Fig. 2.6). The room-temperature value of the **thermal conductivity** $\kappa = 1.3$ W/cm·K is a little smaller than the predicted value of 1.7 W/cm·K [2.26]. The **Debye temperature** θ_D of GaN at 0 K was calculated to be $\theta_D \approx 600$ K [2.26]. Other thermal properties of Wz-GaN have been studied by a number of researchers. The **specific heat** of Wz-GaN at constant pressure (C_p) is given by [2.27]

$$C_p(T) = 9.1 + (2.15 \cdot 10^{-3} T) \quad [cal \cdot mol^{-1} K^{-1}] . \tag{2.1}$$

Thermodynamic properties of Wz-GaN have been reported by *Elwell* and *Elwell* [2.26]. From the reaction

$$Ga(s) + \tfrac{1}{2}N_2(g) = GaN(s) , \tag{2.2}$$

the **heat of formation** of Wz-GaN was calculated to be $\Delta H_{300K} = -26.4$ kcal/mol [2.28] or the standard heat of formation $\Delta H = -37.7$ kcal/mol. The equilibrium vapor pressure of N_2 over solid GaN has been found to be 10 MPa at 1368 K, and 1 GPa at 1803 K [2.29]. A thorough description of the GaN phase diagram including the equilibrium vapor pressure of N_2 over GaN as well as AlN and InN was presented by *Porowski* and *Grzegory* [2.30].

Experimental investigations of the **elastic constants** of Wz-GaN have been carried out by *Savastenko* and *Sheleg* [2.31] by X-ray diffraction in powdered GaN crystals. The estimates from the elastic coefficient $-2C_{13}/C_{33}$ [2.31] and the measured [2.32] values of the Poisson ratio $\nu\langle 0001 \rangle = (\Delta a/a_0)/(\Delta c/c_0)$ of 0.372 and 0.378 are in good agreement. The estimates were from the elastic coefficients while the experiments were due to X-ray diffraction performed on GaN layers on sapphire substrates. However, there is a wide spread in the reported values of elastic stiffness coefficients, as alluded to below.

Chetverikova et al. [2.33] measured the **Young's modulus** and **Poisson's ratio** of their GaN films. From the elastic stiffness coefficients, Young's modulus $E_{\langle 0001 \rangle}$ is estimated to be 150 GPa [2.20, 31]. *Sherwin* and *Drummond* [2.34] predicted the elastic properties of ZB-GaN on grounds of values for those Wz-GaN samples reported by *Savastenko* and *Sheleg* [2.31]. The elastic stiffness coefficients, and Young's and the bulk moduli are compiled in Table 2A.2. Considering the wide spread in the reported data, more reliable figures are underscored by bold letters. The bulk modulus of Wz-GaN was calculated from first principles [2.35], and the first-principle orthogonalized Linear Combination of Atomic Orbitals (LCAO) method [2.36], leading to the values of 195 and 203 GPa, respectively. Another estimate for B is 190 GPa [2.37]. These figures compare well with the value of 194.6 GPa estimated from the elastic stiffness coefficient [2.38] and a measured value for 245 GPa [2.39].

Pertaining to mechanical properties, a group-theoretical approach predicted the optical phonon modes of Wz-GaN, namely one A_1 mode, one E_1 mode, two E_2 modes, and two B_1 modes. Their wave numbers, measured by Raman scattering of E_1-TO, A_1-TO, E_1-LO, A_1-LO and two E_2 phonons, are listed in Table 2A.3 [2.20, 40, 41] along with those obtained from first-principles pseudopotential calculations [2.35]. Also listed are TO and LO optical phonon wave numbers of ZB-GaN [2.42].

2.3 Aluminum Nitride

AlN exhibits many useful mechanical and electronic properties. For example, hardness, high thermal conductivity, resistance to high temperature and caustic chemicals combined with, in non-crystalline form, a reasonable thermal match to Si and GaAs, make AlN an attractive material for electronic packaging applications. The wide bandgap is also the reason for AlN to be touted as an insulating material in semiconductor device applications. Piezoelectric properties make AlN suitable for surface-acoustic-wave device applications [2.43]. However, the majority of interest in this semiconductor stems from its ability to form alloys with GaN producing AlGaN and allowing the fabrication of AlGaN/GaN based electronic and optical devices, the latter of which could be active from the green wavelengths well into the ultraviolet.

AlN is not a particularly easy material to investigate because of the high reactivity of aluminum with oxygen in the growth vessel. Early measurements indicated that oxygen contaminated material can lead to errors in the energy bandgap and, depending on the extent of contamination, in the lattice constant. Only recently achieved contamination-free deposition environments coupled with advanced procedures have allowed researchers to consistently grow improved-quality AlN. Consequently, many of the physical properties of AlN have been reliably measured and bulk AlN synthesized.

When crystallized in the hexagonal wurtzite structure, the AlN crystal has a molar mass of 20.495 gm/mol. The cubic form is hard to obtain and thus will be ignored. The space-group symmetry is C_{6v}^4 (P6$_3$mc) and the point-group symmetry is C_{6v} (6mm) The c/a ratio for this is $(8/3)1/2 = 1.333$. Reported lattice parameters range from 3.110 to 3.113 Å for a, and from 4.978 to 4.982 Å for c. The c/a ratio thus varies between 1.000 and 1.602. The deviation of the c/a ratio from that of the ideal wurtzite crystal is probably due to lattice stability and ionicity. Whereas for the metastable zincblende polytype AlN has a value of a = 4.38 Å, the rocksalt structure has a value of a = 4.043 ÷ 4.045 Å at room temperature. Table 2A.4 summarizes the observed structural and optical properties of AlN.

2.3.1 Thermal and Chemical Properties of AlN

AlN is an extremely hard ceramic material with a melting point higher than 2000°C. The **thermal conductivity** κ of AlN at room temperature has been predicted at ≈ 3.2 W·cm^{-1}K^{-1} [2.44,45]. Values of κ measured at 300 K are 2.5 [2.46] and 2.85 W·cm^{-1}K^{-1} [2.47]. The measured thermal conductivity as a function of temperature is plotted in Fig.2.7.

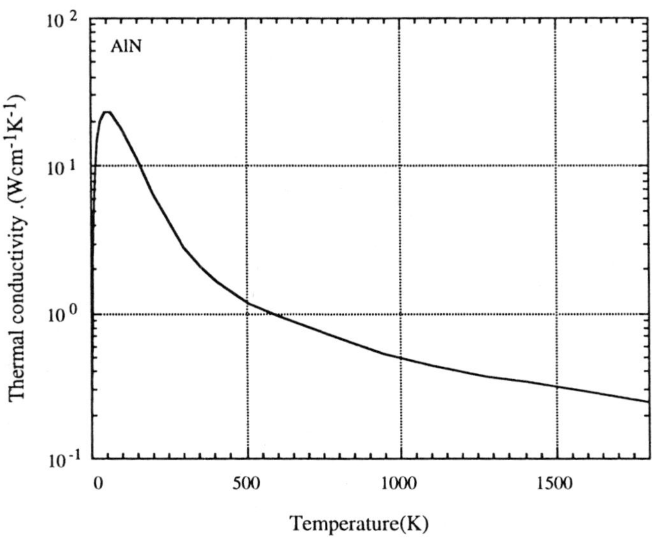

Fig. 2.7. Thermal conductivity of single-crystal AlN. After [2.46]

Using X-ray techniques across a broad temperature range $(77 \div 1269$ K), it was noted by *Slack* and *Bartram* [2.47] that the **thermal expansion** of AlN is isotropic with a room-temperature value of $2.56 \cdot 10^{-6}$ K^{-1}. The thermal expansion coefficients of AlN measured by *Yim* and *Paff* [2.48] have mean values of $\Delta a/a = 4.2 \cdot 10^{-6}$ K^{-1} and $\Delta c/c = 5.3 \cdot 10^{-6}$ K^{-1}. The dependence of the **thermal expansion coefficient** in the c plane and in the c direction is shown in Fig. 2.8, which can be fitted by the following polynomials

$$\Delta a/a_0 = -8.679 \cdot 10^{-2} + 1.929 \cdot 10^{-4} T + 3.400 \cdot 10^{-7} T^2 - 7.969 \cdot 10^{-11} T^3 \tag{2.3}$$

and

$$\Delta c/c_0 = -7.006 \cdot 10^{-2} + 1.583 \cdot 10^{-4} T + 2.719 \cdot 10^{-7} T^2 - 5.834 \cdot 10^{-11} T^3 . \tag{2.4}$$

The equilibrium N_2-vapor pressure above AlN is relatively low compared to that above GaN which makes AlN easier to be synthesized. The calculated temperatures at which the equilibrium N_2 pressure reaches 1, 10 and 100 atmospheres are 2836 K, 3088 K and 3390 K, respectively [2.50]. Details of the thermodynamic properties of AlN can be found in [2.30].

Similar to GaN but even more so, AlN exhibits an inertness to many chemical etches. A number of AlN etches have been reported in the literature. However, none of these etches have been performed on high-quality single-crystal AlN. The surface chemistry of AlN has been investigated by numerous techniques, including Auger electron spectroscopy, X-ray and Ul-

18

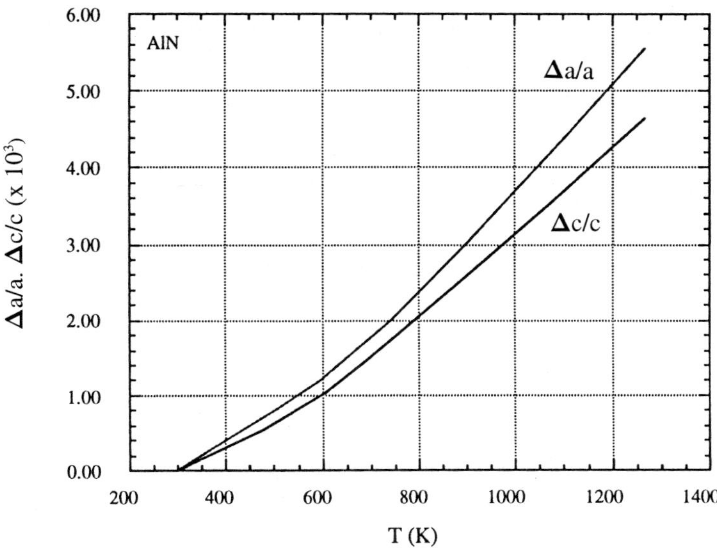

Fig. 2.8. Variation of the thermal expansion coefficient of AlN on temperature in and out of the c-plane. After [2.52]

traviolet Photoemission Spectroscopy (XPS, UPS), ultraviolet photoelectron spectroscopy, and electron spectroscopy. One of these investigations by *Slack* and *McNelly* [2.49] indicated that the AlN surface grows an oxide $50 \div 100$ Å thick when exposed to ambient air for about a day. However, this oxide layer was protective and resisted further decomposition of the AlN samples.

Detailed references to the subject treated in this subsection can be found in [2.19, 50].

2.3.2 Mechanical Properties of AlN

Early investigations of the elastic properties of AlN were carried out on sintered polycrystalline specimens, due to the unavailability of large single crystals. This, however, paved the way to more refined measurements as single crystalline AlN became available. The measured **bulk modulus** B and **Young's modulus** E are compiled in Table 2A.5 along with the entire set of elastic stiffness coefficients. The latter was obtained by fitting the results of surface-acoustic-wave measurements made on epitaxial AlN films and by Brillouin scattering measurements made on an AlN single crystal.

The phonon structure of AlN has been the subject of numerous investigations. The phonon dispersion spectrum of AlN has 12 branches, 3 acoustic and 9 optical ones [2.51]. Transverse Optical (TO) and Longitudinal Optical (LO) phonon energies have been obtained from fits to infrared reflectivity measurements, the results of which are also tabulated in Table 2A.5. Raman-active optical-phonon modes belong to the A_1, E_1, and E_2 group representations. Several Raman-scattering studies on AlN have been conducted, and the measured phonon energies are listed in Table 2A.5.

The hardness of AlN has been measured to be ≈ 12 GPa on the basal plane (0001) using a Knoop diamond indenter [2.52]. Some anisotropy in Knoop hardness has been observed with the indent direction perpendicular to the c axis with measured values in the range of $10 \div 14$ GPa [2.53].

2.3.3 Electrical Properties of AlN

Due to the low intrinsic carrier concentration, and the deep native defect and impurity energy levels, the electrical characterization of AlN has usually been limited to resistivity measurements. One such measurement by *Kawabi* and co-workers [2.54] on transparent AlN single crystals yielded resistivities $\rho = 10^{11} \div 10^{13}$ $\Omega \cdot$cm, a value consistent with other reports [2.55-57]. However, it was found that impure crystals, which exhibited a bluish color possibly due to the presence of Al_2OC, have much lower resistivities $\rho = 10^3 \div 10^5$ $\Omega \cdot$cm. These resistivity values are much lower than those according to *Chu* et al. [2.58] who were able to obtain both n- and p-type AlN by introducing Hg and Se, respectively, but failed to determine the net carrier concentrations due to very high resistivities. The n-AlN films grown by *Rutz* [2.59] had a quite low resistivity ($\rho = 10^3 \Omega \cdot$cm), which is comparable to those of *Kawabi* et al. [2.53]. Although *Rutz* did not determine the source of the electrons, *Rutz* et al. [2.60] observed an interesting transition in their AlN films in which the resistivity abruptly decreased by two orders of magnitude with an increase in the applied bias. This observation found applications to switchable resistive memory elements that are operated at 20 MHz.

The insulating nature of these early films hindered meaningful studies of their electrical transport properties. With the availability of refined growth techniques, AlN is presently grown with much improved crystal quality and shows both n- and p-type conductions. This has rejuvenated efforts to measure both the electron and hole Hall mobilities. *Edwards* et al. [2.61], and *Kawabe* et al. [2.62] carried out some Hall measurements in p-type AlN, which produced a very rough estimate of the hole mobility $\mu_p = 14$ cm^2/Vs at 290 K. The predictions for the Hall mobility in the entire range of AlGaN alloy including the binary end points are treated in Chap. 8. Not all the parameters needed for the calculations are known precisely, which somewhat reduces the confidence in predicted values. As is the case for GaN, the room-temperature mobility is dominated by the polar optical phonon scattering.

2.3.4 Optical Properties of AlN

Since an AlN lattice has a very large affinity to oxygen, it is almost impossible to eliminate oxygen contamination in AlN. Currently, commercially available AlN contains about $1 \div 1.5$ at.% oxygen. Some oxygen is dissolved in the AlN lattice while the remainder forms an oxide coating on the surface of each powder grain. *Harris* and *Youngman* [2.63] have recently reviewed photoluminescence and cathodoluminescence characteristics of AlN. After irradiation with ultraviolet light, AlN doped with oxygen was found to emit a series of broad luminescence bands at the near-ultraviolet frequencies at room temperature, no matter whether the sample was powdered, single-crystal, or sintered ceramic. *Pacesova* and *Jastrabik* [2.64] observed two broad luminescence lines centered in the vicinity of 3.0 and 4.2 eV and more than 0.5 eV wide for samples contaminated at about 1 to 1.5 at.% oxygen. *Youngman* and *Harris* [2.65], and *Harris* et al. [2.66] investigated the luminescence characteristics of polycrystalline sintered AlN samples and noted a continuous shift of the peak position in the ultraviolet luminescence line as a function of oxygen content up to a critical concentration of about 0.75 at.%. The luminescence lines beyond this limit of oxygen concentration remained stationary.

Yim et al. [2.56] characterized high-quality AlN by optical absorption (Fig.2.9) and determined the room-temperature bandgap to be direct with a value of 6.2 eV. Several groups have reported comparable values whereas others have produced questionable values considerably below 6.2 eV, probably due to oxygen contamination. Oxides formed with Ga are not stable at GaN-growth temperatures and are easily desorbed. However, the same fortuity does not hold for AlN. Both MOCVD and MBE films show oxygen contamination. *Yim* et al. [2.48] also observed a broad emission-spectrum range of $2 \div 3$ eV with a peak at 2.8 eV. Other investigations confirmed the presence of a 2.8 eV peak which *Slack* and *McNelly* [2.49] have attributed to oxygen impurities. Samples grown at the present author's laboratory by MBE and measured at the Honeywell Technology Center revealed a broad peak at 4 eV as well as the band-edge or near-band-edge signal in the emission spectrum. These observations are common to MOCVD samples as well. *Perry* and *Rutz* [2.67] performed temperature-dependent optical absorption, measuring a bandgap of 6.28 eV at 5 K compared to their room temperature value of 6.2 ± 0.1 eV. In the only optical study of AlN impurities, *Karel* and coworkers [2.69-72] reported on the luminescence of Mg and rare-earth centers in AlN.

Measurements of the **refractive index** of AlN have been carried out in amorphous, polycrystalline, and single epitaxial thin films. The values of the refractive index are in the n $= 1.99 \div 2.25$ range with several groups reporting n $= 2.15 \pm 0.05$. These values are found to increase with increasing

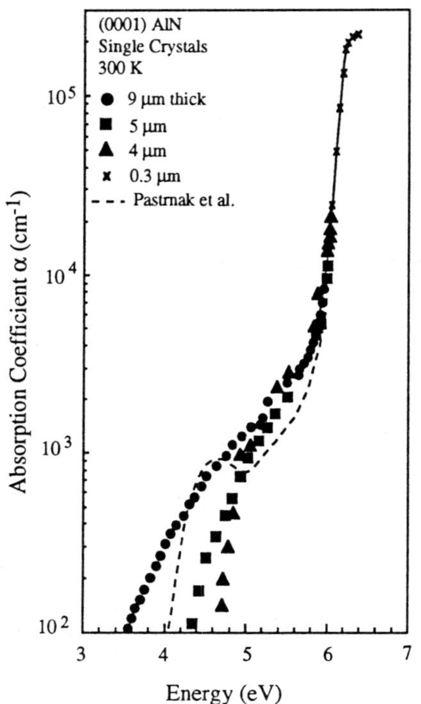

Fig. 2.9. Room-temperature absorption spectra of AlN films of varying thicknesses whose principle absorption edge occurs at 6.2 eV. The bump near $4.5 \div 4.8$ eV in the data of *Pastrnak* and *Souckawa* [2.68] was attributed to oxygen absorption bands. After [2.55-57].

structural order, varying between 1.8 to 1.9 for amorphous films, 1.9 to 2.1 for polycrystalline films, and 2.1 to 2.2 for single-crystal epitaxial films. The spectral dependence and the polarization dependence of the index of refraction has been measured and showed a near-constant refractive index in the wavelength range of $400 \div 600$ nm. Some of these measurements also indicate that, in the long-wavelength range, the **dielectric constant** of AlN (ϵ_0) lies in the range of $8.3 \div 11.5$, and that most of the values fall within $\epsilon_0 = 8.5 \pm 0.2$. Other measurements in the high-frequency range produced dielectric constants of 4.68 and $\epsilon_\infty = 4.84$. AlN has also been examined for its potential for second-harmonic generation.

Synchrotron radiation studies of AlN single crystals out to 40 eV have been performed, which resulted in the observation of a 8 eV luminescence peak. The same peak was also found in vacuum-ultraviolet reflection measurements. Recently, photoluminescence determinations using excimer-laser excitation were performed and the results are displayed in Fig. 2.10 for MBE samples grown at 800°C and 850°C. The samples were grown on sapphire substrates and exhibit broad peaks at a wavelength of 320 nm.

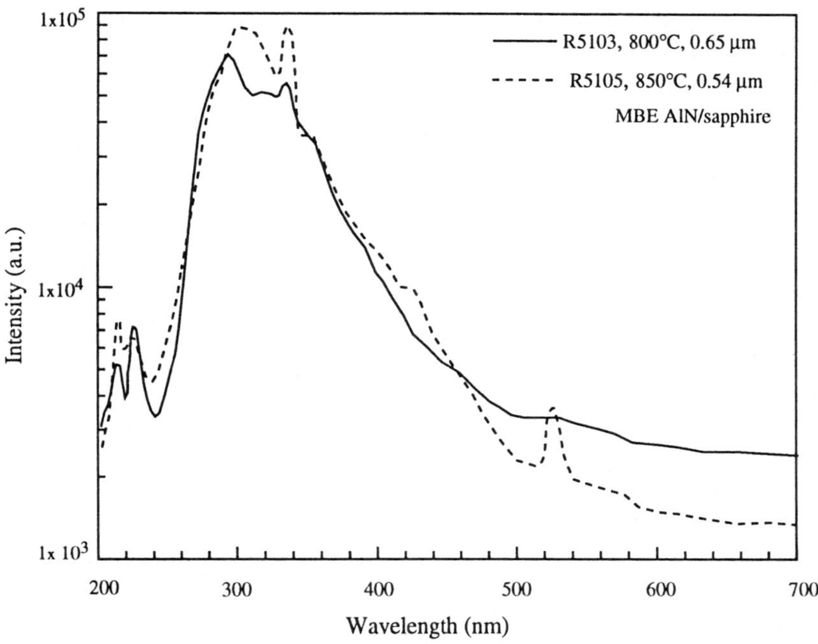

Fig. 2.10. Intensity vs. wavelength curves of the photoluminescence spectra of AlN grown on sapphire

2.4 Indium Nitride

InN has not received the experimental attention given to GaN and AlN. This is probably due to difficulties in growing high-quality crystalline InN samples and because of the existence of alternative, well-characterized semiconductors such as AlGaAs and Zn(Ga, Al)AsP, which have energy bandgaps close to the that of InN (1.89eV). The energy bandgap of InN corresponds to a portion of the electromagnetic spectrum in which alternative and well-developed semiconductor technology is available. Consequently, practical applications of InN are restricted to its alloys with GaN and AlN. The growth of high-quality InN and the enumeration of its fundamental physical properties remain for the present a purely scientific enterprise. InN is not different from GaN and AlN in the sense that it suffers from the same lack of a suitable substrate material and, in particular, a high native defect concentration. All these have hindered its progress. In addition, due to its rather poor thermal stability, InN cannot be grown at the high temperatures required by CVD growth processes. As InN rapidly dissociates at high temperatures, even as low as 600°C, an extraordinarily high nitrogen overpressure would be required to stabilize the material up to the melting

23

point, which is practically impossible. The large disparity of the atomic radii of In and N is another factor that enhances the difficulty in obtaining InN of good quality. Despite the problems inherent to InN growth, some excellent work has been reported. The seminal work of *Tansley* and *Foley* [2.75] first characterized many of the fundamental physical properties of InN. However, InN remains the least understood of the nitrides. Table 2A.4 lists the measured physical properties of InN. No group has yet reported the growth of high-quality single crystalline material which would enable detailed optical, structural and electrical measurements to be performed. Given the fact that the growth of bulk single-crystal InN films using equilibrium techniques is unlikely, attention turned to the deposition of thin films using non-equilibrium techniques. All of the data reported below, unless otherwise specified, were obtained from highly conductive n-type polycrystalline InN grown by non-equilibrium techniques.

There have been several studies which report rapid dissociation of InN at temperatures above $500\,^\circ$C. Since no high-quality InN has yet been grown, the resistance of the high-quality material to chemical etching is unknown. Successful etching of single-crystalline InN films in a hot $H_3PO_4:H_2SO_4$ solution has been measured and so has surface oxidation.

2.4.1 Crystal Structure of InN

Indium nitride normally crystallizes in the wurtzite (hexagonal) structure with the $C_{6V}^4(P6_3mc)$ space-group and C_{6v} (6mm) point-group symmetries. The zincblende (cubic) form has been reported to occur in films containing both polytypes. Due to absence of good-quality single-crystal films, the crystal structure of InN has been measured mainly in non-ideal thin films, particularly the ordered polycrystalline films with crystallites in the range of $50 \div 500$ nm. The end results of the measurements indicate that, although InN normally crystallizes in the Wz (wurtzite) structure, occasionally it also crystallizes in the zincblende (cubic) polytype. Table 2A.6 lists values of the basal-plane and out-of-basal-plane lattice constants, a_0 and c_0, deduced from a series of experiments as well as other basic properties.

The value of c_0/a_0 data from various measurements is about 1.615 ± 0.008. This is close to the more optimistic value of 1.633 determined from layers specially grown under significant precautions, best possible growth conditions, and presumably with reduced nitrogen vacancies. An examination of the tabulated data indicates an unacceptably large scatter. This may possibly be due to nitrogen deficiency, since nitrogen atoms are closely packed in (0001) planes. While the cubic polytype of InN yields a molecular cell volume of 30.9 $Å^3$, the hexagonal polytype gives a molecular cell volume of 31.2 ± 0.2 $Å^3$.

2.4.2 Mechanical and Thermal Properties of InN

The experimental density of InN deduced from Archimedean-displacement measurements is 6.89 g/cm^3 at 25°C [2.76]. This is comparable to 6.81 g/cm^3 estimated from X-ray data. The bulk modulus determined from first-principles calculations by a local-density approximation method and by the linear muffin-tin orbital method is B = 165 GPa. In a hexagonal structure, the second-order **elastic moduli** are C_{11}, C_{12}, C_{13}, C_{33}, and C_{44}, but there are no reports of these parameters yet. Since these figures depend on the lattice constants which are within some 10%, values of other nitrides can be used as a first approximation when absolutely needed [2.77]. The bulk modulus has been calculated from first principles by the local-density approximation [2.78] and by the linear muffin-tin orbital method [2.79] suggesting B = 165 GPa. Most of the properties of InN are tabulated in Table 2A.7

Indium nitride has twelve phonon modes at the zone center (symmetry group: C_{6v}), three acoustic and nine optical ones with the acoustic branches near zero at k = 0. The infrared active modes are of the E_1(LO), E_1(TO), A_1(LO) and A_1(TO) type. Moreover, a transverse optical mode has been observed at 478 cm^{-1} (59.3 meV) by reflectance and 460 cm^{-1} (57.1 meV) by transmission measurements [2.77].

The linear **thermal expansion coefficients** measured at five different temperatures between 190 K and 560 K [2.80], indicate that both along the parallel and perpendicular directions to the c axis of InN these coefficients increase with increasing temperature. Deriving thermal conductivity data from the Leibfried-Schloman scaling parameter and assuming that the **thermal conductivity** is limited by intrinsic phonon-phonon scattering, the thermal conductivity is about 0.80 ± 0.20 W$\cdot$cm^{-1}K^{-1}. While the heat capacity of InN is $(9.1 \pm 2.9) \cdot 10^{-3}$ (cal/mol$\cdot$K) at temperatures between 298 and 1273 K, the entropy is 10.4 cal/(mol$\cdot$K) at 298.15 K. The equilibrium partial pressure of N_2 above InN is about 1 atm at 800 K, and it increases exponentially with temperature to 105 atm at 1100 K.

2.4.3 Electrical Properties of InN

It is fair to state that no reliable experimental data for the electron mobility in InN have yet to be obtained. InN suffers from the lack of a suitable substrate material and high native-defect concentrations that limit its quality. A large disparity of the atomic radii of In and N is an additional contributing factor to the difficulty to obtain InN of good quality. As a result, nitrogen vacancies are thought to lead to large background electron concentrations in InN. Because of all these factors, the electron mobilities obtained from various films have varied very widely, as can be inferred from Table 2A.8. The electron mobility in InN can be as high as 3000 cm^2/Vs at room temperature [2.81]. A recent study of the electron mobility of InN as a function of the growth temperature indicates that the mobility of Ultra-High - Electron Cyclotron Resonance - Radio-Frequency Magnetron Sputtering (UHV-ECR-RMS) grown InN can be as much as four times the mobility of conventionally grown (vacuum depostion) InN [2.82].

2.4.4 Optical Properties of InN

A number of groups have described optical measurements performed on InN [2.83-85]. Values of the room-temperature InN direct-bandgap range from $1.7 \div 2.07$ eV with a value of 1.89 eV measured by optical absorption by *Tansley* and *Foley* [2.86] who also measured the infrared absorption of InN and observed an unidentified donor level approximately $50 \div 60$ meV below the conduction-band edge. *Tyagai* et al. [2.87] performed reflection and transmission measurements. They were able to estimate an effective mass of $m_e^* = 0.11m_0$ and an index of refraction n = 3.05± 0.05. The long-wavelength limit of the **refractive index** was reported to be 2.88± 0.15. The temperature dependence of the bandgap of InN indicate a bandgap temperature coefficient of [2.88]

$$\frac{dE_g}{dT} = 1.8 \cdot 10^{-4} \frac{eV}{K} \, .$$
(2.5)

From reflectance data, the existence of a TO phonon mode at 478 cm^{-1} and an LO mode at 694 cm^{-1} were deduced [2.88].

2.5 Ternary and Quaternary Alloys

Many important GaN-based devices involve heterostructures as the primary means of achieving an improved performance. For an in-depth understanding of the physical mechanisms that underly their operations, the properties of these alloys need to be extensively studied. Many of these properties such as the energy bandgap, effective masses of the electrons and holes, and the dielectric constant are dependent on the alloy composition. Although measured data for these parameters of InGaN and InAlN have been obtained, they are still not very precise, for AlGaN see [2.20] and for InGaN see [2.89]. *Yamasaki* et al. [2.90] have reported on p-InGaN. More research is necessary to confirm these conclusions regarding p-AlN and p-

InGaN. AlN and GaN are slightly lattice mismatched (2.4%). It has been noted that, for many devices, only small amounts of AlN are needed in the GaN lattice to provide sufficient carrier and optical-field confinements.

2.5.1 AlGaN Alloy

An accurate knowledge of the compositional dependence of the barrier and well materials is a requisite in attempts to analyze heterostructures, in general, and in quantum wells and superlattices, in particular. With the nitride system, a plethora of options is available for constructing such structures. The barriers can be formed of AlGaN or GaN; while depending on the barrier material, the wells can be constructed of GaN or InGaN layers. The energy bandgap of $Al_xGa_{1-x}N$ may be expressed by

$$E_g(x) = xE_g(AlN) + (1-x)E_g(GaN) - bx(1-x) , \qquad (2.6)$$

where $E_g(GaN) = 3.4$ eV, $E_g(AlN) = 6.20$ eV, x is the AlN molar fraction, and b is the bowing parameter having controversial values, to be discussed below. Figure 2.11 shows the compositional dependence of the bandgap of

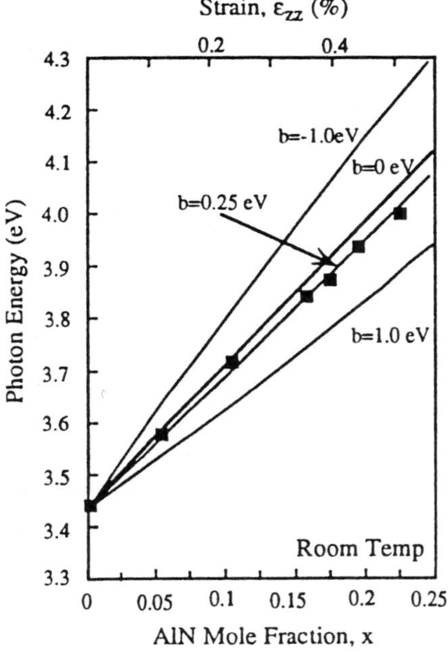

Fig. 2.11. Compositional dependence of the bandgap of AlGaN. Data with negative, and zero bowing parameters are accordig to [2.92 and 93]. Data with positive bowing parameters of 0.25 and 1 eV are from [2.92 and 93], respectively. Compiled in [2.92]

the AlGaN, as compiled by *Amano* et al. [2.91]. The solid line drawn through the square data points of *Amano* et. al. represents the calculated compositional dependence with the effect of strain included. Independent investigators came up with widely varying conclusions. For example, *Yoshida* et al. [2.92] concluded that, as the AlN mole fraction increases, the energy bandgap of $Al_xGa_{1-x}N$ deviates upwards that implies a negative value for the bowing parameter b. This contrasts the data of *Wickenden* et al. [2.93] which support a vanishing bowing parameter b. Moreover, *Koide* et al. [2.94] observed that the bowing parameter is positive and that the bandgap of the alloy deviates downwards that indicates a positive value for the bowing parameter. In the compositional-dependence calculation of *Amano* et al. [2.91] the contribution of the joint density of states and the three-dimensional exciton of the E_0 specific point were taken into consideration. With the ability to see the contribution of both the joint density of states and the three-dimensional excitons close to the bandgap energy, *Amano* et al. [2.91] were able to determine the bandgap of AlGaN within an error of less than 10 meV. Since the tensile strain in an AlGaN layer is less than 0.5%, the strain-induced change in the transition energy is expected to be small.

A recent Hall measurement for n-$Al_{0.09}Ga_{0.91}N$ demonstrated a carrier concentration of $5 \cdot 10^{18}$ cm^{-3} and a mobility of 35 cm^2/Vs at 300 K [2.95]. This measurement did not reveal any temperature-dependent mobility of n-$Al_{0.09}Ga_{0.91}N$. Other Hall measurements [2.96] on Mg-doped p-$Al_{0.08}Ga_{0.92}N$ grown by MOVPE, however, addressed the temperature dependence of the mobility [2.96]. They indicate that the hole mobility decreases with increasing temperature, reaching a value of about 9 cm^2/V·s for a doping density of $1.48 \cdot 10^{19}$ cm^{-3}. This low mobility is ascribed to a high carrier concentration and the inter-grain scattering present in the samples. While the lattice constant was studied, it was observed to be almost linearly dependent on the AlN mole fraction in AlGaN.

Until recently the resistivity of unintentionally doped AlGaN was believed to increase so rapidly with increasing AlN mole fraction that AlGaN became almost insulating for AlN mole fractions exceeding 20%. As the AlN mole fraction increased from 0 to 30%, the n-type carrier concentration dropped from 10^{20} to 10^{17} cm^{-3}, and the mobility increased from 10 to 30 cm^2/V·s. An increase in the native-defect ionization energies with increasing AlN may possibly be responsible for this variation. Our knowledge of the doping characteristics of AlGaN is still incomplete. It is not, for example, known how the dopant atoms such as Si and Mg respond to the variation of the AlN mole fraction in AlGaN. It was, however, suggested that, as the AlN mole fraction increases, the dopant atom moves deeper into the forbidden energy bandgap. AlGaN with an AlN mole fraction as high as $50 \div 60\%$ is dopable by both n- and p-type impurity atoms. The ability to dope a high-mole-fraction AlGaN, especially when low-resistivity p-type material is required, is important because it may otherwise restrict the overall characteristics of devices such as laser diodes. Until now, a low AlN mole fraction in AlGaN is considered to be sufficient for good optical-field confinement. However, it has not yet been confirmed how well it is and, hence, this must be addressed before the potential of AlGaN with respect to the other wide-bandgap semiconductors is fully realized.

2.5.2 InGaN Alloy

The growth of high-quality InN and an enumeration of its fundamental physical properties remain elusive for the present. In contrast, InGaN is already an integral part of important device designs. $In_x Ga_{1-x} N$ (x is the InN mole fraction) is not any less important than $Al_x Ga_{1-x} N$ for the fabrication of electrical and optical devices, such as LEDs and lasers, which can emit in the violet or blue wavelength range. It can be a promising strained Quantum Well (QW) material for these devices, but added complexities such as the phase separation and other inhomogeneities make the determination of the bandgap of InGaN versus composition a very difficult task. Owing to significant progress in the growth and characterization of this material during the last several years, particularly the solid solution, the bandgap dependence of $In_x Ga_{1-x} N$ on the InN mole fraction has been studied by a number of researchers. The energy bandgap of $In_x Ga_{1-x} N$ over $0 \leq x \leq 1$ can be expressed by

$$E_g(x) = (1-x)E_g(InN) + xE_g(GaN) - bx(1-x) \qquad (2.7)$$

where $E_g(GaN) = 3.40$ eV, $E_g(InN) = 1.9$ eV, and $b = 1.0$ eV. The observed and calculated dependencies on the InGaN energy bandgap with the InN mole fraction are displayed in Fig.2.12. The square data points are from [2.91] and so is the solid line which represents the calculations. One gets a positive bowing parameter of 3.2 eV. However, the results of *Nakamura* et al. [2.97] support a bowing parameter of +1.0 eV, which obviously does not agree with those of *Amano* et al.[2.89]. Moreover, a composition-dependent bowing paraameter has also been suggested. Since the InGaN is grown on GaN, there are many complicating factors, such as the piezoelectric effect and the non-uniform strain; the impact of the former can be made negligible by growing thick films. If the strain caused by the lattice mismatch were uniform, it would be compressive, due to the InN lattice constant being 11% larger than that of GaN with an accompanying blue shift. Herein lies the dilemma faced by the experimentalists. The extent of strain and the piezoelectric effect could be minimized by growing thick

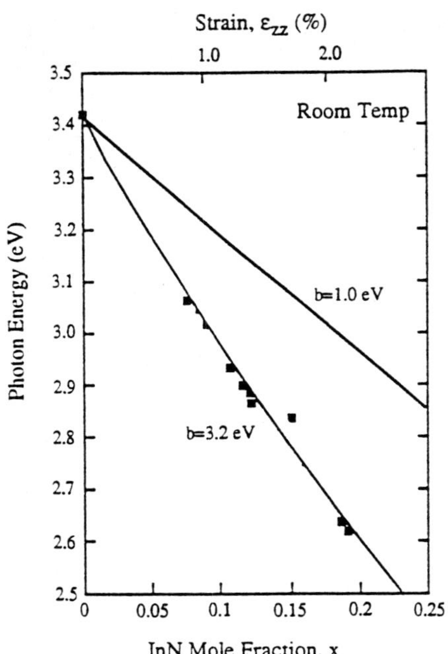

Fig. 2.12. Optical energy gap of $In_xGa_{1-x}N$ layers as a function of alloy composition x. The small solid circles, small open circles, and small open squares represent experimental data reported in the literature. The dashes represent the fitted curve. After [2.92, 98]

films. But this is hard to do because the relaxation value of the lattice constant and the origin of the optical transitions must be known accurately to determine the bandgap vs. composition dependence. Absorption and/or reflection measurements, provided that the absorption edge is sharp, are more useful in determining the bandgap but again require thick and/or good films which are lacking. Detailed X-ray reciprocal-space mapping undertaken by *Amano* et al. [2.91] are purported to indicate that InGaN wells and even somewhat thicker InGaN layers grown on GaN buffer layers are coherently strained; a conclusion reached by the observation that the in-plane lattice constants of GaN and InGaN match. At the same time though, the layer thicknesses well exceeded the calculated critical values. This paradox remains unresolved, possibly until such time when heterostructures with much reduced structural-defect concentrations can be obtained.

There are other reports dealing with the compositional dependence of the bandgap of InGaN. *Nagatomo* et al. [2.98] noted that the $In_xGa_{1-x}N$ lattice constant varies linearly with the In mole fraction up to at least x = 0.42, but it violates **Vegard's law** for x > 0.42. In an investigation with a different set of objectives, *Yoshimoto* et al. [2.99] studied the effect of growth conditions on the carrier concentration and transport properties of $In_xGa_{1-x}N$. They observed that if the deposition temperature is increased from 500° to 900°C, $In_xGa_{1-x}N$, with x ≈ 0.2 and grown on sapphire, will suffer from a reduction in carrier concentration from 10^{20} to 10^{18} cm^{-3}, but will gain from an increase in the carrier mobility from less than 10 cm^2/V·s to 100 cm^2/V·s. The same group later noted that this trend does not change if the films are grown on ZnO substrates instead of sapphire [2.100]. They could achieve good InGaN material with In mole fractions as large as 23%. *Nakamura* and *Mukai* [2.101] discovered that the film quality of In$_x$Ga$_{1-x}$N can be significantly improved if these films are grown on high-quality GaN films. Thus, from the reports cited above it may be concluded that the major challenge for obtaining high-mobility InGaN is to find a compromise in the growth temperature, since InN is unstable at typical GaN deposition temperatures. This growth temperature would undoubtedly be a function of the dopant atoms, as well as the method (MBE, MOCVD, etc.) used for the growth. This is evident from a study by *Nakamura* et al., who have since expanded the study of InGaN employing Si [2.102] and Cd [2.103]. A review of various transport properties of GaInN and AlInN was given by *Bryden* and *Kistenmacher* [2.104], the growth and mobility of p-GaInN is was discussed by *Yamasaki* et al. [2.90].

2.5.3 InAlN Alloy

In$_{1-x}$Al$_x$N is an important compound that can provide a lattice-matched barrier to GaN. The growth and electrical properties of this semiconductor have not yet been extensively studied. *Starosta* [2.105], and later *Kubota* et al. [2.106] grew InAlN alloy by Radio Frequency (RF) sputtering. *Kistenmacher* et al. [2.107], on the other hand, used the RF-Magnetron Sputtering (RF-MS) from a composite metal target to grow InAlN at 300°C. It was observed that the energy bandgap E$_g$ of this semiconductor varies between 2.0 eV and 6.20 eV for x between 0 and 1 [2.106]. The carrier concentration and the mobility of In$_{1-x}$Al$_x$N for x = 0.04 were 2·10^{20} cm^{-3} and 35 cm^2/V·s, respectively, and for x = 0.25 were 8·10^{19} cm^{-3} and 2 cm^2/V·s, respectively [2.104]. Thus, the mobility was found to decrease substantially with an increase in the Al mole fraction because the structure of the InAlN approaches the structure of the insulating AlN.

2.6 Substrates for Nitride Epitaxy

One of the major difficulties which has hindered GaN research is the lack of a suitable substrate material that is lattice matched and thermally compatible with GaN. GaN, AlN and InN have been grown primarily on sapphire, most commonly the (0001) orientation, but also the (21$\bar{3}$1), (110$\bar{1}$), (11$\bar{2}$0). In addition, III-V nitrides have been grown on Si, NaCl, GaP, InP, SiC, W,

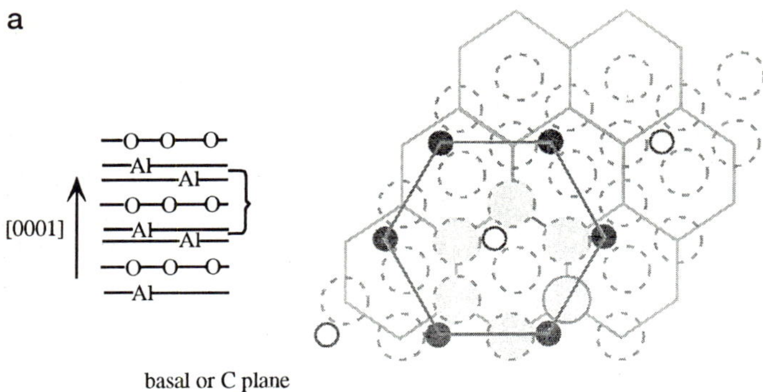

Sapphire (0001) stacking sequence and (0001) surface showing the top Al layer

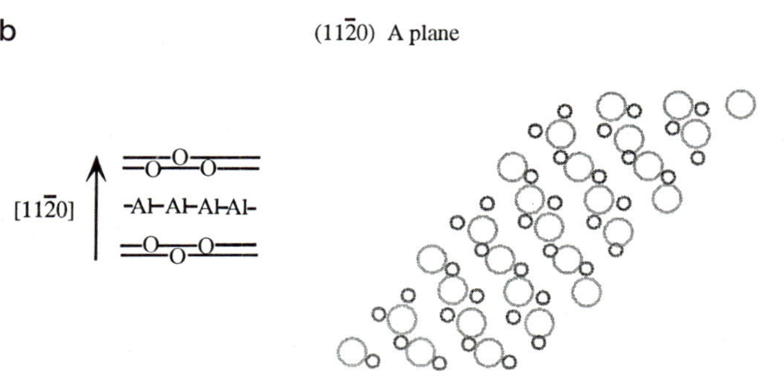

Sapphire $[11\bar{2}0]$ stacking sequence and $(11\bar{2}0)$ surface plane structure

Fig. 2.13. The atomic arrangement and stacking order on the (**a**) c-plane and (**b**) on the a-plane of sapphire

ZnO, $MgAl_2O_4$, TiO_2, and MgO. There are other substrates as well that have been used for nitride growth, which include Hf and $LiAlO_2$ and $LiGaO_2$. Given below is a succinct review of growth attempts on various substrates. However, the curious reader is referred to recent reviews on the topic [2.19, 108]. Table 2A.9 compiles the lattice parameters and thermal characteristics of a number of prospective substrate materials for nitride growth.

By and large, sapphire is the most commonly used substrate for epitaxy of nitrides. Among the faces of sapphire that have been utilised for nitrides are the c, a, and r planes. The atomic arrangement on the c and a planes of

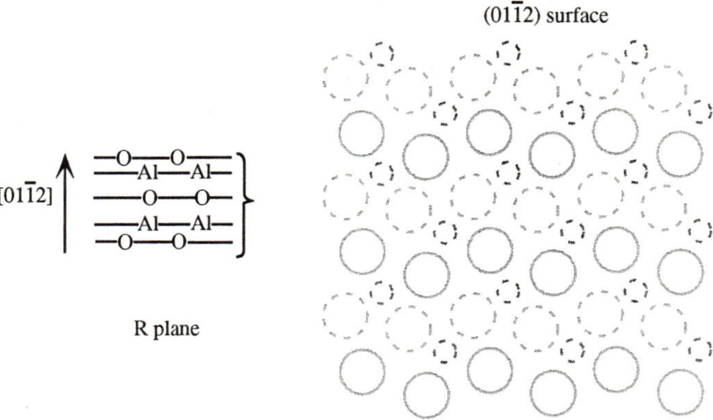

Sapphire [01$\bar{1}$2] stacking sequence and (01$\bar{1}$2) surface showing the top O layer

Fig. 2.14. The atomic arrangement and stacking order on the r-plane of sapphire

sapphire is shown in Fig.2.13a and b, respectively. The **stacking configurations** perpendicular to the c and a planes are also displayed in Fig.2.13a and b. Figure 2.14 depicts the atomic arrangement on the r plane of sapphire and the stacking arrangement perpendicular to the r plane. The **thermal expansion coefficients** of sapphire in c and a planes as a function of temperature are exhibited in Fig.2.15. Also shown are the thermal expansion coefficients of Si and poly-sapphire for comparison. The thermal expansion coefficient of SiC in c and a planes as a function of temperature is displayed in Fig.2.16. The thermal expansion coefficient of ZnO in the c-plane as a function of temperature is depicted in Fig.2.17. Also shown are the thermal expansion coefficients of Si and polycrystalline ZnO.

Most laboratories, in the past and present, have selected sapphire substrates for nitride epitaxy. This is supported by the fact that the layers grown on sapphire have in many cases better quality, and sapphire is available up to 6 inches in diameter, and it is inexpensive. The problems plagueing the SiC approach are the cost, scattered quality and bad surface finish. While these impediments may be temporary, there is a fundamental problem which has to do with the stacking-order mismatch. However, ZnO has the desired stacking order, and a reasonably close lattice match to GaN. The **thermal expansion coefficients** and the lattice constants of ZnO are $\Delta a/a = 4.8 \cdot 10^{-6}$ K^{-1}, $\Delta c/c = 2.9 \cdot 10^{-6}$ K^{-1}, a = 3.2426 Å, and c = 5.1948 Å, respectively.

There are other non-traditional substrates that are beginning to appear; among them are LiAlO$_2$ and LiGaO$_2$. LiAlO$_2$ is tetragonal having the lattice dimensions a = 5.1687 Å and c = 6.2679 Å with a hardness of 8. The **thermal expansion coefficients** are $\Delta a/a = 7.1 \cdot 10^{-6}$ K^{-1} and $\Delta c/c =$

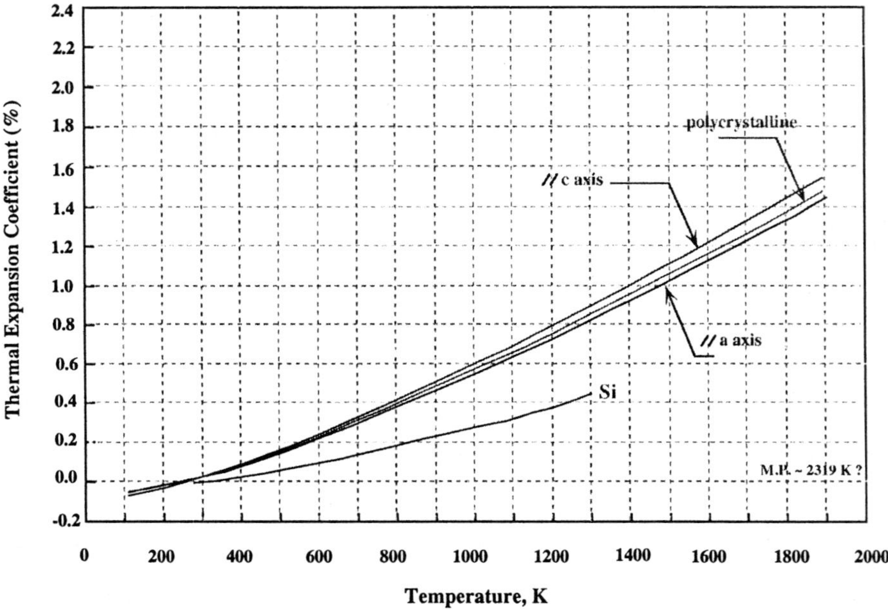

Fig. 2.15. Thermal expansion coefficient of sapphire in the c-plane as a function of temperature. Also shown are the thermal expansion coefficients of Si and poly sapphire for comparison. After [2.109]

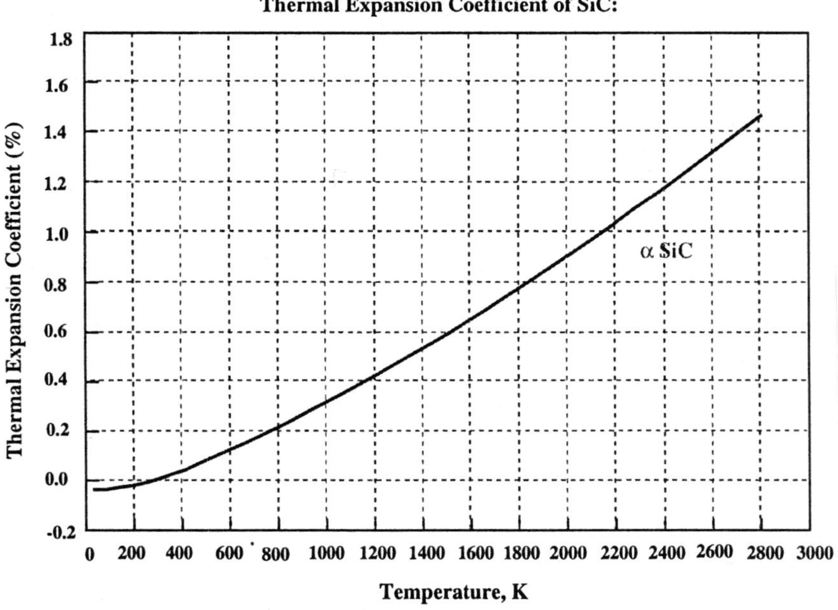

Fig. 2.16. Thermal expansion coefficient of SiC in the c- and a-planes as a function of temperature. After [2.109]

34

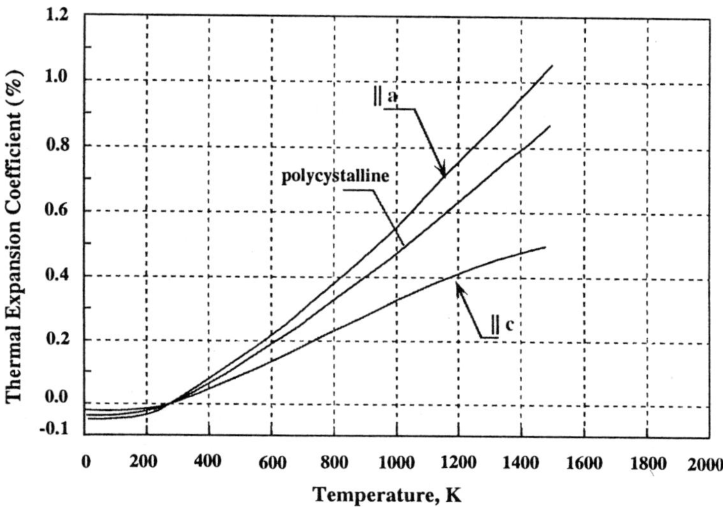

Fig. 2.17. Thermal expansion coefficient of ZnO in the c-plane as a function of temperature. Also shown are the thermal expansion coefficients of Si and polycrystalline ZnO [2.109]

$15 \cdot 10^{-6}$ K^{-1}. Optically, it is uniaxial with a transparency from 0.2 to 4 μm, and refractive indices of $n_c = 1.6014$ and $n_a = 1.6197$ at 633 nm. The crystal has no natural cleavage planes. The structure of interest is the high-temperature form (or g form) of the compound. It melts congruently around 1700°C, and is stable at room temperature. Single crystals can be grown by the conventional Czochralski melt-pulling method.

LiGaO$_2$ is orthorhombic. The lattice dimensions are a = 5.402 Å, b = 6.372 Å and c = 5.007 Å with a density of 4.175 and a **hardness** of 7.5. The thermal expansion coefficients are $\Delta a/a = 6 \cdot 10^{-6}$ K^{-1}, $\Delta b/b = 9 \cdot 10^{-6}$ K^{-1} and $\Delta c/c = 7 \cdot 10^{-6}$ K^{-1}. Optically, it is biaxial with a transparency from 0.3 to 6 μm, and **refractive indices** of $n_a = 1.7617$, $n_b = 1.7311$ and $n_c = 1.7589$ at 620 nm. The crystal has no natural cleavage planes. The structure of interest is the only stable form of the compound from room temperature to its melting point. It melts congruently around 1600°C. Single crystals can also be grown by the conventional Czochralski melt-pulling method.

Since the preferred growth direction of the GaN epitaxial film is the [0001], the match along the a-axis is more critical for the film deposition than the c-axis. Considering the a-axis lattice constant, it is obvious that LiGaO$_2$ is preferred for GaN, and LiAlO$_2$ for Al$_{1-x}$Ga$_x$N. Another important factor is that since these two crystals have exactly the same structure as GaN, the growth orientation may not have to be limited just to the c-axis or the [0001] direction. In fact, epitaxial growth can be achieved at any orientation. The degree of lattice matching may vary slightly depending on the exact orientation.

It has been established that the crystal structure of epitaxial nitrides is strongly influenced by the substrate material and orientation. Like many wide-bandgap semiconductors, each of the nitrides has been shown to exist in at least two polytypes, the most common being the equilibrium wurtzite structure, and the zincblende structure (Fig.2.2). In general, wurtzite material grows on hexagonal substrates, while zincblende material grows epitaxially on cubic substrates. The two polytypes differ only through the stacking order of the planes in the [111] (in the zincblende case) direction. The nearest-neighbor positions are almost identical, which explains the observed similarity in the physical properties of the wurtzite and zincblende GaN phases.

For lack of an ideal substrate, nearly all III-V nitride materials have been grown on sapphire despite its poor structural and thermal match with the nitrides. The preference towards sapphire substrates can be attributed to its wide availability, hexagonal symmetry, and its ease of handling and pre-growth cleaning. Sapphire is also stable at high temperatures ($\approx 1000°C$) that are required for epitaxial growth using the various CVD techniques commonly employed for GaN growth. Due to the thermal and lattice mismatches between sapphire and the III-V nitrides, it is necessary to grow a thick epilayer to obtain good-quality material.

The recent commercial availability of SiC substrates initiated a development which may influence nitride research. The thermal and lattice matches between SiC and AlN are excellent compared to sapphire and other common substrates. However, SiC substrates are hard to prepare by conventional methods, and typically require high-temperature processes incompatible with typical nitride deposition systems. An additional complication inherent in 6H- and 4H-SiC is that it does not have the stacking order of nitrides. Therefore, steps on the surface cause what is dubbed the **Stacking Mismatch Boundary** (SMB) defects, which propagate throughout the film. These propagating defects can be eliminated only when the steps are six, and four bilayers in 6H- and 4H-SiC, respectively. The 2H SiC polytype has the same stacking order as the nitrides, but are not stable and available. The $(0001)_{Si}$ face of 6H-SiC has only Si dangling bonds at the surface and allows one to use techniques developed for Si(111). After controlled oxide growth and stripping cycles, the surface is hydrogen passivated in an HF solution. In-situ hydrogen plasma cleaning produces a high-quality, contaminant-free surface for epitaxial growth. GaN grown on SiC substrates prepared in this manner exhibited sharp X-ray and photoluminescence peaks, whose full width at half maxima were comparable to the best reported values.

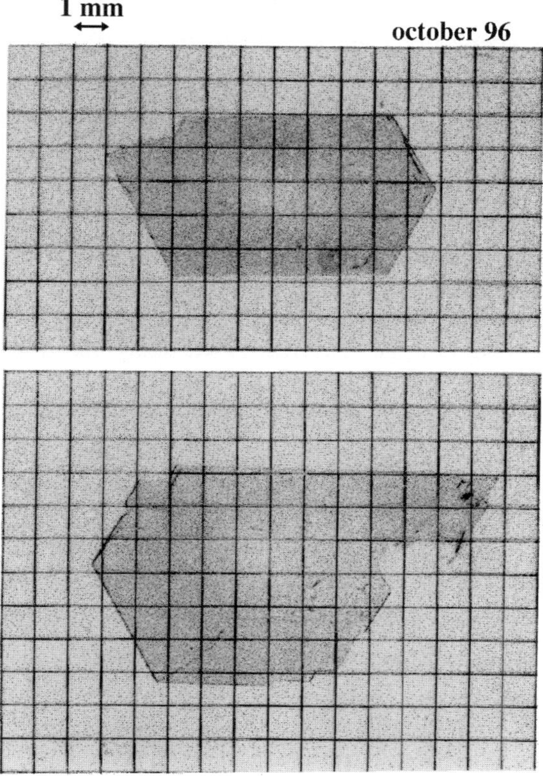

Fig. 2.18. Photograph of a slice of bulk GaN sample prepared at UNIPRES in Warsaw.
Courtesy of S. Porowski and P. Perlin, UNIPRESS

Lasers employing films grown on the c-plane of sapphire utilize etched cavities because sapphire does not cleave well. The cleaving process is further complicated in that the GaN epilayers are rotated (about 30°) with respect to the underlying sapphire substrate making it impossible to align the cleavage plane of GaN and sapphire. For this reason, *Nakamura* et al. [2.110, 111] explored lasers grown on the a plane of sapphire where the c plane of GaN aligns normal to the substrate surface. The facets in GaN and underlying sapphire are oriented on the a-plane and c-plane, respectively. Even though the materials quality does not compare to that on the c-plane, improved cavity formation along the m-plane $(10\bar{1}0)$ or a-plane $(11\bar{2}0)$ outweighs its reduced materials quality. Moreover, laser structures on (111) $MgAl_2O_4$ (**spinel substrates**) which lead to wurtzite GaN along the c-plane have also been explored for injection-laser experiments as well [2.111]. Spinel cleaves along the (100) plane, inclined to the surface with cleavage following the m-plane of GaN about where the epilayers is reached. Even

37

though the facet quality in this scheme is the best among the aforementioned approaches, materials-quality degradation is too severe to pull it ahead of
the other approaches. It is, however, very clear that a substrate with good
cleavage characteristics and on which GaN can be grown without rotation is
desperately needed. While SiC meets some of these criteria, its cost and
poor surface quality of SiC as far as the growth is concerned is making its
implementation difficult although optically pumped stimulated-emission
pulsed RT injection lasers has been reported [2.112, 113]. Until such time
as native substrates are available, ZnO meets the stacking criteria. Again,
this approach, too, has been hampered by the scarcity of high-quality ZnO
substrates prepared by the sublimation technique.

GaN substrates are not available in large quantities. Extremely high N_2
equilibrium pressure on GaN will most likely preclude even pseudo-conven-
tional growth methods to fall short of producing GaN substrates. Bulk GaN
substrates have been prepared under high pressures ($12 \div 20$ kbar) and temperatures ($1200 \div 1600 °C$). A photograph of a slice of one such sample prepared at UNIPRES in Warsaw [2.114] is displayed in Fig.2.18. A much
lower equilibrium pressure of N on AlN, on the other hand, lends itself to
the growth of this alloy utilising the sublimation technique. Samples with
dimensions of some 4 mm in diameter and 12 mm in length have already
been prepared [2.115] and used to determine much of the AlN thermal data
discussed in this chapter. The deposition temperature is around $2250 °C$.
The size of the crystal was determined by the deterioration of the W boat.
Early growth attempts on GaN and AlN substrates are encouraging. Table
2A.10 lists the properties of bulk GaN and AlN specimens. Efforts are
underway to explore the aforementioned techniques as well as new approaches to produce GaN and AlN bulk materials for nitride growth.

2A Appendix: Fundamental Data for Nitride Systems

Table 2A.1. List of the known properties of wurtzite and zinckblende GaN

Wurtzite polytype

Bandgap energy	$E_g(300\,K) = 3.42$ eV	$E_g(4K) = 3.505$ eV
Tamperature coefficient	$dE_g/dT = -6.0\cdot10^{-4}$ eV/K	
Pressure coefficient	$dE_g/dP = 4.2\cdot10^{-3}$ eV/kbar	
Lattice constant	$a = 3.189$ Å	$c = 5.185$ Å
Thermal expansion	$\Delta a/a = 5.59\cdot10^{-6}$ /K	$\Delta c/c = 3.17\cdot10^{-6}$ /K
Thermal conductivity	$\kappa = 1.3$ W/cm·K	
Index of refraction	$n(1eV) = 2.35$	$n(3.42eV) = 2.85$
Dielectric constant	$\epsilon_r = 10.4$	$\epsilon_\infty = 5.47$
Electron effective mass, m_e	$0.22m_0$	
Hole effective mass, m_p	$> 0.8m_0$	

Zincblende polytype

Bandgap energy	$E_g(300\,K) = 3.2 - 3.3$ [eV]
Lattice constant	$a = 4.52$ Å
Index of refraction	$n(3eV) = 2.9$

Table 2A.2. Elastic stiffness coefficients in Wz and ZB GaN [2A.1-3]

	GaN	Wz	ZB
Elastic stiffness coefficient $[10^{11}\,dyn/cm^{-2}]$	C_{11}	29.6	25.3
	C_{12}	13.0	16.5
	C_{13}	12.0	–
	C_{33}	39.5	–
	C_{44}	2.41	6.04

Table 2A.3. Zone-center optical-phonon wave numbers of GaN obtained from Raman scattering at 300 K [2A.4]

Mode	Wz [cm^{-1}]	Mode	ZB [cm^{-1}]
E_1-TO	$556 \div 559$	TO	558
A_1-TO	$533 \div 534$	LO	730
E_1-LO	741		
A_1-LO	710		
E_2 low	$145 \div 146$		
E_2 high	$569 \div 568$		

Table 2A.4. A summary of the observed structural and optical properties of AlN

Wurtzite polytype

Bandgap energy	$E_g(300\,\mathrm{K}) = 6.2$ eV	$E_g(5\mathrm{K}) = 6.28$ eV
Lattice constant	$a = 3.112$ Å, $c = 4.982$ Å	
Thermal expansion	$\Delta a/a = 4.2 \cdot 10^{-6}$ /K	$\Delta c/c = 5.3 \cdot 10^{-6}$ /K
Thermal conductivity	$\kappa = 3.2$ W/cm·K	
Index of refraction	$n(3\mathrm{eV}) = 2.15 \pm 0.05$	$n(3.42\mathrm{eV}) = 2.85$
Dielectric constant	$\epsilon_r = 8.5 \pm 0.2$	$\epsilon_\infty = 4.68 \div 4.84$

Zincblende polytype

Bandgap energy	$E_g(300\,\mathrm{K}) = 5.11$ eV, Theory
Lattice constant	$a = 4.38$ Å

Table 2A.5. Mechanical properties of AlN. The measured bulk modulus and Young's modulus [2A.5-9]

Physical properties	Value	Method of measurement
c_{11}	$345 \div 411$ GPa	SAW velocity, Brillouin scattering
c_{12}	$125 \div 149$ GPa	SAW velocity, Brillouin scattering
c_{13}	$99 \div 120$ GPa	SAW velocity, Brillouin scattering
c_{33}	$389 \div 395$ GPa	SAW velocity, Brillouin scattering
c_{44}	$118 \div 125$ GPa	SAW velocity, Brillouin scattering
Bulk modulus B	$160 \div 201$ GPa	Sound velocity
Young's modulus	$293 \div 308$ GPa	Sound velocity
$\omega_{\Gamma 0}(E)$	657 cm^{-1}	
$\omega_{\Gamma 0}$	$664 \div 667$ cm^{-1}	IR reflectance
$\omega_{\Gamma 0}(A_1)$	$614 \div 667$ cm^{-1}	Raman scattering
$\omega_{LO}(A_1)$	$892 \div 895$ cm^{-1}	
$\omega_{TO}(E_1)$	$667 \div 673$ cm^{-1}	
$\omega_{LO}(E_1)$	$895 \div 921$ cm^{-1}	
$\omega(E_2)$ high	$241 \div 252$ cm^{-1}	
$\omega(E_2)$ low	$655 \div 665$ cm^{-1}	
Refractive index n	1.98	for $\lambda = 250$ nm in PCVD films
Refractive index n	2.30	for $\lambda = 300$ nm in PCVD films
Refractive index n	2.30	for $\lambda = 400$ nm in poly films
Refractive index n	2.40	for $\lambda = 620$ nm in poly films
Refractive index n	2.14	for $\lambda = 633$ nm, calculated
Refractive index n	2.11	for $\lambda = 633$ nm in sputtered films

Table 2A.6. The basal plane and perpendicular-axis lattice constants along with other basic parameters for InN

Wurtzite polytype

Bandgap energy	$E_g(300\,K) = 1.89$ eV	
Temperature coefficient	$dE_g/dT = -1.8 \cdot 10^{-4}$ eV/K	
Lattice constant	$a = 3.548$ Å	$c = 5.760$ Å
Index of refraction	$n = 2.80 \div 3.05$	
Dielectric constant	$\epsilon_r = 15.3$	$\epsilon_\infty = 8.4$

Zincblende polytype

Bandgap energy	$E_g(300\,K) = 2.2$ eV, Theory
Lattice constant	$a = 4.98$ Å

Table 2A.7. Measured physical properties of InN. (LWL: Long-Wave Limit, TSFC: Thermodynamic State Function Changes, NIRSC: Normal Incidence Reflectance of Synchrotron Radiation)

Property	Value	Comments
Density (hcx)	$6.89 \cdot 10^3$ kg/m^3	Measured by displacement
Density (cubic)	$6.97 \cdot 10^3$ kg/m^3	Derived from X-ray data
Molar mass	128.827 gm/mol	
Mole volume (hcx)	31.2 Å^3	From lattice constant
Mole volume (cubic)	30.9 Å^{33}	From lattice constant
$C_\perp$	$4.42 \cdot 10^{11}$ dyn/cm^2	Estimate
C_1	$2.65 \cdot 10^{12}$ dyn/cm^2	Estimate
$h_{14}{}^2 (4C_\perp + 3C_1)/12$	$7.78 \cdot 10^{12}$ V^2/dyn	Estimate
Defortmation potential	7.10 eV	Estimate
$\hbar\omega_{TO}$	59.3 meV (478 cm^{-1}	Reflectance method
	57.1 meV (478 cm^{-1}	Transmission method
$\hbar\omega_{TO}$	86.2 meV (478 cm^{-1}	Est. Brout. sum rule
	89.2 meV (478 cm^{-1}	Est. Brout. sum rule
Thermal conductivity	80 ± 20 W/m$\cdot$K	Estimate
Heat capacity C_p	$9 \pm 3 \cdot 10^{-3}$ (cal/mol$\cdot$K)	$298 \div 1273$ K
Entropy S^0	10.4 cal/mol$\cdot$K)	298.15 K
N$_2$ equil. vapor pressure	1 atmosphere	800 K
	10^5 atmosphere	1100 K
TSFC* at formation $\Delta H_f{}^0$	-34.3 kcal/mole	Experimental 298.15 K
TSFC at formation $\Delta S_s{}^0$	-25.3 kcal/mol$\cdot$K	Experimental 298.15 K
TSFC* at formation $\Delta G_f{}^0$	-22.96 kcal/mole	Experimental 298.15K
TSFC* at fusion ΔH_m	14.0 kcal/mole	Theoretical
TSFC* at fusion ΔS_m	10.19 kcal/mole	Theoretical
Refractive index at LWL	2.88 ± 0.30	Theoretical
""" at $600 \div 800$ nm	2.90 ± 0.30	—
""" at $900 \div 1200$ nm	2.90	Transmission interference
""" at $900 \div 1200$ nm	3.05 ± 0.30	Transmission interference
""" at 620 nm	2.65	NIRSR
""" at 120 nm	1.0	NIRSR
Dielectric constant	8.4	LWL

Table 2A.8. A compilation of electron mobilities obtained for InN on different substrates and under various deposition conditions [2A.10]. (ZB: Zincblend polytype)

n-type carriers concentration [cm^{-3}]	Carrier mobility [cm^2/v·s]	Substrate	Depostion technique
$(5 \div 8) \cdot 10^{18}$	250 ± 50	Sapphire, silicon, various metals	Reactive sputtering
10^{20}	20	Sapphire	Reactive evaporation
$(3 \div 10) \cdot 10^{18}$	$20 \div 50$	Glass, fused	Reactive sputtering
$(1 \div 200) \cdot 10^{18}$	3	Fused quartz	Reactive sputtering
$(2 \div 80) \cdot 10^{20}$	$35 \div 50$	Sapphire	CVD
$5 \cdot 10^{18}$	20	Glass, NaCl	Reactive sputtering
$6 \cdot 10^{20}$	2	Fused quartz	Cathodic sputtering
$6 \cdot 10^{16}$	2	Glass, silicon 304 stainless steel	RF ion plating
$(7 \div 70) \cdot 10^{16}$	$730 \div 3980$	Glass, silicon	Reactive sputtering
$3 \cdot 10^{16}$ at 150 K	5000	Glass, silicon	Reactive sputtering
10^{20}	10	Glass	Reactive DC magnetron sputtering
$4.8 \cdot 10^{20}$	38	Sapphire	Magnetron sputtering
$(1 \div 8) \cdot 10^{20}$	50	Sapphire	Plasma-assisted MOVPE
$(1 \div 10) \cdot 10^{20}$	50		MOVPE
$(2 \div 3) \cdot 10^{20}$	$20 \div 60$	Sapphire, silicon, mica	Reactive RF magnetron sputtering
$2 \cdot 10^{20}$	≈ 100	GaAs	ECR-assisted MOMBE
10^{20} (ZB)	220 (ZB)	GaAs	Plasma-assisted MBE

Table 2A.9. Lattice parameters and thermal characteritics of a number of the prospective substrate materials for nitride growth, and their lattice mismatch with GaN [2A.11, 12]. For $LiGaO_2$ is the lattice constant $b = 6.372$ Å

Crystal		a [Å]	c [Å]	Mached a [Å]	κ [W/cm·K]	$\Delta a/a$ $\Delta c/c$ ($\times 10^{-6}$/K)	Orientation	Mismatch (to GaN)
AlN	Hexagonal	3.104	4.966	3.104	3.2	4.2 5.3	4.966	−2.7%
GaN	Hexagonal	3.18	5.16	3.18	1.3	5.59 3.17	5.16	0%
Sapphire	Hexagonal	4.758	12.991	2.747	0.3÷0.5	7.5 8.5	4.330	49% (≈13%)
6H-SiC	Hexagonal	3.0817	15.112	3.0817	4.9	5.0374	−3.1%	
ZnO	Hexagonal	3.2496	5.2065	3.2496	0.3÷0.4	5.2065	+2.1%	
$LiAlO_2$	Tetragonal	5.1687	6.2679	3.1340		7.1	5.1687	1.4%
$LiGaO_2$	Orthorhom.	5.402	65.407	6.372		a:6,b:9	5.007	0.18%
$MgAl_2O_4$	Cubic	8.083				7.45		
Si	Cubic	5.4301			1.5	3.59		
GaAs	Cubic	5.6533			0.5	6		
3C-SiC	Cubic	4.3			4.9			
MgO	Cubic	4.216				10.5		

Table 2A.10. Some Properties of bulk GaN and AlN [2A.13, 14]

	Lattice parameter	
	a ± 0.0003 [Å]	c ± 0.0001 [Å]
GaN bulk		
Clear	3.1879	5.1856
Zn doped	3.1884	5.1860
Dark	3.1884	5.1861
Temperature	1200°÷1600°C	
Pressure	12÷20 kbar, 1% nitrogen	
Size	1÷2 mm	
FWHM	20÷30 arcseconds	
n [cm^{-3}]	10^{19}	
AlN bulk		
Size	4 mm$^\phi$, 12 mm long, 0.3 mm/h growth rate	
Temperature	2250°C in a tungsten boat	

3. Electronic Band Structure
of Bulk and QW Nitrides

The band structure of a given semiconductor is pivotal in determining its potential utility. Consequently, an accurate knowledge of the band structure is critical if the semiconductor in question is to be incorporated in the family of materials considered for serious investigations and device applications. The group III-V nitrides are no exception and it is their direct-bandgap nature and the size of the energy gap what spurred the recent activity. A number of researchers have published band-structure calculations for both Wurzite and zincblende GaN, AlN, and InN. The first Wz GaN band structure found through a pseudo-potential method led to a 3.5 eV direct bandgap. The band structure for ZB GaN has been obtained by a first-principles technique within the local-density functional framework with a direct bandgap of 3.40 eV and a lattice constant of 4.50 Å. A treatise of the bad structure in bulk and quantum wells with and without strain will be given below.

3.1 Band-Structure Calculations

The wide-bandgap group III-V nitrides can exist in Wurtzite (Wz) and Zincblende (ZB) crystal polytypes with the Wz phase being the stable and widely used form. Calculations of electronic and optical properties of Wz GaN and related structures have been undertaken over the years [3.1, 2] with more emerging recently as nitride-based devices became popular. *Wimmer* et al. utilized the Full Potential Linearized Augmented Plane Wave (FLAPW) method for calculating the electronic band structure of Wz semiconductors within the Local Density Approximation (LDA) [3.3]. Methods such as the ab-initio, tight-binding, Linear Muffin-Tin Orbital (LMTO) [3.4], Linear Combination of Atomic Orbitals (LCAO), Linearized Augmented Plane Wave (LAPW), and pseudopotential methods have been employed to calculate the energy bands for both wurtzite and zincblende GaN, InN and AlN bulk materials. These methods change in their capabilities to varying degrees. Of importance is that the LDA method can handle the overlapping between the Ga 3d semi-core bands with the N 2s bands, which leads to two separate bands. In contrast, the overlapping can not be handled with the pseudopotential method where the Ga 3d bands are treated as core states. The resulting energy resonance causes the Ga 3d electrons to strongly hybridize with both the upper and lower valence-band s and p levels. Such a hybridization is predicted to have a profound influence on the GaN properties, including such quantities as the bandgap, the lattice constant, acceptor levels and valence-band heterojunction offsets. Since Al has no 3d core states, there is no hybridization between the cation d states and the N 2s states. It has been predicted in the cases of ZnS and ZnSe that potential acceptors such as Cu, whose d electrons are resonant with the lower valence band, are repelled by the d-hybridized upper valence band resulting in a deep level. Impurities without d-electron resonance form shallow acceptors. Mg has no d electrons and turns out to be sufficiently shallow for room-temperature p-type doping of GaN. On the other hand, Zn, Cd, and Hg, which all have d electrons, form deep levels in GaN [3.5]. Further insight is warranted before conclusive statements can be made with certainty as, for example, photoemission data show the N 2s to be well below the Ga 3d band. A review of the band-structure calculation and the methodologies used for both polytypes of all group-III nitrides, BN, AlN, GaN, and InN can be found in [3.6]. The calculated band structures of Wz GaN, AlN and InN are exhibited in Fig.3.1. The structure and the first Brillouin zone of a wurtzite crystal is displayed in Fig.3.2.

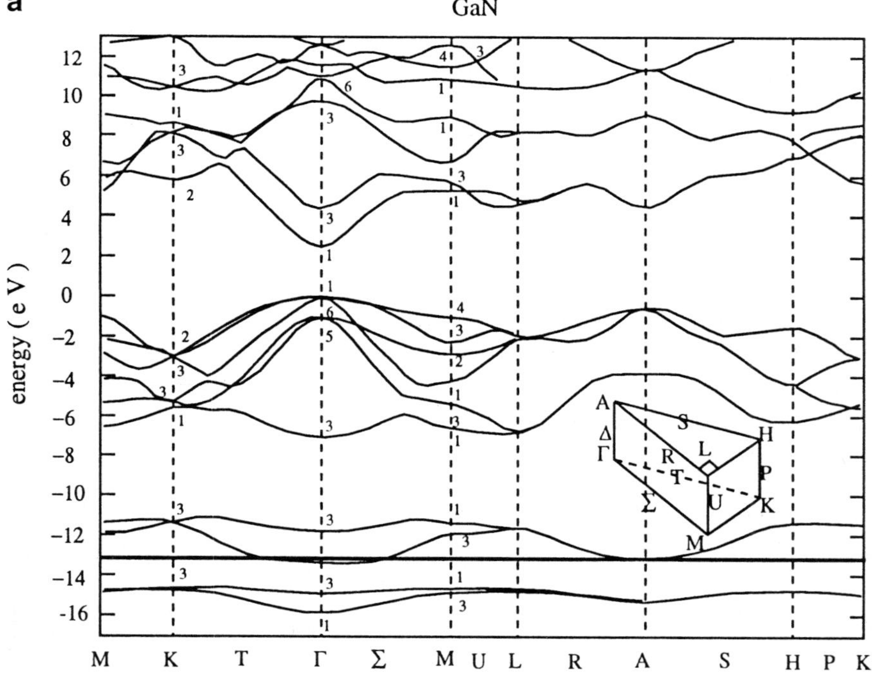

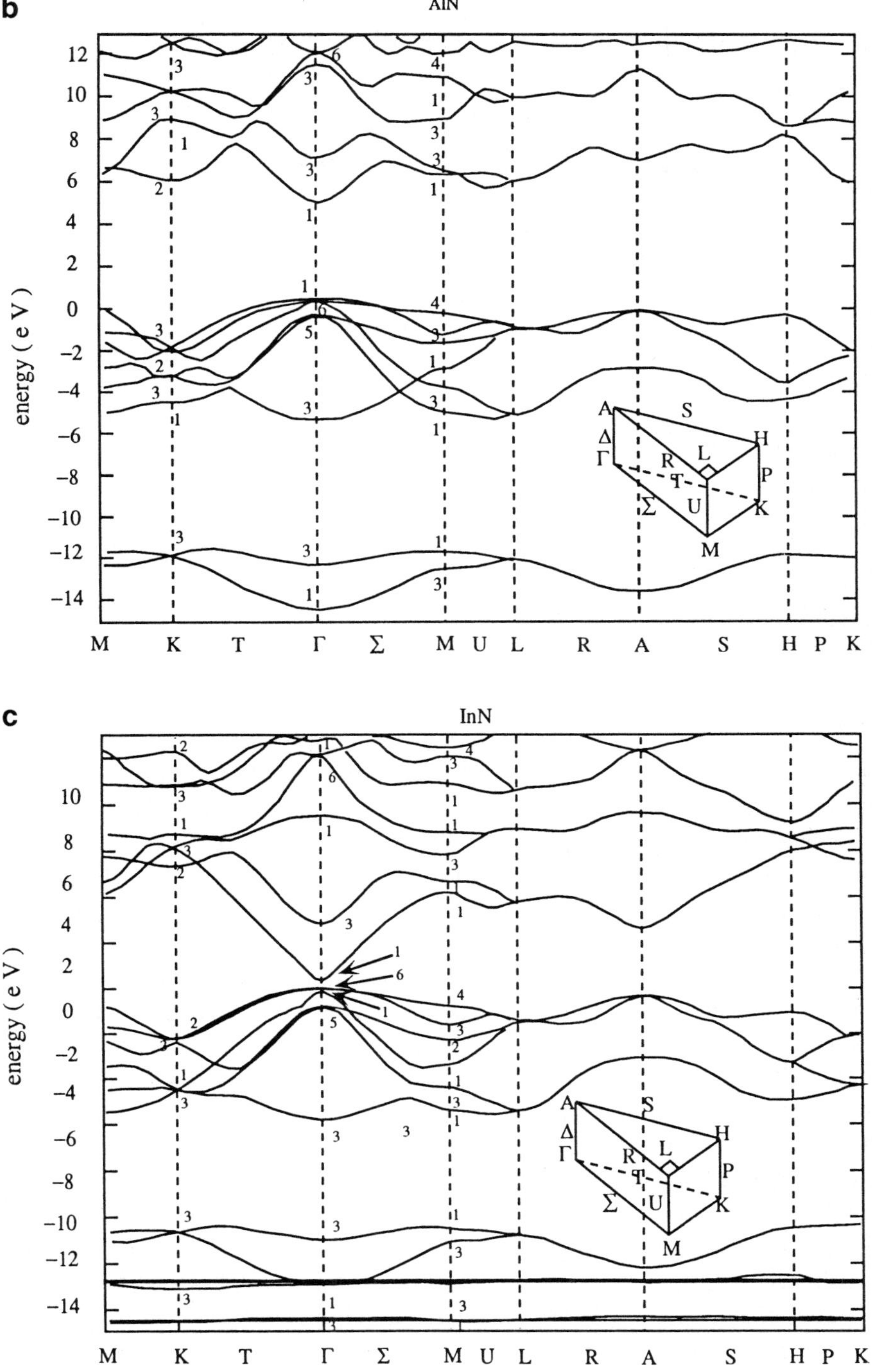

Fig. 3.1. Calculated band structures of (**a**) wurtzite GaN, (**b**) AlN, and (**c**) InN. The first Brillouin zone is also shown for convenience. After [3.6]

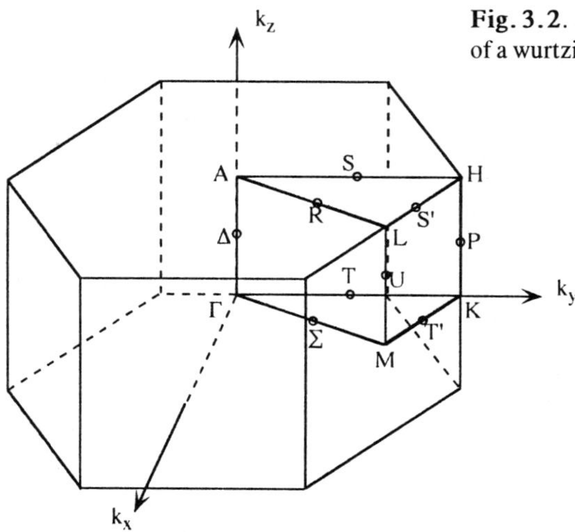

Fig. 3.2. Structure and the first Brillouin zone of a wurtzite crystal

It should be noted at the outset that all these binary materials, including alloy compounds obtained by combinations of these binaries are wide direct-bandgap semiconductors in both crystal phases, except zincblende AlN which is expected to have an indirect gap with the conduction-band minimum being at the X valley. Due to the lack of reliable experimental data, many details of these studies must be improved in order to provide an accurate band description. Approaches, such as the $\mathbf{k} \cdot \mathbf{p}$ model [3.7], for calculating the valence-band structure of Wz GaN including strain have been applied to band-structure calculations [3.7, 8]. First-principles calculations of effective-mass parameters, and valence-band structures in bulk and confined systems with and without strain, utilising the FLAPW method [3.9-13] and envelope function formalism for valence bands in wurtzite quantum wells have been undertaken [3.14].

The wurtzite structure has a hexagonal unit cell and thus two lattice constants, c and a. It contains 6 atoms of each type. The space group for the wurtzite structure is $P6_3mc$ (C_{6v}^4) [3.15]. The wurtzite structure consists of two interpenetrating Hexagonal Close Packed (HCP) sublattices, each with one type of atoms, offset along the c axis by 5/8 of the cell height (5c/8). The zincblende structure has a cubic unit cell, containing four group-III elements and four nitrogen elements. The space group for the zincblende structure is $T_d^2:F\bar{4}3m$. The position of the atoms within the unit cell is identical to the diamond crystal structure. Both structures consist of two interpenetrating face-centered cubic sublattices, offset by one quarter of the distance along a body diagonal. Each atom in the structure may be viewed as positioned at the center of a tetrahedron with its four nearest neighbors defining the four corners of the tetrahedron. The zincblende and wurtzite structures are similar. In both cases, each group-III atom is coordinated by four nitrogen atoms. Conversely, each nitrogen atom is coordinated by four group-III atoms. The main difference between these two structures lies in the stacking sequence of closest-packed diatomic planes. For the wurtzite structure, the stacking sequence of the (0001) planes is ABABAB in the $\langle 0001 \rangle$ direction. For the zincblende structure, the stacking sequence of the (111) planes is ABCABC in the $\langle 111 \rangle$ direction.

The structure and the first Brillouin zone of a Wz crystal is depicted in Fig.3.2. In a crystal with Wz symmetry, the conduction-band wave functions are formed of the atomic s orbitals and transform the Γ point congruent with the Γ_7 representation of the space group C_{6v}^4. The upper valence-band states are constructed out of appropriate linear combinations of products of p^3-like (p_x, p_y, and p_z-like) orbitals with spin functions.

Under the influence of the crystal field and spin-orbit interactions, the hallmark of the wurtzite structure, the six-fold degenerate Γ_{15} level associated with the cubic system, splits into Γ_9^v, upper Γ_7^v, and lower Γ_7^v levels (Fig.3.3). Figure 3.4 shows the dispersion of the uppermost valence- and conduction-band structures in Wz GaN. The influence of the crystal-field splitting, which is present only in the wurtzite structure, transforms the semiconductor from ZB to Wz, which is represented in the section on the left-hand side in Fig.3.3. The crystal field splits the Γ_{15} band of the ZB structure into Γ_5 and Γ_1 states of the wurtzite structure. These two states

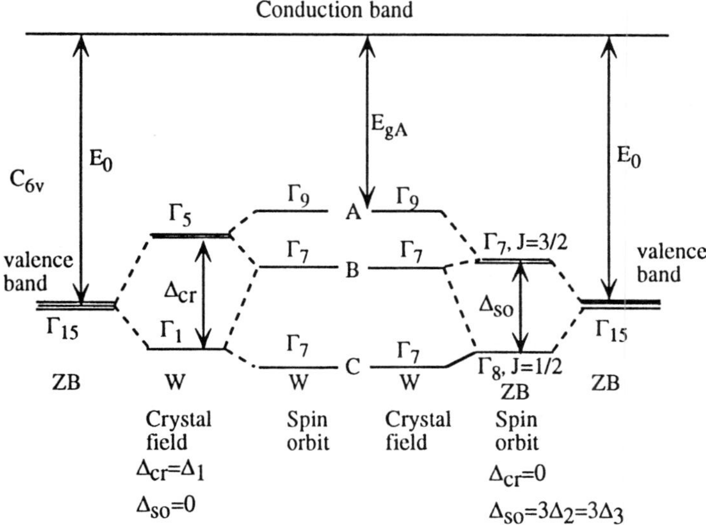

Fig. 3.3. Splitting of the valence band in Wz crystals due to crystal-field and spin-obit interaction. *From left to right*, the crystal-field splitting is considered first. *From right to left*, the spin-orbit splitting is considered first. Regardless of which is considered first, the end result is the same in that there are three valence bands which are sufficiently close to one another for band mixing to be non-negligible

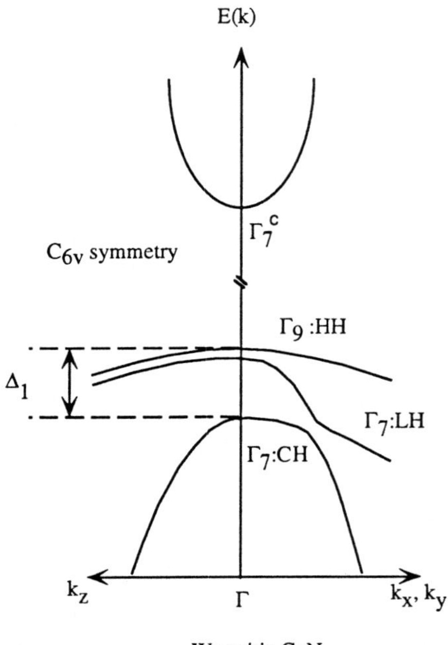

Fig. 3.4. Valence and conduction bands in crystal with wurtzite symmetry such as GaN where the spin-orbit splitting leads to the bands labeled as HH and LH. That caused by splitting due to crystal field is labeled as CH. After [3.9, 10]

are further split into Γ_9^v, upper Γ_7^v and lower Γ_7^v levels by spin-orbit interactions. Application of the spin-orbit splitting, from right to left, splits the Γ_{15} band of the ZB crystal into Γ_8 and Γ_7 states while the crystal possesses the zincblende symmetry. Application of a crystal field further splits these states into Γ_9^v, upper Γ_7^v and lower Γ_7^v levels, and the crystal now possesses the wurtzite symmetry. As shown in Fig. 3.4, the hole effective masses of the three uppermost valence bands Γ_9^v, Γ_7^v, and Γ_7^v exhibit a large k dependence. The bands are labeled as **HH** (heavy Γ_9^v), **LH** (light, upper Γ_7^v spin orbit split) and **CH** (lower Γ_7^v, crystal-field split). The mass of the Γ_9^v band is heavy in all k directions whereas that of the upper Γ_7^v is relatively light in the x,y plane but heavy in the z direction. That of the lower Γ_7^v is light in the x,y plane, but it is heavy along the z direction (c direction). Two different definitions are prevalent in the literature. The Γ_6 , Γ_1 pair has been used in [3.15, 14] and the Γ_5, Γ_1 pair in [3.9, 13, 14, 17-19]; for background information on group theory and symmetries in physics, see [3.20, 21].

3.2 Effect of Strain on the Band Structure of GaN

In recent years, strain in conventional group-III-V semiconductors has been
a much coveted feature for its beneficial effects [3.22]. In the world of
GaN, however, it is not necessarily a desirable commodity but a nemesis
brought upon by the lack of lattice- and thermal-matched substrates and
uncomfortably large lattice and thermal mismatch with its ternaries. It is
therefore imperative that strain effects be considered. Figure 3.5 exhibits
the valence-band structure of GaN in the x,y plane under biaxial compressive strain and uniaxial strain in the c plane with the direction of strain as in
Fig. 3.5c. There are no major changes in the HH, LH and CH bands, other
than crystal splitting becoming larger, with the hole effective mass remaining heavy and the density of states staying high, and the crystal symmetry
remaining the same, C_{6v}^4. In contrast, the uniaxial strain in the c plane
causes an anisotropic energy splitting in the x,y plane which leads to a symmetry lowering from C_{6v}^4 to C_{2v}. When a compressive uniaxial strain is induced along the y direction, the HH band in the x direction and the LH band
in the y direction move to higher energies. This causes a reduction in the
density of states. A tensile uniaxial strain along the x direction has the same
effect. On the other hand, when a tensile uniaxial strain is induced along
the x direction, the HH band in the x direction and the LH bands along the y
direction move to lower energies; this causes a reduction in the density of
states. A tensile compressive strain along the x direction has the same effect.

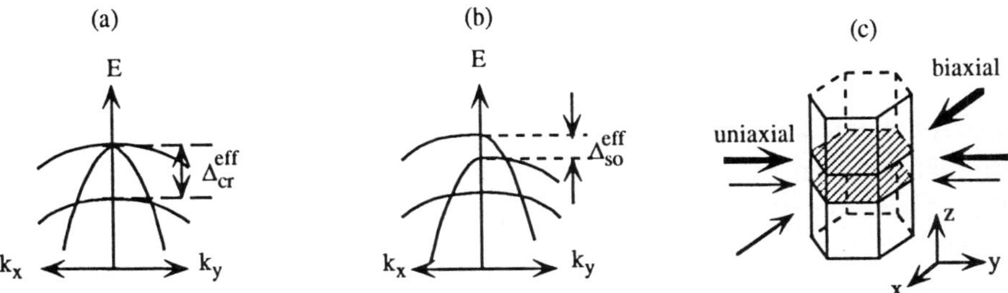

Fig. 3.5. The valence-band structure of GaN under (**a**) biaxial strain in the c-plane, (**b**) uniaxial strain in the c-plane, and (**c**) schematic of the particulars of the strain. After [3.9]

3.3 k·p Theory and the Quasi-Cubic Model

The conduction and valence bands of nitride semiconductors are comprised of s- and p-like states, respectively. Unlike the conventional ZB III-V semiconductors and the lack of a high degree of symmetry, the crystal field present removes the degeneracy at the top of the conduction band. Moreover, unlike the ZB case, the spin-orbit splitting is very small and makes all three bands in the valence band closely situated in energy. Consequently, the three valence bands and the conduction band must be considered in unison and this makes the use of an 8×8 **k·p** Hamiltonian imperative. Since the bandgaps of nitrides are very large, the coupling between the conduction and valence bands can be treated as a second-order perturbation which allows the 8×8 Hamiltonian to be split into one 6×6 Hamiltonian dealing with the valence band and another 2×2 dealing with the conduction band [3.9]. As indicated above, the conduction band is made of s-like states which means that it can be treated as parabolic with the dispersion relation

$$E(k) \; = \; E_{c0} + \frac{\hbar^2 k_z^2}{2m_c^{\parallel}} + \frac{\hbar^2 (k_x^2 + k_y^2)}{2m_c^{\perp}} + a_c^{\perp}(\epsilon_{xx} + \epsilon_{yy}) + a_c^{\parallel}(\epsilon_{zz}) \qquad (3.1)$$

where $a_c^{\parallel}$ and $a_c^{\perp}$ represent the in-plane and out-of-plane deformation potentials, respectively.

For an isotropic parabolic conduction band Eq.(3.1) reduces to

$$E(k) \; = \; E_{c0} + \frac{\hbar^2 k^2}{2m_c} + a_c \epsilon \; . \qquad (3.2)$$

E_{c0} is the conduction-band energy at the $k=0$ point, ϵ is the strain, and a_c is the deformation potential for the conduction band. The other terms have their usual meanings. It should be pointed out that we are dealing with a linear system.

The strain-stress relationship for a hexagonal crystal with C_{6v} symmetry is expressed as [3.9]

$$
\begin{vmatrix} \sigma_{xx} \\ \sigma_{yy} \\ \sigma_{zz} \\ \sigma_{xy} \\ \sigma_{yz} \\ \sigma_{zx} \end{vmatrix}
=
\begin{vmatrix}
C_{11} & C_{12} & C_{13} & 0 & 0 & 0 \\
C_{12} & C_{22} & C_{13} & 0 & 0 & 0 \\
C_{13} & C_{13} & C_{33} & 0 & 0 & 0 \\
0 & 0 & 0 & C_{44} & 0 & 0 \\
0 & 0 & 0 & 0 & C_{55} & 0 \\
0 & 0 & 0 & 0 & 0 & C_{66}
\end{vmatrix}
\begin{vmatrix} \epsilon_{xx} \\ \epsilon_{yy} \\ \epsilon_{zz} \\ \epsilon_{xy} \\ \epsilon_{yz} \\ \epsilon_{zx} \end{vmatrix}
\tag{3.3}
$$

where C_{ij} denotes the elastic stiffness coefficient with $C_{66} = (C_{11}-C_{12})/2$. If the crystal is strained in the (0001) plane, and allowed to expand and constrict in the [0001] direction, $\sigma_{zz} = \sigma_{xy} = \sigma_{yz} = \sigma_{zx} = 0$, $\sigma_{xx} \neq 0$ and $\sigma_{yy} \neq 0$, and the strain tensor has only three non-vanishing terms, namely,

$$
\epsilon_{xx} = \epsilon_{yy} = \frac{a-a_0}{a_0} , \quad \epsilon_{zz} = \frac{c-c_0}{c_0} = -\frac{C_{13}}{C_{33}}(\epsilon_{xx}+\epsilon_{yy}) .
\tag{3.4}
$$

If the in-plane strains in the x and y directions are identical, $\epsilon_{xx} = \epsilon_{yy}$.

When the crystal is uniaxially strained in the (0001) c plane and free to expand and constrict in all other directions, σ_{zz} is the only non-vanishing stress term, and the strain tensor is reduced to

$$
\begin{vmatrix} \epsilon_{yy} \\ \epsilon_{zz} \end{vmatrix}
=
\frac{1}{C_{13}^2 - C_{11}C_{33}}
\begin{vmatrix} C_{12}C_{33} - C_{13}^2 \\ C_{11}C_{13} - C_{12}C_{13} \end{vmatrix}
\epsilon_{xx} .
\tag{3.5}
$$

The values of the elastic stiffness coefficients for GaN have been measured by *Sheleg* and *Savastenko* [3.23] and reproduced in [3.9]. These data are listed in Table 2A.2.

For the band structure near the top of the valence band maximum, the 6×6 Hamiltonian with biaxial strain (compression or extension) in the c-plane can be expressed as

$$
\begin{bmatrix}
F & -K^* & -H^* & 0 & 0 & 0 \\
-K & G & H & 0 & 0 & \Delta \\
-H & H^* & \lambda & 0 & \Delta & 0 \\
0 & 0 & 0 & F & -K & H \\
0 & 0 & \Delta & -K^* & G & -H^* \\
0 & \Delta & 0 & H^* & H & \lambda
\end{bmatrix}
\tag{3.6}
$$

on the base

$$
\left\{
\begin{array}{ll}
|11\uparrow\rangle & = -(X\uparrow+iY\uparrow)/2^{\frac{1}{2}} \\
|1\bar{1}\uparrow\rangle & = (X\uparrow-iY\uparrow)/2^{\frac{1}{2}} \\
|10\uparrow\rangle & = Z\uparrow \\
|1\bar{1}\downarrow\rangle & = (X\downarrow-iY\downarrow)/2^{\frac{1}{2}} \\
|11\downarrow\rangle & = -(X\downarrow+iY\downarrow)/2^{\frac{1}{2}} \\
|10\downarrow\rangle & = Z
\end{array}
\right.
\tag{3.7}
$$

where

$$
\lambda = \frac{\hbar\Delta}{2m_0}[A_1 k_z^2 + A_2(k_x^2+k_y^2)] + D_1\epsilon_{zz} + D_2(\epsilon_{xx}+\epsilon_{yy}) , \tag{3.8a}
$$

$$
\theta = \frac{\hbar\Delta}{2m_0}[A_3 k_z^2 + A_4(k_x^2+k_y^2)] + D_3\epsilon_{zz} + D_4(\epsilon_{xx}+\epsilon_{yy}) , \tag{3.8b}
$$

$$
F = \Delta_1 + \Delta_2 + \lambda + \theta . \tag{3.8c}
$$

$$
G = \Delta_1 - \Delta_2 + \lambda + \theta , \tag{3.8d}
$$

$$
K = \frac{\hbar^2}{2m_0}A_5(k_x+jk_y)^2 , \tag{3.8e}
$$

$$
H = \frac{\hbar^2}{2m_0}A_6(k_x+jk_y)k_z , \tag{3.8f}
$$

and

$$
\Delta = 2^{\frac{1}{2}}\Delta_3 , \tag{3.8g}
$$

with Δ_1 and $\Delta_{2,3}$ being the crystal-field and spin-orbit splitting energies, respectively, and A_i the valence-band parameters corresponding to the Luttinger parameters in the ZB system. The D_i parameters represent the deformation potentials for the valence band, and ϵ_{ij} $(i,j = 1,2,3)$ are the strain terms where the diagonal terms are positive for tensile strain.

3.4 Quasi-Cubic Approximation

The genesis of the quasi-cubic approximation relies on the fact that the Wz and ZB structures are both tetrahedrally coordinated and hence are closely related. The nearest-neighbor coordination is the same for Wz and ZB structures, but differs at the next-nearest neighbor positions. The basal plane {0001} of the Wz structure corresponds to one of the {111} planes of the ZB. When the in-plane hexagons are lined up in Wz and ZB structures, the Wz [0001], [11$\bar{2}$0], and [1$\bar{1}$00] planes are parallel, in order to ZB [111], [10$\bar{1}$], and [1$\bar{2}$1] planes. This, in turn, leads to correlations between the symmetry direction and the k-points for the two polytypes. There are, however, twice as many atoms in the Wz unit cell as there are in the ZB one. In addition to the band-structure similarities between the doubled ZB and Wz structures, one can establish a correlation between the Luttinger parameters in the ZB system and parameters of interest in the Wz system by taking the z axis along the [111] direction, and the x and y axes along the [11$\bar{2}$] and [$\bar{1}$10] directions. For details regarding the symmetry relations between the ZB and Wz polytypes, refer to *Lambrecht* and *Segall* in [3.24]. Doing so leads to

$$
\begin{aligned}
\Delta_2 &= \Delta_3 \, , \\
A_1 &= A_2 + 2A_4 \, , \\
A_3 &= -2A_4 = 2^{1/2}A_6 - 4A_5 \, , \\
A_7 &= 0 \, , \\
D_1 &= D_2 + 2D_4 \, , \\
D_3 &= -2D_4 = 2^{1/2}D_6 - 4D_5 \, .
\end{aligned}
\tag{3.9}
$$

The A parameters can be related to the classical Luttinger parameters γ_i through

$$
\begin{aligned}
A_1 &= -(\gamma_1 + 4\gamma_3) \, , \\
A_2 &= -(\gamma_1 - 2\gamma_3) \, , \\
A_3 &= 6\gamma_3 \, , \\
A_4 &= -2\gamma_3 \, , \\
A_5 &= -(\gamma_2 + 2\gamma_3) \, , \\
A_6 &= -2^{1/2}(2\gamma_2 + \gamma_3) \, .
\end{aligned}
\tag{3.10}
$$

The calculated values of the spin-orbit and crystal-field splitting parameters along with those deduced from the observation of A, B and C excitons are listed in Table 3A.1. The calculations agree well in terms of the spin-orbit splitting, but the theoretical crystal-field splitting is much too large compared to experimental data. The discrepancy may be due to the unaccounted residual strain and strain inhomogeneities present in GaN films, as well as the inaccuracy of the parameter values. The debate will probably continue until strain-free or homogeneously strained films can be prepared.

The calculated and, when available, effective hole masses for all three valence bands are compiled in Table 3A.2. The values seem reasonable in light of the available experimental data except that the heavy-hole mass deviates from the commonly accepted value of 0.8. However, there are suggestions that the heavy-hole mass may be as large as 2, in which case, the issue is open. One can also argue that all three valence bands are very close in energy, and there may be mixing among them. In averaging the masses, one gets an average value of about 1.

Tabulated in Table 3A.3 are the calculated and experimental electron effective masses. The large bandgaps of nitrides lead to large effective masses of electrons. As can be seen, the anisotropy is negligible for both GaN and AlN.

The Luttinger-like parameters along with deformation potentials for the valence band are listed in Table 3A.4. The A parameters transformed from the Luttinger parameters obtained with the help of the quasi-cubic approximation are also given; they are in good agreement with the calculated values. This goes on to prove that the quasi-cubic approximation, which greatly simplifies the problem, is a good one.

The deformation potentials determined by the PLAPW calculations of *Suzuki* and *Uenoyama* [3.9] are compiled in Table 3A.5 for GaN and AlN. Biaxial and uniaxial strains have been introduced and reduced shifts in the Γ point energy to which a linear fit in terms of strain, was obtained. From the linear fit, the deformation potential values for the valence band were deduced. The figures obtained from the quasi-cubic approximation are listed, too. The good agreement between the calculated values and those determined from the quasi-cubic model is strikingly obvious, which is indicative of the excellence of the quasi-cubic approximation.

Assuming an approximately spherical potential in the neighborhood of the N atoms, the higher energy of the two spin states is the one in which the electron spin and the orbital angular momentum are parallel. This result is also anticipated on the basis of the atomic spin-orbit splitting in which the $P_{3/2}$ state is known to have an energy higher than the $P_{1/2}$ state. The contributions of spin-orbit interaction and the crystal-field perturbation to the experimentally observed splittings ($E_{1,2}$ and $E_{2,3}$) have been calculated with different LCAO approximations [3.25, 26].

The large effective mass and the small dielectric constant of GaN, relative to more conventional group-III-V semiconductors, lead to relatively large exciton binding energies and makes excitons, together with large exciton recombination rates, clearly observable even at room temperature. The bottom of the conduction band of GaN is predominantly formed from the s-levels of Ga, and the upper valence-band states from the p-levels of N. *Hopfield* and *Thomas* [3.26] have treated the wurtzite energy levels as a perturbation of those in the zincblende structure. Using the quasi-cubic model of *Hopfield* [3.28], we have

$$E_1 = 0, \tag{3.11}$$

$$E_2 = \frac{\Delta_{cr} + \Delta_{s0}}{2} + \left[\left(\frac{\Delta_{cr} + \Delta_{s0}}{2}\right)^2 - \frac{2}{3}\Delta_{cr}\Delta_{s0}\right]^{1/2} , \qquad (3.12)$$

$$E_3 = \frac{\Delta_{cr}\Delta_{s0}}{2} - \left[\left(\frac{\Delta_{cr} + \Delta_{s0}}{2}\right)^2 - \frac{2}{3}\Delta_{cr}\Delta_{s0}\right]^{1/2} , \qquad (3.13)$$

where Δ_{cr} and Δ_{s0} represent the contributions of uniaxial-field and spin-orbit interactions, respectively, to the splittings $E_{1,2}$ and $E_{2,3}$.

3.5 Confined States

When the size of the well is comparable to the de Broglie wavelength in the semiconductor forming the well, the conduction and valence bands are modified in that the density of states in both the conduction and valence bands are discretized. This picture is depicted schematically in Fig. 3.6a and b for a wurtzite semiconductor quantum well. Moreover, equally important is the modification of these energies in the semiconductor caused by strain effects. In short, eigenstates of strained-layer superlattices require the consideration of both strain and quantum size effects. Eigenstates applicable

a   GaN/AlGaN Quantum Well

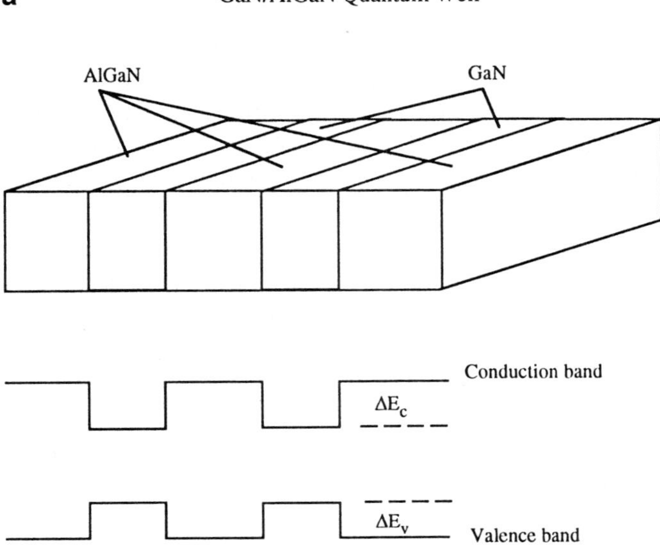

Fig. 3.6. Caption see next page

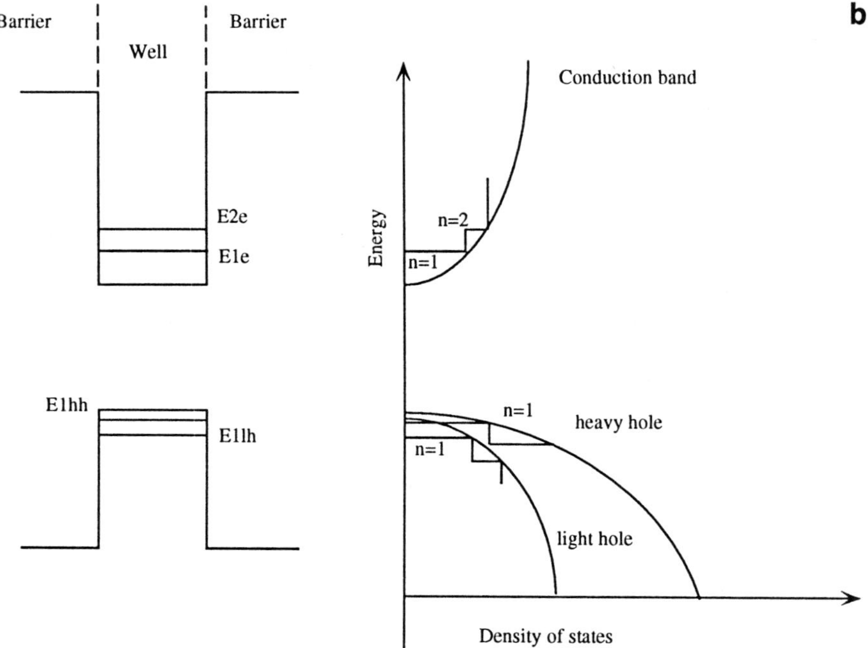

Fig. 3.6. Schematic representation of (**a**) a GaN/AlGaN multiple quantum-well structure and the conduction- and valence-band edges at the Γ point, and (**b**) valence- and conduction-band diagrams of the same with the confined states indicated

to the early versions of compound-semiconductor-based quantum wells without strain were modeled by the envelope-function formalism of *Bastard* [3.29]. If this method is to be used, one has to calculate the strain-dependent bandgap first. *Bastard's* formalism can then be utilised to determine the transition energies in quantum wells, which must be added to the strain component, as shown by *Marzin* [3.30]. He developed a method for calculating the bandgap of a strained cubic semiconductor which, when employed in conjunction with the envelope-function approximation, leads to eigenstates in quantized structures of strained systems.

In *Marzin's* revision of *Bastard's* method, an 8×8 Kane Hamiltonian matrix [3.31] is employed to give an accurate description of a strained quantum structure. The size of the Hamiltonian is justified since there are three valence bands, in addition to the conduction band, with spin up and spin down for each band. Similar to cubic semiconductors, the wurtzite phase also has one conduction band and three valence bands, heavy and light hole states as well as spin-orbit bands, each with spin up and down necessitating the use of an 8×8 Hamiltonian. The strain effects in the GaN system has been treated by *Gil* et al. [3.32, 33] and others, as reviewed in [3.9]. However, until the time when uniformly strained films can be grown, correlations to experiments will remain weak and reduce the level of confidence in the predictions especially and the analysis as a whole. This is particularly true for InGaN wells as there is also phase separation to deal with, in addition to non-uniform strain.

We shall once again emphasize that there exist many discrepancies concerning the determination of some important parameters for the relevant bulk material properties such as hole masses, bandgap-energy bowing parameters, shear and deformation potentials, and band offsets, which are necessary for determining and understanding the properties of confined structures. Calculations, using ab-initio methods, have been applied to estimate the unknown but necessary parameters for the band-structure calculations. More recently, *Sirenko* et al. [3.14] have performed envelope-function calculations of the valence band in wurtzite quantum wells following the formalism of Rashda-Sheka-Pikus (RSP) developed for bulk wurtzite semiconductors. Employing a 6×6 Luttinger-Kohn model, *Ahn* [3.34] studied the effect of a very strong spin-orbit split-off band coupling on the valence-band structure of GaN-based materials. Considering that the spin-orbit band is extremely important for GaN because of its very narrow spin-orbit splitting (10 meV), the spin-orbit split-off coupling was taken into consideration in the calculations. In addition, it was assumed that the electrons in QW are confined by the conduction-band offset (ΔE_c) and the holes by the valence-band offset (ΔE_v), the values of which are also of some controversy.

When the size of the well is comparable to the Bohr radius of an exciton in the semiconductor forming the well. The exciton transition energies are modified. The Bohr radius is given by

$$a_0 = \frac{4\pi \epsilon_s \hbar^2}{m^* q^2} , \tag{3.14}$$

where ϵ_s is the dielectric constant of the semiconductor. In GaN, due primarily to its large effective mass, this radius is about 28 Å necessitating very small physical dimensions before noticeable quantization can occur. In an experimental quantum well, the wave function is constricted along the growth direction which we shall term the z direction. In the orthogonal directions, the system is free.

The problem is similar to that of a vibrating string with the ends held stationary. The vibration wavelengths are given by

$$\lambda_n = \frac{2L}{n} \quad \text{with} \quad n = 1, 2, 3 \tag{3.15}$$

where L represents the length of the string.

In a quantum well, the rapidly varying Bloch waves will be affected by the barriers in the z direction, and the effect can be lumped into an envelope function which is slowly varying. The wave function can be expressed as [3.9, 31]

$$\Psi_n(k_\perp, z) = \sum_v f_n(k_\perp, z) u_v \exp(ik_\perp r_\perp) \qquad (3.16)$$

where $f_n(k_\perp, z)$ and u_v represent the envelope and Bloch functions, respectively, and n is the subband index. The summation is performed for spin up and down of the conduction band where the value of six is assumed for the three valence bands. The wave-function expression must be solved for the conduction band and the valence band with the envelope function satisfying Schrödinger's equation for the particular potential-barrier height [3.9]. For the case where the conduction band is almost s-like, the Γ_7 state suffices. However, for the valence band, due to band mixing, a 6×6 Hamiltonian including all the three uppermost valence bands must be used. Moreover, if strain is present which is the case in almost all nitride-based structures, the Hamiltonian must include the effect of strain as well. To complicate matters further, the piezoelectric effect induces large electric fields at the hetero-interfaces, particularly in samples utilizing InGaN wells.

The envelope function must satisfy

$$\sum_{v'} \left[H_{vv'} \left(k_\perp, \frac{\hbar}{i} \frac{\partial}{\partial z} \right) + V_v(z)\delta_{vv} + H_{vv'}(\epsilon) \right] f_{v,v'}(k_\perp, z)$$

$$= E_n(k_\perp) f_{v,v'}(k_\perp, z) \qquad (3.17)$$

where $\nu = 1, 2$ for the conduction band and $\nu = 1, ..., 6$ for the valence-band states. The details of how the dispersion of the conduction- and valence-band states are computed and can be found in [3.9]. Suffice it to say that the conduction band is formed of nearly s-like states and can be considered nearly parabolic ameliorating the confined state calculations, as will be discussed in the next section. In finding the total energy between the confined conduction- and valence-band states, one may assume that the majority of the contribution is due to the confinement energy in the conduction band because of the large disparity between the electron and hole effective masses in favor of the conduction band. However, when gain in semiconductor lasers is considered, the dispersion of valence-band states must be taken into consideration.

3.6 Conduction Band

If the potential barrier is infinitely high, the wave vector in the z-direction will be quantized and assumes the discrete values of

$$k_{zn} = \frac{2\pi}{\lambda_n} = \frac{n\pi}{L}, \quad n = 1,2,3 \tag{3.18}$$

where L is the thickness of the quantum well. Assuming a parabolic band structure which satisfactorily describes the s-like conduction band, the confinement energy can be expressed as

$$\Delta E_{conf} = \frac{\hbar^2 k_{zn}^2}{2m^*} = \frac{\hbar^2}{2m^*}\left(\frac{n\pi}{L}\right)^2, \quad n = 1,2,3 \ . \tag{3.19}$$

Taking into account the energy dispersion relationship in the x and y direction for a parabolic conduction band, we have

$$E = E_c + \frac{\hbar^2}{2m^*}\left[\left(\frac{n\pi}{L}\right)^2 + k_x^2 + k_y^2\right] \tag{3.20}$$

with n = 1, 2, 3, and E_c representing the conduction-band edge.

If the barrier potential is large, but not infinite, the wave function outside the well decays exponentially, which is called the **evanescent wave**. When the potential barrier is not infinitely high, no analytical solution exits for the subband energies. Graphical solutions treated in many textbooks on quantum mechanics and numerical solutions as well do exist. The problem is made even more complicated in semiconductors in that not only is the barrier not infinite, but also the barrier and well materials do not have the same carrier mass. In this case, the boundary condition must be changed from the continuity of the derivative of the wave function in the z direction to the continuity of the particle flux in the z direction, i.e.,

$$\frac{1}{m_B^*}\frac{\partial f_{nB}}{\partial z} = \frac{1}{m_W^*}\frac{\partial f_{nW}}{\partial z} \tag{3.21}$$

at z = $\pm$L/2, assuming the origin of the z axis to be in the middle of the well. The terms m_B^* and m_W^* represent the effective masses in the barrier and well materials, respectively. Likewise, f_{nB} and f_{nW} depict the envelop wave functions in the barrier and well materials, respectively. The solution for the subband energies can be computed numerically.

The conduction-band minimum for GaN as well as AlN is at the zone center and two-fold degenerate. The confinement energies for the GaN/ AlGaN quantum wells can reasonably be estimated by means of the envelope-function approximation in the same manner as that extensively used for the GaAs/AlGaAs material system [3.35]. Following the *Weisbuch* and *Vinter* notation, the low-lying conduction electron state can be represented by [3.35, 36]

$$\Psi(r) = \sum_{j=W,B} \exp(ik_\perp r) u_c^i(r) f_n(z) \qquad (3.22)$$

where W and B represent the well and barrier materials, $u_c^j(r)$ is the conduction-band zone-center Bloch wave function of GaN or AlGaN, and $f_n(z)$ is a slowly varying envelope function, $k_\perp$ is the transverse (in-plane) wave vector, $f_n(z)$ is the envelop wave function, and the growth direction is along the z axis. Since $u_c^j(r)$ is the same for GaN and Al(Ga)N, Schrödinger's equation reduces to

$$\left[\frac{-\hbar^2}{2m^*(z)} \frac{\partial^2}{\partial^2 z} + V(z) \right] f_n(z) = E_n f_n(z) \ . \qquad (3.23)$$

In the above equation, $m^*(z)$ is the corresponding effective mass of the conduction electron, $V(z)$ represents the profile of the minimum of the conduction band along the growth direction, and $E_n(z)$ is the confinement energy. Assuming no doping in either regions, $V(z)$ has a rectangular well-like profile. The solution of the confinement energies is similar to a particle in-a-box problem. The boundary conditions are that $f_n(z)$ and $1/m^*(z) \cdot df_n(z)/dz$ are continuous across the interface. The latter is necessary to ensure the conservation of the particle current. In Fig.3.7, plots of the conduction-band ground-state (solid line) and excited-state (dashed line) confinement energies for a GaN/$Al_{0.15}Ga_{0.85}N$ quantum well as a function of well width are displayed. GaN/AlGaN quantum wells may be prominent in that use as the active layer in GaN-based short-wavelength injection mode lasers. In generating the plot, a linear interpolation was used with the bandgap energies of AlN and GaN being 6.2 eV and 3.42 eV, respectively, to calculate the AlGaN bandgap. The effective conduction-electron masses of $0.22m_e$, and $0.27m_e$ have been adopted for GaN and $Al_{0.15}Ga_{0.85}N$. The effective mass of $Al_{0.15}ga_{0.85}N$ was deduced through an interpolation of AlN for 0.27eV which can be found in Table 2A.4.

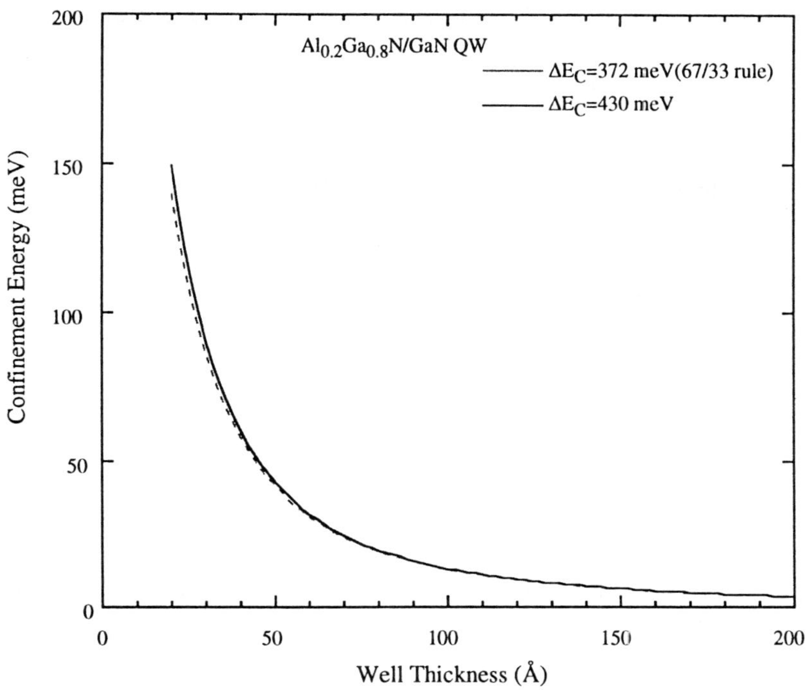

Fig. 3.7. Plot of the conduction-band ground state (*solid line*) and excited state (*dashed line*) confinement energies for a GaN/Al$_{0.15}$Ga$_{0.85}$N quantum well as a function of well width (the masses for GaN and AlN have been assumed the same at 0.19). Hole effective masses of 0.22 for GaN and 0.27 for AlN have been as listed in Table 2A.4. The value for AlGaN has been obtained by means of a linear interpolation between the binary end points

3.7 Valence Band

In an attempt to determine the valence-band subband structure, *Suzuki* and *Uenoyama* [3.9] calculated the band discontinuities from first-principles calculations and found them to be 0.11 and 0.43 eV for the valence and conduction bands of GaN/Al$_{0.2}$Ga$_{0.8}$N. The elastic-stiffness constants, taken from prior experimental data, employed for GaN were, in units of 10^{11} dyn·cm^{-2}, 29.6, 13.0, 15.8, 26.7, and 2.41 for C_{11}, C_{12}, C_{13}, C_{33}, and C_{44}, respectively (see Table 2A.2 for details). For illustrative purposes, the valence-band structure in unstrained 30 and 50 Å GaN/Al$_{0.2}$Ga$_{0.8}$N quantum wells is exhibited in Figs.3.8a and 3.9a where the strain due to the lattice and thermal mismatch are neglected. Bandgap discontinuities of 0.11 and 0.43 eV were adopted for the valence and conduction bands, respectively. The confinement energies in deep wells are inversely proportional to

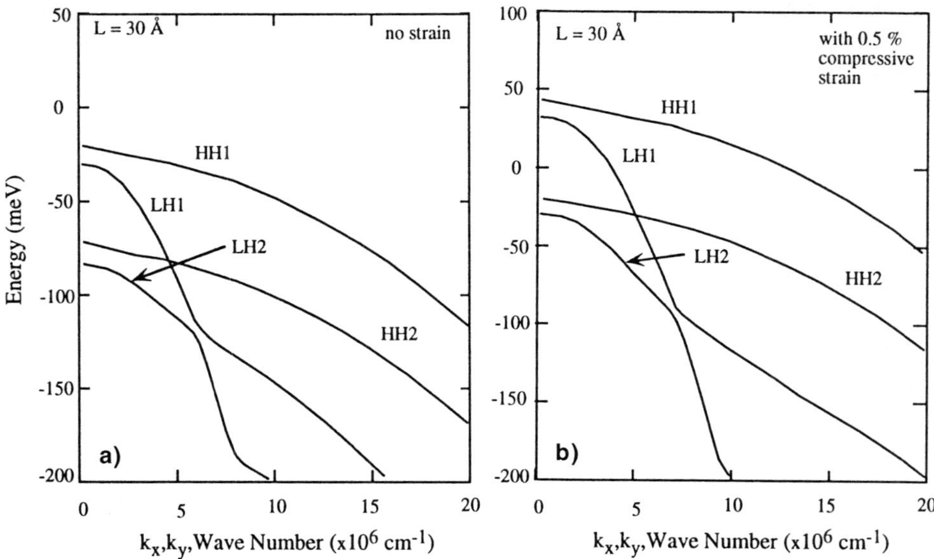

Fig. 3.8a,b. Plot of the upper valence-band structure (HH, LH and CH bands) in a 30 Å GaN/Al$_{0.2}$Ga$_{0.8}$N quantum well with bandgap discontinuities of 0.11 and 0.43 eV for the valence and conduction bands, respectively; (**a**) without strain and (**b**) with 0.5 compressive strain in the c-plane. After [3.10]

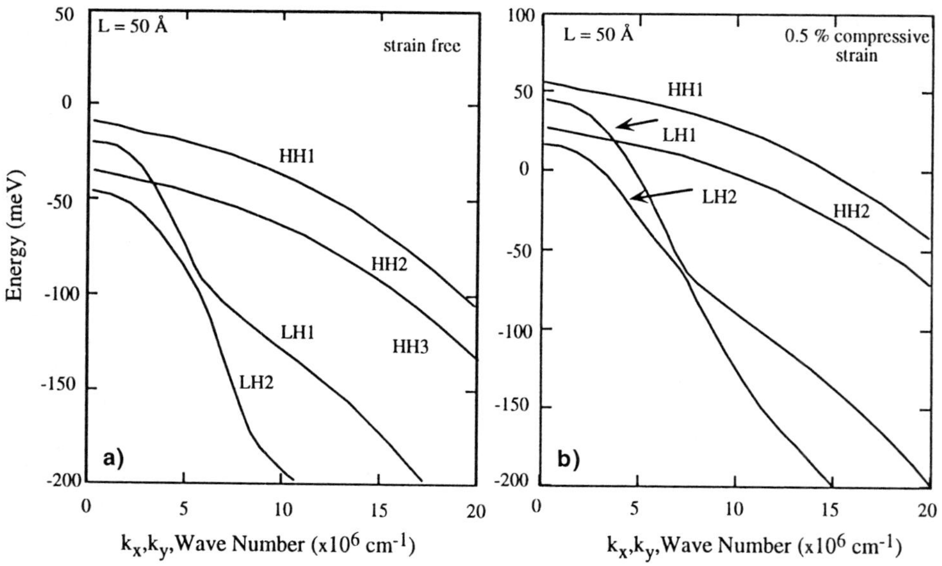

Fig. 3.9a,b. Plot of the upper valence-band structure (HH, LH and CH bands) in a 50 Å GaN/Al$_{0.2}$Ga$_{0.8}$N quantum well with bandgap discontinuities of 0.11 and 0.43 eV for the valence and conduction bands, respectively; (**a**) without strain and (**b**) with 0.5 compressive strain in the c plane. After [3.10]

the effective mass in the growth direction and directly proportional to the square of the well length. The HH and LH bands are not coupled. Consequently, the HH band can be construed as parabolic with little, if any, change with strain. This is because the C_{6v} crystal symmetry of the bulk remains. The upper (LH) and lower (CH) bands are coupled with the constant coupling coefficient $2^{1/2}\Delta_3$, which means that coupling of these bands is independent of k_z and strain. The effective masses of the HH and LH bands are too heavy to cause substantial confinement energy whereas the CH band with a lighter mass causes more split due to quantization, and, in a sense, makes the crystal splitting (Δ_{cr}) larger. For comparison, in ZB structures the coupling between the light hole and the spin-orbit band is dependent on k_z and thus the bands change considerably with strain, and the in-plane heavy-hole mass becomes light and the in-plane light-hole mass becomes heavy.

The effect of strain on the quantum well was also considered by *Suzuki* and *Uenoyama* [3.10] assuming coherently strained quantum wells. Consequently, the in-plane lattice constant of the layers is assumed to be that of the substrate with a substantial effect on the subband structure of the quantum wells. The beneficial effects of strain in electronic and optoelectronic devices based on ZB crystals have been documented well [3.5]. This and the fact that there is some lattice mismatch between GaN and its ternar-

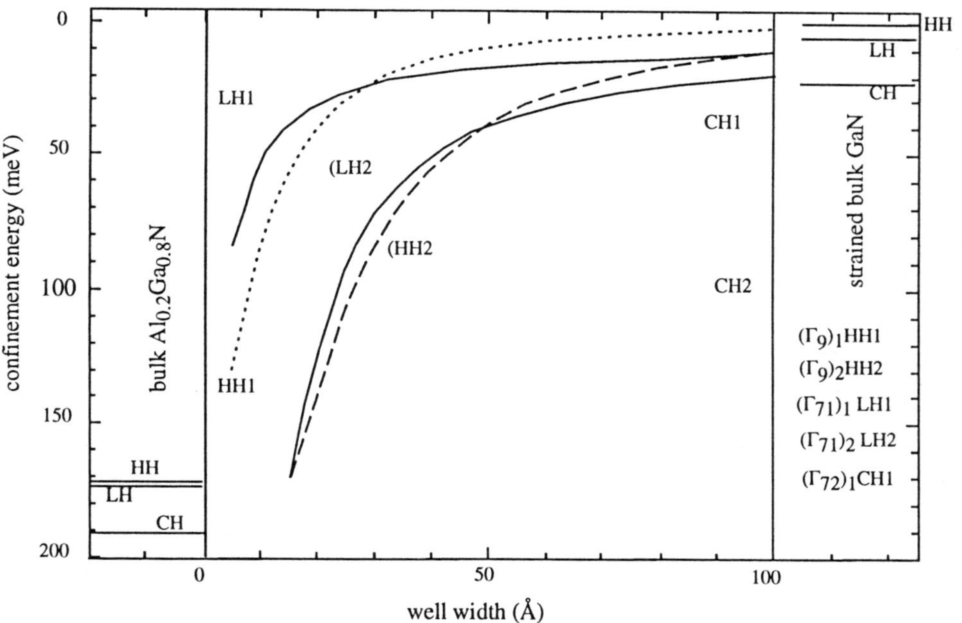

Fig. 3.10. Valence-band confinement energies vs. the wells thickness for GaN/Al$_{0.2}$Ga$_{0.8}$N quantum wells. After [3.38]

ies make it imperative that the effect of strain on the properties of Wz quantum wells be considered. Figures 3.8b and 3.9b exhibit the valence-band subband structure for 30 and 50 Å GaN/Al$_{0.2}$Ga$_{0.8}$N quantum wells. The biaxial strain was assumed to be 0.5% in the (0001) c-plane. Superimposed is the valence-band structure for 30 and 50 Å Wz GaN/Al$_{0.2}$Ga$_{0.8}$N quantum wells with 0.5% tensile strain in the c-plane. Results indicate that the wells get deeper for compressive strain and shallower for tensile strain. The density of states at the valence-band maximum gets smaller for the compressive and larger for tensile strain. However, the change is very small as the symmetry remains as in the bulk with no further removal of the degeneracy. It must be mentioned though that uniaxial strain along the x or the y direction only causes the HH band to move to a higher energy and leads to a reduced density of states. Consequently, the effects of this type of strain resemble those in bulk GaN. In a sense the Q well alone does not result in any special characteristic that would lead to much improved results for lasers [3.38].

Valence-band confinement energies for relaxed and 0.5% compressively strained, in the c-plane, GaN/Al$_{0.2}$Ga$_{0.8}$N quantum wells are depicted in Fig.3.10. The data have been deduced from the calculations of *Suzuki* and *Uenoyama* [3.9] for 30, 40, 50 and 60 Å well thicknesses and band discontinuities of 0.11 and 0.43 eV for the valence and conduction bands, respectively. The other parameters utilised can be found in the tables presented in Appendix 2A.

3.8 Exciton Binding Energy in Quantum Wells

Exciton binding energies in reduced-dimensional systems vary from those of the bulk [3.39]. In the GaN/AlGaN system, one would expect the binding energy to go up as the confinement gets stronger. If the well thickness is continually reduced, at some point the overlap with AlGaN becomes very noticeable. Toward zero well thickness, the binding energy should approach the value of AlGaN. Considering the strong localization and possible role of excitons in optical processes even at room temperature, it is imperative that the binding energy be known. *Bigenwald* et al. [3.40] considered the very problem of exciton binding energy in a GaN/Al$_{0.2}$Ga$_{0.8}$N system with results leading to the obvious conclusion that the effect of confinement is large and can not be ignored. They calculated the exciton binding energies and oscillator strengths with the formalism previously developed [3.41] by applying a two-parameter trial function. Due to the anisotropy of the structure, the dielectric constant was globalized and the particle masses were

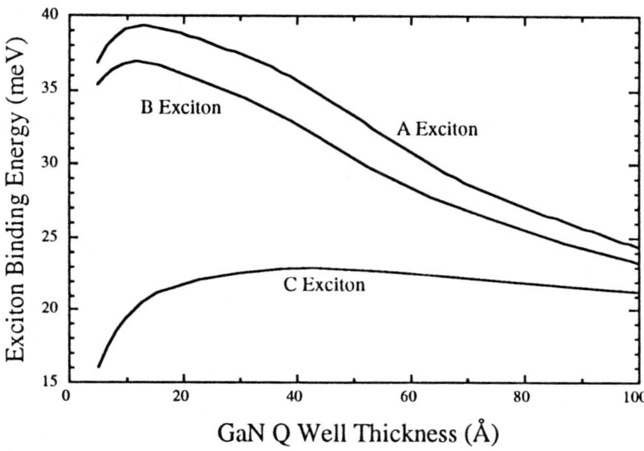

Fig. 3.11. A, B, and C exciton binding energies as a function of the well width in a GaN/Al$_{0.2}$Ga$_{0.8}$N system. After [3.40]

weighted with the probability densities. The A, B, and C exciton binding energies so computed as a function of GaN well thickness are depicted in Fig. 3.11. Two essential points stand out:

The first point is that the Γ_{9v} and Γ_{7v}^1 states are confined to the well as is the electron state. The largest binding energy corresponds to a well thickness of L ≈ 15 Å for which the spreading of both electron and hole functions in the barrier are minimal.

The second point is that the relatively light Γ_{7v}^2 hole state leaks out of the well (for L < 100 Å) that the electron-hole pair has a small binding energy and is almost constant for 20 Å $< $ L $ < 100$ Å. Caution should be exercised in applying the calculations of [3.41] for well thicknesses larger than about 100 Å, being three times the Bohr radius.

In GaN-based systems, the terms quantum well and superlattices have been used very liberally in that structures with well thicknesses well in access of the Bohr radius are referred to as quantum wells. The term superlattice requires that barriers are penetrable by the wave function; and, further, the wave functions in the adjoining wells overlap and form the superlattice minibands. If these standards are strictly applied at this point in time, there may not be much to discuss. Consequently, a conscious decision was made to treat many heterostructures with reduced dimensions, in at least one direction, as **Quantum Confined Structures**.

3.9 Polarization Effects

Group III-V nitride semiconductors exhibit highly pronounced polarization effects. Semiconductor nitrides lack inversion symmetry and exhibit piezoelectric effects when strained along $\langle 0001 \rangle$. In addition, wurtzite GaN has a unique axis which allows spontaneous polarization even in the absence of strain. This manifests itself as a polarization charge at hetero-interfaces. Spontaneous polarization was only recently understood fully by *King-Smith* and *Vanderbilt* [3.42], and *Resta* et al. [3.44]. The piezoelectric effect has two components. One is due to lattice mismatch (misfit) strain while the other is due to thermal strain caused by the thermal expansion-coefficient difference between the substrate and the epitaxial layers. In relative terms, spontaneous polarization is larger than piezoelectric polarization in AlGaN/GaN-based structures. In the case of InGaN/GaN structures, spontaneous polarization is relatively small, as spontaneous polarizations in GaN and InN are not that different from one another. However, the strain-induced piezoelectric polarization can be sizable. If and when defects reduce the strain in the films, the strength of the piezoelectric polarization is lowered. Spontaneous polarization and piezoelectric polarization affect the band structure of heterostructures. The effects are very large and can easily obscure the engineered modifications.

Polarization is dependent on the polarity of the crystal, namely whether the bonds along the c-direction are from cation sites to anion sites, or visa versa. The convention is that the [0001] axis points from the face of the N plane to the Ga plane, and marks the positive z-direction. In other words, when the bonds along the c-direction (single bonds) are from cation (Ga) to anion (N) atoms, the polarity is said to be the **Ga polarity**, and the direction of the bonds from Ga to N along the c-direction marks the [0001] direction which is generally taken to be the +z-direction (Fig.3.12). By a similar argument, when the bonds along the c-direction (single bonds) are from anion (N) to cation (Ga) atoms, the polarity is said to be the **N polarity**, and the direction of the bonds from N to Ga along the c-direction marks the direction which is generally taken to be the −z-direction. To shed further light, the Ga polarity means that if one were to cut the perfect solid along the c-plane where one breaks only a single bond, one would end up with a Ga-terminated surface.

Substrates upon which nitride films are grown lack the wurtzitic symmetry of nitrides. Consequently, the polarity of the films may not be uniform, as schematically depicted in Fig.3.13, where the section on the left is of Ga polarity and the section on the right is of N polarity. These regions

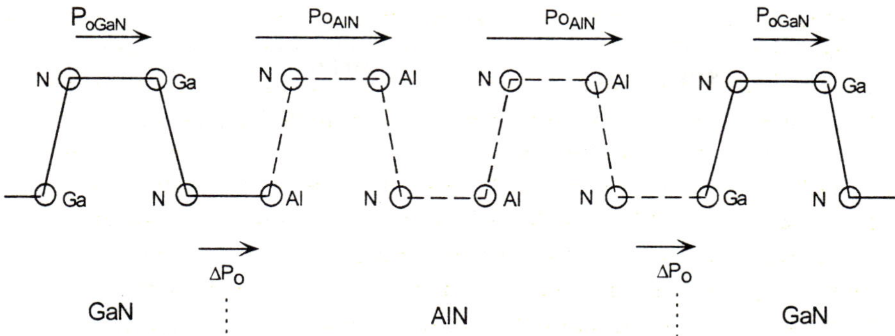

Fig. 3.12. Schematic depicting the convention used for determining the polarity and crystalline direction in wurtzitic nitride films. The diagram shows the case for a Ga-polarity film with its characteristic bonds parallel to the c-axis (horizontal in the figure) going from the cation (Ga or Al) to the anion (N). The spontaneous polarization components P_{0GaN} and P_{0AlN} for a periodic GaN/AlN structure are also indicated with that for AlN having a larger magnitude. The spontaneous polarization is negative and thus points in the [000$\bar{1}$] direction. The cation must be exercised here as there is no long-range polarization field, just that at the interface. The polarization in AlN is larger in magnitude than in GaN. There exist a difference in polarization at the interface, ΔP_0 pointing in the [000$\bar{1}$] direction for both GaN/AlN interfaces

are termed as **inversion domains** and the boundaries between them are called **inversion-domain boundaries**. When inversion domains occur, the alternating nature of anion-cation bonds can not the fully maintained throughout. Inversion domains combined with any strain in nitride-based films lead to flipping PiezoElectric (PE) fields with untold adverse effects on the characterization of nitride films in general and the polarization effect in particular, and on the exploitation of nitride semiconductors for devices. Such flipping fields would also cause much increased scattering of carriers as they traverse in the c-plane.

The spontaneous polarization P_{spont} (also commonly referred to as P_0) in a solid is not well defined. Only those differences in P between two phases which can be linked by an adiabatic transformation that maintains the insulating nature of the system throughout are well defined. For example, one phase can be considered unstrained and the other strained. *Vanderbilt* proved that the polarization difference ΔP between the wurtzite and zincblende phases can be calculated by considering an interface between the two phases and by defining P_{spont} to be zero in zincblende. In short, by calculating the integral of a quantum mechanical Berry phase along a line in Brillouin Zone (BZ) from one end to the other in the bulk wurtzite leads to the polarization P with respect to that in zincblende (which is zero by defin-

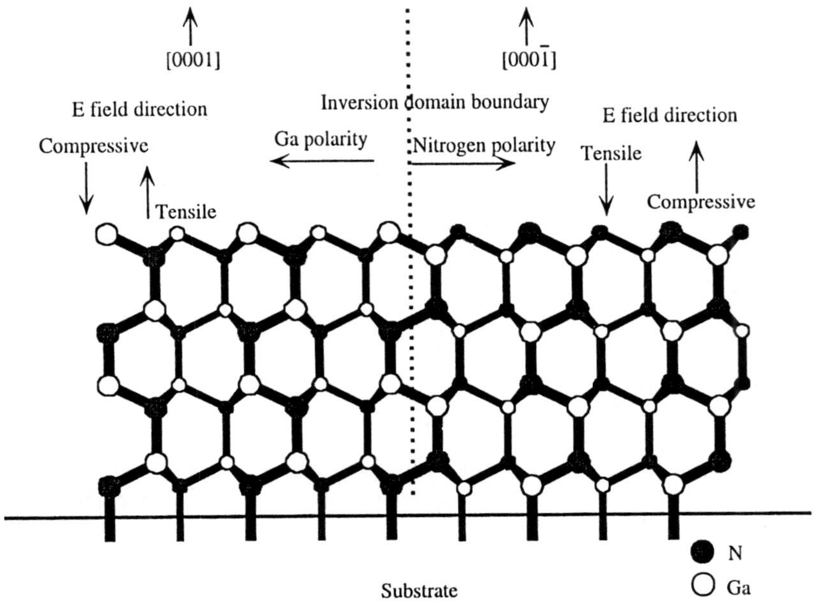

Fig. 3.13. A schematic representation of an inversion-domain boundary. On the left, the film has the Ga-polarity which is designated as the [0001] direction. On the right, the film has the N-polarity which is designated as the [000Ī] direction. Also shown are the directions of the piezoelectric polarization under tensile and compressive strain on both sides

ition because zincblende is cubic and cannot have a spontaneous polarization in an infinite bulk periodic crystal). The Berry phase actually represents an overlap integral between the periodic part of the Bloch function at k and a neighboring k-point, k'. *Bernardini* et al. [3.45] showed that the charges accumulating at each interface in a self-consistent calculation can be obtained from the ΔP of the two bulks. The relation between charge and P follows basically Gauss' law. The spontaneous-polarization charge P_0, as calculated by *Bernardini* et al. [3.44] are -0.081, -0.029, and -0.032 [C/m^2] for AlN, GaN, and InN, respectively. Note that these are all in C/m^2 and that $1 \text{ C/m}^2 = 6.25 \cdot 10^{14}$ e/cm^2.

The same anisotropy which cause piezoelectric polarization can also lead to pyroelectric effects. Such phenomena can be rather important in nitride-based devices as the junction temperature is high by the nature of the applications such as lasers and high-power amplifiers. Consequently, the thermally induced electric field, **pyroelectric effect**, would most likely be present [3.46] with consequences similar to those ascribed to polarization effects.

In order to gain a quantitative understanding of the piezoelectric polarization, the piezoelectric tensor which is defined as the derivative of the polarization with respect to strain, must be considered. First, the displacement vector in a dielectric can be expressed as [3.47]

$$\mathbf{D} = \epsilon \mathbf{E} + \mathbf{P} \tag{3.24}$$

where $\mathbf{E}$ and $\mathbf{P}$ represent the electric-field and polarization vectors. Considering only the piezoelectric effect, the polarization vector is given by

$$\mathbf{P} = \underset{\sim}{\mathbf{d}} \cdot \underset{\sim}{\mathbf{T}} \tag{3.25}$$

where $\underset{\sim}{\mathbf{d}}$ and $\underset{\sim}{\mathbf{T}}$ are the piezoelectric and stress tensors. The number of independent components of d reduces to three: e_{15}, e_{31} and e_{33} due to the wurtzite symmetry. The index 3 corresponds to the direction of the c-axis. Hence, the piezoelectric properties of the Wz structures are somewhat more complicated. However, we restrict ourselves to structures with growth along the [0001] direction when only the e_{31} and e_{33} components are important.

In a heterojunction system, the amplitude of the displacement vector is related to the total interfacial charge by

$$|\mathbf{D}| = qn_s \, , \tag{3.26}$$

where n_s is the total sheet charge, both fixed and mobile.

In heterojunctions containing donors and acceptors, and shallow defects, the associated free carriers within the Fermi statistics diffuse to the semiconductor with the smaller bandgap where they are confined, due to potential barriers, to potential minima. The resulting charge separation due to free carriers causes an internal electric field, *screening field*, which is represented by the first term in (3.24). In addition, an electric field can also be induced by the application of an external voltage such as done through the use of Schottky barriers, metal-oxide semiconductor structures, and p-n junctions. Spontaneous and strain-induced piezoelectric polarization can influence the final status of the interfacial free-charge density in these heterostructures. Any shallow defects (induced fields would change the ionization ratio), free carriers, surface contacts must be included for a complete treatment.

Under the assumption that the layers are coherently strained, the projections of the distorted primitive translation vectors on the interface plane, which are defined by the unit vector $\hat{\mathbf{n}}$ orthogonal to the interface, are equal [3.48]. This condition leads to five independent strain parameters with six independent strain components:

$$\epsilon_{xx} = \epsilon_{xx}^0 + \frac{n_x}{n_z}\epsilon_{xz} \,,$$

$$\epsilon_{yy} = \epsilon_{xx}^0 + \frac{n_y}{n_z}\epsilon_{yz} \,,$$

$$\epsilon_{zz} = \epsilon_{zz}^0 + \frac{n_z}{n_x}\epsilon_{zx} \,, \qquad (3.27)$$

$$\epsilon_{xy} = \epsilon_{xx}^0 + \frac{n_x}{n_y}\epsilon_{xy} \,,$$

$$\epsilon_{yz} = \frac{n_y}{n_x}\epsilon_{xz} \,,$$

where $\epsilon_{xx}^0 = (a-a_0)/a_0$, $\epsilon_{zz}^0 = (c-c_0)/c_0$, with a and c being the lattice constants of the relaxed buffer layer, and a_0 and c_0 representing the lattice constants of the strained layer. The coordinate z is chosen along the c-axis, and x and y lie in the plane normal to c with n_x, n_y, n_z representing the Cartesian components of the unit vector $\hat{\mathbf{n}}$. Minimizing the strain energy with respect to the unknown strain provides the basis for calculating all strain components, with the aid of (3.27), for an arbitrary surface orientation [3.48]. For the (0001) growth direction ($n_z = 1$), we have for the strain:

$$\epsilon_{xx}^0 = \epsilon_{yy}^0 \,, \quad \text{and} \quad \epsilon_{zz}^0 = \frac{c-c_0}{c_0} = -\frac{C_{13}}{C_{33}}(\epsilon_{xx}^0 + \epsilon_{yy}^0) \,. \qquad (3.28)$$

Once the strain components are found, the strain tensor can be determined and with the piezoelectric constants, the strain-induced polarization can be determined. The condition of mechanical equilibrium yields

$$T_3 = 0 \,. \qquad (3.29)$$

Using the relationship between the stress and strain tensors in the Voigt notation,

$$T_1 = \sigma_{lk}\epsilon_k \qquad (3.30)$$

one can find the polarization in the growth direction. For a biaxially strained layer, the magnitude of the strain induced polarization along the c direction is given by

$$P_z^{\text{piezo}} = [e_{31} - (C_{31}/C_{33})e_{33}]\epsilon_\perp \,, \qquad (3.31)$$

where $e_\perp = \epsilon_{xx} + \epsilon_{yy}$ is the in-plane strain, and C_{31} and C_{33} are elastic constants (Chap.2). By using the values for the elastic constants of *Kim* et al. [3.49] and the piezoelectric-polarization data from *Bernardini* et al. [3.45] numerical values for $[e_{31} - (C_{31}/C_{33} e_{33}]$ can be found as -0.86, -0.68, and -0.90 C/m^2 for AlN, GaN, and InN, respectively. For Al$_x$Ga$_{1-x}$N pseudomorphically strained on a relaxed GaN substrate, the strain $\epsilon_\perp$ is expected to be proportional to x and given by $\epsilon_\perp = 2x(a_{GaN} - a_{AlGaN})/a_{AlGaN}$, which is 0.0495x (using 3.112Å for the AlN c-plane lattice constant and 3.189Å for the GaN c-plane lattice constant, with x being in the range of $0 \div 1$. It depicts the AlN mole fraction in the alloy) and is tensile. The piezoelectric polarization is then $P_{piezo} = -4.26x \cdot 10^{-2}$ C/cm^2 or $2.66x \cdot 10^{13}$ e/cm^2 and points in the [000$\bar{1}$] direction.

The corresponding difference in the spontaneous polarization between Al$_x$Ga$_{1-x}$N and GaN is also expected to be proportional to x and is given by ($P_{spon} = -0.052x$ C/m^2 or $3.25x \cdot 10^{13}$ e/cm^2). Note that the two polarization effects have the same sign for Ga polarity and tensile strain in the alloy, and point in the [000$\bar{1}$] direction. For an In$_x$Ga$_{1-x}$N layer, the situation is rather different in that the differential spontaneous polarization between In$_x$Ga$_{1-x}$N and GaN is much smaller, $\Delta P_{spon} = -0.003x$ which translates to $1.88x \cdot 10^{12}$ e/cm^2. Furthermore, the In$_x$Ga$_{1-x}$N layer on GaN would be under compressive stress $\epsilon_\perp = -0.195x$ using 3.533 Å and 3.189 Å for the c-plane lattice constants of InN and GaN. Here x is in the range of $0 \div 1$ and depicts the InN mole fraction, and $P_{piezo} = +0.176x$ or $1.1x \cdot 10^{14}$ e/cm^2. For an InN mole fraction of 0.15, $P_{piezo} = 1.65 \cdot 10^{13}$ e/cm^2. Here, the piezoelectric polarization dominates and is opposite in direction but even larger in absolute magnitude. In the AlGaN case, the sign of the polarization is such that it produces a potential energy for electrons, sloping down from the Ga face towards the N face.

Polarization effects and devices are inextricable. In devices with a large concentration of free carriers, the polarization charge would be screened. In devices where the modulation of charge is the basis of operation such as the MOdulation Doped Field Effect Transistors (MODFETs), a detailed accounting of all polarization charge must be undertaken. In considering a Normal MODFET (N-MODFET) structure where the larger bandgap AlGaN donor layer is deposited on top of a GaN channel layer, both the spontaneous polarization and the piezoelectric polarization must be accounted for. For an N-MODFET structure with Ga polarity, the potential will slope down from the surface of the AlGaN layer towards the AlGaN/ GaN interface. It will help to drive free electrons towards the interface forming a 2DEG (Fig.3.14). For example, if there is an ohmic metal contact on the AlGaN surface, electrons will flow towards the 2DEG below the AlGaN layer from the contacts. Since nitride semiconductors in question

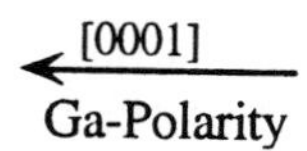

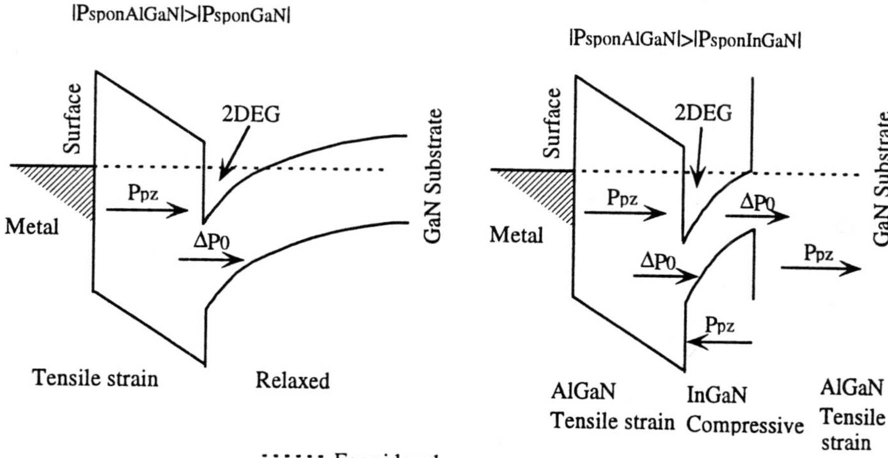

Fig. 3.14. Schematic band structure for an AlGaN/GaN heterostructure having Ga-polarity and with a metal on the AlGaN, which is similar to that used in modulation-doped field-effect transistors. This is shown on the left. The AlGaN layer is under tensile strain and piezoelectric polarization and spontaneous polarization support each other. The same for an AlGaN/InGaN/AlGaN where AlGaN and InGaN layers are assumed under tensile and compressive strains, respectively, is exhibited on the right. The piezoelectric polarization produced by compressive strain supports the bend bending in the well. Consequently, the structure can accommodate large concentrations of electrons if they are present. Perfectly insulating semiconductors are assumed, and free carrier screening is neglected

have large bandgaps, thermal generation rates are minuscule and the role played by thermally generated carriers can be ignored.

The most favorable situation for enhancing sheet carrier concentration would occur for an InGaN (under compressive strain) quantum well on top of a relaxed n-GaN and below an AlGaN barrier (under tensile strain) with the entire structure having cation (Ga) polarity, as shown on the right-side in Fig. 3.14. In that case, the field will slope down towards the InGaN/AlGaN interface in the quantum well, and will help to localize the carriers in the 2DEG. Note that the piezoelectric polarizations estimated here are based on the theoretical values for perfectly insulating material. The field will be screened by the carriers present in each layer. For example, if carriers flow from a metal contact towards the 2DEG, then this will set up a counteracting field. The equilibrium self-consistent field is ultimately determined by the requirement that the chemical potential for electrons (i.e., the Fermi level) must be constant throughout the structure. It thus depends on the doping and band bending in the buffer layer and possibly in each of the layers. At the very least, one may expect that these fields will be reduced by

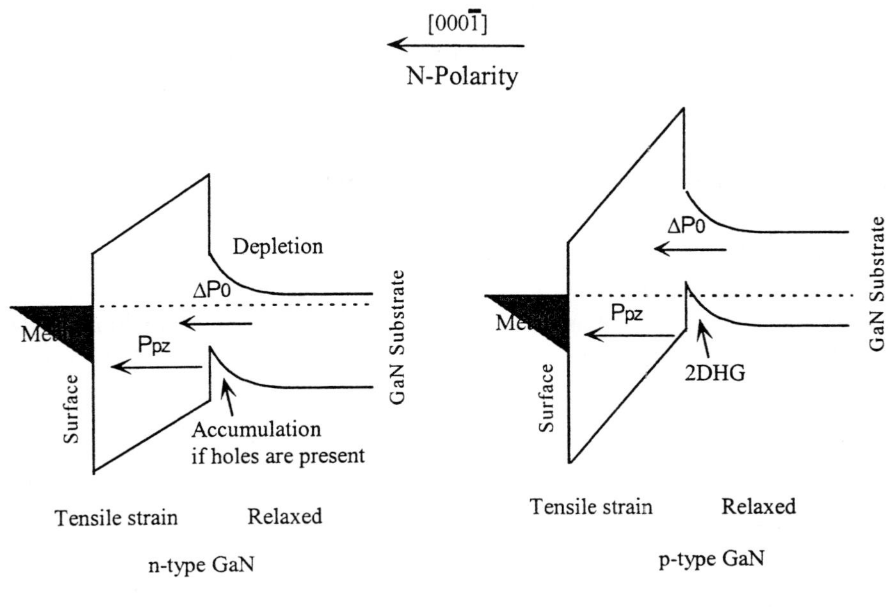

Fig. 3.15. Schematic band diagrams associated with n-GaN and p-GaN layers overlaid with AlGaN for n-polarity. As in the case of Fig. 3.14, the piezoelectric polarization and the spontaneous polarization support one another. Even in the n-type GaN buffer layer case, the interface is favorable for hole accumulation if they are present in the system. In reality, hole accumulation would take place in p-type GaN and not in n-type GaN as the thermal generation rate is miniscule. Perfectly insulating semiconductors are assumed, and free carrier screening is neglected

a factor that corresponds to the macroscopic dielectric constant, i.e. a factor of the order of 10, even possibly larger if the layers acquire conductivity by free carriers. Consequently, a more realistic expectation for the effects on the sheet carrier concentration is of the order of $10^{11} \div 10^{12}$ e/cm^2. A fully consistent solution of the Poisson equation must be found with the spontaneous polarization component as the interface charge in order to get the accurate charge distribution.

A case of an [AlGaN (tensile strained)]/[GaN (relaxed)] heterostructure with nitrogen polarity is shown in Fig.3.15 for n-type and p-type GaN buffer layers. As in the cases depicted in Fig.3.14, the piezoelectric polarization and the spontaneous polarization support one another. Even in the n-type GaN case, the interface is favorable for hole accumulation if they are present in the system. In reality this would happen in a p-type GaN and not in an n-type GaN as the thermal generation rate is miniscule. In inverted MODFETs with Ga-polarity, not shown, where the GaN is situated on top of AlGaN, the spontaneous polarization vector and the piezoelectric polarization vector (AlGaN is assumed to be under tensile strain) point away from the surface. Consequently, the screening charge would be of positive polarity which means that the hole accumulation at the interface is favored providing that holes are present in the system. If, on the other hand, the film were n-polarity, both polarization vectors would point toward the surface, and the screening charge would be negative. This means that electron accumulation at the interface is favored.

The electric field resulting from the piezoelectric polarization in strained GaN can cause a large Stark shift and reduce the effective bandgap which is of paramount importance in optoelectronic devices. The Stark shift can be screened on a length scale of the order of the Debye length $(\epsilon kT/q^2 n)^{1/2}$ by injecting free carriers in the GaN layer(s) as is the case in LEDs, lasers and PL experiments with large excitation intensities. In such a case, the strain-induced field causes carrier separation which, in turn, causes a field which opposes the strain field. Strain-induced polarization in such a heterostructure can also lead to a net electric field which can be measured as a voltage drop across the sample.

The large Stark shifts in quantum wells lead to reduced effective bandgaps. This, in turn, exacerbates efforts to determine the mole fraction-bandgap relationship, particularly in InGaN quantum wells. This is, in part, due to the fact that the preparation of high-quality bulk InGaN is complicated because of phase segregation and re-evaporation of In from the growing surface. Consequently, high-quality films are limited to thin films straddled by InGaN of low mole fraction or GaN or AlGaN. Even bulk layers, being not completely relaxed on non-native substrates, could pave the way to piezoelectric polarization. The polarization-induced field, unless accounted for, causes errors in the determination of mole fractions of the barrier and well layers in quantum wells, and of band discontinuities. In fact, the field induced by the piezoelectric and spontaneous polarization effects could, in many cases, be larger than the effect of engineered charge separation. Such may be the case in the modulation doped-structures depicted in Figs.3.14 and 15. Unless the strain present in the films is coherent and/or the inhomogeneities can be modeled, an accurate description will not be possible.

Polarization can be converted to electric field by dividing it with the dielectric constant. The voltage drop due to the polarization effects can then be found by multiplying the field by the thickness of the layer, such as that of the AlN layer. Underlying assumptions here are that the strain is coherent and uniform in the c-direction, which gives rise to a constant electric field, and the sample does not contain defects and impurities except possibly for misfit dislocations. If the misfit dislocations are present, the piezoelectric polarization would be averaged over the period of misfit dislocations. The field due to any carrier transfer to the interface, free carriers and weakly bound states, is opposite to the strain-induced field. Thus, in calculating the voltage drop, this filed must be subtracted from the strain induced field.

Let us choose the GaN/AlGaN system to demonstrate quantitatively the effect of piezoelectrically induced polarization. Measured piezoelectric constants of AlN have been reported [3.50]. However, this paper deals with ceramic-grade AlN. Since the distortion of atomic orbitals [3.33] causes the piezoelectric polarization under discussion, one would expect non-negligible errors in adopting figures associated with ceramic-grade AlN for single crystals. The piezoelectric constants of GaN have been estimated from ab-initio calculations [3.45] and from the data on electromechanical coupling coefficients (k^2) of GaN [3.51] and AlN for the same orientations [3.52] (the results of which are listed in Table 3A.4). As tabulated, the data on nitrides indicate that the piezoelectric constants are about 4 times larger than those for conventional III-V semiconductors. As expected, the data also indicate that the constants for GaN and ZnO are very close.

A theoretical estimate of e_{14} for cubic GaN [3.53] was used along with the transformation procedure described in [3.54] to get e_{14} for the wurtzite structure (third column in Table 3A.4). A somewhat lower $e_{14} = 0.375$ C/m^2 was extracted from the GaN mobility data [3.55] (see the second column in Table 3B.6). Since the e_{14} values vary between 0.375 and 0.6, the value of 0.5 has been adopted in Chap. 8. For comparison, the data for AlN, InN, ZnO, and GaAs are also listed Table 3A.6.

For $In_xGa_{1-x}N$ with small x, the piezoelectric constants can be estimated as xe_{ij}, where e_{ij} are the piezoelectric constants for GaN. Thus, this estimate gives a lower bound for the piezoelectric effect in $In_xGa_{1-x}N$. For a superlattice with the layer thicknesses L_1 and L_2, the strain-induced electric fields can be calculated as follows [3.56].

$$E_1 = \frac{P_2 - P_1}{\epsilon_0(\epsilon_1 + \epsilon_2 L_1/L_2)} , \qquad (3.32)$$

$$E_2 = - E_1(L_1/L_2) , \qquad (3.33)$$

where ϵ_1 and ϵ_2 are the relative dielectric constants, ϵ_0 is the dielectric permittivity of vacuum, P_1 and P_2 are the corresponding piezoelectric polarizations, and L_1 and L_2 are the thicknesses of the regions 1 and 2.

Shown in Fig. 3.16 are the strain-induced polarizations in GaN and AlN layers of $(GaN)_n(AlN)_n$ superlattices of equal monolayer thicknesses as functions of the thickness L. The AlN layers are under positive tensile strain whereas the GaN layers are under negative compressive strain. The anion (B face) is assumed to be on the surface and the field in AlN is out of

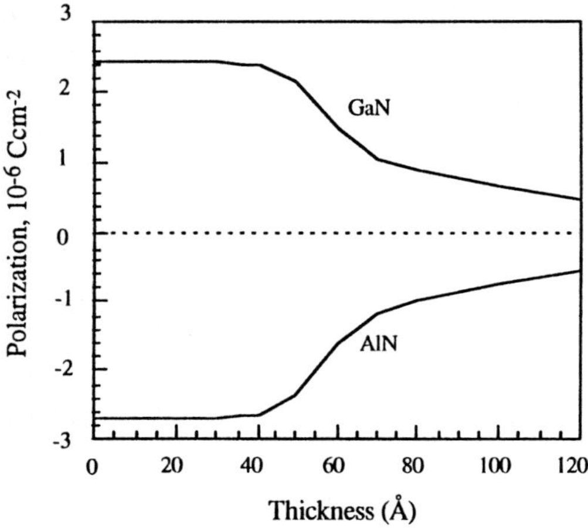

Fig. 3.16. Strain-induced polarizations in GaN and AlN layers of $(GaN)_n (AlN)_n$ superlattices of equal monolayer thicknesses as functions of the thickness L. The AlN layers are under tensile (positive) strain, whereas the GaN layers are under compressive (negative) strain. The anion (B face) is assumed to be on the surface and the field in AlN is into the surface whose polarity is chosen as negative. After [3.51]

the surface whose polarity is chosen as negative. Obviously, the field direction in the GaN layer is opposite to that in AlN and thus positive [3.51]. The reduction in the magnitude of fields in both layers at around 50 Å is indicative of approaching/exceeding the coherency limit and onset of relaxation, sending the strain-induced field toward zero. For these calculations, the set of values for the GaN piezoelectric constants from the first column of Table A3.6 was used. For comparison purposes, the typical polarization in these nitride semiconductors is approximately 10 times larger than those for the GaAs-$Ga_{0.8}In_{0.2}$As system [3.56]. This leads to a sheet-charge concentration of about 10^{13} cm^{-2}. Once again, this is a bound charge in the absence of free carriers, impurities and shallow defects.

The strain-induced electric fields in GaN and AlN layers of $(GaN)_n \cdot (AlN)_n$ SL can be very high. For example, for an 8×8 monolayer superlattice, the electric field in AlN and GaN can be on the order of $3.6 \cdot 10^6$ V/cm and $3 \cdot 10^6$ V/cm for the AlN and GaN layers, respectively. The corresponding reduction in the effective bandgap of a 50 Å thick GaN well can be about 1.5 eV. Since the nitride system is not insulating and contains defects and impurities, the strain-induced field and thus the effective lowering of the bandgap would be screened to the extent allowed by the availability of carriers supplied to the interfaces. If the layer is coherent and within the critical thickness limit, the strain and thus the polarization field is constant and independent of the layer thickness within the confines of the linear approximation. However, if the critical thickness is exceeded and the film begins to relax, the polarization field would decrease accordingly. Assuming that about $5 \cdot 10^{12}$ cm^{-2} carriers are provided by free carriers and weakly bound states, the strain-induced charge/field would be screened by 50%. Recently, there have been some efforts to consider the impact of strain-induced polarization on the performance of injection lasers [3.59]. However, if one uses a transparency carrier density in the vicinity of 10^{19} cm^{-3} for a 100 Å well thickness, which is conservative, one gets a sheet-charge density of $1 \cdot 10^{13}$ cm^{-2}, which is comparable to that induced by strain polarization. Clearly, in this case the free carriers would screen the polarization field. Thus, it is fair to state that the strain-induced polarization effect in and of itself would not be expected to play a very important role in injection lasers, particularly when the injected carrier densities are high.

3A Appendix

Table 3A.1. Spin-orbit and crystal field splitting parameters which are given in terms of meV [3.9-11]

	Δ_2	Δ_3	Δ_{cr}	Δ_{so}	
Wz GaN	5.2	5.2	72	15.6	calculated
			25	17	observed [3.12]
Wz AlN	6.8	6.8	−58.5	20.4	calculated
ZB GaN				20	
ZB AlN				20	calculated

Table 3A.2. The valence-band effective-mass parameters for Wz and ZB GaN and AlN: m_{hh}, m_{lh}, m_{ch} (m_{sh}) denote the heavy hole, light hole, crystal field (spin-orbit) split-off hole masses, respectively [3.9-11]

	Wz	m_{hh}	m_{lh}	m_{ch}	ZB	m_{hh}	m_{lh}	m_{sh}
GaN	$k_x(\Sigma)$	1.61	0.14	1.04	[100],Δ	0.86	0.86	0.17
(calculated)	$k_y(T)$	1.44	0.15	1.03	[110],Σ	4.89	0.84	0.15
	$k_z(\Delta)$	1.76	1.76	0.16	[111],Λ	1.74	1.74	0.15
observed	$m_h > 0.8$							
AlN	$k_{x(\Sigma)}$	10.42	0.24	3.81	[100],Δ	1.40	1.40	0.32
(calculated)	$k_y(T)$	5.02	0.25	3.61	[110],Σ	>10	1.41	0.27
	$k_z(\Delta)$	3.53	3.53	0.25	[111],Λ	3.72	3.72	0.25

Table 3A.3. The effective electron masses for Wz and ZB GaN and AlN. Calculated numbers have been taken from [3.9-11]

Wz	m_e^*	$m_e^{\perp(\Sigma)}$	$m_e^{\perp(T)}$	$m_e^{\parallel(\Delta)}$
GaN calculated	0.18	0.18	0.18	0.20
observed	0.22			
AlN calculated	0.27	0.25	0.25	0.33

ZB	m_e^*	$m_e^{100(\Delta)}$	$m_e^{110(\Sigma)}$	$m_e^{111(\Lambda)}$
GaN calculated	0.17	0.17	0.17	0.18
AlN calculated	0.30	0.30	0.30	0.30

Table 3A.4. Values of the A parameters and the Luttinger parameters. The calculated A parameter values as well as those deduced from quasi-cubic approximation are also shown. The A parameters are in terms of (Ry·cm) in units of $\hbar^2/2m_0$. After [3.9-11]

			GaN	AlN
Luttinger-like parameter [Ry·cm]	calculated	A_1	−6.56	−3.95
	transformed		−6.98	−3.98
	calculated	A_2	−0.91	−0.27
	transformed		−0.56	−0.26
	calculated	A_3	+5.65	+3.68
	transformed		+6.42	+3.72
	calculated	A_4	−2.83	−1.84
	transformed		−3.21	−1.86
	calculated	A_5	−3.13	−1.95
	transformed		−2.90	−1.63
	calculated	A_6	−4.85	−2.92
	transformed		−3.66	−1.98
	both	A_7	0	0
Zincblende		γ_1	2.70	1.50
(calculated)		γ_2	0.76	0.39
		γ_3	1.07	0.62

Table 3A.5. Calculated values of the deformation potentials (D) for GaN and AlN. Those that are deduced from the quasi-cubic approximation are also listed. After [3.9-11]

			GaN	AlN
Deformation potential [eV]	calculated	D_1	-13.87	-12.34
	quasi-cubic		-15.35	-12.93
	calculated	D_2	-10.95	-9.36
	quasi-cubic		-12.32	-8.46
	calculated	D_3	2.92	4.76
	quasi-cubic		3.03	4.46
	calculated	D_4	-5.84	-2.08
	quasi-cubic		-1.52	-2.23
	calculated	D_5	-2.05	-2.57
	quasi-cubic		-2.05	-2.57
	quasi-cubic	D_6	-3.66	-4.12
zincblende		a	-13.33	-9.95
		b	-2.09	-2.17
		c	-1.75	-2.57
Splitting energy [meV]		Δ_1	72.9	
		Δ_2	5.17	

Table 3A.6. Piezoelectric constants in GaN, AlN, InN, ZnO, and GaAs

ϵ_{ij} [C/m^2]	GaN[c,d]	GaN[b]	GaN[f]	GaN[c]	AlN[b]	InN[b]	ZnO[b]	GaAs[d]
ϵ_{33}	1	0.53	0.44	0.65	0.81	0.87	0.68	-0.185
ϵ_{31}	-0.36	-0.36	-0.22	-0.33	-0.31	-0.51	-0.39	0.093
ϵ_{15}	-0.3		-0.22	-0.33	-0.48[a]		-0.48[a]	0.093
ϵ_{14}			0.375	0.56[e]				-0.16

[a] [3.50], [b] [3.45], [c] [3.51], [d] [3.52], [e] [3.53], [f] [3.55]

4. Growth of Nitride Semiconductors

Although the synthesis of GaN goes back more than a half century, there are several pivotal developments which, in the opinion of the author, are responsible for laying the technological framework and paving the way for the tremendous commercial and scientific interest in nitrides. They are as follows: The synthesis of AlN by *Tiede* et al. [4.1], the synthesis of GaN through the reaction of Ga and ammonia to produce GaN by *Johnson* et al. [4.2], the synthesis of InN by *Juza* and *Hahn* [4.3], the epitaxial deposition of GaN using the hydride VPE technique by *Maruska* and *Tienjen* [4.4], the employment of nucleation buffer layers by *Amano* et al. [4.5] and *Yoshida* et al. [4.6], the achievements of p-type GaN by *Akasaki* et al. [4.7]. A more recent development which paved the way for all the commercial activity is the preparation of high-quality InGaN by *Nakamura* et al. [4.8] which followed the synthesis of InGaN by *Osamura* et al. [4.9]. Nearly every crystal-growth technique, substrate-type and orientation, has been tried in an effort to grow high-quality group-III-V nitride thin films. In recent years, various researchers have successfully taken advantage of the Hydride Vapor Phase Epitaxy (HVPE), Metal Organic Vapor Phase Epitaxy (MOVPE), and Molecular Beam Epitaxy (MBE) techniques, which have yielded greatly improved film quality.

A major drawback of GaN is that native substrates are not yet available in large quantities. This is, in part, due to the low solubility of nitrogen in Ga and the high vapor pressure of nitrogen on GaN. As a result, the bulk growth must resort to high temperatures and high pressures. AlN fares better in both of the aforementioned aspects and early work on producing bulk AlN looked promising [4.10]. The problem of nitrogen is endemic in epitaxial deposition techniques as well. Regardless of the growth method employed, the major difficulty in growing group-III nitrides arises from having to incorporate stoichiometric quantities of nitrogen into the film. This is accomplished in vapor-phase processes at high substrate temperatures by decomposing a nitrogen containing molecule, such as ammonia, on the substrate surface. It can also be accomplished at lower temperatures in MBE growth by increasing the reactivity of nitrogen or ammonia through remote plasma excitation or ionization.

4.1 Bulk Growth

It is well known that nitride semiconductors do not enjoy native substrates of their own, notwithstanding considerable efforts to produce them. The main impediment is the large vapor pressure of N on AlN, GaN and InN, in an ascending order, coupled with a low solubility of N in the metal melts at reasonable temperatures and pressures. It is thus imperative to consider the phase diagrams of these binaries. Shown in Fig.4.1 are the partial pressures of N_2 over AlN and Al liquid as a function of temperature, as determined by *Slack* and *McNelly* [4.10], and on GaN by *Karpinski* et al. [4.11] and InN by *Porowski* and *Grzegory* [4.12]. The calculated values for the N_2 vapor pressure on AlN by *Slack* and *McNelly* are 1, 10, and 100 atm at about 2550, 2800, and 3120°C, respectively. The GaN data, however, point to a different story in that the partial pressure of N_2 is very high, necessitating high-pressure experiments to collect data. *Karpinski* et al. employed a tungsten carbide anvil cell and pressures of up to 60 kbar to collect

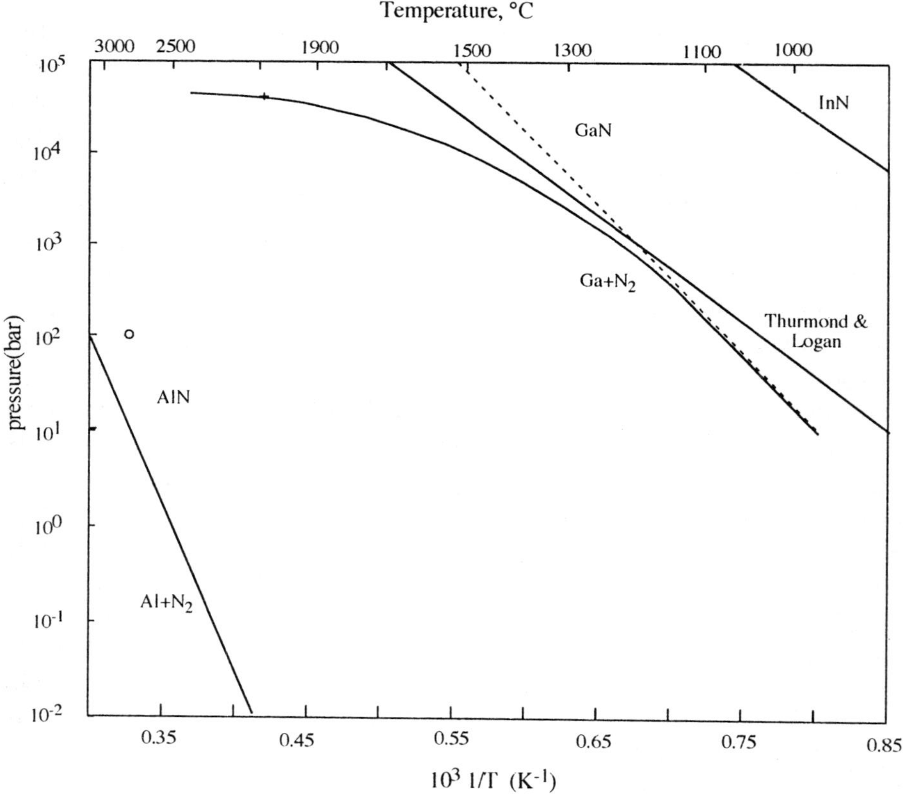

Fig. 4.1. Partial pressure of N over AlN, GaN and InN. After [4.10-12]

the data presented in part in Fig.4.1. The GaN data deviate from the calculations of *Thurmond* and *Logan* [4.13], most noticeably at high temperatures. The equilibrium N_2 pressure data were calculated from the measurement of the equilibrium ammonia pressure over GaN assuming N_2 to be an ideal gas, and thus would predict a linear dependence in the log vs. 1/T scale. The data for InN unequivocally indicate that InN is not really stable even at moderate temperatures, as evidenced by the extremely high vapor pressure of N_2 over InN. The partial pressure data alone are indicative that AlN would be easiest to synthesize of all the three nitride binaries under discussion. In fact, *Slack* and *McNelly* obtained growth rates in the millimeters per hour range under very moderate pressures with a duration of growth determined by the reaction rate of Al with a tungsten crucible. The same partial pressure data also indicate the difficulties associated with an epitaxial deposition going from AlN to GaN and then to InN that increases the amount of nitrogen necessary to avoid decomposition.

The data of the phase diagrams of GaN, AlN and InN are limited and contradictory by reason of the high melting temperatures (T_M) and the high nitrogen dissociation pressures (P_{N2}^{dis}). Dissociation pressure of MN, where M represents metal species such as Al, Ga, or In, is defined as the nitrogen pressure at the thermal equilibrium of the reaction [4.14]

$$MN(s) = M(\ell) + (1/2)N_2(g) \qquad (4.1)$$

where s, ℓ and g indicate solid, liquid and gas, respectively. Reported values for P_{N2}^{dis} of GaN are plotted in Fig.4.1. Examining the available data, *Sasaki* and *Matsuoka* [4.14] concluded that the data of *Madar* et al. [4.15] and *Karpinski* et al. [4.11] are the most reliable ones. The data of *Karpinski* et al. are therefore shown. As can be deduced from Fig.4.1, the nitrogen dissociation pressure equals 1 atm at approximately 850°C, and 10 atm at 930°C. At 1250°C GaN is unstable and decomposes even under a pressure of 10,000 bar of N_2. The case of InN is even more problematic at the decomposition temperature of about 700°C [4.12]. It should therefore come as no surprise that the incorporation of nitrogen at high temperatures is a non-trivial problem at best. For pressures below equilibrium at fixed temperatures, the thermal dissociation occurs at a slow and apparently constant rate, suggesting a diffusive process of dissociation. Despite the fact that equilibrium growth of InN appears nearly impossible, InGaN can be grown at temperatures in the range of 700° to 900°C with very high amounts of nitrogen on the surface.

Owing to N dissociation, the data on the melting points (T_M) of binaries vary susbtantially. Landolt and Börnstein tables [4.16] list T_M of AlN to be 2400°C at a pressure of 30 bar, T_M of GaN to be higher than 1700°C at a pressure of 2000 bar, and T_M of InN to be $1100 \div 1200\,°C$ at a pressure above 10^5 bar. The aforementioned nitrogen dissociation pressure of AlN is several orders of magnitude smaller than that of GaN and InN. *Massalski* [4.17] cited T_M of AlN to be $2800\,°C$. There are reports estimating T_M of AlN at $2200\,°C$ [4.18] and $2450\,°C$ [4.19]. *Van Vechten* [4.20] theoretically estimated T_M of AlN to be 3487 K. *Porowski* and *Grzegory* [4.12] predicted T_M of AlN to be over 3000 K at a dissociation pressure of a few hundred bars, and T_M of GaN to be larger than 2500 K at tens of kbar. Taking into account the high melting temperatures and the high dissociation pressures of group-III-N compounds, none of the conventional bulk crystal-growth methods can be successfully applied.

Judging from the N pressure at equilibrium, it can be discerned that AlN is the closest to successful production of all the group-III-V nitride semiconductors under discussion. The most satisfactory method of growing high-purity AlN is that in which AlN itself is used as the starting material. *Slack* and *McNelly* [4.10, 21] employed a technique in which high-purity AlN is produced through an intermediate AlN powder formed by utilizing AlF_3. The AlN powder is converted into single crystals by sublimation in a closed tungsten crucible or in an open tube with a gas flow. The main problem with this growth technique is perhaps the surface oxidation of the powder due to the strong reactivity of oxygen and aluminum. If this oxidation is minimized, then AlN could be produced only with 100 ppm of oxygen and lower amounts of other impurities. The purest AlN prepared by *Slack* and *McNelly* employing a tungsten crucible had 350 ppm oxygen. However, when W or Re was used for the crucible, very little contamination of the AlN with metal impurities was found. Furthermore, the crystals grown in W or Re crucibles generally showed uniform amber color indicating that, indeed, both oxygen and carbon contamination were scrupulously minimized by using W and Re crucibles. Much of the single-crystal AlN data obtained and reported throughout this book, have been extracted from material produced by this method.

Leszczynski et al. [4.22], *Porowski* et al. [23], and *Grzegory* et al. [4.24] performed nitride crystal growth from the solution under high N_2 pressure. The experiments were carried out in a gas pressure chamber of 30 mm internal diameter with a furnace dimension of 14 mm ($1500\,°C$) or 10 mm ($1800\,°C$) and with a boron nitride crucible containing Al, Ga or In. The temperatures were stabilized to a precision better than $1\,°C$. Attempts were made to optimize the pressure range for growth. With this optimization, the crystals were grown only at a pressure for which the nitride was stable over the whole temperature range. In the case of GaN, single crystals were grown from a solution in liquid Ga under an N_2 pressure of $8 \div 17$ kbar at temperatures ranging between $1300 \div 1600\,°C$. The quasi-linear temperature gradient in 5 to 24 h processes was $30 \div 100\,°C/cm$. The nucleation and growth of single-crystal GaN took place through the process of thin poly-crystalline GaN film on a Ga surface, and through its dissociation and transport into the cooler part of the crucible. At high N_2 pressure, the synthesis rate of AlN is high, and the synthesis rate of InN is extremely low. The rate of AlN is so high that, at a pressure lower than 6.5 kbar, thermal explosion takes place during heating of a bulk Al sample. Due to a low stability, the crystallization rate of AlN at $1600 \div 1800\,^\circ C$ is marginally low. On the other hand, due to kinetic (low-temperature) and thermodynamical (low-stability) barriers, crystal-growth experiments of InN result in very small crystallites ($5 \div 50$ μm), particularly when grown by slow cooling of the system from the temperatures exceeding the stability of InN.

An examination of the crystal morphology indicated that the crystal shape and size depend on the pressure, the temperature range, and the super-saturation during growth. For pressures and temperatures lying deeply in the GaN stability field (e.g., higher pressure and lower temperature), the crystals are hexagonal prisms elongated in the c-direction. Under conditions close to the equilibrium curve, the dominating shape of the crystals is a hexagonal plate. The crystals grown slowly (slower than 0.1 mm/h) at smaller temperature gradients, exhibited high crystal quality. These are transparent, slightly yellowish, and have flat mirror-like surfaces. Typical Full Widths at Half Maximum (FWHM) of the X-ray rocking curves for (004) CuK_α reflections are $23 \div 32$ arcsec. Note that these curves are significantly narrower than the corresponding curves of heteroepitaxial GaN films grown by MBE or MOCVD techniques.

Probably due to the non-uniform distribution of nitrogen in the solution across the growing crystal face, it was observed that the quality of the GaN crystals deteriorates with increasing growth rate (high supersaturation) and with increasing dimensions of the crystals. This is apparent especially when the size of the face becomes comparable to the size of the crucible. The deterioration of quality of $5 \div 10$ mm crystals grown with a rate of $0.5 \div 1$ mm/h was evidenced by the broadening of the rocking curve.

4.2 Substrates Used

Lacking a commercial native substrate, a plethora of substrates have been employed in the growth of GaN films. Recognizing the bleak thermodynamical bottleneck with regard to native substrates, unconventional methods such as compliant substrates and grapho-epitaxy, van der Waals susbtrates, have been explored. The most promising results so far have been obtained on sapphire, SiC and ZnO substrates. Below, I shall give succinct descriptions of each of these approaches.

4.2.1 Conventional Substrates

GaN, as the most studied member of the semiconducting group-III nitrides, has been grown on many substrates. Many of the major problems which have hindered the progress in GaN and related semiconductors can be traced to the lack of a suitable substrate material which is lattice matched to and thermally compatible with GaN. The semiconductors GaN, AlN and InN have been grown primarily on sapphire, most commonly the (0001) orientation, but also on the a and r planes [4.25]. In addition, the group-III-V nitrides have been grown on SiC, ZnO, $MgAl_2O_4$, Si, GaAs, MgO, NaCl, W, and TiO_2. Some of the suitable substrate materials have only recently become commercially available. Almost all the group-III-V nitride semiconductors have been deposited on sapphire despite its poor structural and thermal match to the nitrides. The preference for sapphire substrates can be ascribed to its wide availability, hexagonal symmetry, and its ease of handling and pre-growth cleaning. Sapphire is also stable at high temperatures ($\approx 1000\,^{\circ}C$) required for epitaxial growth using the various CVD techniques commonly employed for GaN growth. Due to the thermal and lattice mismatches between sapphire and the group-III-V nitrides, it is necessary to grow a thick epilayer to obtain good quality material.

4.2.2 Compliant Substrates

The concept behind this approach is to force the defects due to mismatch to propagate into the substrate as opposed to the epilayer. This approach works in the deposition of nearly structural defect-free SiGe on Si substrates. In the nitride case, the initial approach proposed was that Si on an insulator be utilized for GaN growth. In this case, a thin Si layer on silicon dioxide which is, in turn, on Si would be employed. The downside is that the quality of GaN on Si has been poor at best. The modified approach to overcome this barrier is to carbonize the Si to convert it to SiC. If one utilizes the (111) orientation, one would then get the wurtzitic phase of GaN. To be specific, *Yang* et al. [4.26] offered a blue print in which they suggested that before the beginning of the carbonization process, SiO_2, Si and C need to be deposited successively on a Si substrate. By exposing the new composite substrate to a flux of acetylene or carbon particles at 900 °C, a thin layer (less than 50 nm) of Si (on SiO_2) will be partially or completely converted into SiC. GaN will then be grown on this SiC. Again, the problem here, setting aside the problems associated with the growth on SiC, is that SiC so formed is not contiguous and is extremely defective both in terms of bulk and surface structural properties. In addition, air gaps also form beneath the layer surface. Consequently, this technique has not yet showed that it lives up to the original proposal and expectations. Efforts still continue to exploit this approach despite the lack of progress so far.

4.2.3 Van der Waals Substrates

To get around the lattice-mismatch problem, a new growth method called **Van der Waals epitaxy** has been proposed [4.27], which delivers strain-free films. In this approach, the substrate and the epitaxial film are separated by an intermediate epitaxial two-dimensional buffer material such as MoS_2, WS_2 or other materials such as II-VI (ZnTe) or III-VI compound (GaSe, InSe, etc.) having weak Van der Waals bonding to the substrate and the film. Strain from lattice mismatch between the epitaxial film and the substrate is completely relieved between the layer and the buffer region. As in the case of the compliant substrate scheme, this approach has not been very successfuly applied to nitrides.

4.3 Substrate Preparation

Substrate preparation is like the foundation of a building. As such, substrate preparation deserves the most attention. Though the details of the procedures employed vary from one growth method to the next, a chemical preparation before loading into the growth reactor is common. This is followed by either a simple heat treatment or a combination of a heat treatment with gas-phase etching required by the OMVPE technique where temperatures in the vicinity of $1200\,^\circ C$ are possible. In the case of vacuum-deposition techniques where it is not always possible to achieve sufficiently high temperatures, dry processing techniques are employed. Among them is one utilizing the ECR remote plasma etching with a mixture of hydrogen and helium. The purpose of He gas is to take advantage of its energetic metastables with their long mean-free paths. In addition to a clean surface, the goal is to get as flat a surface since possible as the nitride stacking order, ABAB..., is different for most of the substrates under investigation. The exception is ZnO whose stacking order matches that of the nitrides. Since the atomic steps on the (0001) surface are of the bi-layer type, the surface terraces would have the same surface polarity so that stacking mismatch boundaries can be avoided.

The chemical (solvent) degreasing procedure for sapphire, SiC, and ZnO substrates is similar and can be described as follows: The substrate of choice is first dipped in a solution of TriChlorEtane (TCE) and thereafter kept at 300°C, for 5 minutes. It is then rinsed for 3 minute each in acetone as well as methanol. This procedure is followed by a 3 minute rinse in DeIonized (DI) water. The process sketched is repeated three times to complete the degreasing process. The substrates are then etched, the details of which are substrate dependent, as will be discussed next.

Following the degreasing procedure, surface damage must be removed either by solution etching and/or dry etching. In the solution-etching approach, approximately 3 μm of the SiC material are removed from the substrate surface in a hot KOH solution (300° $\div$ 350°C). This is followed by a DI rinse for 3 minutes and the wafer is blown to dry in N_2. The SiC substrate then undergoes an oxidation and passivation procedure. The substrate is immersed for 5 minutes in a 5:3:3 solution of HCl: H_2O_2 :H_2O at 60°C followed by a 30 second DI rinse. The resulting oxide layer is then removed by dipping the substrate, for 20 seconds, in a 10:1 solution of H_2O:HF. This procedure is repeated several times after which the substrate should not be exposed to the atmosphere for longer than 30 minutes or another oxidation-passivation procedure would be required. Although very smooth surfaces can be obtained on the C face of (0001) SiC, the Si surface is more conducive for growth. A successful growth of nitrides on SiC hinges on the ability to remove the damage caused during mechanical lapping/polishing. In addition to the KOH etch, a high-temperature H or HCl treatment is employed to selectively remove damaged or partially damaged portions from the surface. In the case of hydrogen, the removal most likely takes the form of gas-phase transport of silane and methane. In the case of HCl, in addition to the processes involving H, chlorides of Si and C are also volatile and can be removed by the gas phase.

Vacuum deposition equipment is not compatible with the high-temperature H and HCl treatments, but remote plasma etching techniques can be employed to at least remove the damaged surface layer. In what follows, a SiC substrate will be used to describe the plasma process which is almost identically applicable to sapphire which cannot be etched easily and forms a very stable native oxide; it is highly suited for hydrogen plasma cleaning. To circumvent the need for high temperatures and exotic treatments incompatible with conventional MBE setups, a preparation procedure adapted from conventional Si technology, and augmented by H plasma cleaning, has been shown to work for SiC. In the first step of this *Lin* et al. [4.28] procedure, the surface is hydrogen passivated using an HF dip before introduction into vacuum. Second, the substrate is treated with a hydrogen plasma which reduces the oxygen-carbon bonding to a value below the X-ray photoemission detection limit. Upon heating in the MBE chamber, the SiC substrates were observed to have a sharp (1×1) surface reconstruction. The (0001)Si face of 6H SiC has only single Si dangling bonds at the surface and is chemically analogous to (111) Si. After several controlled oxide growth and stripping cycles, the surface can be hydrogen passivated by an HF dip which successfully removes the oxygen from the surface Si atoms. Still, X-ray photoemission spectroscopy (XPS) reveals a significant amount of C-O bonds remaining on the surface. After an exposure to a low-energy H_2 plasma at 650°C, the surface O contamination is reduced to below the XPS detection limit, producing a high-quality, stoichiometric, nearly contaminant free surface for epitaxial growth.

It has long been suspected that nitridation of the Si face of (0001) SiC could lead to a SiN formation which would prevent successful growth of GaN. Consequently, growth on SiC commences with nearly simultaneous exposure to the metal and nitrogen in the MBE growth case. While nitrides are grown on SiC by OMVPE (MOCVD), the growth generally is initiated with AlN or AlGaN at temperatures as high as 1200°C, which requires much lower nitrogen flows than GaN. Convergent-beam electron-diffraction investigations do indeed indicate that growth on the Si face of SiC proceeds with Ga first followed by N [4.29].

For Al_2O_3 substrates, a 3:1 solution of hot H_2SO_4:H_3PO_4 is employed as the etchant. The substrate is dipped in this solution at 300°C for 20 minutes. This is followed by a rinse in DI water for 3 minutes. Prior to growth, the substrates undergo an H_2 plasma treatment for about one hour. Sapphire substrates can also be heat-treated in an oxygen environment at elevated temperatures, i.e. 1200°C for durations in excess of 10 hours. Shown in Fig.4.2 are AFM images of two sapphire surfaces, one before and the other after such an anneal. Clearly, the sapphire substrate subjected to this high-temperature heat treatment exhibits terrace-like features with about 2 μm long terraces. The GaN films grown on annealed substrates show smoother surfaces as compared to the unannealed ones, but their electronic properties are inferior to those grown on unannealed substrates. The likely cause for this observation is that ledges may act as nucleation sites and guides for the extension of islands that prevent columns (Sect.4.7) from arbitrarily forming with the freedom to twist and minimize strain energy and thus defect formation.

Sapphire substrates which have undergone MBE-like pre-growth in situ heat treatment followed by an exposure to ammonia exhibit the polishing damage in the form of random scratches [4.30]. Considering the stacking-order mismatch between sapphire and nitrides, these features are undesirable as atomically smooth surfaces are ideally required. Figures 4.3a and b depict the AFM images of sapphire substrates after thermal desorption at 850°C for 3 min and an exposure to NH_3 gas at 800°C for a period of 15 min. Substrate-surface damage is clearly visible on the surface of each of the samples due to the substrate-polishing process. A low density ($\approx 10^8$ cm^{-2}) of surface outgrowths was observed after 30 minutes of nitridation,

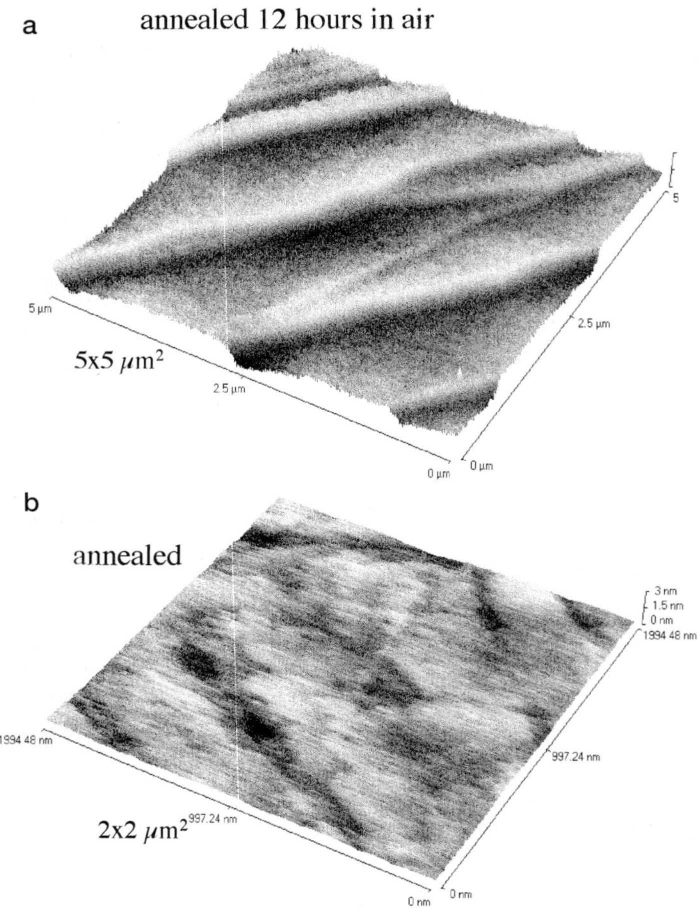

Fig. 4.2. AFM images of sapphire surface (**a**), as prepared in hot H_2SO_4 : H_3PO_4 at $300\,^{\circ}C$ for 20 min, and (**b**) after anneal in air at $1200\,^{\circ}C$ for 2 hours. Courtesy of M. Yeadon, University of Illinois

as illustrated in Fig. 4.3b. The presence of surface damage does not appear to have influenced the formation of protrusions. There is no clear correlation between the positions at which the protrusions have formed and the local surface topography. *Uchida* et al. [4.31] observed similar protrusions after 5 minutes of nitridation at $1050\,^{\circ}C$ in an MOCVD system, however, the density was approximately three orders of magnitude larger than that found in Fig. 4.3b. It is likely that a combination of a higher substrate temperature and the background ammonia pressure promotes a more rapid nitridation reaction leading to a higher density of protrusions in MOCVD-grown samples.

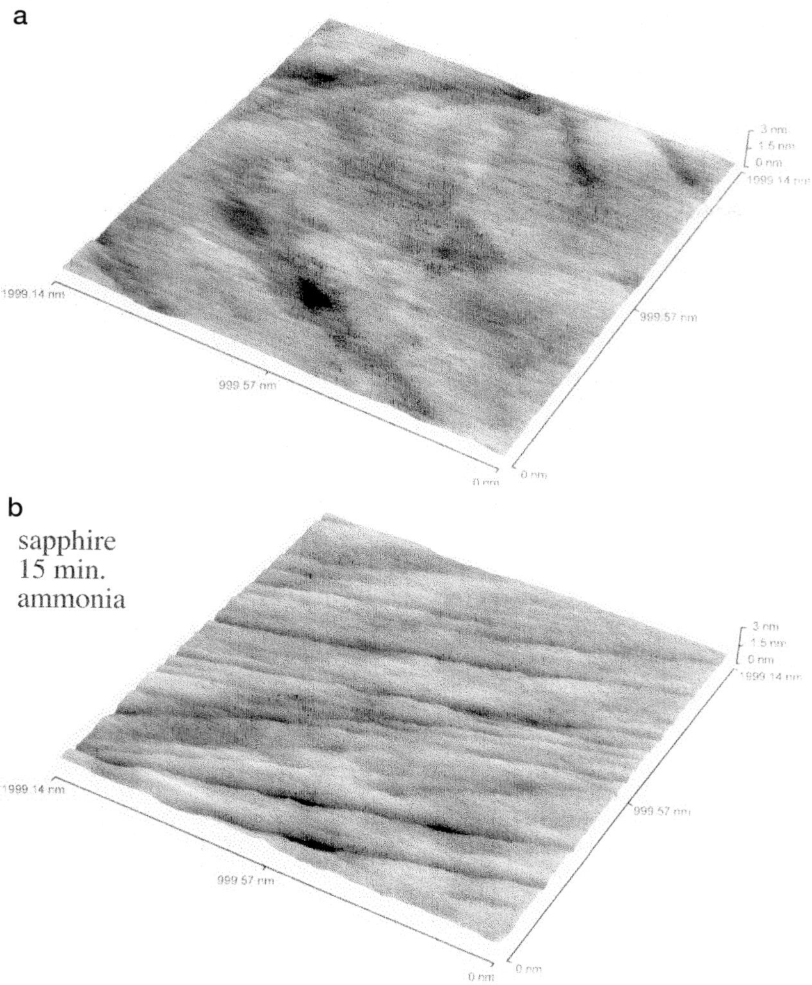

Fig. 4.3a, b. AFM images of sapphire before and after exposure to ammonia at $800°C$. (**a**) After degassing at $850°C$ for 3 min, and (**b**) after exposure to ammonia for 15 min at $800°C$. Polishing damage is clearly visible in (**a**)

The nitridation reaction is believed to yield AlN_xO_{1-x} via the exchange of oxygen atoms from sapphire and nitrogen atoms from ammonia. XPS analyses were carried out to confirm the incorporation of nitrogen atoms into the sapphire substrate during the nitridation process. A check of the sapphire substrates did not exhibit a 1s N peak, but those exposed to ammonia even for as little as 1 min did indeed show the nitrogen peak. This is an indication that nitrogen atoms react with the surface. The down-shifting of the oxygen 1s line by about 0.25 eV, relative to the spectra obtained from a bare sapphire sample, and the sample nitrided for 1 min suggests the generation of a significant number of new bonds between oxygen and nitrogen atoms in the nitrided layer [4.30].

High-quality ZnO substrates have recently become available in laboratory quantities. While stable in air and O environments at temperatures as high as 900°C, perhaps even higher, exposure to ammonia etches ZnO even at temperatures as low as 600°C. It is believed that atomic hydrogen reacts with O forming volatile water vapor. The Zn metal is removed from the surface by evaporation. The chemical preparation of ZnO involves a 5 min acetone bath followed by the same procedure in methanol while using ultrasound agitation to remove particulates. The wafers are then rinsed in deionized water, followed by blow drying with filtered nitrogen prior to introduction into the growth chamber. The details of the growth processes will be given in Sects.4.8 and 4.19.

4.4 Substrate Temperature

The substrate temperature is a very important parameter as it determines much of the reaction processes, the properties of the low-temperature buffer layer, the sizes of islands and the manner in which they coalesce. As such, it must be measured as accurately as possible and certainly controlled reproducibly. The simplest method is to monitor the output of a thermocouple situated as closely as possible to the subtrate. However, the reproducibility is dependent on the ability to mount the wafer in the same manner at all times, which proves to be problematic. Fortuitously, the large refractive index difference between the nitride layers and the substrate provides the necessary index step to allow the formation of constructive and destructive interactions of the exiting radiation with varying degrees of reflection events which leads to interference fringes. Since the refractive index is a function of temperature, the periodicity of the oscillations contain not only the thickness information but also the temperature information as well. A few calibrations are sufficient to provide the necessary parameters needed to deconvolve the data. Moreover, the occurrence of these oscillations is a sign of smooth layer growth, which is also useful to know by the grower. The interference oscillations can be measured in the reflection mode by letting impinge a laser beam on the substrate and record the reflected light as the film growth proceeds. Alternatively, the evolution of the IR radiation, emitted from the substrate, which goes through the layers can be monitored. This method is applicable not only to MOCVD but also to MBE as well. Other highly useful information can also be garnered. For example, if a contiguous layer is not formed and the surface contains many islands of different shape and thickness, the phase coherence is lost, which prevents interference fringes from forming. If the surface of the growing film smoothes out, the intensity of the radiation would be relatively high compared to the rough case. In short, there is a wealth of information to be gleaned by this unsophisticated measurement which is as simple as reading a radiation thermometer.

4.5 Epitaxial Relationship to Sapphire

Sapphire is a ubiquitous substrate on which to grow any semiconductor, and GaN is no exception. Sapphire remains the most frequently used substrate for group-III nitride epitaxial growth owing to its low cost, the availability of three-inch-diameter crystals of good quality, its transparent nature, its stability at high temperatures, and a fairly mature technology for nitride growth on it. The orientation order of GaN films grown on principal sapphire planes – c (basal)-plane (0001), a-plane (01$\bar{1}$20), and r-plane (1$\bar{1}$02) – was studied in great detail by Electron Cyclotron Resonance - MBE (ECR-MBE) [4.32] and reviewed in [4.33]. The same epitaxial relationship holds for growth by OrganoMetallic Vapor Phase Epitaxy (OMVPE).

The calculated lattice mismatch between the basal GaN and the basal sapphire plane is larger than 30 %. However, the actual mismatch is smaller ($\approx$ 16 %) because the small cell of Al atoms on the basal sapphire plane is oriented 30° away from the larger sapphire unit cell (Fig.4.4). This smaller lattice mismatch can be calculated by adopting the model explained in Fig.4.4, i.e.,

$$\frac{\sqrt{3}\, a_{wGaN} - a_{sapphire}}{a_{sapphire}} = 0.16 \,. \tag{4.2}$$

It is on this plane that the best films have been grown with relatively small in- and out-of-plane misorientations. In general, films on this plane show either none or nearly none of the cubic GaN phase.

GaN grown on the (11$\bar{2}$0) a-plane turns out to be (0001) oriented and anisotropically compressed. The effect of this uniaxial stress on the laser performance will be discussed in Chap. 12. The in-plane relationship of GaN and sapphire is depicted in Fig.4.5 where the [11$\bar{2}$0] direction of GaN is aligned with the sapphire [1$\bar{1}$00] direction. In this orientation, the bulk positions of the substrate and the GaN cations lie along the sapphire [0001] direction. The mismatch between the substrate and film is given by

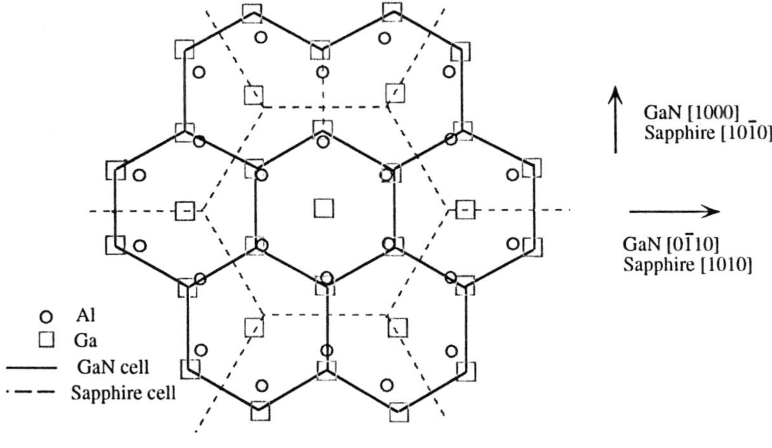

Fig. 4.4. Projection of bulk basal-plane sapphire and GaN cation positions for the observed epitaxial growth orientation. The *circles* mark Al-atom positions and the *dashed lines* show the sapphire basal-plane unit cells. The *open squares* mark Ga-atom positions and *solid lines* show the GaN basal-plane unit cell. The Al atoms on the sapphire plane sit at positions approximately 0.5 Å above and below the plane position. After [4.32]

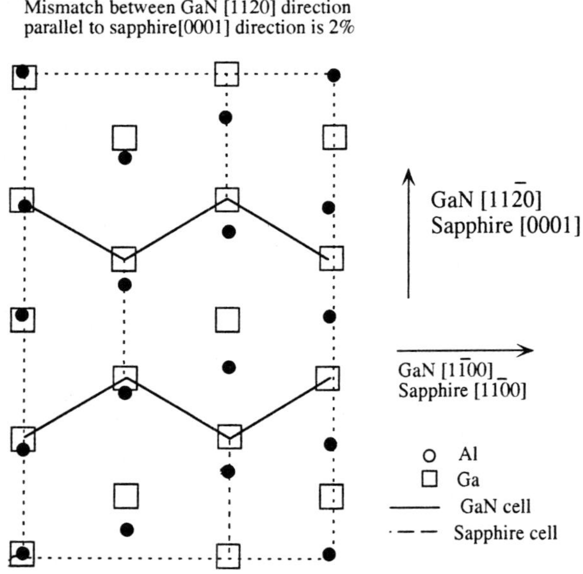

Fig. 4.5. Projection of bulk a-plane sapphire and basal-plane GaN cation positions for the observed epitaxial growth orientation. The *dots* mark Al-atom positions and the *dashed lines* show the sapphire a-plan unit cells. The *open squares* mark Ga-atom positions and the *solid lines* show the GaN basal-plane unit cell. After [4.32]

96

$$\frac{c_{sapphire} - 4a_{wGaN}}{c_{sapphire}} = 0.02 \qquad\qquad (4.3)$$

and for the GaN $[1\bar{1}00]$ direction parallel to the sapphire $[1\bar{1}00]$ direction by

$$\frac{c_{sapphire} - 1.5a_{wGaN}}{c_{sapphire}} = -0.005 . \qquad\qquad (4.4)$$

In one investigation [4.34], the range of the growth conditions leading to good films on the a-plane was found to be wider than that on the c-plane. Another impetus for exploring growth on this plane is the relative ease with which the sapphire could be cleaved along the weak single bond in the c-plane. The GaN film on the top would cleave along the $(11\bar{2}0)$ a-plane or the $(10\bar{1}0)$ m-plane.

GaN films have also been grown on the r face $(1\bar{1}02)$ of sapphire purportedly to achieve a lattice mismatch smaller than on the c-plane sapphire. Films grown on the r-face assume the $(2\bar{1}\bar{1}0)$ orientation. The arrangement in the case of the $(1\bar{1}02)$ face of sapphire and $(2\bar{1}\bar{1}0)$ of GaN is depicted in Fig.4.6. The lattice mismatch between the $[\bar{1}101]$ direction of sapphire and the $[0001]$ direction of GaN parallel to the sapphire $[\bar{1}101]$ direction is equal to

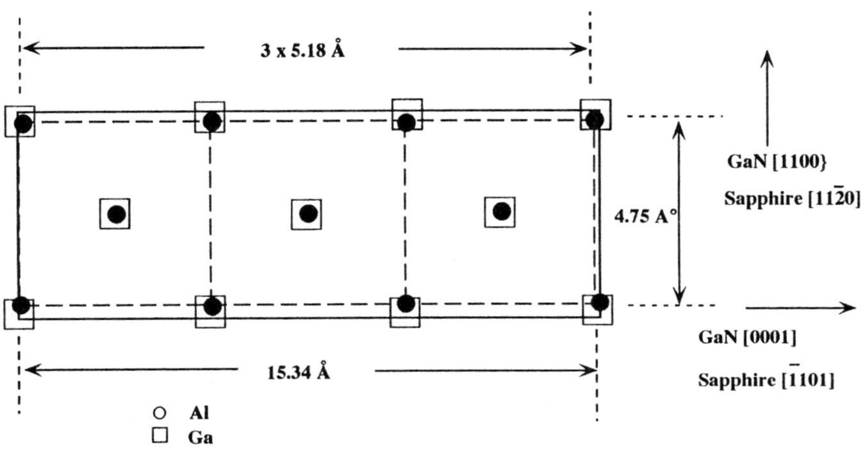

GaN grown on sapphire $(11\bar{2}0)$, A plane is oriented in the [0001] direction.

Fig. 4.6. Projection of bulk r-plane sapphire and a-plane GaN cation positions for the observed epitaxial growth orientation. The *dots* mark Al-atom positions and the *dashed lines* show the sapphire r-plane unit cells. The *open squares* mark Ga-atom positions and *solid lines* show the GaN a-plane unit cell. After [4.32]

$$\frac{3c_{GaN} - \sqrt{3a^2_{sapphire} + c^2_{sapphire}}}{\sqrt{3a^2_{sapphire} + c^2_{sapphire}}} = 0.01 \ . \tag{4.5}$$

In the case when the $[1\bar{1}00]$ direction of GaN is parallel to the sapphire $[11\bar{2}0]$ direction, the lattice mismatch is

$$\frac{a_{wGaN} - a_{sapphire}/\sqrt{3}}{a_{sapphire}/\sqrt{3}} = 0.16 \ . \tag{4.6}$$

The mismatch along the $[0001]$ direction of GaN parallel to the $[\bar{1}101]$ direction of sapphire is 1%, which is much smaller than the 16% along $[1\bar{1}0]$ direction of GaN parallel to the $[11\bar{2}0]$ direction of sapphire. Growth on the r-face exhibits ridge-like features with allow relaxation of the mismatch.

4.6 Growth by Hydride Vapor Phase Epitaxy (HVPE)

Maruska and *Tietjen* grew the first single-crystal epitaxial GaN thin films by vapor transport [4.4]. In their method, HCl vapor flowing over a Ga melt, cause the formation of GaCl which was transported downstream. At the substrate, the GaCl mixed with NH_3 leads to the chemical reaction:

$$GaCl + NH_3 \Rightarrow GaN + HCl + H_2 \ . \tag{4.7}$$

The growth rates were quite high ($0.5\,\mu m/min$) that allows extremely thick films to be deposited. Although early GaN grown by vapor transport had a very high background of n-type carrier concentration (typically 10^{19} cm^{-3}), this level has recently dropped to values of about the 10^{17} cm^{-3} range with room-temperature mobilities approaching 900 cm^2/V·s, as detailed in Chap. 8. Figure 4.7 displays a schematic diagram of a horizontal HVPE reactor which is, in principle, similar to the vertical types with vertical varieties that yield better uniformity.

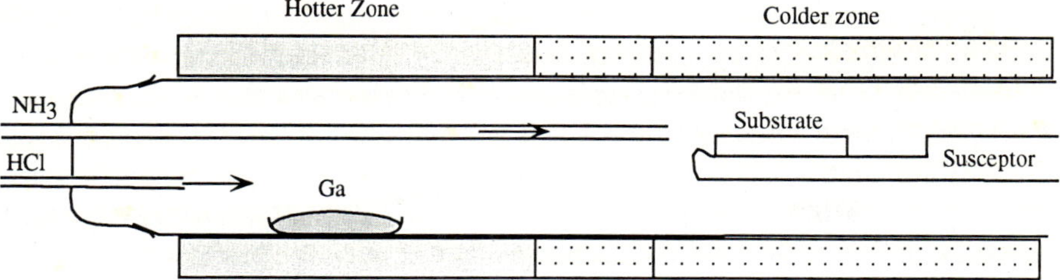

Fig. 4.7. A schematic representation of a horizontal HVPE reactor

4.7 Growth by OMVPE (MOCVD)

The approach of *Maruska* and *Tietjen* [4.4] was a precursory to the modern-day OMVPE GaN growth reactor, which is capable of depositing films with AlN as well [4.35]. In OMVPE or Metalorganic Chemical Vapor Deposition (MOCVD), TriMethylGallium (TMG), TriMethylAluminum (TMA), and TriMethylIndium (TMI) react with NH_3 at a substrate which is heated to roughly 1000°C. A schematic representation of an atmospheric-pressure reactor, with the optional low-pressure capability used in the laboratory of I. Akasaki is shown in Fig. 4.8. Once again there are vertical varieties with pros and cons on both sides. The common feature among all the reactors is that the reactants must be allowed to interact with one another only on the surface of the substrate. Any prior interaction, particularly Al and ammo-

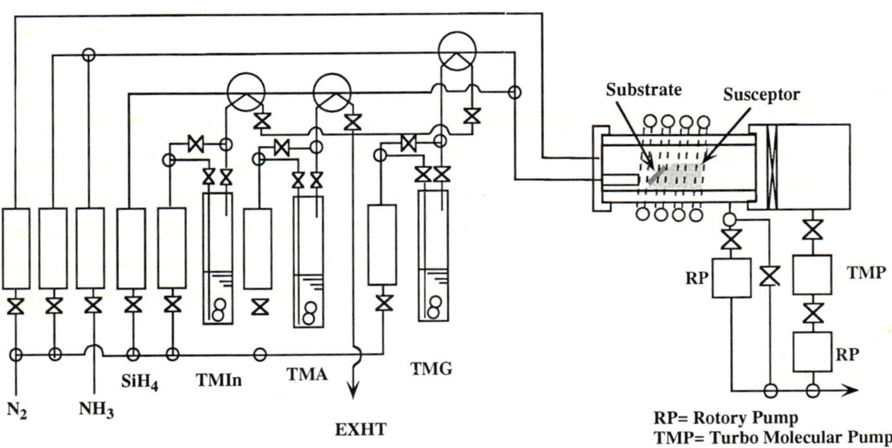

Fig. 4.8. A schematic representation of a horizontal OMVPE reactor. Courtesy of I. Akasaki, Meijo University

nia, could cause adduct formation in the gas phase with the adverse effect of suffocating the growth. *Nakamura* et al. [4.36] designed an atmospheric-pressure MOVPE reactor specifically for nitride growth which has been highly successful. Reactant gases, diluted with H_2, enter the growth area through a quartz nozzle which directs the flow across the rotating substrate. The key aspect of this design is a downward subflow of H_2 and N_2, which was claimed to improve the interaction of the reactant gases with the substrate.

MOCVD reactors for group-III nitride film growth incorporate laminar flow at high operating pressures and feature separate inlets for the nitride precursors and the ammonia to minimize predeposition reactions. In the Nichia reactor, the main flow carries the reactant gas parallel to the substrate. The second subflow perpendicular to the substrate forces, on the other hand, the reactant gas toward the substrate, and suppresses thermal convections. A rotating susceptor was used to enhance the uniformity of the deposited films. Commercial systems tend to be of the vertical type, some low-pressure varieties with a rotating substrate for uniform coverage, and others operating at atmospheric pressures. One of the new CVD technologies utilizes an activated form of nitrogen in order to lower the deposition temperatures of group-III nitrides. That these technologies are interesting becomes apparent, for example, from the deposition of polycrystalline and amorphous GaN films at temperatures lower than 350°C by plasma-enhanced CVD. Epitaxial GaN and AlN have been grown by variants of methods that activate nitrogen, such as **laser-assisted CVD**, remote plasma-enhanced CVD, atomic-layer epitaxy with NH_3 cracked by a hot filament or decomposed by a reaction on the substrate surface, **photo-assisted CVD**, and **ECR plasma-assisted CVD**.

The MOVPE growth follows a sequence of substrate outgassing, low-temperature AlN or GaN buffer-layer growth followed by a high-temperature anneal and, finally, growth of the desired layers, the details of which can be found elsewhere [4.37-39]. As alluded to earlier, the MOCVD growth sequence begins by the growth of a GaN or AlN buffer of a thickness in the range of $20 \div 50$ nm at a low temperature in the $480° \div 600°C$ window. The growth is then stopped and the substrate temperature is raised to the main growth temperature under an ammonia flow and allowed to stabilize. The growth process is then restarted to build up the main GaN epilayers.

4.7.1 Sources

Trimethylgallium for metal and high-purity ammonia for nitrogen are the most commonly used ingredients. They are transported to the growth chamber using hydrogen carrier gas and are injected into the reaction chamber. The gas manifold typically features fast switching of the group-III elements and dopants, and permits the separate injection of ammonia. Nominally c-plane [0001] oriented sapphire substrates are used, although other orientations of sapphire and alternative substrates are sometimes employed, too. For a hydrogen flow of 40 sccm through the TMGa bubbler ($-10°C$), an NH_3 flow of 3.0 standard liter per minute (slm) and hydrogen make up the flow of 1.5 slm; the GaN growth rate was 2 μm/h. Sources commonly employed for Al and In are TMAl and TMI whereas for n- and p-type dopants they are silane and $C_{p2}Mg$.

Before the employment of low-temperature buffer layers, to be discussed in Sect. 4.7.2, it was commonly accepted that the inability to provide sufficient quantities of nitrogen was the bottleneck problem to which the reader should pay particular attention [4.40]. Consequently, much of the early efforts dealt with exploring ways to address this issue of the high vapor pressure of N_2 on GaN. Several investigators substituted the more reactive hydrazine in favor of ammonia [4.41, 42]. Moreover, attempts have been made to use $Ga(C_2H_5)_3 NH_3$, which already has a Ga-N bond, for the GaN source [4.43]. Although no evidence to justify the approach exits, plasma excitation of ammonia in a CVD-growth environment was attempted. Suitable precursors are those possessing good reactivity, thorough pyrolisis, and transportability. Ideally, the precursors should be nonpyrophoric, water and oxygen insensitive, noncorrosive, and nontoxic. TriMethylGallium (TMG) and TriEthylGallium (TEG) are very popular for Ga, though GaCl has been tried. TriMethylIndium (TMI) and TriMethylAluminum (TMA) are the commonly employed sources of In and Al, respectively. Ammonia (NH_3) is the unchallenged source of nitrogen, as it is reasonably pure and stable. TMI and TMA are the commonly used sources of In and Al, respectively. TriMethylAluminum (TMA) and NH_3 have been reacted in the presence of a hydrogen plasma to grow AlN. TMI has been utilized with microwave activated N_2 to grow InN. Even metallic Ga transported with nitrogen to the reaction zone where it is allowed to react with active nitrogen has been tried. During growth of nitrides employing trialkyl precursors, adduct formation between ammonia and TMA and TMG has well been documented. Usually when mixing at room temperature, the adduct formation between TMG or TEG and ammonia is completed in less than 0.2 s. The resulting adduct $Ga(CH_3)_3$-NH_3 has a vapor pressure of 0.92 Torr at room temperature, while the vapor pressure of $Ga(C_2H_5)_3$:NH_3 is much lower. For a review, see [4.40].

As mentioned earlier, plasma excitation of the nitrogen species has not proved to be necessary in CVD growth. Although trialkyl compounds (TMA, TMG, TMI, etc.) are pyrophoric and extremely water and oxygen sensitive, and ammonia is highly corrosive, much of the best material grown today has been produced by conventional MOCVD via reacting these trialkyl compounds with NH_3 at substrate temperatures in the neighborhood of 1000°C. Temperatures in excess of 800°C are required to obtain single-crystalline high-quality GaN films, with the best GaN films grown at 1050°C. At substrate temperatures exceeding 1100°C, the dissociation of GaN results in voids in the grown layer. A similar situation has also been observed for AlN film growth. To overcome this difficulty, N_2H_4, instead of NH_3, has been employed with the conclusion that a significantly smaller amount of N_2H_4 was required to maintain the same growth rate.

4.7.2 Buffer Layers

Until the advent of buffer layers, the quality of GaN films grown directly on sapphire were very poor, and the background concentrations were in some cases as high as 10^{20} cm^{-3} [4.5, 6]. The poor crystalline quality was characterized by wide X-ray rocking curves of typically 10 min and uneven surface morphologies that contain hillocks, and a considerable yellow emission in the luminescence spectra. As will be detailed below, a low-temperature AlN or GaN buffer layer improves the GaN-layer quality substantially. Although, the low-temperature AlN and GaN buffer layers were initially used solely for MOCVD, their modified implementation in the MBE-growth process proved highly beneficial, too. A short period of nitridation of sapphire generally precedes the buffer-layer growth, during which time a thin $AlO_{1-x}N_x$ film grows on the substrate surface which transforms Al_2O_3 into AlN through $AlO_{1-x}N_x$.

The time sequence employed during growth by the group of I. Akasaki is depicted in Fig. 4.9 where the low-temperature buffer layer has been grown at 600°C. This must be contrasted to a temperature range of 480 ÷ 600°C employed at various laboratories for the low-temperature AlN and GaN buffer-layer growth on sapphire substrates. The purpose of the low-temperature buffer layer is to optimize the transition between the sapphire substrate and the GaN device layers. A high nucleation density, and not having to follow the surface arrangement of the sapphire substrate are among the benefits. The latter is to allow the nitride layer to assume its own structure with as few defects as possible by appropriate twisting, tilting and prismatic growth; hardly the goal of homoepitaxy. The crystalline structure of this low-temperature buffer layer is still shrouded with controversy. What can be stated is that this layer contains cubic and wurtzitic phases as well as much disorder. Although very different in nature from the MOCVD case, some sort of buffer layer is also employed in the MBE approach. Unlike the MOCVD case, in the MBE approach no matter how low the substrate tem-

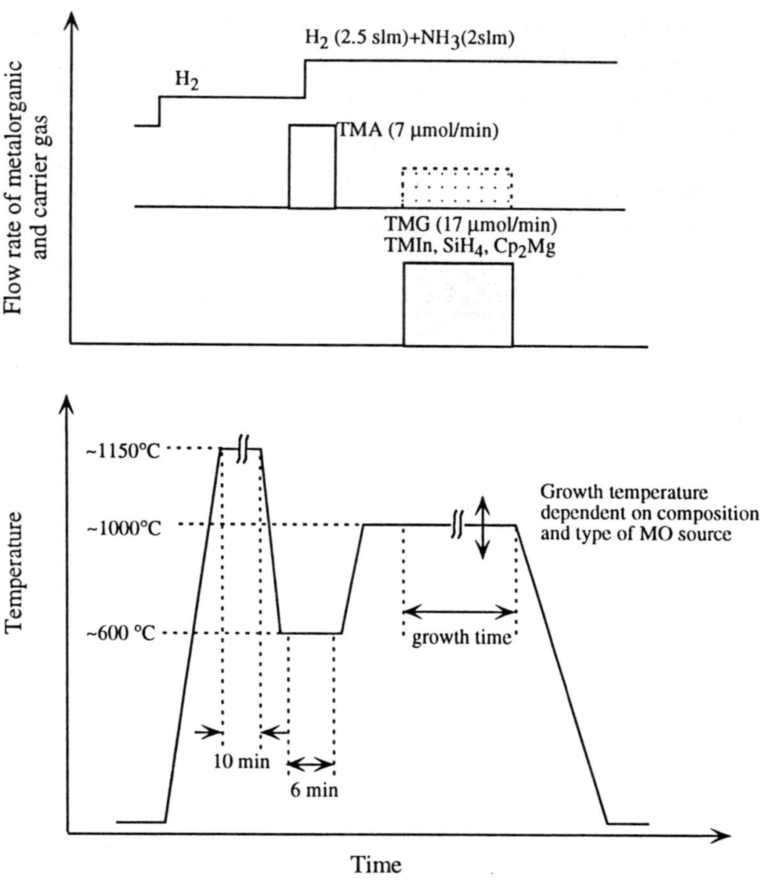

Fig. 4.9. The time sequence employed in the growth of group-III nitride heterostructures at Meijo the University. Courtesy of I. Akasaki, Meijo University

perature is, the registry between the buffer and the underlying substrate is strong, which is not desirable in nitride heteroepitaxy.

The initial stage of growth is very important for obtaining heteroepitaxy and a good quality of the resulting film. In general, the three cases of epitaxy are the two-dimensional (2D) layer-by-layer mode, the three-dimensional (3D) island mode, and the mixed (M) mode, where layer-by-layer growth is followed by an island formation. The first mode (2D) results in smooth surfaces, while the last two (3D and M) lead to rough surfaces and to low-quality epitaxial layers unless one deals with quantum-dot-like structures. The mode of growth is determined by many parameters, such as the interfacial energy of the solid and vapor phases, and the interfacial energy of the vapor phase and substrate, which, in turn, depends on the growth temperature, the bond strength and bond lengths of the substrate and the overgrowth material, the rate of impingement of species, surface migration

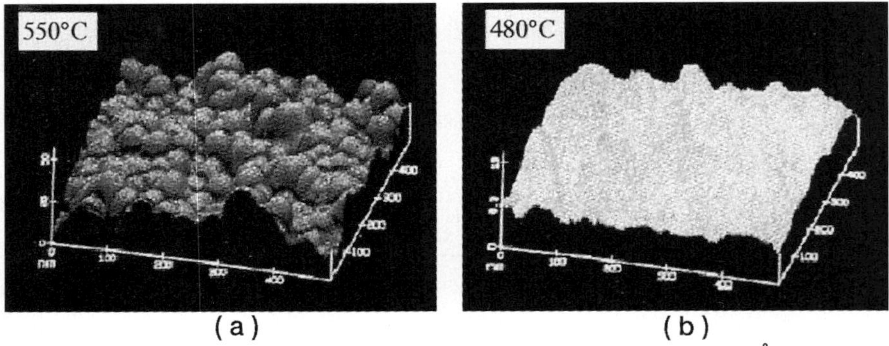

Fig. 4.10a,b. Atomic Force Microscopy (AFM) images of nominally 200 Å thick, low-temperature GaN buffer layers at (**a**) 550°C and (**b**) 480°C. Each image represents a 0.5 × 0.5 mm^2 area of the sample. The scales are in nm. After [4.39]

rates of reactants, super saturation of the gas phase, the size of critical nuclei, and others. Since the roughness of the nucleation layer is larger at higher growth temperatures, the epitaxial growth is as a rule, divided in two steps: a smooth thin (≈ 20 nm) buffer layer grown at low temperatures and the main layer grown at higher temperatures.

Atomic Force Microscopy (AFM) images (Fig. 4.10) of nominally 200 Å thich low-temperature GaN buffer layers demonstrate the island-growth characteristics of the AlN buffers [4.39] wherein the islandic nature becomes diffused for the very low substrate temperature of 480°C (Fig. 4.10b). To gain an insight into the benefits of high-temperature annealing and the nature of the surface prior to the deposition of the main layers, AFM images of the buffer layers grown at 480° and 580°C were taken after ramping the substrate temperature to 1025°C. The 550°C buffer (Fig. 4.11a) shows larger, well developed islands that are approximately 200

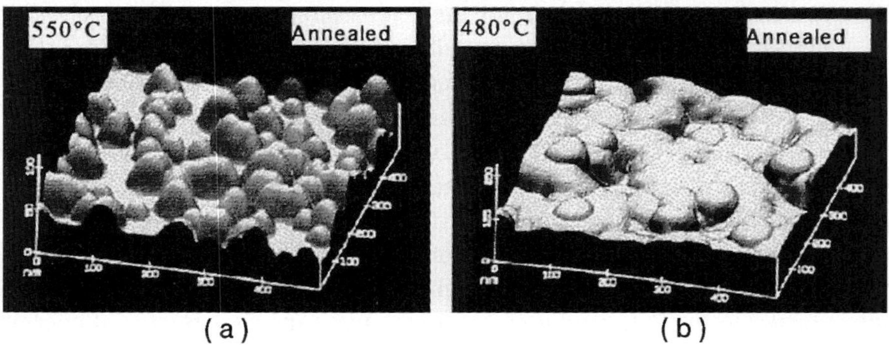

Fig. 4.11a,b. AFM images of the buffer layers of Fig. 4.10 annealed for 20 min during the ramp to the main growth temperature. (**a**) Buffer initially grown at 550°C. (**b**) Buffer initially grown at 480°C. (The images are at the same scale as in Fig. 4.10). After [4.39]

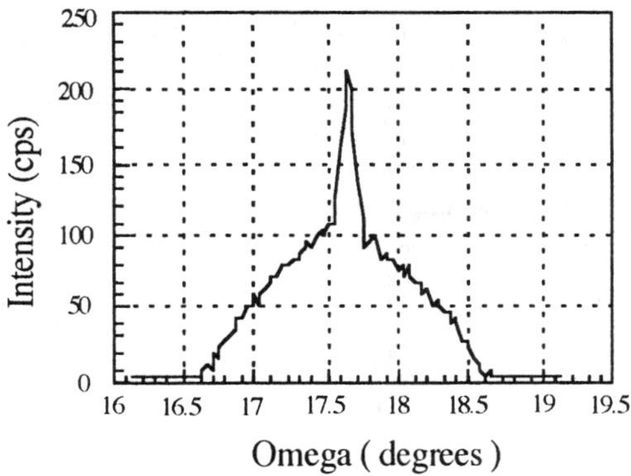

Fig. 4.12. Double-XRD peak showing a broad low-intensity peak that is topped by a narrow high-intensity peak indicating a preferred orientation among the buffer islands. After [4.39]

to 300 Å high. The 480°C buffer (Fig. 4.11b) is also modified by the anneal but retains a much smoother morphology after ramping with a typical peak-to-valley roughness of about 100 Å. A comparison of buffer layers that had been deposited with 30 and 70 Å/min indicates that larger islands are formed for the lower rate. This implies a mechanism that is kinetically limited by island growth, where a higher temperature and a lower growth rate produce the larger islands.

The crystallinity of these thin buffer layers can be probed with X-ray diffraction (XRD). The GaN (0002) reflection reveals broad X-ray rocking-curve peaks (Fig. 4.12) with a sharper peak superimposed on it. The broadening is consistent with the misorientation of the islands. The sharp peak at the center indicates a preferred orientation which can, to some extent, be influenced by the rate of the temperature ramp and the anneal time. Using AFM images and XRD analyses as probes, the effect of the ramping procedure with durations between 0 and 20 minutes indicates that the island size increases with the ramping duration. The XRD measurements exhibit a narrow high-intensity peak component which grows in lieu of the broader part of the peak indicating a more closely matched orientation. These measurements reveal that the GaN buffer layers initially consist of a dense array of islands (as in the case of AlN buffer layers [4.37]) whose sizes and crystallinities are limited by the kinetics, and are affected by the deposition temperature, the growth rate, and the ramp duration.

Cross-sectional Transmission Electron Microscopy (TEM, see [4.44] for the technique) images reveal that about a 50-nm region of the GaN layer immediately on top of the AlN buffer layer contains a large concentration of defects. Above this defect-laden region, another zone which has a number of trapezoidal crystals can be delineated. The density of defects in this 150-nm zone is much lower than in the interfacial zone. The remaining region on top exhibits a rapid decline in defects, eventually down to defect densities in the range of 10^8 to 10^{10} cm^{-2} depending on the substrates in use and the growth parameters applied.

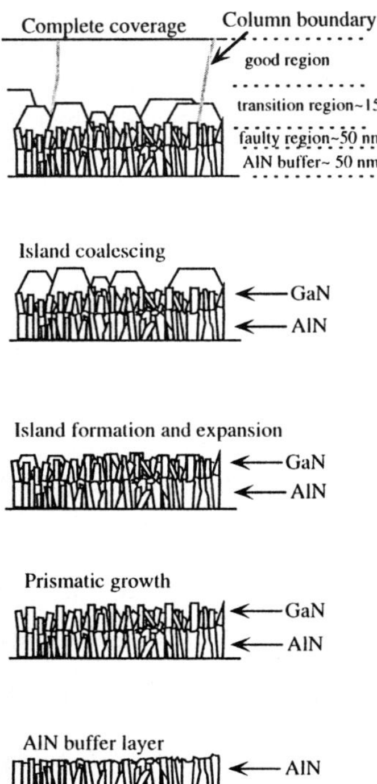

Fig. 4.13. Artistic rendition of the evolution of salient structural features of GaN grown on a low-temperature AlN buffer layer as the substrate temperature is ramped to the growth temperatures of about $1000 \div 1100°C$. The schematic is based on the analysis of the cross-sectional TEM images. Patterned after [4.37]

The AlN buffer layer has an amorphous-like structure at the deposition temperature of $600°C$ [4.37]. But during the ramping process, it is crystallized and exhibits the well-known columnar structure, as shown in the first panel of Fig. 4.13. Cross-sectional TEM images of the highly defective initial GaN layer display features similar to those of the AlN buffer layer, which suggests columnar fine crystals, as sketched by the artist's view in Fig. 4.13 (second panel). Each GaN column is probably grown from a GaN nucleus formed on top of each columnar AlN region. Consequently a high-density nucleation occurs owing to the high density of the AlN columns. The columns have disordered orientations and as the base of the columns gets larger, the number of columns emanating at the growth surface gradually decreases. The crystalline structure supports the prismatic growth which leads to a general alignment along the c-direction with some twist and tilt. The relative *tilt* and *twist* (Fig. 4.14) between the columns decreases as the layer thickness increases, expressing that the film continues to evolve towards a more ordered structure as growth proceeds. The sub-grain boundaries are more likely than the primary defected regions of the epilayer when the dislocation density has stabilized. The small value of strain broadening observed for the (0002) and (1102) reflections indicates that within each GaN columns, the crystal quality is high.

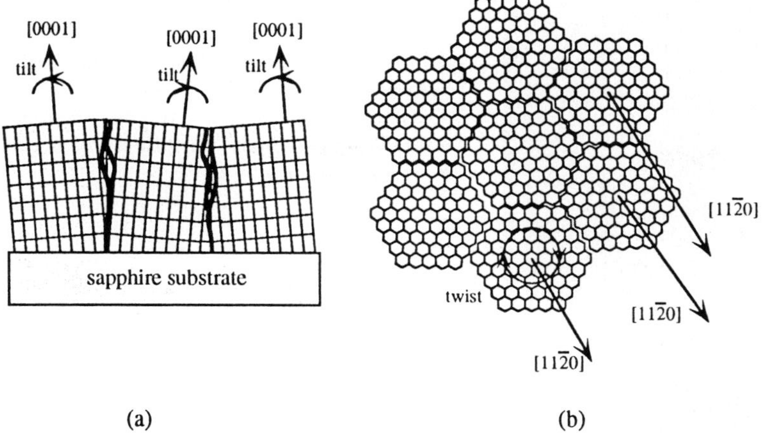

Fig. 4.14a,b. Artistic view of the microstructure of GaN on sapphire. (**a**) Side view showing the relative *tilt* of the [0001] directions between columns. (**b**) Plan view showing the relative *twist* of columns in the [112̄n] direction. After [4.39]

During the stage where the trapezoidal crystals are formed and the front area of the column increases by a geometric selection, the growth front follows the c-face (Fig.4.15, 3rd from top). This figure illustrates the changes in surface morphology during the early growth stage of GaN deposition [4.37]. The stage depicted in Fig.4.15a corresponds to the generation of trapezoidal islands where the preferred orientation begins to be established. Subsequently, lateral growth and coalescing of the islands occur in stages (Fig.4.15b and c). The trapezoidal crystals grow at a higher rate in a transverse direction, as sketched in the Fig.4.13 with a schematic of the growth front (panel 5), and the islands coalescence. Finally, since the crystallographic directions of all islands agree well with each other within a few

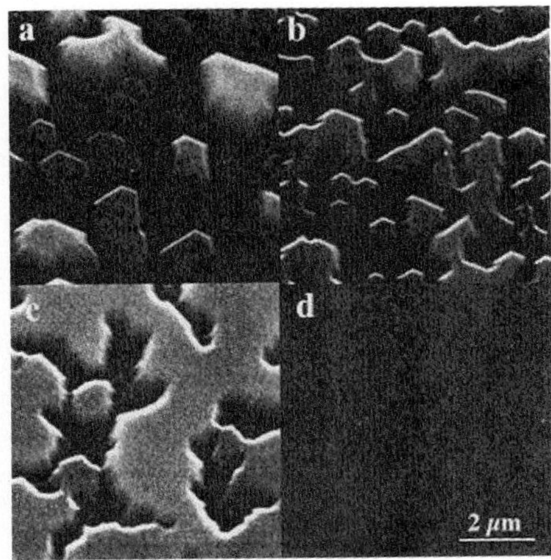

Fig. 4.15a-d. Surface morphology of GaN layers, as observed by Scanning Electron Microscopy (SEM), grown on annealed low temperature AlN buffer as the growth evolves. The panels **a, b, c,** and **d** correspond to 3, 5, 10 and 60 min, respectively. Patterned after [4.37]

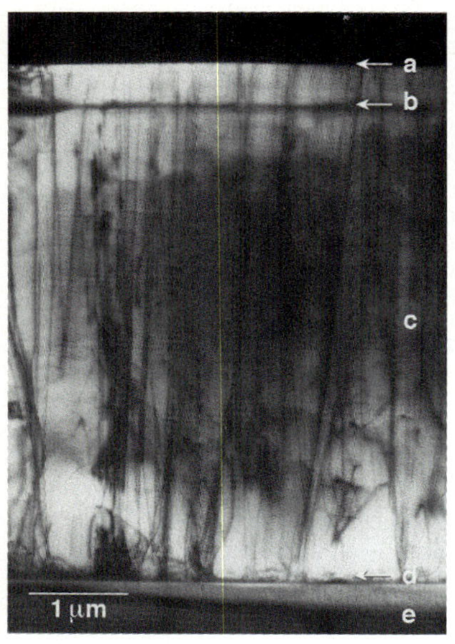

Fig. 4.16. Cross-sectional transmission microscopy image of a Nichia LED structure showing the columnar structure with nearly vertical defects traversing the entire structure. The amazing observation is that the heterointerfaces are strikingly flat. Courtesy of F. Ponce, Xerox Corp.

degrees of disorder, a smooth GaN layer with a small number of defects results, as seen in Fig.4.13 (5th from top) and in Fig.4.15d. Consequently, the pseudo-uniform growth is due to what may be construed as almost layer-by-layer growth. A cross-sectional TEM image of a Nichia LED structure is exhibited in Fig.4.16. Further away from the interface, decent hetero-interfaces are possible, but with its success hinging on the ability to grow the nucleation layer and the low-temperature buffer layer in a way that would promote GaN with good optical quality. As will be discussed in Chaps.11 and 12, the recombination process is a localized one in semiconductor nitrides, and the luminescence efficiency alone may not be a universal figure of merit. This is true for any one measurement alone. A combination of parameters, such as the in-plane X-ray diffraction, the in-plane mobility, and the luminescence efficiency all may have to be measured with weighting factors closely representative of the device application in mind.

The critical importance of the deposition rate of the low-temperature buffer layer is further illustrated by the FWHM of the (0002) XRD peak for 5 μm thick GaN epilayers grown on these buffer layers. In one experiment [4.39], the measured FWHM values were 180, 250, and 320 arc seconds for 30, 70, and 90 Å per min growth rates. These data clearly demonstrate a direct correlation between the crystallinity of the buffer layer and the crystallinity of the 5-μm GaN films grown on top of these buffer layers.

To gain an appreciation for the implications of the data, it is imperative that the origin of the XRD-peak broadening is clarified. It should be pointed out that this is applicable to any layer grown by any method. To a

108

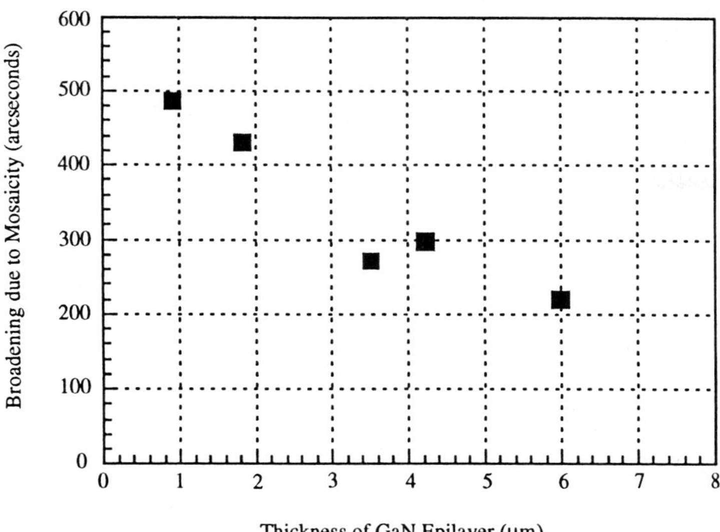

Fig. 4.17. Contributions to broadening of the (0002) XRD peak made by mosaicity. After [4.39]

first extent, the broadening is a measure of the disorder on the alignment of the c-axis among the columns and the inhomogeneous strain. High-resolution XRD can discriminate between broadening of an X-ray peak due to strain (a range of crystal-plane spacing) and broadening due to the mosaicity (a range of misorientation for a particular set of crystal planes). Broadening due to strain turns out to be less than about 30 arcseconds leaving the major influence of the broadening to the mosaicity, as displayed by the plots of Fig.4.17 where the GaN (0002) XRD peak data are presented for epilayers of different thicknesses. The mosaicity broadening decreases with increasing layer thickness. Thus, the trend towards a common crystal orientation, that was observed during buffer-layer annealing, continues to be operative even after several micrometers of GaN growth. The relative constancy of the strain broadening is attributed to an inclined slip plane in GaN grown on [0001] sapphire, which is purported to preclude annihilation of dislocations.

4.7.3 Lateral Growth

Lateral overgrowth which was successfully utilized for conventional III-V semiconductors can also be applied to nitrides for slightly different motivations. In compound semiconductors, this concept has been exploited for a variety of applications. In one such approach, periodic stripes of photoresist

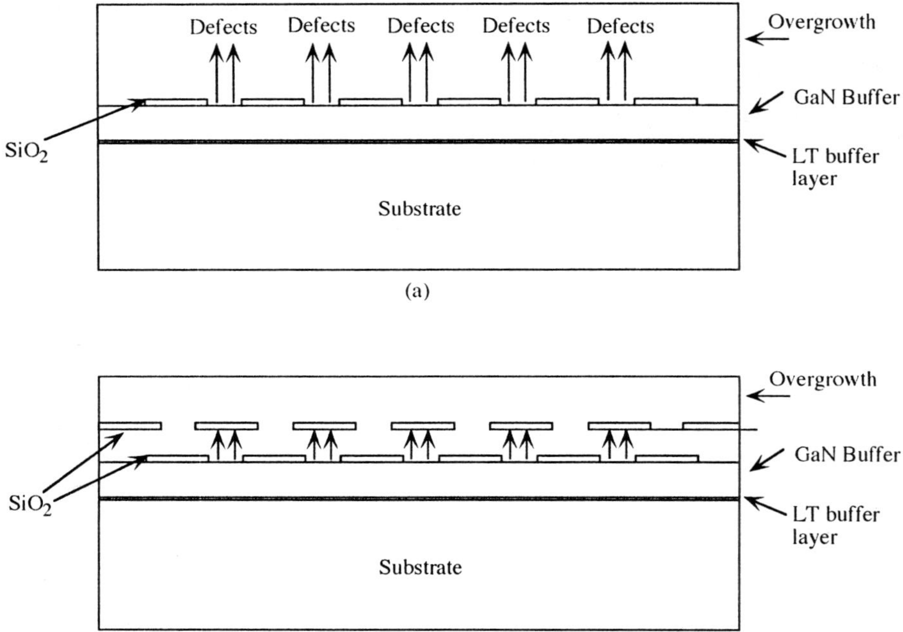

Fig. 4.18a,b. Schematic representation of lateral over-growth of GaN on SiO_2 stripes. (**a**) One lateral growth process which has been successfully demonstrated by several research groups already. (**b**) Extension of (**a**) to a two-step lateral over-growth scheme which could potentially eliminate almost all the structural defects, but the picture is not that simple. Part (**a**) is based on a figure provided by S. Nakamura, Nichia Chemical Ltd.

patterns on a GaAs substrate was used by researchers at Lincoln Laboratory to grow GaAs where in the growth initiates on the bare substrate surface followed by overgrowth on the areas covered by photoresist. Once the growth process is completed, the epitaxial layers were successfully removed from the substrate. In another Lincoln-Laboratory approach, the overgrowth was again successfully performed over W stripes as part of the **permeable-base transistor** processing. In the case of nitrides, overgrowth using SiO_2 stripes on GaN has been successfully employed to reduce structural defects in the overgrown regions (Fig.4.18a). If the growth parameters, such as the substrate temperature, the TMG flow rate and pressure are optimized, the GaN growth first initiates on bare GaN stripes and is followed by a lateral extension of the growth over the SiO_2 stripes, culminating in a continuous GaN coverage. As in the case of Si lateral growth over SiO_2, the interface where the Si layers merge, may contain many structural defects. A confirmation of the defect reduction is obtained by reactive ion-etching experiments in a gas chemistry containing some form of Cl gas on samples utilizing the lateral overgrowth method and conventional means

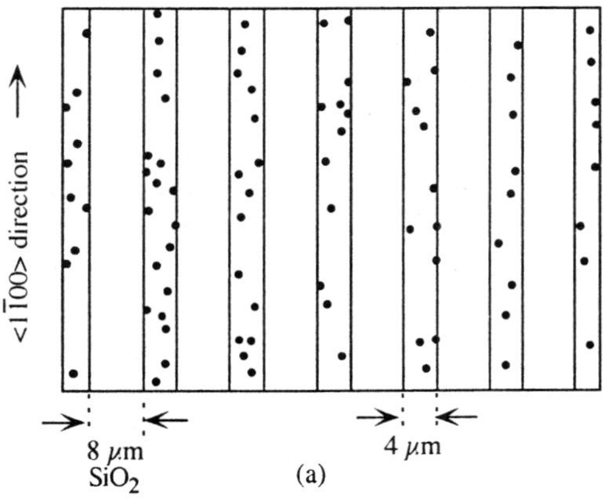

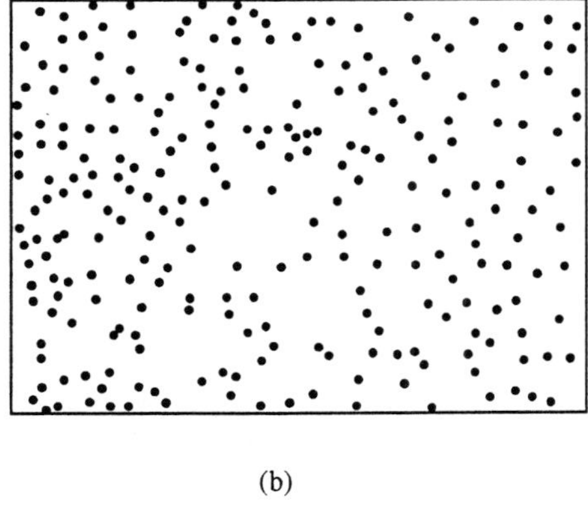

Fig. 4.19. Reactive ion etched GaN layer surfaces grown by the (**a**) lateral over-growth method, and (**b**) conventional growth scheme. The etch-pit distribution support the argument that structural defects are reduced or completely eliminated in the regions over the SiO_2 stripes. Based on figures provided by S. Nakamura, Nichia Chemical Ltd.

(Fig. 4.19a and b). The solid circles represent the etch pits where the etch rate is higher. It is clear that in the sample with lateral growth (Fig. 4.19a) the etch pits are limited to the bare GaN stripes. On the contrary, in a standard sample the etch pits are distributed throughout the sample (Fig. 4.19b). If one can extend the idea of lateral growth further, the two-step Lateral Epitaxial Overgrowth (LEO) process shown in Fig. 4.18b would eliminate just about all the structural defects. Since the process is independent of the substrate employed, issues such as thermal conductivity, electrical conductivity, and availability, would be the primary consideration in choosing a substrate. The dielectric SiO_2 is a poor thermal conductor, which must be taken into considerations in applications where the thermal budget is an issue. Moreover, each one of these steps requires a growth interruption followed by a photolithography step even if all is picture perfect. As is the case for conventional III-V semiconductors, the LEO method would require a gaseous Ga source, as metallic Ga reacts with SiO_2. This necessitates the use of MBE systems equipped with organometallic sources.

4.7.4 Growth on Spinel (MgAl$_2$O$_4$)

Cubic MgAl$_2$O$_4$ has a spinel-type structure (Fd3m) with the oxygen atoms forming a face centered cubic sublattice, and Mg and Al atoms occupying the tetrahedral and octahedral sites, respectively. Lattice mismatch with GaN, $\Delta d/d$, is 10%. The crystals are stable at the GaN growth temperature. The spinel substrate has an advantage over the sapphire substrate for obtaining mirror laser facets by cleaving [4.45-47]. GaN crystals were grown on (100)- and (111)-oriented MgAl$_2$O$_4$ substrates by MOCVD. As in the case of sapphire substrates, low-temperature buffer layers are employed followed by a few micrometers thick GaN films grown at about 1000°C. As expected, GaN films grown on (111) substrates are wurtzitic single crystals. The crystallinity of the films on spinel is not comparable to that on sapphire. The only attractive feature of spinel is the cleaved facet prospect which with the recent advances on sapphire and SiC is becoming a mute point.

4.8 Molecular Beam Epitaxy

Molecular Beam Epitaxy (MBE) with its innate refined control of growth parameters and in-situ monitoring capabilities is well suited for depositing heterostructures [4.48]. The control of growth parameters with this technique is such that any structure can be grown in any sequence. In this vein, the structures for surface-emitting vertical cavity lasers, IR lasers, ZnSe-based lasers, and high-performance electronic devices such as Pseudo-morphic Modulation-Doped Field-Effect Transistors (PMODFETs) are all produced very successfully by MBE. The nitride growth, however, is somewhat different because temperatures encountered are much higher than those customarily used for conventional III-V semiconductors, for which the systems have been designed, and active N species at sufficiently high rates have proved to be troublesome. Despite these mechanical limitations and its relatively late entry, MBE has already played a key role on a number of fronts such as high-performance GaN-based MODFETs and fast solar-blind detectors. Initial MBE techniques relied solely on the use of solid sources for both the group-III and -V elements. Later, gas sources were utilized in some cases, for example, for P, while still using the metal source for the group-III elements. This variation was dubbed the Gas-Source MBE (GSMBE) [4.49]. Further along, metalorganic sources were introduced and employed in parallel with the hydride group-V sources. Yet additional terms such as Chemical Beam Epitaxy (CBE) or Metal Organic Molecular Beam

Epitaxy (MOMBE) were deemed necessary to describe the process. Utilization of gas sources has the obvious advantage that the systems do not have to be opened to air for recharging. However, solid-source manufacturers were not to be outdone that easily. They produced two-stage sources for group-V elements and incorporated large cells for group-III elements to successfully combat the necessity to brake the vacuum. In the case of nitride growth by MBE, while solid sources are commonly practiced for the group-III species, a wide range of options are available for nitrogen. Among them are ammonia, supersonic jet sources, Electron Cyclotron Resonance (ECR), Radio Frequency (RF) sources and ion sources. Details of the energetic sources have been discussed elsewhere and are beyond the scope of this monograph [4.50]. Below, following a brief discussion of an MBE system, the use of the MBE method for nitride growth will be described.

4.8.1 MBE Growth Systems

An MBE system [4.51] can be considered a refined form of an ultrahigh vacuum (UHV) evaporator, as illustrated in Fig.4.20. The arrangement shown utilizes ammonia for nitrogen. In the case of plasma-like approaches, the ammonia injector is simply replaced with a plasma or ion source. Elements are heated in Knudsen cells and directed beams of atoms or molecules are condensed onto a heated single-crystal substrate where

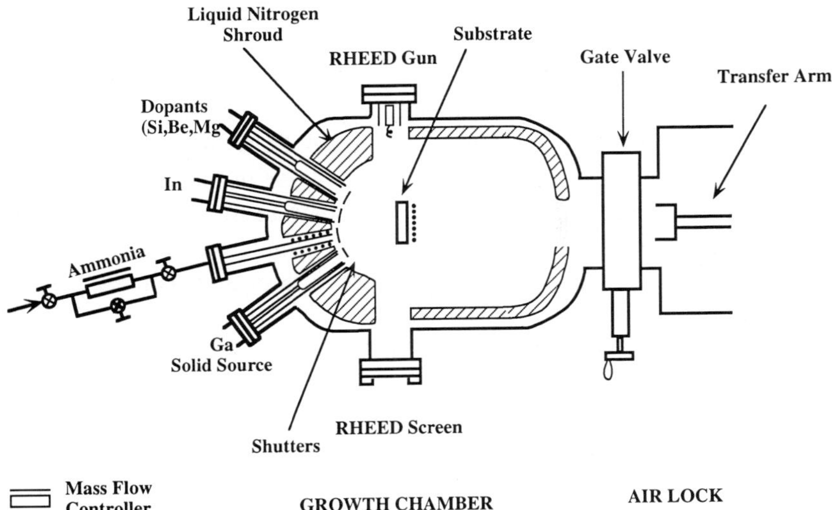

Fig. 4.20. Schematic cross sectional diagram of a MBE system. This particular one situated in the author's laboratory features ammonia as the reactive nitrogen source. To depict other varieties requires the removal of the ammonia injector and replacing it with an RF, ECR or an ion source

they react. Because it is a UHV-based technique, it has the advantage of being compatible with a wide range of surface analysis techniques, such as Reflection High-Energy Electron Diffraction (RHEED), quadrupole mass spectrometry for gas analysis, and Auger electron spectrometry [4.52]. Typically, the growth rate employed is one to three monolayers per second, approximately $0.3 \div 1$ μm/h, although much higher growth rates can be attained. The MBE growth is usually performed at relatively low temperatures of $650 \div 800\,^\circ$C. The molecular nitrogen is inert and does not chemisorb on a GaN surface below $950\,^\circ$C due to the strong N-N bond of the nitrogen molecule. Atomic nitrogen or nitrogen containing molecules with weaker bonds should therefore be provided. Several modifications of conventional MBE methods have been implemented for III-N growth. Among them, RF or ECR plasma sources are most commonly employed to activate the nitrogen species. A solid source is usually utilized for Ga. Mg or Si are usually evaporated during the growth for p- and n-doping, respectively. Electron concentrations in the range of $10^{17} \div 10^{19}$ cm^{-3} and the dependable hole concentration in the range of $10^{17} \div 10^{18}$ cm^{-3} with good Hall mobilities have been successfully produced without any post-growth treatment.

4.8.2 Plasma-Enhanced MBE

Plasma excitation of nitrogen or nitrogen containing molecules for group-III nitride growth has been studied for over twenty years [4.50-54]. Early models of RF source plasmas used by researchers suffered from a number of disadvantages including high ion energies and unfavorable plasma-source geometries. Only recently, there has been a revitalization of plasma-based nitride research, which has been stimulated by the development and commercial availability of MBE compatible compact ECR microwave plasma sources, and RF sources. In an RF or ECR microwave plasma-assisted MBE approach, the GaN films are formed on the substrate through the reaction of atomic Ga provided by a Knudsen cell, and atomic and ionic nitrogen formed by passing molecular nitrogen at a pressure of approximately 10^{-4} Torr.

There are, at least, two commercial MBE-compatible ECR designs available. These compact sources operate at 2.45 GHz, enabling the plasma to be confined to a smaller volume. To better meet the low-pressure requirements for MBE, ECR sources employ a magnetic field designed to match the electron cyclotron frequency of 2.45 GHz. Resonant absorption by the electrons greatly enhances the coupling of the microwave input to the plasma. Each source is compact, allowing it to be inserted into the source flange and through the cryoshroud of an MBE chamber. The plasma enjoys a direct line of sight to the substrate so that collision losses are minimized.

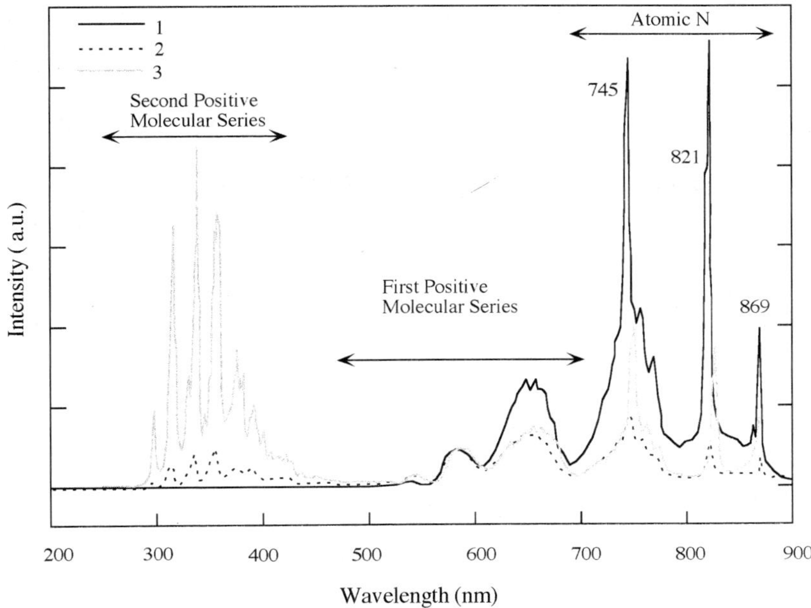

Fig. 4.21. Optical emission spectra from three different nitrogen RF sources operating at 400 W. Courtesy of J. Schetzina, North Carolina State University

In an ECR source, approximately 10 % of the molecular nitrogen gas is converted into atomic nitrogen.

As in the case of ECR sources, there are several manufacturers of RF sources. RF plasma sources take advantage of relatively-low-frequency excitation such as 13.56 MHz through an RF coil to couple energy into the plasma-discharge region. The plasma-sheath effect confines ions and electrons within the plasma discharge regions allowing only low-energy ($<$10eV) neutral species to escape. RF MBE permits growth rates up to about one micrometer per hour. For high-quality crystals, however, a relatively low growth rate of about 0.2 μm/h should be employed. This figure is up to about 0.5 μm/h in more recent designs.

Optical emission spectra of RF sources are very useful in determining, to a good extent, the nitrogen species responsible for growth. Shown in Fig.4.21 are the emission spectra from three different RF-plasma sources operating at an input RF power of 400 W and a system pressure of $5 \cdot 10^{-5}$ Torr. Unlike that for ECR sources, the RF-source emission is primarily neutral and consists of peaks associated with molecular (first-positive and second-positive) and atomic species. The first- and second-positive molecular-nitrogen emissions occur near orange-red and UV wavelengths, respectively. The atomic-nitrogen-related emission, however, occurs near-red (InfraRed, IR) wavelengths. One particular source, labeled as 1 in Fig.4.21, exhibits the strongest emission lines associated with atomic species as well as the first-positive molecular species. This source is capable of producing GaN growth rates of more than 1 μm/h at substrate temperatures as high as 800°C. At first glance, this observation would imply that the atomic species are the primary participants in the growth process.

Several outstanding issues regarding plasma sources must be resolved before a validation of their eventual utility can be undertaken. These are exacerbated by the lack of availability of sources with a larger volume-to-surface ratio which require among others, 15-cm (6 inch) ports on MBE systems which are not readily available. There is also the inadequate knowledge base regarding the molecular dynamics of the nitride-growth surface. It is not clearly understood what type of species are most conducive to nitride growth, namely atomic nitrogen, ions, or another form of excited nitrogen. Fluences are determined by Langmuir probes, which measure only ions, making it difficult to evaluate neutral-flux rates. MBE source flanges with larger dimensions are under development to allow the use of larger plasma sources which can operate under more optimized geometries and allow increased growth rates to be obtained with ion energies lower than that is possible with the presently available models. Finally, contamination must be minimized by eliminating the exposure of metal, quartz and alumina components to the plasma discharge. The influence of the nitrogen-ion bombardment on the layer quality is controversial, and not fully understood. It has widely been observed that optical and electrical properties of films grown at high powers can degrade. The damage threshold of GaN was estimated to be 24 eV [4.55]. Ion removal magnets have also been employed to reduce the damage caused on the growing surface.

In an MBE process utilizing ECR-activated nitrogen, a typical growth procedure is as follows: The treated substrates, as outlined in Sect. 4.3, for the growth are transferred to the growth chamber for epitaxial growth. The substrate temperature is slowly increased to 750°C if the substrate is directly mounted on a Mo-block; it is increased to 850°C for substrates mounted on Si. This process normally takes 35-40 minutes. A streaky bulk RHEED pattern is observed for H_2/He plasma-treated substrates. The substrate temperature is then lowered to $550° \div 560°$C for the growth of $350 \div 500$ Å thick AlN buffer layer. The ion-pump gate valve is closed, and N_2 is introduced into the chamber. The chamber pressure increases from $(3 \div 5) \cdot 10^{-10}$ Torr to $8.5 \cdot 10^{-5}$ Torr. The plasma is turned on with the ECR power set to 60 watts, and the current magnets set to $18.5 \div 19.5$ A. The following conditions are adopted for AlN growth: The N_2 flowmeter is set to lead to a growth-chamber pressure of about $(8.5 \div 8.7) \cdot 10^{-5}$ Torr. The nitrogen-ion flux is kept at $J_g = (0.8 \div 1.0) \cdot 10^{-6}$ cm^{-2}s^{-1}, and Al cell temperatures are set to 1150°C and 1040°C for the top and bottom heaters, respectively. AlN growth on Al_2O_3 substrates is initiated by opening the Al and ECR shutters at the same time. For growth on SiC substrates, the ECR shutter is opened $20 \div 30$ seconds after opening the Al shutter. The growth rate using the above-stated conditions was found to be $200 \div 220$ Å/h.

After the desired AlN thickness has been reached, the Al shutter is closed and the Ga shutter opened. Various conditions, such as the substrate temperature and the nitrogen-ion flux, have been tried to obtain GaN growth. For each specific setting of the substrate and the Ga cell temperature, there is an equilibrium N_2-flux condition. Below this particular value, the RHEED pattern becomes streaky but weak, and the resulting surface is covered with Ga droplets. With the N_2 flux above this equilibrium value, the RHEED pattern becomes spotty and bright, and the resulting surface is rough. So far, the optimum condition is one utilizing a high substrate temperature and a low nitrogen-ion flux, corresponding to the order of $(1 \div 5) \times 10^{-6}$ A.

The growth with an RF source differs from the case of an ECR source in the settings of the source itself. As in the case of ECR plasma-assisted growth, the substrates are cleaned in a H:He plasma in a vacuum-connected chamber. This process takes place at a substrate temperature of $1000\,°C$ in the sapphire. Nitridation is performed at a substrate temperature of $700 \div 800\,°C$, determined by the substrate-temperature limitation of the substrate heater, for 3 min at a nitrogen flow rate of 2 sccm and an RF power of 300 W. Following the nitridation process a buffer layer is grown. Various buffer layers incorporated are 200 Å thick AlN layer, 6 periods of an AlN (20 Å) / GaN (20 Å) superlattice buffer layer, and in the case of ZnO, either a GaN or InGaN buffer layer. The RF source settings for the growth of the remainder of the structure can be the same, but those resulting in growth rates of about $0.15 \div 0.2$ μm/h are optimal. Higher growth rates on the order of 0.4 μm/h are accomplished at the expense of n-type background concentrations of about 10^{18} cm^{-3}. Increasing the RF power to 500 W reduces this background somewhat. With an RF source, the entire mole-fraction range of InGaN can be obtained with the substrate temperature reduced to $500\,°C$ for pure InN, while applying a 500 W RF power with a growth rate of 0.16 μm/hour. In the case of InGaN/GaN multi-layered structures, RF power levels of about 350 W seem to work.

4.8.3 Reactive-Ion MBE

Several laboratories have attempted MBE growth of the group-III nitrides using non-plasma-based growth techniques. The most successful one of these utilized low-energy nitrogen ions from a Kaufmann ion source to grow GaN of a quality comparable to that grown with ECR sources. Several laboratories in the past have tried reactive-ion molecular beam epitaxy in which N_2 or NH_3 was decomposed on the substrate surface [4.56, 57]. However, decent growth rates and material quality were not obtained at MBE substrate temperatures. Two possible extensions of the original RMBE approach, which were thought to reduce the required substrate temperature, are the application of hydrazine, and cracked NH_3 as the source gas, which could be accomplished by a commercial hydride cracker. Hydrazine is believed to increase the reactivity of the nitrogen containing species. Much effort on thermal cracking of ammonia turned out to be misguided for several reasons. Thermal cracking requires high temperatures with its expected complications which reduce this approach to a laboratory curiosity. In addition, the little atomic N produced recombines readily through collisions forming very inert and stable N_2. Needless to say, this approach is not seriously pursued anymore. The use of hydrazine and the likes thereof raise safety concerns and water vapor contamination. Ammonia cracked only at the surface of the substrate is the successful approach and has been employed in various laboratories; as will be outlined below.

4.8.4 Reactive MBE

The technique utilizing ammonia as a nitrogen source has been termed the **Reactive MBE** (RMBE) in which ammonia is decomposed only on the surface of the substrate by pyrolysis. As such, the lower end of the temperature range where this technique can be utilized is about 600°C, albeit with very low growth rates. Growth rates of about $1 \div 3$ μm/h can be obtained at substrate temperatures between 750° and 850°C while maintaining a high quality. This growth temperature is much lower than the growth temperature in MOCVD. *Kim* et al. [4.58] showed that RMBE can produce films of a very good quality. Though applicable to all substrates, this particular study was performed on sapphire substrates. The actual growth is preceded with a nitridation and a thin AlN buffer layer. In this approach, the substrate temperatures varied between 610° and 820°C. Growth rates up to 2.9 μm/h were obtained for temperatures between 725° and 820°C. These growth rates are quite comparable to those produced by conventional MOCVD. Studies of the growth kinetics have been carried out by employing various V/III ratios and substrate temperatures. It was found that the layer quality increases and compressive stress decreases as the N/Ga ratio increases. It should be pointed out that incorporation of residual impurities as well as the creation of a high density of native defects are made less efficient at higher N/Ga ratios.

The growth mechanism of GaN in RMBE involves thermal cracking of ammonia and surface reaction of N with Ga atoms to form GaN. Therefore, if cracked-ammonia flux on the growing surface is insufficient as compared

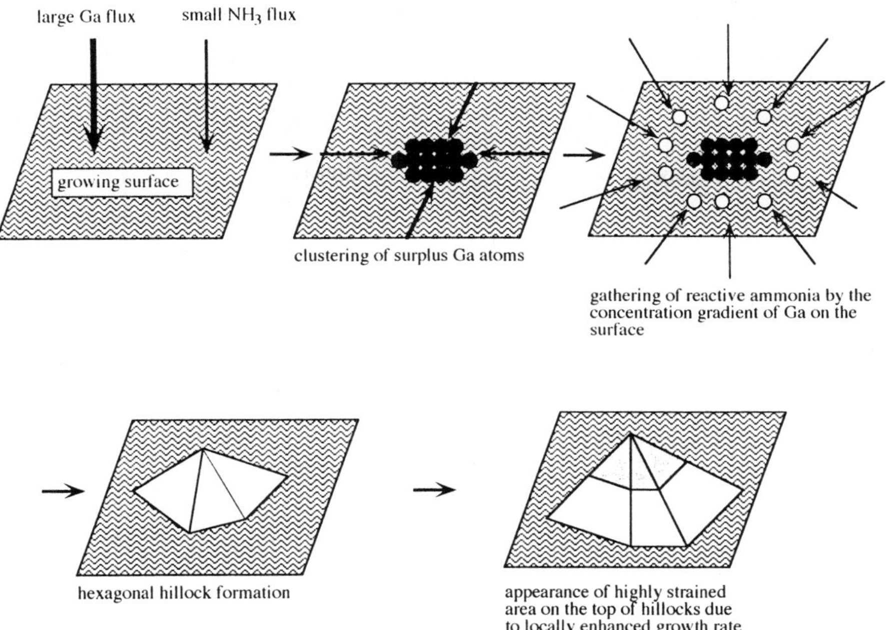

Fig. 4.22. A proposed pathway to hillock formation in the growth of GaN

to the Ga flux, the surplus Ga would thermally desorb as substrate temperatures are high, in the vicinity of $800°C$. Consequently, those Ga atoms which do not participate in the formation of GaN tend to stick closer to the surface with minimum surface energy. This gives rise to Ga clusters on the surface which, in turn, appear to serve as preferred nucleation sites. It is then reasonable to argue that the larger population of Ga atoms near the cluster sites causes a locally accelerated growth. The immediate result of this is the appearance of huge hexagonal hillocks on the film surface. The proposed process is depicted schematically in Fig.4.22. The conclusion that can be drawn is that if the Ga-to-ammonia flux ratio is high, hillocks are likely to form on the surface. These features are confirmed with scanning electron images of GaN surfaces. A noticeable feature of hexagonal hillocks is that their tops appear to be composed of many overlapping triangles and etch pits.

To investigate the growth kinetics [4.58], the substrate temperatures and ammonia flow rates were varied in the ranges of $700 \div 820°C$ and $4 \div 25$ sccm, respectively, while keeping the Ga flux at about $3 \cdot 10^{15}$ cm^{-2}s^{-1}. The variation of growth rate with the ammonia flow at a substrate temperature of $800°C$ is displayed in Fig.4.23. Although the scatter in data is quite substantial, a weak trend of increasing the growth rate with increased ammonia flow rate before reaching a saturation is apparent. It can be argued that, at low ammonia flow rates, the Ga desorption is responsible for

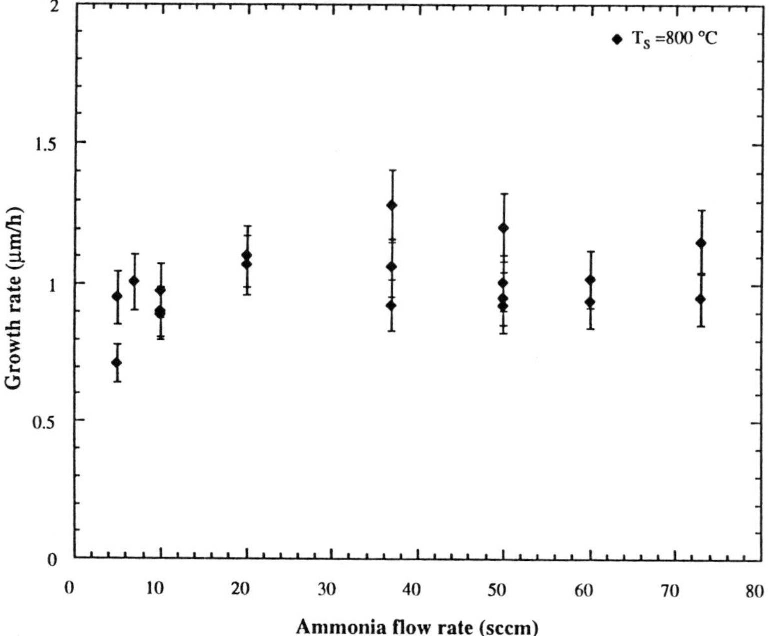

Fig. 4.23. Variation of the growth rate with the ammonia flow at a substrate temperature of 800 °C. Although these experimental results are very preliminary in nature, the growth rate apprears to increase with ammonia flow rate initially and before saturation. This is indicative of the suppression of Ga disorption with increased ammonia flow

the reduced growth rate. At high ammonia flow rates, the growth rate may be limited by the Ga arrival rate, which is responsible for the apparent saturation. Many more experiments must be conducted before a quantitative argument can be made. At much lower growth temperatures, e.g. 700 °C, not shown, the growth rate is much smaller because of the lower ammonia cracking efficiency, supporting the Ga-desorption argument.

Based on the assumption of no desorption from chemisorbed states, a growth kinetics study of GaN grown by Reactice Ion Molecular Beam Epitaxy (RIMBE), which is also applicable to RMBE with some modifications, was carried out by *Powell* et al. [4.59]. In this model, competitive contributions of collisionally induced dissociative chemisorption of nitrogen and of ion-stimulated Ga desorption were considered to be the primary mechanisms that affect the growth. The model was successful in demonstrating that, at a constant substrate temperature and constant nitrogen-to-Ga flux ratio ($J_{N_2^+}/J_{Ga}$) the growth rate decreases with increasing ion energy. However, the model could not account for the experimentally observed drop in the growth rate at high substrate temperatures. For an RMBE study, the same model can be used with some modifications of parameters and of activation energies for the surface-state transitions. The equations describing the rate of change of the number of physisorbed and chemisorbed Ga atoms on the surface may be given as

$$\frac{dn_{Ga}}{dt} = J_{Ga} - \frac{n_{Ga}}{\tau} - kn_{Ga}(1-q) + k'N_s q \, , \qquad (4.8)$$

$$N_s \frac{dq}{dt} = kn_{Ga}(1-q) - k'N_s q - fqJ_{NH_3} \, , \qquad (4.9)$$

where k is the forward reaction rate constant from the physisorbed state to the chemisorbed state, k' is the backward reaction rate constant from the chemisorbed state to the physisorbed state, τ is the average residence time of free Ga prior to thermal desorption, J_{NH_3} is the ammonia flux, and f is the fraction of effective ammonia that is thermally cracked and reacts with Ga to form GaN on the growing surface. To correctly describe the nature of growth, the last term on the right-hand side of (4.9) is modified to fqJ_{NH_3} for RMBE from $2fqJ_{N_2}$, originally formulated for RIMBE. The multiplication factor of 2 was reduced to 1 because each ammonia molecule in RMBE can provide only one nitrogen atom. The parameters used in (4.8 and 9) are defined as

$$k \quad = n_0 e^{-E_p/k_B T} \, , \qquad (4.10)$$

$$1/\tau = n_0 e^{-E_d/k_B T} \, , \qquad (4.11)$$

$$k' \quad = n_0 e^{-E_c/k_B T} \, , \qquad (4.12)$$

$$f \quad = e^{-E_{cr}/k_B T} \, . \qquad (4.13)$$

All parameters other than those defined by (4.10-13) are the same as those used in Powell's model. The solution of (4.8 and 9) at steady state with the same values for the various parameters, as given in Powell's model, results in a decreasing growth rate as the substrate temperature increases from 650° to 850°C for a constant Ga flux and a constant V/III ratio. Another calculation for a constant Ga flux ($3 \cdot 10^{15}$ cm^{-2} s^{-1}) and a constant substrate temperature (800°C) reveals an increasing growth rate as the V/III ratio becomes larger. These results are exhibited in Figs. 4.24a and b.

Evans et al. [4.60] reported on an empirical value for Ga desorption in the growth of GaN with gas-source MBE, which follows essentially the same growth scheme, as used in the present experiment. The desorption energy for Ga atoms obtained by TPD (Temperature-Programmed Desorption) was published to be 1.9 ± 0.2 eV. In addition, *Burns* et al. [4.61] measured the thermal desorption energy of Ga on Al_2O_3 by means of a Thermal

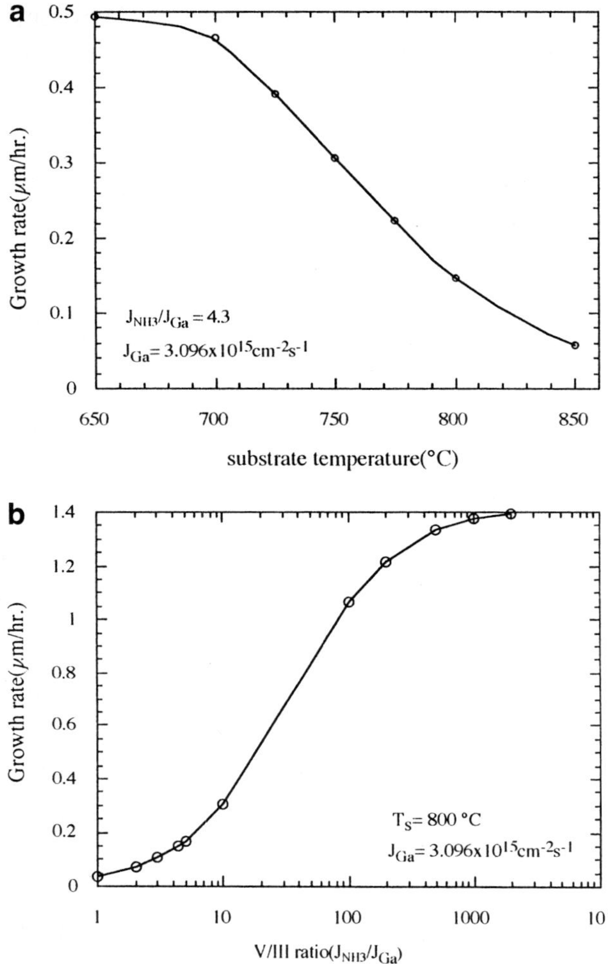

Fig. 4.24. Calculated growth rate from the simple precursor-mediated chemisorption model. (**a**) Growth rate change with substrate temperature (V/III = 4.3, J_{Ga} = 3.096 × 10^{15} cm^{-2} s^{-1}) (**b**). Growth rate change with V/III ratio (T_s = 800 °C, J_{Ga} = 3.096 × 10^{15} cm^{-2} s^{-1}). Courtesy of W. Kim, LG Corporate Institute of Technology, Seoul, Korea

Desorption Spectroscopy (TDS) technique and deduced a value of 2.05 eV. Among these two measured values, The 1.9 eV according to *Evans* et al. is deemed more applicable because of similarities in the growth schemes employed in the RMBE investigation of [4.58]. The growth rate has been calculated with 1.0 and 0.25 eV, respectively, for the E_c and E_{cr} parameters. The activation energy for the transition from the physisorbed state to the chemisorbed state was chosen to be zero because the sticking coefficient of Ga is generally considered to be one. In Fig.4.24, the calculated growth rate data for 12 and 25 sccm ammonia are compared with experiments. Even though there are differences in magnitude between the experimental data and the calculated values for the 12 sccm ammonia case, the calculated data follow the same qualitative trend of growth rate versus substrate temperature as the experimental data. More specifically, the calculated results also present growth rates increasing with temperature due to more efficient thermal cracking of ammonia in the temperature range where Ga desorption is not the limiting factor.

Considering that the effective amount of ammonia molecules is proportional to the ammonia flux and assuming that the cracking rate is identical for a particular temperature, it is quite reasonable to expect that the growth rate of the sample grown with the lowest ammonia flux (4 sccm) would decrease rather drastically with decreasing substrate temperature. That this is the case is apparent in Fig. 4.24 for ammonia flow rates of 4 and 12 sccm, respectively. If the substrate temperature is not sufficiently high to produce as much or more reactive ammonia than that necessitated by the given surface concentration of Ga at that substrate temperature, the growth rate would not be determined by the ammonia flow rate. Instead, it would be determined by the Ga population on the surface, and would remain constant for a given temperature, unless there is a change in Ga flux. Additional experiments carried out at 800°C point to the growth rate remaining more or less constant, no matter what the ammonia flux. However, one can conclude that the growth rate is determined by the ammonia flow rate if the substrate temperature is not sufficiently high to crack substantial amounts of the ammonia supplied. In the present case, this is a temperature below 750°C.

It is to be noted that, once the substrate temperature becomes higher than that leading to the highest growth rate, Ga desorption from the surface becomes the limiting factor. Consequently, the growth rate decreases despite more efficient ammonia cracking. For example, a significant drop of growth rate was found at 820°C for the sample grown with 25 sccm ammonia (Fig. 4.24). Furthermore, in this growth regime, the growth rate is constant at a specific temperature regardless of the ammonia flux unless it is lower than the Ga flux. The same trend was also observed for growth rates corresponding to other flow rates of ammonia. When the substrate temperature was increased to 850°C, the growth rate practically became nil.

The incorporation kinetics of gallium during GSMBE of GaN using elemental Ga and NH_3 gas as source materials was studied by *Jenny* et al. [4.62]. Desorption Mass Spectrometry (DMS) was applied to perform in-situ quantitative measurements of GaN formation, Ga desorption and Ga surface accumulation during growth. The rate of formation of GaN was reported as a function of incident Ga flux ($0.1 \div 0.75$ ML/s) and incident NH_3 flux ($1 \div 300 \cdot 10^{-7}$ Torr beam equivalent pressure) at a growth temperature of 725°C. Three distinct growth regimes (Fig. 4.25) were observed:

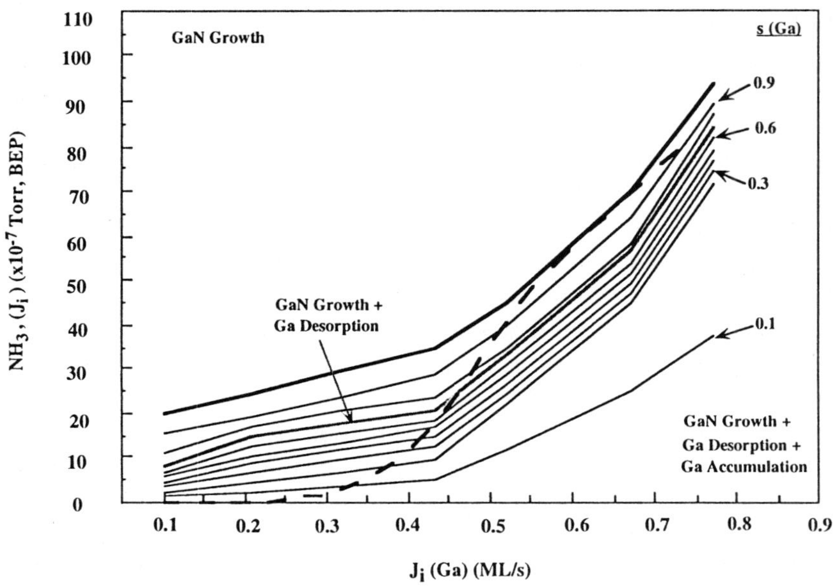

Fig. 4.25. Plot of the Ga incorporation ratio, s(Ga), contours as a function of Ga and NH$_3$ fluxes. Labeled in the figure are the three primary regimes for Ga adatoms. After [4.62]

(1) GaN is formed where all of the incident Ga flux is consumed, (2) part of the incident Ga atoms are consumed to form GaN while the excess is desorbed by the surface, and (3) GaN formation coexists with desorption and surface accumulation of Ga. It was generally found that the Ga surface accumulation is inhibited by increasing the rate of incidence of NH$_3$ and/or by decreasing the rate of incidence of Ga at this temperature. In addition, the order of the reaction between Ga and NH$_3$ is determined to be unity, and this supports the validity of the Ga + NH$_3$ → GaN + 3/2H$_2$ reaction.

The MOCVD and MBE grown films with very narrow out-of-plane-rocking curves with FWHM as low as 37 arcsec have been reported. Conventional wisdom would imply that the X-ray rocking curve FWHM of the GaN (002) peak is a good barometer of quality. This conclusion is based on a good deal of experience with conventional semiconductors grown homoepitaxially. In conventional semiconductor technology it is typical that the rocking curves become significantly broadened at high densities of dislocations (generally $> 10^5 \, \mathrm{cm}^{-2}$). However, the GaN films with narrow rocking curves contain threading dislocations which are predominantly pure edge types along the c axis. The specific threading dislocation geometry leads to distortions of only the specific crystallographic planes. Edge dislocations will distort only the (hkℓ) planes with either h or k being non-zero. Rocking curves on off-axis (hkℓ) planes will be broadened while symmetric (00l) rocking curves will be insensitive to the pure edge threading dislocation content in the film. Screw threading dislocations with [001] directions have a pure shear strain field that distort all (hkℓ) planes with ℓ non zero. Therefore, the rocking-curve widths for the off-axis reflections are more reliable indicators of the structural quality of GaN films. The properties of GaN film grown on sapphire differ for the in-plane and out-of-plane in terms of structural features. The measured in-plane coherence lengths are smaller and the rocking-curve widths are larger than in the plane-normal direction. It implies that optical and electrical properties are anisotropic along the film plane and plane-normal directions. The in-plane structure measurements are more closely related to the electronic mobility and optical properties than those in the plane-normal direction. This premise is supported a high-resolution X-ray investigations [4.63]. Films with very sharp (45 arcsec) out-of-plane rocking curves show poor electrical and optical properties, which is consistent with an in-plane X-ray analysis. On the other hand, the films with about 5 arcmin out-of-plane rocking curves reveal much better optical and electrical properties; this is again consistent with the excellent in-plane X-ray diffraction. Judging from the out-of-plane XRD data and the cross-sectional TEM images of the same films, it appears that a sharp X-ray rocking curve is likely a result of planar defects and is not due to any enhanced overall quality.

4.8.5 Modeling of the MBE-Like Growth

The hallmark of the MBE approach, as applied to conventional group-IV and group-III-V semiconductors, is that it provides very reliable laboratory investigations of basic growth processes. The case of nitrides contradicts this long-standing status somewhat in that there is little work published on modeling of group-III-N growth. The difficulty of quantitative estimates of growth parameters resides in the lack of reliable data on surface migration rates, nucleation parameters, and sticking coefficients of group-III and N atoms to the different substrate surfaces, as well as other thermodynamic and kinetic parameters. There are some indications that the grown surfaces are likely to be Ga stabilized which is purported to be the case in some MBE- and MOCVD-grown layers. As depicted in Fig. 4.26, the growth rate increases with the increasing V/III ratio and reaches a saturation value determined by Ga flux [4.64]. At a given temperature, the lower the V/III ratio the better the growth front, but the growth rates are correspondingly lower. The results indicate that there is a temperature window and a N/Ga ratio window in which a good growth front could be obtained while preserving a reasonable growth rate.

Kinetic modeling of microscopic processes during ECR microwave plasma-assisted molecular beam epitaxial growth of GaN/GaAs based heter-

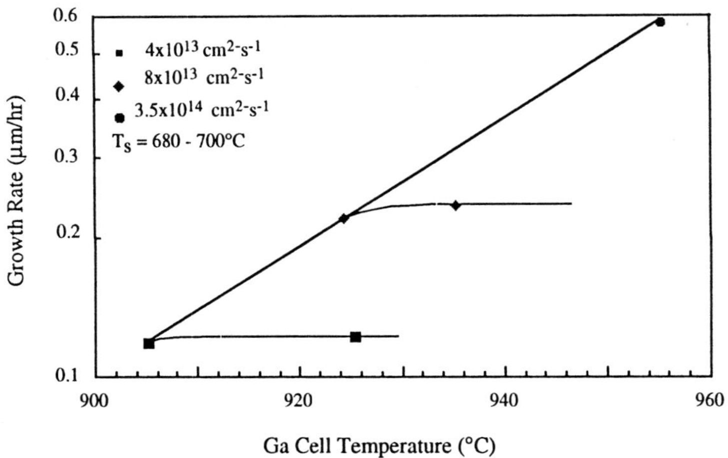

Fig. 4.26. Growth rate vs. the Ga cell temperature in MBE growth of GaN with nitrogen flux as the parameter. At low N fluxes and the higher end of the Ga temperatures, the growth is limited by N flux, presumably leading to Ga stabilized surfaces with smooth surfaces and 2×2 surface reconstruction

ostructures was done by *Bandic* et al. [4.65]. Microscopic growth processes associated with GaN/GaAs molecular beam epitaxy were examined through the introduction of a first-order kinetic model. The model has been applied to the ECR microwave plasma-assisted MBE (ECR-MBE) growth of a set of delta $GaN_y As_{1-y}$/GaAs strained-layer superlattices that consist of nitrided GaAs monolayers separated by GaAs spacers, and that exhibit a strong decrease of y with increasing T over the range $540 \div 580°C$. This y(T) dependence is quantitatively explained in terms of microscopic anion exchange, and thermally activated N surface-desorption and surface-segregation processes.

A theoretical model for the MBE growth which accounts for a physisorption precursor of molecular nitrogen was proposed for the analysis of group-III nitrides, independently suggested by *Averyanova* et al. [4.66]. The kinetics of nitrogen evaporation was found to be an essential factor influencing the MBE growth process. The high thermal stability of nitrides was explained to be related to the desorption kinetics, it results in a low value of the evaporation coefficient. Evaporation coefficients as functions of temperature were extracted from the experimental Langmuir evaporation data of GaN and AlN. The process-parameter-dependent growth rate and the transition to extra liquid-phase formation during the GaN MBE were calculated.

High-quality layers grown on sapphire substrates by MBE permitted the fabrication of a variety of devices, such as Schottky diodes [4.67] and MODFETs [4.68, 69], multiple GaN/AlGaN and InGaN/AlGaN quantum wells [4.70, 71], and the observation of photo-pumped stimulated emission at 300 K with a low pumping intensity [4.72]. Although the layer quality on sapphire is almost unmatched, sapphire (0001) substrates are not necessarily most suitable for GaN-based lasers, because the cleavage planes in sapphire substrates are not perpendicular to the surface. Thus, the use of a sapphire (0001) substrate requires the fabrication of laser mirror facets by reactive ion etching or other more complicated cleavage techniques. Attempts have been made to get around this problem by employing the a-plane of sapphire, but the film quality is inferior to that on c-plane sapphire.

4.9 Growth on 6H-SiC (0001)

The development of the growth technology on SiC is not as well advanced as that on sapphire because of high cost and difficult surface preparation. As a substrate, 6H-SiC offers the advantage in that the lattice constant and the thermal expansion coefficient are almost identical to those of AlN but having a lattice mismatch of 3.5% with GaN for the basal plane. The large thermal conductivity of SiC and the ability to form facets with somewhat more ease as compared to sapphire, are additional advantages in favor of SiC. Common wisdom would imply that growth on this substrate should be better. However, an examination of the stacking shows that 6H-SiC is two thirds cubic and one third wurtzite, while nitrides are fully wurtzite with stacking order of ABAB along the c-direction, and in the cubic form the stacking order is ABCABC [4.73]. The researchers who have so far been successful with growth on SiC are those with access to good SiC substrates and surface cleaning and damage removal capabilities. LEDs and injection lasers as well as modulation-doped structures with high mobilities have been reported on SiC, as detailed in Chapters 8, 11 and 12.

Lattice mismatch between nitride films and the most frequently used substrates (SiC, Al$_2$O$_3$) is often cited as a major obstacle in nitride growth and is a cause of the observed defects. Although lattice mismatch in wurtzite nitride films grown in the close-packed (0001) direction can be relieved by the formation of a network of edge dislocations at the substrate/film interface and can be effectively confined to the vicinity of this interface, other structural defects, such as Inversion Domain Boundaries (IDB) and Double Positioning Boundaries (DPB), have been identified as defects spreading into the bulk GaN layers. If special care were taken to initiate growth with only one species and the substrate surface did not contain any steps, then IDBs would not form. However, substrates invariably contain steps and, even with single-species initiation, the lattice inverts across each single step substrate. The solution for avoiding IDBs is by making double-stepped substrates and choosing a substrate with a stacking order identical to the epitaxial layer. For example, in 6H-SiC the stacking sequence is ABCACB so three stacking sequences would be possible for wurtzite GaN: (1) substrate ...ABC leading to BCBC GaN; (2) substrate ...BCA leading to CACA GaN; and (3) substrate ...CBA leading to BABA GaN, and all lead to vertical defects, between differently stacked domains in wurtzite GaN; they are named **Stacking Mismatch Boundaries** (SMB).

Consider a 6H-SiC substrate region terminated as ...ABC and an adjacent region separated by a bilayer step terminated as ...ABCA. With subsequent wurtzite GaN growth, the stacking sequence BCBC would form over the first region while the stacking sequence CACA would form over the second region and inevitably lead to an SMB, as illustrated by the arrow labeled S1 in Fig. 4.27. Two important points must be made. First, a surface bilayer step on a wurtzite substrate would not lead to SMB because subsequent wurtzite growth would have the same stacking sequence on both sides of the bilayer step. Second, 4H- and 6H-SiC are a mixture of cubic and hexagonal stackings, and not every substrate bilayer step would lead to SMB. For 6H-SiC, consider the substrate terminations ...ABCA and ...ABCAC separated by a bilayer step. The first would lead to CACA GaN, whereas the second would lead to ACAC GaN. Clearly, there would be no SMB at this substrate step, as illustrated by the arrow labeled S2 in Fig.

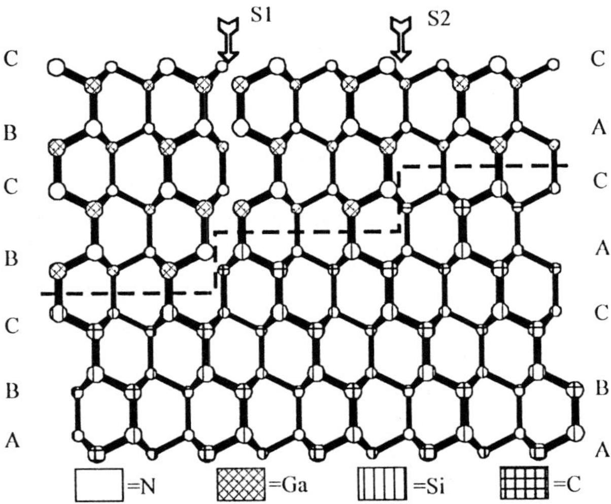

Fig. 4.27. Cross-sectional atomic model of wurtzite GaN grown on 6H-SiC in the (0001) direction. Steps on the SiC surface are likely to create SMBs, as indicated by the arrow S1, although certain steps do not lead to SMBs as indicated by the arrow S2. The circle sizes and line widths are used to give a three-dimensional effect, and have no relation to atomic size or bond strength. The cross-section is a bilayer where the large circles and lines are raised out of the plane above the small circles and lines

4.27. Wurtzite nitride growth on sapphire is more complex because the exact interface arrangement of atoms is not known and because sapphire has a coordination of atoms completely different from GaN. However, if we suppose that nitrogen atoms attach in the same way to each uppermost Al atom of the substrate during the so-called **nitridation of the sapphire surface**, then every step on the sapphire substrate has a high probability of producing an SMB due to the position change of Al atoms in successive basal planes. One may then suggest that every bilayer surface step, on a close-packed non-isomorphic substrate of wurtzite GaN, has a high probability of creating an SMB threading defect during wurtzite GaN epitaxy.

The above-discussed viewpoint provides a fresh perspective on the notorious problem of finding a good substrate for GaN growth. For the reason of stacking-order match and close lattice match to GaN, it is tempting to state that a ZnO substrate would be superior to a ZnO buffer layer on sapphire for subsequent GaN growth. Since none of the presently known growth methods on available substrates completely eliminates the problem of SMBs in wurtzite GaN, we will briefly consider possible deleterious effects of these defects on the properties of epitaxial films. SMB originating from a linear step is essentially two-dimensional and an equivalent density of defects averaged over the GaN volume could be as high as $10^{18} \div 10^{19}$ cm^{-3} if we estimate the linear density of these defects from TEM micrographs. The geometry of atoms and unsaturated bonds at SMB is not presently known. Extensive theoretical work is needed in order to understand the electronic character of these defects.

4.10 Growth on ZnO

ZnO is considered a promising substrate for group-III N since it has a close match for the GaN c- and a-planes and the close stacking order [4.74, 75]. ZnO was used as a buffer layer for GaN growth but the early films were inferior in quality to films with AlN and GaN buffer layers. *Matsuoka* et al. [4.74] have used ZnO substrates to grow lattice matched $In_{0.22}Ga_{0.78}N$ alloys. The most comprehensive study to date of GaN growth on ZnO was accomplished by MBE [4.75, 76]. High-quality films have been confirmed by photoluminescence and reflectance measurements. The width of excitonic transitions (≈ 8meV) were in the same range as those obtained on sapphire by MBE. The yellow luminescence was nearly absent from PL spectra. X-ray measurements as well as polarized optical data confirm that the GaN epilayers are better oriented with respect to the substrate axis than those on sapphire and SiC. More work is needed to delineate intrinsic pro-

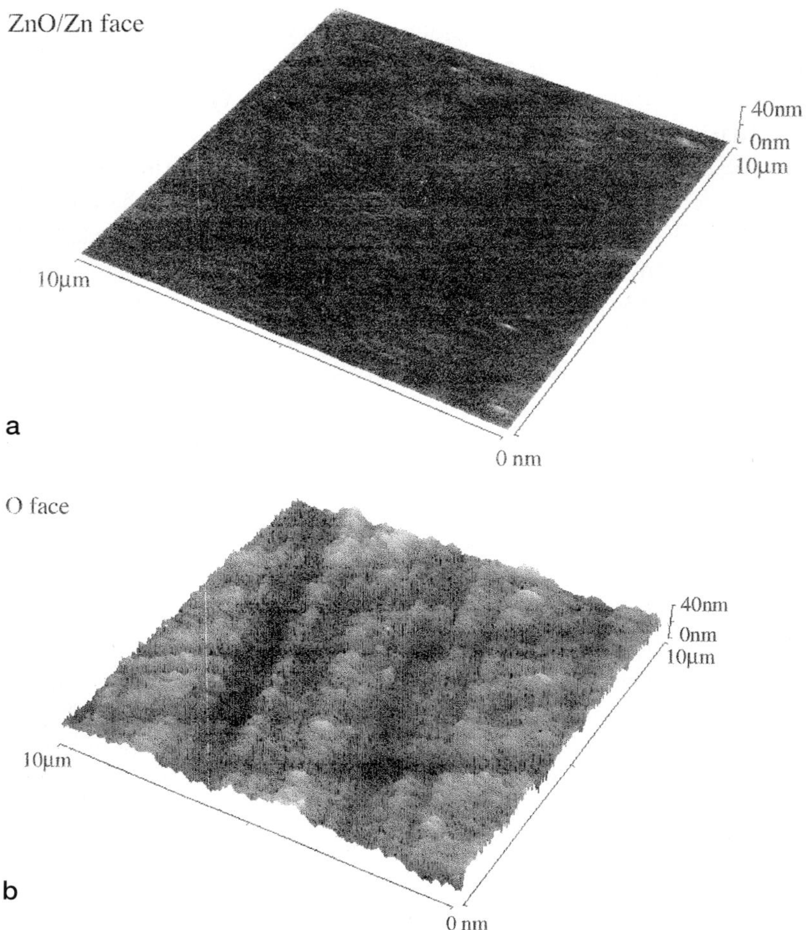

Fig. 4.28. An AFM images obtained on (**a**) a Zn surface, and (**b**) an O surface of ZnO (0001) substrates

perties from extraneous issues such as the defective substrate surface morphology which may be overcome in time.

The c-face termination of ZnO consists of either group-II (Zn) atoms termed as the **zinc face** or the group-VI (O) atoms called the **oxygen face**. Although the zinc face does not have unpaired electrons, the oxygen face does have them, making it more chemically active. The growth of GaN on polar substrates is strongly dependent on the interface polarity between the GaN epilayer and the substrate surface. As pointed out earlier, the Si-terminated 6H-SiC (0001) surface is more suitable for the GaN growth compared to the C-terminated surface, although the latter can be made smoother. Chemical etching in nitric acid and electrical conductivity can be employed to determine the surface polarity of ZnO prior to growth. Owing to the absence of dangling bonds, the zinc face is less affected by the etchant, and the surface electrical conductivity is three orders of magnitude lower compared to the oxygen face. As for chemical etching, the Zn-terminated surface becomes smoother after a 2 minute etch. In contrast, the O-terminated surface becomes rough after etching in the same solution, and the etching rate is an order of magnitude faster compared to the Zn-terminated surface. The first problem encountered in the growth of GaN on ZnO using RMBE is the reactivity of ZnO with ammonia. This reactivity is dependent on both the growth temperature (T_g) and the polarity of the ZnO surface. Figure 4.28 exhibits the surface morphology of oxygen and zinc faces of a ZnO substrate that had been obtained after etching in a 20% nitric acid solution (HNO_3) for 2 minutes. The Zn face presents a smoother surface compared to the O face. The root mean square (rms) of the O-surface roughness is one order of magnitude larger than that of the Zn surface which has a surface roughness comparable to that of the sapphire substrate roughness, as may be inferred from Table 4.1.

To observe the effect of the substrate surface polarity, GaN layers on O- and Zn-terminated faces of ZnO have been grown under otherwise identical conditions, i.e. the same III/V flux ratio, sample thickness and growth temperature (T_g) [4.76]. Figures 4.29a and b present the Atomic Force Microscopy (AFM) images for GaN samples grown at $T_g = 760\,°C$ with different surface polarities and without using any buffer layer. Despite the

Table 4.1. AFM data obtained for the 2×2 mm^2 scans, on different samples grown by RF-MBE. Ra denotes the average roughness, RMS is the roughness root-mean square, Av-Ht is the average height, and Max Range is the maximum range

Sample label	buffer	Ra [nm]	RMS [nm]	Av-Ht [nm]	Max Range [nm]
ZnO substrate					
O face		1.7	2.2	11.3	20.7
Zn face		0.24	0.3	1.2	2.3
Sapphire substrate	0.08	0.11	0.5	0.8	
GaN/ZnO					
5511/O-face		18	23	66	124
5514/Zn-face		23	29	105	178
5516	AlN buffer	24	29	78	157
5351	AlN buffer	15	18	66	126
5518	GaN buffer	9	11	33	71
GaN/sapphire					
5361	AlN buffer	11	13.6	70	140

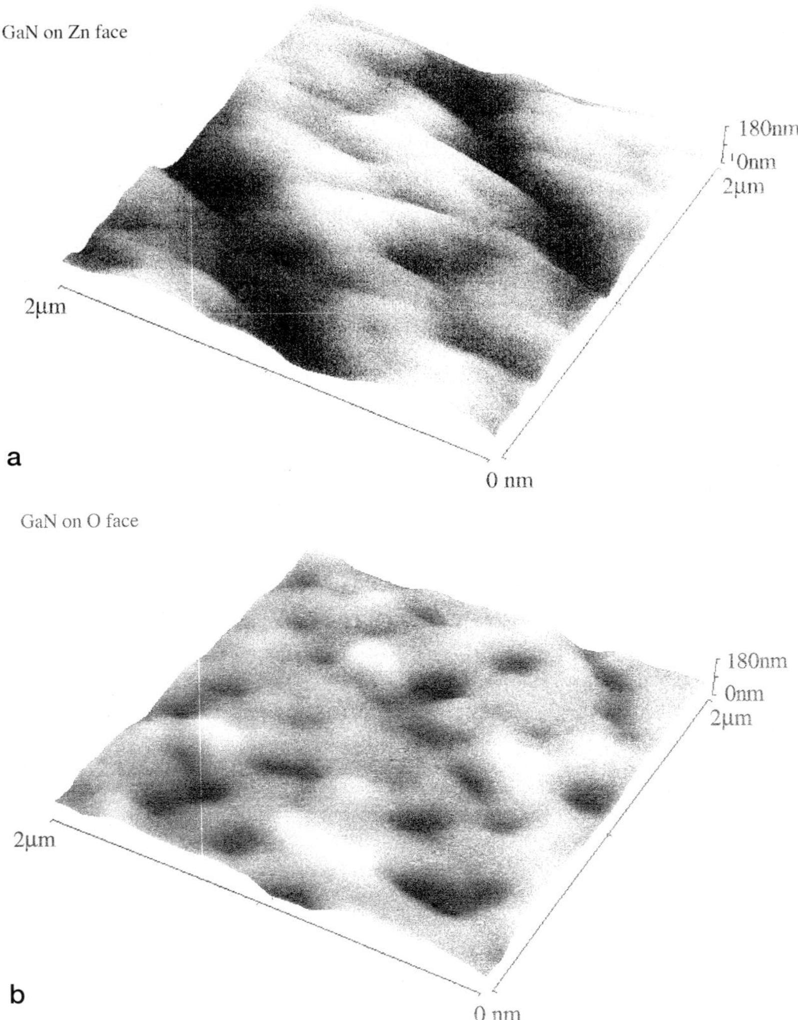

Fig. 4.29a-d. Caption see next page

fact that the Zn faces present smoother substrate surfaces, the samples grown on Zn-terminated surfaces present an increased surface roughness compared to those grown on O-terminated surfaces. For substrate temperatures above 750°C, the GaN layers are grown on ZnO without any buffer peeled off. This effect is due to the reactivity of ammonia with the ZnO surface during the early stages of growth. In order to minimize this interaction and to avoid any layer peeling, a low-temperature buffer layer has been incorporated as an intermediate step. Presented in Fig.4.29c and d are the AFM images obtained on samples grown with two different buffer layers.

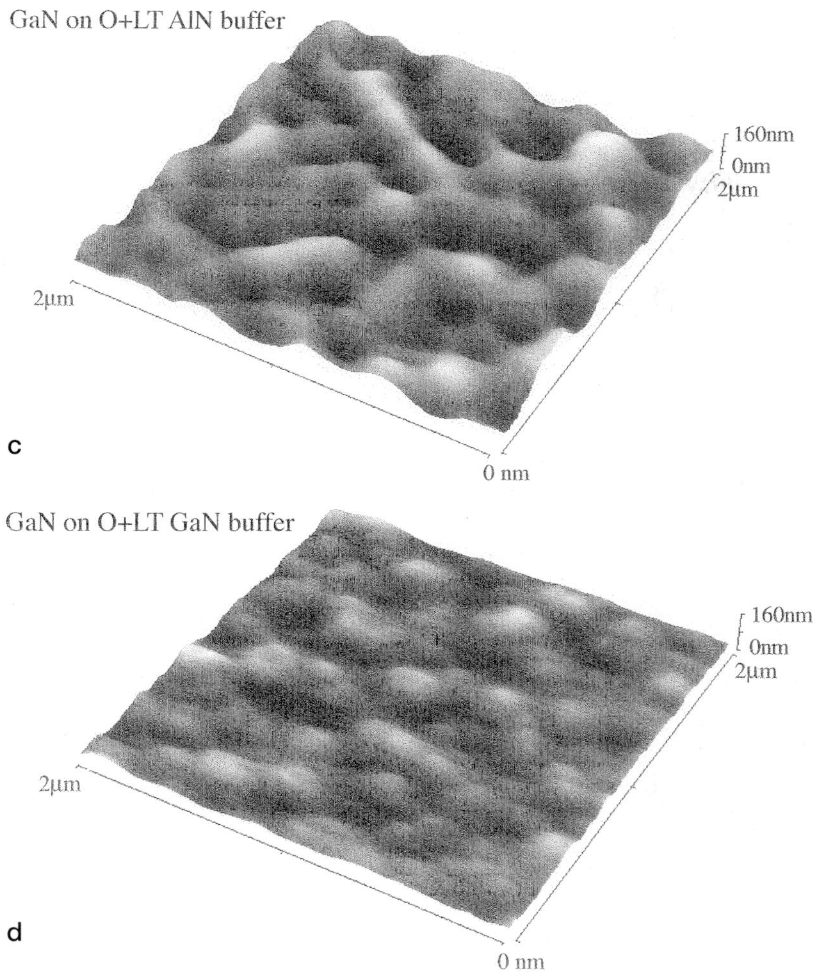

Fig. 4.29a-d. AFM images obtained on (**a**) GaN grown on O-face ZnO at $T_g = 760\,^{\circ}$C, (**b**) GaN grown on Zn-face ZnO at $T_g = 760\,^{\circ}$C, (**c**) GaN grown on O-face ZnO with low-temperature AlN buffer grown at $700\,^{\circ}$C, and (**d**) GaN grown on O-face ZnO with a GaN low-temperature buffer grown at $700\,^{\circ}$C

Figure 4.29c shows that the surface roughness is increased due to the AlN buffer on the ZnO surface. This is expected since the lattice mismatch is higher between these two materials compared to that between GaN and ZnO. Figure 4.29d, obtained on the sample grown with the GaN buffer layer at $T_g = 700\,^{\circ}$C, depicts a surface morphology which was the smoothest obtained in the author's laboratory with ZnO or sapphire substrates (Table 4.1).

TEM images have also been employed to characterize the growth on ZnO substrates. For example, a 1-μm thick GaN grown on the O-face ZnO

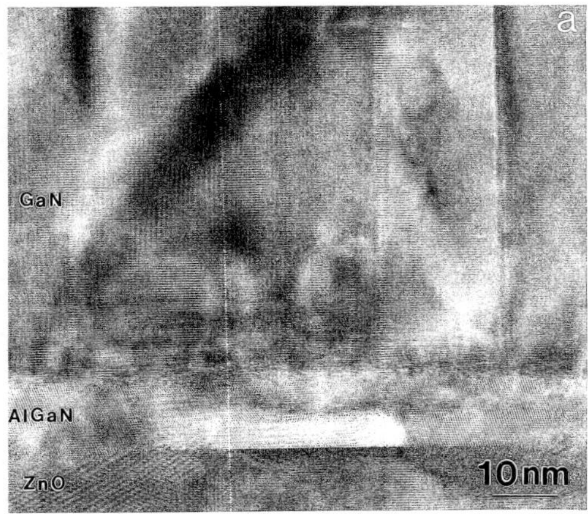

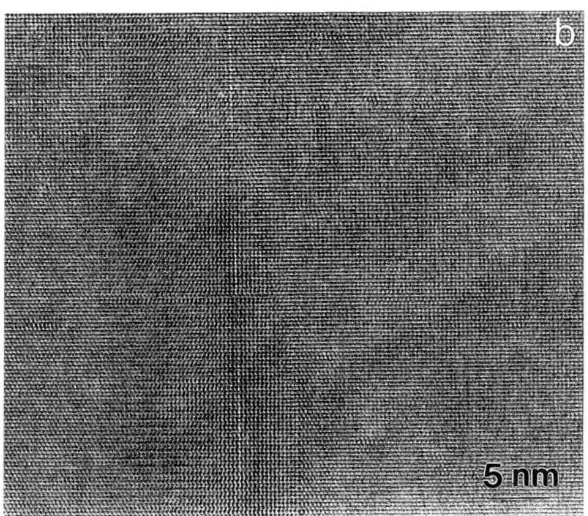

Fig. 4.30a,b. Cross-sectional transmission electron micrograph of GaN grown with GaN low-temperature buffer layer on O-face ZnO substrate. (**a**) Interfacial region and (**b**) upper GaN surface

substrate, without any buffer layer, revealed a lateral defect density much lower than films on other substrates. It has also been observed that most of the observable dislocations did not appear to originate from the substrate, and many of them start and terminate within the field of view, indicating a short propagation length. The facets seen on the upper surface indicate that the surface is N-terminated and consequently the interface polarity is formed by the O-Ga bonding. TEM images obtained in 4 μm-thick GaN grown at 720°C on an O-face ZnO substrate, with 200 nm GaN buffer layer grown at 650°C, indicate the buffer layer to be heavily disordered (Fig.4.30a). However, at the surface the structural order and the defect density are at least three orders of magnitude lower compared to those of the buffer-layer region (Fig.4.30b).

From the luminescence point of view, low-temperature buffer layers, grown at 650°C with RF-MBE, lead to an improved optical quality compared to the same buffer layers grown by means of RMBE due to the reactivity of ammonia and ZnO. The details of the optical properties of GaN on ZnO are discussed in Chap. 10. Although, the origin of the yellow luminescence is multi-faceted and its presence or absence in isolation may not mean much unless considered with all the other properties, it is warranted to note that the yellow signal is not observed for GaN on ZnO with low-temperature PL spectra. The room-temperature PL spectrum of GaN/ZnO exhibits a very small yellow signal with an intensity being three to four orders of magnitude smaller compared to the band-edge PL peak.

The intensity of free exciton transitions depend on the crystallographic direction due to the anisotropy of the wurtzite structures. As a result, with linearly polarized PL and reflectivity measurements, one can detect the misorientation of the epilayer planes with respect to the substrate. Results on GaN grown on $(0001)Al_2O_3$ show the directionality, indicating that GaN layers are tilted with respect to the substrate axis. Unlike the films on sapphire, PL and reflectivity of GaN grown on $(0001)ZnO$ are invariable with the polarization direction and indicate the absence of large tilting effects. This would imply that any tilting must have some degree of long-range coherence as the probe is much larger than the size of these columns. Double crystal X-ray diffraction measurements performed by rotating the

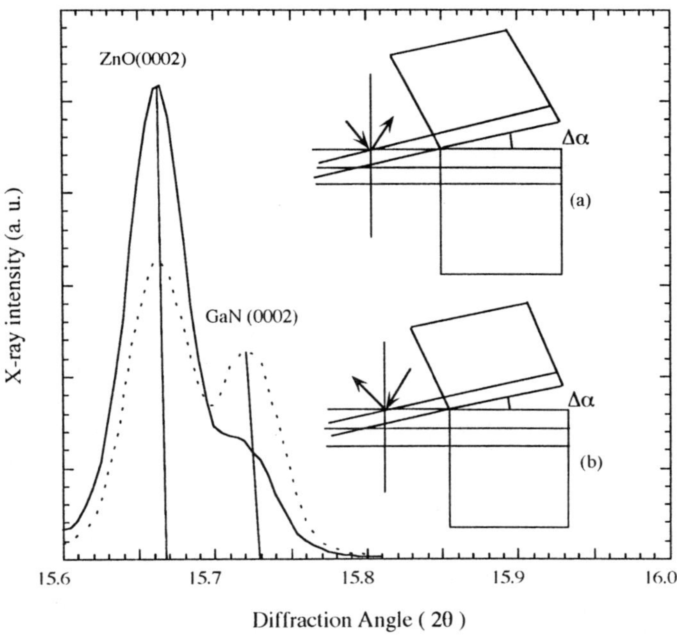

Fig. 4.31. Schematically depicting the arrangement for polarization-dependent X-ray diffraction rocking-curve measurements. The deviation of the relative angle with respect to that for the substrate, at the same spot at which the diffraction occurs is an indication of the tilt present in the film with respect to the substrate and is a measure of tilting

samples by 180° (Fig.4.31) lead to a diffraction peak appearing at the same angle and confirming the conclusions drawn from the polarized optical data, i.e., GaN epilayers are not as tilted with respect to the ZnO substrate axis, as they would be on other substrates used. Conversely, X-ray measurement performed on GaN grown on Al_2O_3 conclude that the presence of tilting, consistent with the picture of Fig.4.11, is due to the lack of coherence in the atomic arrangement between sapphire and GaN.

4.11 Growth on GaN

In the absence of large-area GaN substrates, attempts at homoepitaxial growth were done by both MOCVD [4.77] and MBE [4.78] only on GaN platelets grown from the liquid phase under a high pressure. Such studies serve to establish benchmark values for the optoelectronic properties of thin GaN films. The homoepitaxial films grown by the MBE method showed a superior quality of PL and structural characteristics with respect to heteroepitaxial layers grown on sapphire. The dislocation density was estimated to be $10^7 \div 10^8$ cm^{-2} for homoepitaxial film, as compared to $10^9 \div 10^{11}$ cm^{-2} for films grown on sapphire. Homoepitaxy does not automatically guarantee dislocation-free material as the technology for preparing the surface for epitaxy and removing surface damage has not been developed.

There are, in general, two possible approaches for suitable substrates for homoepitaxial growth. The first possibility is to grow a large bulk crystal, cut it and polish slices from it, as desired. This method is widely employed for conventional semiconductors. It cannot be applied easily to GaN bulk growth since only small pieces of GaN could be grown and only at high temperatures and very high pressures (tens of kbar). The difficulty increases because GaN begins to decompose possibly at temperatures as low as 800°C. The second possibility is to grow very thick (tens of micrometers) GaN on a foreign substrate (sapphire, SiC or Si) . The most suitable method for growing thick GaN layers at rapid rates (up to $100\,\mu$m/h) is the inorganic CVD or HVPE. A similar approach was employed by *Johnson* et al. [4.79]. The MBE homoepitaxial growth of GaN and AlGaN films and of GaN/AlGaN multiple quantum wells was performed on high-quality 3-μm-thick GaN films grown by MOVPE on SiC. The film grown by MBE had an FWHM of the X-ray rocking curve of 156 arcsec. The room-temperature PL spectrum was dominated by band-edge emission at 3.409 eV, with a FWHM of 29.7 meV. In short, conventional approaches do not lend themselves well to nitrides. Minor improvements resulting from these thick buffer layers will most likely run out of steam. The proposal to grow thick GaN layers by

HVPE on Si, followed by the removal of Si substrates, has the same short-comings as the approach on sapphire with the additional complication that the layer quality on Si is inferior. The potential advantage is that Si can selectively be removed, and it comes in large sizes. Again, no successful report on this approach is available. Along the same lines, GaAs has also been explored as a sacrificial substrate with no reports of success.

4.12 Growth of p-Type GaN

Until 1989, efforts to achieve p-type GaN have led to compensated high-resistivity material. Activation of Mg doping for p-type conductivity has been responsible for placing GaN and related materials in a commanding position as light emitters [4.80, 81]. An inordinate number of early investigations were directed toward the potential of Zn as a p-type dopant in GaN. These investigations demonstrated that Zn impurities effectively compensate the semiconductor and result in high-resistivity material. All Zn-doped samples exhibited the commonly observed 2.8 eV emission with heavily doped samples which exhibit room-temperature peaks at $1.8 \div 1.9$ eV, 2.2 eV, and $2.5 \div 2.6$ eV. A number of researchers have investigated Mg doping which is known to effectively compensate GaN. As the Mg concentration is increased, the Mg-related emission peak broadens and shifts to lower energies. Attempts to dope GaN with Cd failed to produce p-type conductivity even though a Cd-related peak at 2.85 eV was observed. As for Be, until very recently, it led to compensated material as well, while yielding high-resistivity GaN. Still, hole concentrations obtained in the most recent samples were too low to be measured by the Hall effect, although p-n junctions have been made, leaving Mg the only practical p-type dopant. Hg doping was also investigated with no electrical measurements. Hg, too, leads to optical transitions at energies much lower than the band edge, about 0.4 to 0.8 eV above the valence-band edge. Carbon has also been tried and produced only deep-level transitions reaching even to the yellow region with no electrical conductivity. There is one report of GaN with CCl_4 doping on GaAs substrates which has not been confirmed by others. Li atoms have gone the same way, and produced only compensated material with optical transitions about 750 meV below the band edge [4.40].

A common point is that all p-type dopants, including Mg at high concentrations, can lead to deep-level transitions where the actual level is dependent on the doping level. Available data appear to suggest that unsuccessful attempts end up generating defects, and that only when the undoped layers have a high resistivity, at least in the case of MBE growth, can the available data point to active p-type doping. The deep transitions have, in fact, been exploited to produce blue emitters.

First achieving and then controlling p-type doping have been, and continues to be, a formidable challenge for semiconductor-nitride researchers. The first breakthrough came when *Amano* et al. [4.80] converted compensated Mg-doped GaN into conductive p-type material by Low Energy Electron Beam Irradiation (LEEBI). *Nakamura* et al. [4.81] improved upon those results using LEEBI to achieve GaN with $p = 3 \cdot 10^{18}$ cm^{-3} and a resistivity of $0.2\ \Omega \cdot$cm with the follow-up of thermal annealing at 700°C under an N_2 ambient, which converted the material to p-type equally well. Due to the high binding energy ($150 \div 200$ meV) of Mg, acceptor activation ratios of only 10^{-2} to 10^{-3} are typically achieved and require Mg chemical concentrations in the 10^{20} cm^{-3} range. Although, OMVPE samples require a post-growth activation, the MBE-grown samples do not. Hydrogen has long been suspected of passivating Mg atoms and is driven off in the subsequent annealing treatment. The general issue will be discussed in Chap. 5.

Recent theoretical insights have provided a plausible explanation for the success of the Mg acceptor in GaN compared to the other group-II materials which continue to compensate GaN even after LEEBI or annealing. Due to the strong binding of the nitrogen anion, group-III nitrides are considerably more ionic than typical group-III semiconductors. Their calculated band structures resemble II-VI semiconductors in many ways, including a large splitting between the upper and the Lower Valence Bands (LVBs). The Ga (3d) core-level energies have been predicted and observed to overlap in energy with the N(2s)-like LVB states as a result of the LVBs being deeper in GaN compared to GaAs, which is a more typical group-III semiconductor. The resulting energy resonance causes the Ga(3d) electrons to strongly hybridize with both the upper and lower valence-band s- and p-levels. Such a hybridization is predicted to have a profound influence on GaN properties including such quantities as the energy bandgap, lattice constant, acceptor levels and valence heterojunction offsets. It is known, in the cases of ZnS and ZnSe, that potential acceptors such as Cu have the p-d acceptor level raised due to anti-bonding repulsion between Cu(3d) and Se(4p) [S(3p)] resulting in a deeper level, while impurities without d-electron resonances form shallow acceptors. Mg has no d electrons and turns out to be sufficiently shallow for room-temperature p-type doping of GaN. On the other hand, Zn, Cd, and Hg, all of which have d electrons, form deep levels in GaN. These observations are consistent with theory.

4.13 Growth of n-Type InN

InN is pivotal in the triad of the group-III nitride system and holds the key
for a full exploitation of the group-III nitride system for optical emitters.
Because of the low dissociation temperature (beginning at temperatures as
low as 550°C at low pressures), growth of high-quality InN films has prov-
en difficult. The nitrogen equilibrium vapor pressure over InN is orders of
magnitude higher than over both AlN and GaN (Fig.4.1). To circumvent
this difficulty, a number of options, including low-temperature (less than
600°C) deposition, various growth techniques such as reactive evaporation,
ion plating, reactive RF sputtering, reactive magnetron sputtering, vapor
phase epitaxy, microwave-excited MOCVD, laser-assisted CVD, halogen
transport, and MBE have been tried [4.40].

Most of the InN growths were performed at about 500°C. The best
films were obtained with high indium and nitrogen fluxes. Short thermal an-
neals (30 min, 450÷500°C) have been found useful for improving the crystallinity of the as-deposited InN films. InN films are generally polycrystalline with agglomerates of small columnar grains having various degrees of
texture and epitaxy. The improvement in the crystalline quality is due to a
rearrangement of these crystallites in the film. The electrical properties of
the as-deposited InN films are dominated by conduction between the grains.
Buffer layers used for the growth also play a role. For example, application
of an AlN buffer layer on a sapphire substrate significantly alters the granu-
lar polycrystalline growth mode of InN and results in substantial improvements in the growth morphology and in electrical properties. However, the
electrical transport properties are still dominated by inter-granular interactions.

4.14 Growth of n-Type Ternary and Quaternary Alloys

By alloying InN together with GaN and AlN, the bandgap of the resulting
alloy(s) can be increased from 1.9 eV to 6.2 eV, which is critical for
making high-efficiency visible light sources, ultraviolet detectors, and high-
performance electronic devices. The quaternary alloys (InGaAlN) have the
added advantage that their lattice constants can be tuned to, at least, ZnO
while its bandgap can be adjusted to meet the need set forth by the device
design. Detection of missiles requires **solar-blind detectors**, meaning
those that do not get affected by the sun's radiation. Nitride-based ultraviolet detectors employing high fractions of AlN would not be blinded by the

large infrared, visible, and UV components of the sun light, but would be able to detect missiles by detecting their plumes.

Incorporation of indium in these alloys is not easy even though the InN molar fraction in InGaN has been pushed almost to 70%. To prevent InN dissociation, the InGaN crystal was originally grown by vacuum deposition at low temperatures (about 500°C). However, the use of a high nitrogen flux rate allowed the high temperature (800°C) growth of high-quality InGaN and InGaAlN films on (0001) sapphire substrates. It was noted that the incorporation of indium is strongly dependent on the flow rate ratio of TMI, TMA and TMG (or TEG), with larger flow rates of TMG, producing larger InN mole fractions. The indium incorporation efficiency decreases quite rapidly with increasing growth temperatures to values above 500°C.

The crystalline quality of InGaN was observed to be superior when grown on the well-matched ZnO substrate than when grown on a bare (0001) sapphire substrate. Employment of a buffer layer provides further improvements in the crystalline quality of InGaN. As in the case of GaN, InGaN films grown on sapphire substrates with low-temperature GaN buffer layers exhibit much better optical properties. This has to do with the nucleation issues on the substrate.

The pathways to deposition of high-quality AlGaN layers are similar to those for GaN with the exception that higher deposition temperatures can be employed. One concern is the reactivity of AlN and oxygen. The higher the temperature, the lower the incorporation of oxygen, the source of which is attributed to ammonia. Despite the above-mentioned complications high-quality InGaN/AlGaN, InGaN/GaN, and $In_x Ga_{1-x} N/In_y Ga_{1-y} N$ superlattices have been grown.

4.15 Growth of p-Type Ternary and Quaternary Alloys

As the nitride activitiy intensified, the p-type nitride-semiconductor family expanded to include AlGaN with AlN mole fractions of about 30% and InGaN with InN mole fractions a good deal lower (9%). To reduce the parasitic reactions between metalorganic sources, particularly with TMA and NH_3, a dual-flow channel reactor was employed [4.82]. For activating Mg in AlGaN, flash-lamp annealing under atmospheric-pressure nitrogen flow at 1140°C for 60 s was used. One problem with p-type nitrides, progressively getting worse as the AlN mole fraction increases, is the poor ohmic contact which makes measurements difficult and less accurate. Measured temperature dependencies of the hole concentration indicate that the hole concentration drops off with decreasing temperature from about 10^{18} cm^{-3} at 500 K to about 10^{13} cm^{-3} at 125 K.

Mg-doped In$_x$Ga$_{1-x}$N (x = 0.09) films were grown at 800°C by *Yamasaki* et al. [4.83], again by employing MOVPE and low-temperature AlN buffer layers. TMG, TMA, TMI, and C$_{p2}$Mg were used for Ga, Al, In, and Mg source gases, respectively, and NH$_3$ for nitrogen. After annealing at 800°C for 5 minutes in a nitrogen ambient gas at atmospheric pressure, the layers showed p-type conduction. Hall-effect measurements in the temperature range of 90 ÷ 500 K indicate that the hole concentration initially decreases rather rapidly, and then almost linearly with decreasing temperature down to T = 143 K, before saturating. The measured hole concentration, if dependable, was about $5 \cdot 10^{18}$ cm^{-3} at 500 K and dropped to $8 \cdot 10^{16}$ cm^{-3} at 100 K. The measured Hall mobility μ_p first increased and then decreased with increasing temperature. It shows a peak value of about 40 cm^2/V·s at 150 K and a value of 10 cm^2/V·s at 300 K. By fitting the theoretical expression of the hole concentration as a function of the measurement temperatures, the activation energy of the Mg acceptor, E_A, was estimated to be 204 meV above the valence band. This is in the same ball park as that in GaN.

4.16 Critical Thickness

Lattice and thermal mismatch lead to strain which, beyond a certain level, is relaxed by generating defects. In conventional group-III-V systems and SiGe, compressive strain has been exploited because the valence-band curvature changes that causes an increase of the in-plane mobility and a reduction in the valence band density of states. Moreover, the use of coherently strained films expands the usable compositional range [4.84]. Unlike conventional group-III-V semiconductors, the strain is unavoidable in nitrides due to both lattice and thermal mismatches, and the wurtzite symmetry does not lead to any beneficial effect other than the piezoelectric effect which may find some applications. It is imperative that we, at the very least, have a cursory exposure to strain issues in nitrides.

The elastic strains and stresses of the GaN-AlN layered structures has generally been calculated by minimizing the deformation energy of the total system. An integral part of this exercise is the inclusion of the elastic strain-relaxation mechanisms such as misfit dislocations. *Bykhovski* et al. [4.85] used an isotropic elasticity approximation to treat the problem while neglecting the difference between the elastic constants of GaN and AlN to simplify the treatment. The method can be applied to the investigation of the elastic strain relaxation in $(GaN)_m(AlN)_n$, $GaN_m(Al_xGa_{1-x}N)_n$ and $Ga\cdot N_m(In_xGa_{1-x}N)_n$ superlattices, m and n indicating the number of atomic layers. Let us first consider the GaN-AlN superlattices with equal numbers of GaN and AlN layers in the superlattice cell according to *Bykhovski* et al. [4.85]. They used an approach similar to that developed in [4.83]. The value of the critical thickness L_c for n/m = 1 represents the upper bound for the critical thickness of the thinner layer in a superlattice with arbitrary n and m, whereas the critical thickness for a very small n/m (or m/n) ratio is the corresponding lower bound. Essentially, there are two contributions to the elastic energy, one due to uniform strain and the other due to misfit dislocations. In the linear theory of elasticity, the total stress is the sum of these contributions. Some assumption about the distribution of misfit dislocations must be made in order to properly treat them. *Bykhovski* et al. [4.86] assumed that the dislocation grids are situated at each AlN-GaN interface, as depicted in Fig.4.32. Furthermore, the dislocation grid at each interface contains two perpendicular slip systems with the slip directions along the x- and z-axes [the y-axis is directed along (0001) and is perpendicular to the GaN/AlN interface]. Each misfit dislocation is supposed to be straight with its Burgers vector **b** in the slip direction and $\mathbf{b} = \langle 11\bar{2}0\rangle/3$. The magnitude of the Burgers vector is equal to the lattice period [4.87]. Typically, hexagonal crystals have three predominant $\langle 11\bar{2}0\rangle$ slip systems. The interval between the two neighboring dislocations is denoted as ℓ and the shift along the Burgers vector in the positions of two nearest dislocations placed on neighboring GaN/AlN interfaces (separated by L) as Δx (Fig.4.32).

With the assumption of elastic isotropy, the elastic energy density f_ℓ has the following form [4.85]

$$f_\ell = \frac{1}{2E}(\sigma_{zz} + \sigma_{xx} + \sigma_{yy})^2$$

$$+ \frac{1+v}{E}(\sigma_{xz}{}^2 + \sigma_{yz}{}^2 + \sigma_{xy}{}^2 - \sigma_{xx}\sigma_{yy} - \sigma_{xx}\sigma_{zz} - \sigma_{yy}\sigma_{zz}) \qquad (4.14)$$

where σ_{km} denote the stress terms, E is Young's modulus which is the ratio of the simple tensile stress to strain, and v is the Poisson's ratio expressing the ratio of the transverse contraction to the elongation in simple tension. The stress contributions from a dislocation array can be calculated [4.86]. This paves the way to the calculation of the total stress components as sums over the contributions from all dislocation arrays at all interfaces of the superlattice plus the unrelaxed misfit stress, which can be expressed as

$$\sigma_{km} = \sum \sigma^j_{km}, \quad \text{with} \quad km = x, y, z \qquad (4.15)$$

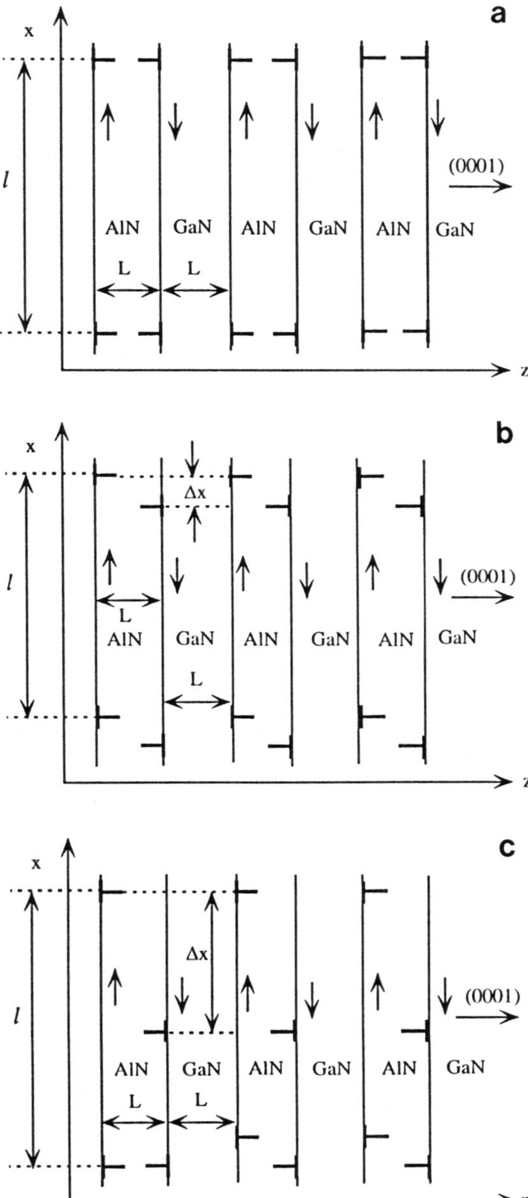

Fig. 4.32a-c. GaN-AlN superlattices: (**a**) low dislocation densities, (**b**) moderate dislocation densities, (**c**) high dislocation densities. *Arrows* show the directions of the Burgers vectors. The straight dislocations shown have a slip direction in the interfaces along x. Structure looks the same way in the yz-plane. After [4.85]

where $j = 0, 1, 2...$ ($j = 0$ denotes a uniform contribution from a misfit, and non-vanishing values of j stand for the corresponding slip systems). To obtain the energy of the layer per unit area, F_1, the term f_ℓ is numerically integrated over the volume with the exclusion of the positions of the dislocation cores. Integrating f_ℓ of (4.14) over the volume and dividing the result by the interface area S, one obtains F_1, neglecting the contribution from the dislocation cores,

$$F_1 = \frac{1}{S} \int_{v.p.} dv f_\ell .$$

(4.16)

The core energy of a single straight dislocation per unit length is equal to [4.83]

$$f_c = \frac{\mu b^2 \ln \xi}{4\pi(1 - v)}$$

(4.17)

where μ is the shear modulus, and b is the absolute value of the Burgers vector. The value of the dimensionless parameter ξ is estimated by means of atomic calculations for different crystals [4.87]. A value of 4 for the parameter ξ is reasonable as it successfuly accounts for a screw dislocation in covalent systems. Turning now our attention to the core energy of the dislocations per unit area, F_c, with $F = F_c + F_1$, multiplying f_c by the dislocation density leads to

$$F_c = \frac{2m_s}{\ell} f_c$$

(4.18)

where m_s is the number of slip systems ($m_s = 2$ for two perpendicular systems, and $m_s = 3$ for hexagonal slip systems). In the isotropic approximation, $m_s = 2$.

Minimizing the deformation energy of the system numerically with respect to ℓ and Δx, one can find ℓ and Δx as functions of L. The average elastic strain components can then be expressed via stress components. In an unrelaxed structure, the misfit strain-tensor components are equal to $u_{xx}^{(0)} = u_{zz}^0 = \frac{1}{2}(a_1/a_2 - 1)$, where a_1 and a_2 are the in-plane lattice parameters. In AlN on GaN, the indices 1 and 2 correspond to GaN and AlN, respectively. In GaN on AlN, the definition is opposite, so that the xx, zz strains are negative.

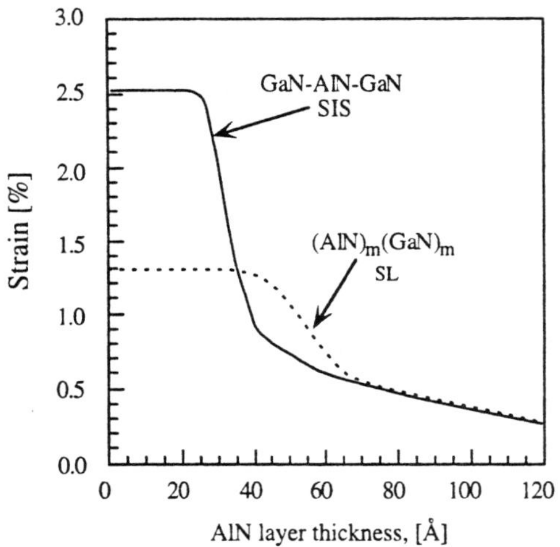

Fig. 4.33. Strain ($u_{xx} = u_{zz}$) as a function of L in $(GaN)_n$-$(AlN)_n$ superlattices (*dashed line*) and GaN-AlN-GaN SIS structures (*solid line*). After [4.85]

Let us now turn our attention to the discussion of the critical thickness plotted in Fig. 4.32 are typical patterns which correspond to thin layers (L $\leq$ 40 Å, dislocation spacing >40 nm) (Fig. 4.32a), to the transition region (L = 50 ÷ 60 Å and the dislocation spacing 25 ÷ 40 nm) (Fig. 4.32b) and to thicker layers (L > 60 Å and the dislocation spacing <25 nm) (Fig. 4.29c). The multipole structure with $\Delta x = 0$ (Fig. 4.32a) is replaced gradually by a more uniform pattern with $\Delta x = 0.5 \dots \ell$ as the relaxation advances.

Figure 4.33 plots the strain $u_{xx} = u_{zz}$ as a function of L for symmetrical $(GaN)_n$-$(AlN)_n$ superlattices (solid curve) and a Semiconductor Insulator Semiconductor (SIS) structures (broken line). The SIS structure represents the case where a thin layer of AlGaN is straddled by two bulk GaN. These two examples represent two exteremes in that the former depicts the case of minimum strain while the latter that of the high strain. The reason for the critical thickness (onset of relaxation) being somewhat larger for the symmetrical SuperLattice (SL) is because the misfit strain is distributed equally between the GaN and AlN layers, halving the strain as compared to an SIS structure.

For a symmetrical superlattice, the critical thickness L_c increases only by 40%, from 25 to 37 Å, while the initial misfit strain, u_{xx} (L = 0) decreases by a factor of 2, which is less than one might expect. This is because the contributions to the stress from different interfaces partially cancel each other. This lowers the deformation energy and makes the dislocation generation more beneficial for thinner layers. To elucidate on this feature, *Bykhovski* et al. [4.85] calculated the strain in a SL layer by considering the dislocations at the two nearest interfaces and neglecting the strain contributions from all other interfaces. This approximation yields, for the symmetrical SL, the value of L_c (single layer) = 65 Å. This value is almost

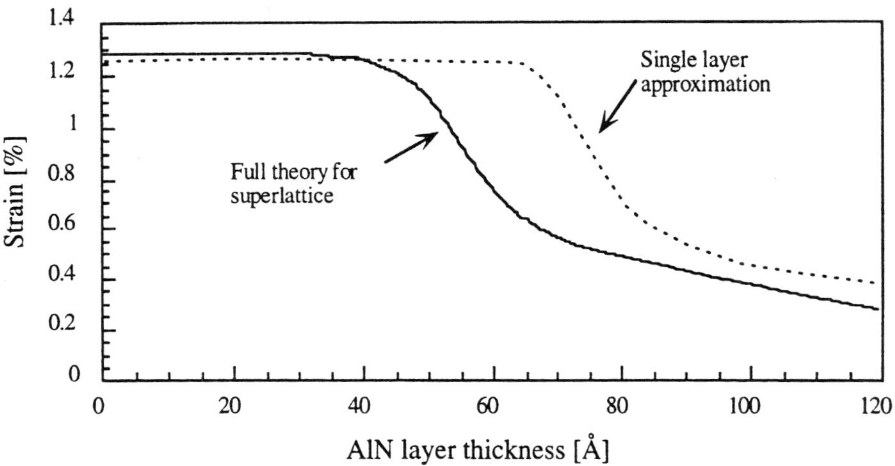

Fig. 4.34. Strain ($u_{xx} = u_{zz}$) as a function of L in $(GaN)_n$-$(AlN)_n$ superlattices. The full calculation includes contributions from all interfaces (*solid line*), only two interfaces are taken into account (*dashed line*). After [4.85]

twice larger than the correct result (Fig.4.34). The correct result ($L_c = 37$ Å) has been obtained by the integration over many layers (10 to 20) with all the dislocations included so that the deformation energy converged. The decrease of the deformation energy is caused by the symmetric dislocation pattern and by the opposite signs of the Burgers vectors for the neighboring interfaces (Fig.4.32).

The critical thickness in a $GaN_m AlN_n$ superlattice with the m/n > 1 (m/n < 1) can not be smaller than in a GaN-AlN-GaN (AlN-GaN-AlN) SIS structure; and it can not be larger than in a $GaN_n AlN_n$ superlattice. Consequently, the thickness L of a dislocation-free layer in a $GaN_m AlN_n$ superlattice with an arbitrary m/n ratio can be estimated to fall in the range

$$25\text{Å} < L < 37\text{Å} . \tag{4.19}$$

The upper bound corresponds to $GaN_n AlN_n$ superlattices with n equal to 15 monolayers and the lower bound to $GaN_n AlN_\infty$ (representing thin GaN and thick AlN) or $GaN_\infty AlN_n$ (representing thick GaN and thin AlN) superlattices with n = 10 monolayers. This leaves one only with a 5-monolayer difference between the worst and the best cases. On the experimental side, the heterolayers are grown on partially relaxed and defective buffer layers which preclude the preparation of coherently and unifromly strained structures for a meaningful comparison to experiments.

The critical thickness as a function of Al concentration, x, for $(GaN)_n$ · $(Al_x Ga_{1-x} N)_n$ superlattices is displayed in Fig.4.35. Within the mole fractions under consideration (0.12 < x < 1) the dislocation interactions be-

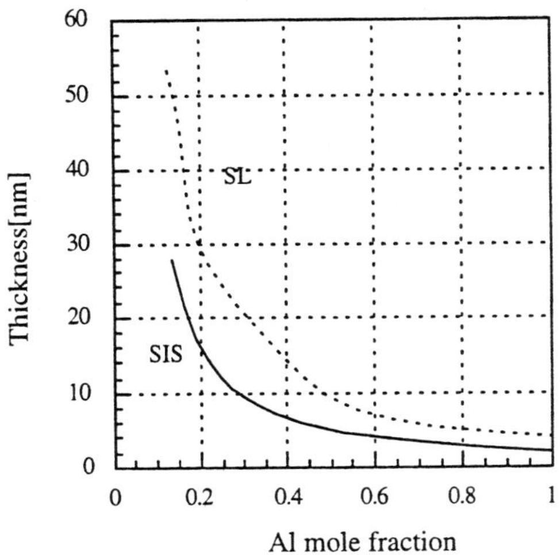

Fig. 4.35. Critical thickness as a function of Al concentration x for superlattice (*dashed line*) and Semiconductor Insulator Semiconductor (SIS) structure (*solid line*). After [4.85].

tween the different arrays reduce the critical thicknesses by about $30\% \div 50\%$. Results of the relaxation of the $(GaN)_n$-$(AlN)_n$ superlattice and GaN-AlN SIS structure, presented in Fig. 4.33, are also applicable to the $(GaN)_n (In_x Ga_{1-x} N)_n$ superlattice with $x \approx 0.2$, and the inequality (4.15) is valid for $(GaN)_n (In_{0.2} Ga_{0.8} N)_m$. More generally, the results obtained for the critical thickness of the $(GaN)_n (Al_x Ga_{1-x} N)_n$ superlattice (Fig. 4.35) are applicable to the $(GaN)_n (In_y Ga_{1-y} N)_m$ superlattice with $y = 0.223x$. It is possible that other defects may cause the strain to be somewhat larger for intermediate AlN film thicknesses in SIS GaN-AlN-GaN structures. For example, the relaxation could be slowed down by point defects and stacking faults. In many films, defects such as threading dislocations which do not relieve strain are present. The dislocation density in nitride films on sapphire is in the range of 10^8 to 10^{10} cm^{-2}. For a dislocation density of $2.5 \cdot 10^9$ cm^{-2}, the average spacing for dislocations is about 65 nm. In a GaN-AlN SL, the theoretical spacing ℓ varies from 100 nm to 60 nm for L from 100 Å to about 60 Å, and, in an SIS structure, the calculated spacing is $2 \div 3$ times smaller within the same range of L. One may conclude that in GaN-AlN-based multilayer structures grown on buffer layers with individual layer thicknesses that exceed the critical thickness, misfit dislocations are the primary source of strain relaxation. In lower-quality films with dislocation spacings under 100 nm, a correction to account for other extended defects should be made in (4.15), a task which can be accomplished by adding a stress term. If the density of threading and other non-misfit defects is known, one can simply repeat the procedure outlined with the corrected Equation (4.15). In high-quality group-III nitride films grown on buffer layers, the correction of the resulting strain will be on the order of only 10%.

In conclusion, the growth technology of nitride semiconductors is, far from being fully understood; consequently, it is not yet as controllable as would be required from a mature technology. Prospects for large-area GaN substrates to become available in large quantities are not bright, and prospects for bulk InN are nil. However, the case for AlN substrates might be different. It should be underscored that even if suitable substrates were available, the prismatic nature of nitride growth promises a good deal of problems with the growth. Challenges of this kind, though, have been the genesis of major breakthroughs in the past.

5. Defects and Doping

Point defects, unless they are neutral, manifest themselves as background doping or autodoping, and complicate attempts to dope the semiconductor in order to control its conductivity. Moreover, defects influence the radiative-recombination efficiency with adverse impact on the LED and laser performance. Unfortunately, GaN and related materials are rich in structural defects as well as point defects such as vacancies. It is therefore imperative that we discuss doping together with native defects.

Due to a variety of reasons, chiefly the non-availability of native substrates in the commercial sense, GaN epitaxial layers have been grown on substrates such as sapphire, SiC, GaAs and Si with sapphire being the dominant substrate. In order to improve the quality of layers, much attention has been devoted to understand the particulars of hetero-epitaxy. In this vein, various planes of sapphire as well as its vicinal planes have been investigated as detailed in Chap. 4. The epitaxial relationship, or rather the lack of it, between the various substrate orientations and the nitride films grown on them has been investigated. There is also the thermal mismatch in hetero-epitaxy which manifests itself as additional structural defects which are created during cool-down from the growth temperature or as strain if any defect formed during cool-down is not sufficient to cause complete relaxation. Strain causes a modification of the band structures, and if controlled can be used to advantage. GaN is a hard material and a sufficiently thick film can actually crack the substrate.

Transmission Electron Microscopy (TEM) has been employed to directly observe structural defects in GaN films. To investigate the effect of misorientation on the crystal morphology, *Hiramatsu* et al. [5.1] grew GaN on (0001) sapphire with a number of different misorientations towards both the [10$\bar{1}$0] and the [1$\bar{2}$10] directions, with the exact (0001) basal plane leading to the best surface morphology. TEM analysis of wurtzite GaN grown on sapphire substrates [5.2, 3] indicates that the major defects in GaN are those related to stacking mismatch and dislocations. GaN grown on 6H-SiC would also suffer from the same types of defects, whose density depends on the step density. Owing to the stacking-order match between GaN and ZnO substrates, stacking-mismatch defects are not expected to occur if proper surface preparation procedures could be developed.

5.1 Dislocations

As stated above and in Chap. 4, the use of non-lattice matched substrates for the growth of thin-film nitrides leads to three-dimensional growth and the necessity to use low-temperature buffer layers. The large difference in lattice mismatch and in thermal expansion rates among the materials involved result in a large number of dislocations, some of which alleviate the strain. *Ponce* et al. [5.3] studied the crystallinity of AlGaN films in the neighborhood of the substrate. The film under investigation was grown on a (0001) sapphire substrate by MOCVD with an AlN buffer layer. Transmission electron lattice images, such as Fig. 5.1, indicate that the sapphire/AlN interface is atomically abrupt and follows a single basal plane. The film appears to be completely relaxed, within the resolution of the technique, which implies a uniform array of **misfit dislocations**. The misfit dislocations lie along $\langle 1\bar{1}00 \rangle$ AlN directions, one of which coincides with the direction of the lattice projection in the image. They are separated from each other by about 2 nm which is close to the 2.03 nm expected for a 12.46% lattice mismatch between AlN and Al_2O_3. When AlN is grown on Al_2O_3, eight AlN planes match nine planes at the interface, and the termination of the extra $\{1100\}$ Al_2O_3 plane results in a misfit dislocation. The critical

Fig. 5.1. Lattice image of the interface region of an AlGaN epilayer grown on sapphire using a AlN buffer layer. Courtesy of F. Ponce of Xerox PARC

thickness for the generation of this misfit dislocation is about one monolayer. The exact nature of chemical bonding responsible for this dislocation is not yet known. One possibility is that the Al atoms at the interface bond with three oxygen atoms in the sapphire direction and with two nitrogen atoms in the AlN direction. Dislocations at the $AlN/Al_{0.5}Ga_{0.5}N$ interface occur along the $\langle 1\bar{1}00\rangle$ direction. The separation between these misfit dislocations is about 22 nm, as compared to 22.6 nm expected for bulk values. The experimental position of these dislocations is also not very distinct. Unlike dislocations at the AlN/Al_2O_3 interface, the plane for these dislocations can vary by ±6 monolayers (±1.5nm). The buffer layer is basically a polycrystalline film which is closely epitaxial with the sapphire substrate but with a distribution in orientation that corresponds to very small angles (<1 degree); the film coalesces within the first 500 nm to form a columnar structure [5.4, 5]. In materials used for high-performance emitters (such as Nichia LEDs) the columns are closely oriented, as extracted from X-ray diffraction measurements with a FWHM of 5 to 6 arcmin (Fig.4.16). The epilayer orientation can schematically be described by the model depicted in Fig.4.14 where the tilt of the c-axis and the rotation (twist) along the c-axis are shown. Most dislocations in GaN are located at the low-angle grain boundaries of the columnar structure. Due to the large Burgers vectors associated with these dislocations, many of these have an open core structure (also referred to as **nanopipes**).

5.2 Stacking-Fault Defects

Stacking faults are a common form of strain relief in face-centered cubic crystal structures since their formation energy is fairly low. The schematic diagram of a typical stacking fault is exhibited in Fig.5.2. From this figure, it may be seen that, when the crystal growth is not in the stacking direction, the formation of a stacking fault requires the termination of an atomic plane at a dislocation. In lattice mismatched hetero-epitaxy of group-III nitrides, stacking faults serve as a form of strain relief, too. If the substrate orientation is chosen in such a way that the growth direction is parallel to the stacking direction, the polytype boundary can be coherent as in the case of (111) 3C and (0001) 2H-GaN. *Davis* et al [5.6] grew zincblende GaN on 3C-SiC and by means of a TEM analysis observed many defects, mainly microtwins and stacking faults, propagating into the epitaxial layer. Likewise, films grown on other substrates, such as ZnO, GaAs, Si, $MgAl_2O_4$, MgO, $LiGaO_2$, and $LiAlO_2$, are not free from these defects either. As discussed in Chap.4, lateral growth over SiO_2 stripes appears to be nearly free of structural defects.

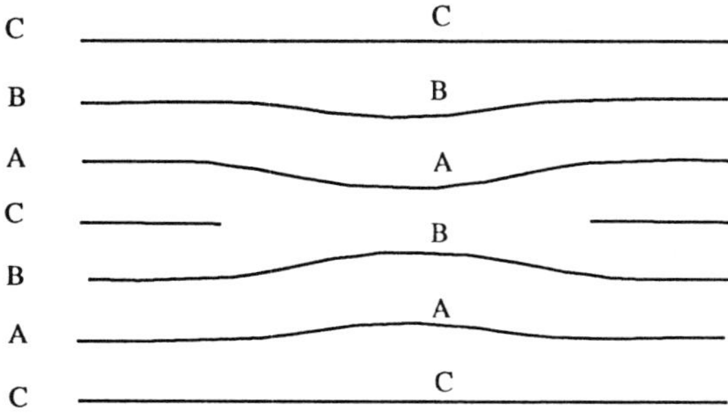

Fig. 5.2. Schematic diagram of a typical stacking fault defect. When growth is not along the stacking direction, the formation of a stacking fault requires the termination of an atomic plane at a dislocation. ABC-type stacking is chosen as the example

Recent investigations have attempted to elucidate the role of stacking faults in group-III nitrides. Based on the observation of a small zincblende component in a bulk wurtzite film, *Seifert* and *Tempel* [5.7] suggested the existence of these faults in GaN. *Powell* [5.8] has since reported the existence of small regions of wurtzite GaN in bulk zincblende GaN crystals. *Lei* et al [5.9] employed X-ray diffraction to suggest that wurtzite and zincblende polytypes coexist widely in GaN thin films. It was concluded that all of the films except those on Si(001) have domains. These domains have their (111) axis parallel to the (0001) wurtzite axis, and the zincblende domains are nucleated at stacking faults. These allow the wurtzite stacking to shift to the zincblende stacking. This view was verified by TEM images showing the nucleation of a wurtzite InN domain from a stacking fault in zincblende InN [5.10]. The impact of stacking faults on the carrier mobility is not precisely known. Away from the fault edges their primary effects would most probably be to cause a local variation of the energy bandgap.

Almost all epitaxial GaN-based structures have so far been grown on substrates that do not have a wurtzite structure. As a result of this and non-ideal growth, inversion and rotation domain boundaries form at each step unless each terrace has the same bonding configuration. In short, unless the steps in 6H-SiC are six double-steps high, these boundaries would form. Consequently, it is highly desirable to deposit GaN on substrates with stacking order common with nitrides. ZnO is one such substrate with the added advantage that the lattice mismatch between GaN and ZnO is $\epsilon = 0.017$, which leads to a critical thickness between 80 and 120 Å. This implies that coherently strained layers of GaN could be grown with thicknesses up to about 100 Å. Some compositions of the InGaAlN alloy can be made to lattice-match ZnO. Although the opinion is divided, GaN grown on the O face of (0001) ZnO exhibit emission characteristics similar to the best ones on (0001) sapphire. More work is underway to harness the possibilities of-

fered, including easy cleavage planes which align with GaN. For details, we
refer the reader to Chap. 4.

5.3 Point Defects and Autodoping

Point defects, also known as native defects or intrinsic defects, are the
most common defects occurring in semiconductors [5.11, 12]. As in all
semiconductors, these defects play an important role in the electrical and
optical activity of nitride semiconductors. A case in point is the carrier lifetime which is often dependent on the type and density of these defects.
Consequently, they play a pivotal role in the radiative quantum efficiency
and the longevity of GaN-based lasers and light-emitting diodes. On the
electrical activity side, experimental evidence and calculations support the
premise that N vacancies form a shallow donor level in InN and GaN.
Though of a secondary importance, this level gets deeper as the Al mole
fraction is increased in AlGaN or the pressure on GaN is increased, as
sketched in Fig. 5.3. Such native donor and trap levels, and unintentional
impurities determine the carrier concentration and the ability to control the
doping of GaN-based device structures. In fact, high n-type background
concentrations in early samples were in part responsible for the lack of p-
type conductivity in GaN.

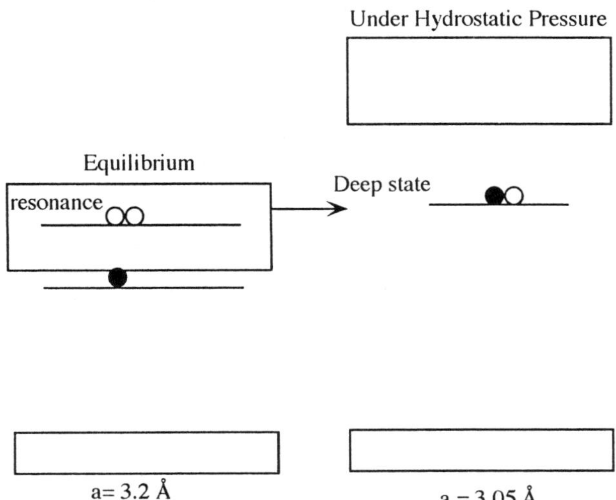

Fig. 5.3. Nitrogen vacancy level in GaN and AlGaN or GaN under sufficient hydrostatic
pressure showing the deepening of the nitrogen vacancy level as the lattice constant is
reduced. After [5.20, 21]

There are three basic types of point defects: vacancies, self-interstitials, and antisites. **Vacancies** may be interpreted as the atoms missing from lattice sites. **Self-interstitials** then are the additional atoms in between the lattice sites. **Antisites** are the cations sitting on anion sites, and vice versa, in compound semiconductors. **Native defects** result when bonds in a semiconductor are either broken or distorted; they often give rise to deep levels within the forbidden gap. *The Fermi level determines* the charge state of a particular defect. Defects can be either donor-type, acceptor-type, or amphoteric.

5.3.1 Vacancies, Antisites and Interstitials

Defects, in general, and point defects, in particular, have been synonymous with GaN research that began with the initial reports of an epitaxial growth. Characteristically, the early films suffered from large n-type backgrounds. *Maruska* and *Tietjen* [5.13] argued that autodoping is due to native defects, probably nitrogen vacancies, because the impurity concentration was, at least, two orders of magnitude lower than the electron concentration in their samples. Consequently, *unintentional n-type doping in GaN is probably due to nitrogen vacancies* became the crutch statement in nitride-related reports, because there had not been an unequivocal experiment to prove or disprove it. It should be pointed out that the uncertainty in the experimental determination of impurity concentration was large. Based on their first-principles calculations, *Neugebauer* and *Van de Walle* [5.14-16] suggested that in n-type material the thermodynamically stable formation of nitrogen vacancies in appreciable quantities is highly improbable. Instead, they pointed out that contaminants such as silicon or oxygen may be responsible for the large electron concentrations observed in some unintentionally doped samples. On the other hand, *Perlin* et al. [5.17] interpreted their first-principles calculations as favoring the notion that nitrogen vacancies are responsible for the large electron concentrations, albeit these calculations were oriented toward the electronic energy levels.

In a study, *Wetzel* et al [5.18.19] attempted to determine the position of localized states in GaN, assumed to be related to nitrogen vacancies, with respect to the band edge they employed infrared-reflection and Raman-spectroscopy analyses under a large hydrostatic pressure. An independent observation of the carrier localization by different experimental techniques was reported, too. From the observed reduction of the free-carrier concentration to 3% of the atmospheric pressure value at 27 GPa, they concluded the defect concentration to be as high as 10^{19} cm^{-3}. This defect is strongly localized and has a state 126 ± 20 meV below the conduction band at a pressure of 27 GPa. This is responsible for both the high electron concentration at atmospheric pressure and for the capture and localization at 27 Gpa. A resonant level of the neutral localized level at ambient pressure was predicted to be 0.40 $\pm$0.10 eV above the conduction-band edge.

Jenkins et al. [5.20, 21] calculated the energies of the N and Ga vacancies and antisite defects in GaN by taking advantage of a tight-binding approach. For the calculations, the wave functions of strongly localized defects, like nitrogen vacancies, V_N, were built up from contributions of the whole Brillouin zone. The calculations indicated small variations with respect to the alloy composition. They suggest that V_N would produce an s-like level containing two electrons near the conduction-band edge and a p-like level containing one electron above the conduction-band edge. Since the p-like level is resonant, its electron is auto-ionized, it decays to the conduction-band edge and dopes GaN n-type (one electron per vacancy). It is also possible that the s-like deep states lie slightly higher than predicted (not in the gap) and are resonant with the conduction band, donating their electrons to the conduction band. In this case the nitrogen vacancy would be a triple donor. Experimentally, nitrogen vacancies lie roughly 40 meV below the conduction band, and Ga vacancies form shallow acceptors. Both types of antisite defects, Ga on a N site, Ga_N and N on a Ga site, N_{Ga} lie deep within the energy bandgap. In the case of Ga_N, the doubly occupied A_1 state is approximately 0.11 eV below the conduction band, and a T_2 level is about 0.61 eV above the conduction-band edge. In the case of AlN, the A_1 and T_2 levels were predicted to be positioned 1.60 eV and 0.68 eV below the conduction-band edge, respectively. Calculations lend support to the assertion that the nitrogen vacancy is most likely a single donor and responsible for the n-type behavior in undoped GaN. Furthermore, if the nitrogen vacancy is a simple donor with its s-like deep level in the gap just below the conduction-band edge, the nitrogen vacancy would provide a natural explanation of the 0.2-eV feature in the Tansley-Foley optical absorption data [5.22]. It should be kept in mind that tight binding calculations include only the first-neighbor interactions, and likely result in erroneous energy levels. For a more accurate description of the energy levels, we refer the reader to [5.23].

Using a state-of-the-art supercell approach with 32 atoms per cell, a plane-wave basis set with an energy cutoff of 60 Ry, and soft Troullier-Martins pseudopotentials, *Neugebauer* and *Van de Walle* [5.24] investigated defect-formation energies and electronic structures for native defects in wurtzite and cubic GaN. It was predicted that 3d electrons are important both for the formation energy and the atomic relaxation. Breaking the additional bonds between the Ga 3d orbitals and the N orbitals to create a vacancy costs more energy. This leads to a significant increase in the formation energy. Taking the Ga 3d electrons as core electrons results in a large relaxation. The introduction of 3d electrons causes the system to be stiffer.

As a result, the outward displacement of the surrounding nitrogen atoms is reduced to 0.1 Å and the energy gain to 0.26 eV. The 3d electrons thus prevent the Ga_N bond length from becoming too short. Essentially, all antisites and self-interstitials are high in energy and hence less likely to occur in reasonable concentrations, and then only as compensating defects [5.15].

Another yet extensive theoretical study of native defects in hexagonal GaN was carried out by *Boguslawski* et al. [5.25]. For this study, the effects of cation and anion vacancies, antisites, and interstitials were considered. The computations were carried out utilizing an ab-initio molecular dynamics approach. The supercells contained 72 atoms. Although the numerical results are reasonably comparable, their interpretation differs from the conclusions of *Neugebauer* and *Van de Walle*. The interpretation of the results is that the residual donors responsible for the n-type character of as-grown GaN are nitrogen vacancies. However, the concentration of Ga interstitials under equilibrium conditions in the usual Ga-rich material can become comparable to that of the vacancies. Both n- and p-type doping efficiencies are substantially reduced by the formation of gallium vacancies, nitrogen vacancies, and gallium interstitials. In the zincblende structure, a substitutional defect, for example a vacancy, has four equivalent nearest neighbors. In the wurtzite structure, an atom along the c-axis relative to the defect becomes inequivalent to the three remaining neighbors. This leads to a lowering of the point symmetry that causes defect states, which are three-fold degenerate in the zincblende structure, to split into singlet and doublet pairs in the wurtzite structure. The resulting energy levels, which are comparable to those reported in [5.23], are shown in Fig. 5.4 where V_{Ga} and V_N represent the Ga and N vacancies, respectively, and Ga_N and N_{Ga} depict Ga on N and N on Ga antisites, respectively. N_I is the nitrogen interstitial with Ga_O and Ga_T representing the ideal and relaxed Ga interstitials.

From symmetry arguments, the vacancies would be in the form of singlets and quasitriplets. Since the electronic states of V_N are mainly composed of Ga dangling bonds, the energy of the quasitriplet is quite high. The singlet state of the quasitriplet is about 0.8 eV above the bottom of the conduction band. The hexagonal splitting is 0.5 eV, and the quasitriplet level contains one electron. Since a Ga vacancy causes a N dangling bond, its level is close to the top of the valence bands. The quasitriplet is located about 0.3 eV above the valence-band edge with a hexagonal splitting of only 0.1 eV. Since the quasitriplet is populated by three electrons in the neutral charge state, V_{Ga} can trap both electrons [5.25].

The gallium antisite, Ga_N, induces a quasitriplet near the middle of the band gap. In the neutral charge state, the singlet at $E_v + 1.4$ eV and the doublet at $E_v + 2.1$ eV contain two electrons each. The outward relaxation around Ga_N is large and causes the bond lengths with neighbors to increase. As for the nitrogen antisite, N_{Ga}, it introduces a doubly occupied singlet at

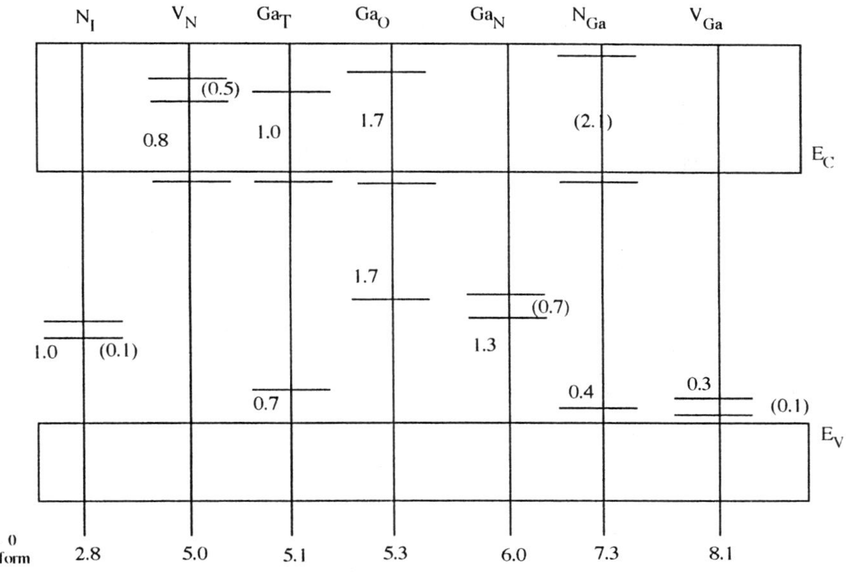

Fig. 5.4. Energy levels of neutral native point defects in GaN. After [5.25]

$E_v +0.4$ eV, and an empty doublet at $E_C -0.2$ eV in its neutral charge state. N_{Ga} strongly distorts along the c-axis, which causes the remaining bond lengths to increase. The empty doublet at $E_C -0.2$ eV cannot be occupied even by one electron, due to the large value of the electron-electron repulsion parameter for this level (0.95 eV) [5.25].

As for the interstitials, there are two highly symmetrical interstitial positions (T and O) shown in Fig. 5.5. The T site is situated in the middle of the line connecting non-bonded Ga and N atoms, and has two nearest neighbors and six next-nearest neighbors. The O site has six nearest neighbors, with distances larger by 28% than at the T site, and 7% larger than the equilibrium bond length. Due to the lack of a reflection plane perpendicular to the c-axis, neither site can represent the equilibrium position, leading to relaxation along the c-axis. In short, both the T and O positions are highly unstable [5.25].

The equilibrium position of the interstitial Ga is strongly charge dependent. Figure 5.5 displays both the initial and the relaxed configurations of the neutral Ga_T. The relaxation leads to an upward displacement of three atoms: Ga_T, the host Ga, and the host N. After the relaxation, the Ga-Ga distance increases nearly to the average Ga-Ga distance for the GaN anti-site. With an increasing charge the complex moves down, assuming an off-axis position for Ga_T^{3+}. Ga_T^0 introduces a deep level at $E_C -1.8$ eV, occupied by two electrons, and a resonance state at $E_C +1.0$ eV. The one electron that should occupy the resonance state auto-ionizes and becomes

157

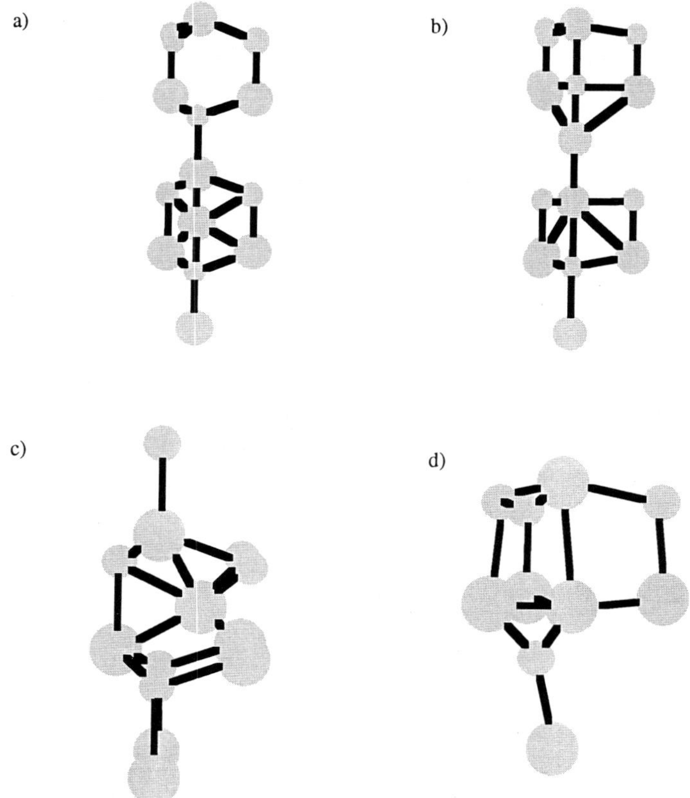

Fig. 5.5a-d. Atomic configurations of various interstitials in GaN: (**a**) An ideal and (**b**) a relaxed Ga(T) interstitial, (**c**) an ideal Ga(O) interstitial, and (**d**) a relaxed N interstitial. Large and small spheres represent Ga and N atoms, respectively. After [5.25]

trapped in an effective-mass level. Finally, Ga_T exhibits a negative U behaviour, since the $+2$ charge state is unstable for all positions of the Fermi level. The energy gain for the reaction $2\,Ga_T^{2+} \approx Ga_T^{3+} + Ga_T^{+}$ is 1.8 eV. The $(+/3+)$ level is located at 1.7 eV above the top of the valence band [5.25].

In the case of the ideal Ga interstitial, Ga_O, the relaxation is much weaker (Fig. 5.5c). The interstitial is displaced upwards and the first three Ga neighbors move down. The electronic structures of Ga_T and Ga_O are quite similar despite the very different surroundings. Ga_O also introduces a resonance in the conduction bands and a deep level at $E_C - 0.8$ eV. The $(+/3+)$ level is located at about $E_v + 1.7$ eV and the U parameter is nearly zero. Finally, the computed formation energy of the neutral Ga_O is only 0.2 eV larger than that of Ga_T and is within the margin of errors of the calculations of *Boguslawski* et al. [5.25].

The N interstitial at either the T or O site leads to the final configuration illustrated in Fig.5.5d. The displacement from the ideal T site is such that the final N-N bond length is very close to the N-N distance in the case of N_{Ga}. considering the electronic structure, N_T introduces a nearly degenerate pair of deep levels at about $E_v + 1.0$ eV which are separated by 0.1 eV and occupied by three electrons [5.25].

The formation energies of stoichiometric defects and growth temperatures must be considered together. Bulk samples are grown at $T_g = 1300 \div 1500°C$ and under Ga-rich conditions while MBE and VPE take place nominally at $600 \div 950$ and $900 \div 1100°C$, respectively. Considering growth temperatures in the $600 \div 1300°C$ range and equilibrium thermodynamics, *Boguslawski* et al. [5.25] treated three scenarios: (i) no doping, (ii) n-type, and (iii) p-type samples with concentrations of the external carries of $10^{18} \div 10^{21}$ cm^{-3}. The formation energies of neutral N and Ga vacancies are 3.2 and 8.1 eV, respectively, and under Ga-rich conditions they lead to a higher likelihood of V_N over V_{Ga} although the formation energies of the defects in highly charged states may be reduced by an energy of up to $2 \div 3$ times the bandgap, i.e., of the order of 10 eV. In n-type samples, the energy gain associated with transfering three electrons from the Fermi level to the low-lying acceptor states of the vacancy makes V_{Ga}^{-3} the prevalent native defect. In p-type samples, this phenomenon occurs for Ga_I^{3+} where three donor electrons are transferred to acceptor levels. Consequently, p-doping of wide-bandgap semiconductors under thermal equilibrium leads to very strong self-compensation effects. An attempt to dope a crystal will result in an increasing concentration of the appropriate compensating defects rather than in an increased hole concentration. The task of unraveling the machanisms of p-type doping is made more complex by the role of hydrogen, as will be discussed in Sect.5.3.2.

The origin of n-type conductivity of the as-grown undoped bulk and heteroepitaxial GaN has been attributed to nitrogen vacancies [5.13-19] and the impurities Si and/or O [5.14]. In an attempt to unveil the origin of auto-doping, undoped GaN films with different ammonia flow rates were grown by RMBE while keeping other experimental conditions the same [5.26]. Hall-effect measurements and Secondary Ion Mass Spectrometry (SIMS) measurements were carried out to obtain free-carrier concentration, and thus a glimpse of background doping levels and impurity levels, respectively. Figure 5.6 exhibits the variation of the background electron level with the ammonia flow rate (V/III ratio). The general trend is that the background electron concentration decreases as the ammonia flow rate increases. It is expected that the concentration of N vacancy in the film should decrease as the V/III ratio increases. Thus, the trend in Fig.5.6 is such that the background doping level and the N vacancy level follow one another, supporting the nitrogen vacancy argument. While this observation

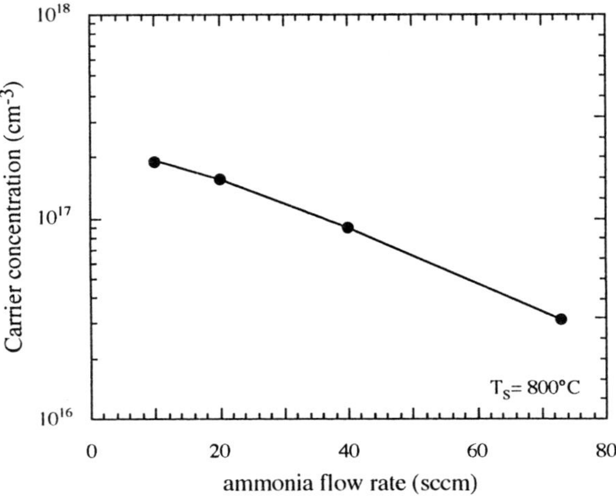

Fig. 5.6. The variation of the background carrier concentration for a series of undoped GaN films grown with different ammonia flow rates. The growth rate and film thickness were kept at 1.2 μm/h. and 2.4 μm, respectively. After [5.26]

alone would not rule out the O_N the argument, other data, which will be discussed in Sect.5.3.2, suggest that oxygen impurities do not really account for the electron concentration in unintentionally doped samples with high electron concentrations. However, the case for lightly doped samples is not yet clearly understood because oxygen is found at some concentrations, and high oxygen as determined by SIMS may also imply less oxygen incorporation as donors.

5.3.2 Role of Impurities and Hydrogen

To probe whether the autodoping is caused by native defects or impurities, SIMS measurements were carried out for two samples grown with 10 (B) and 73 (C) sccm of ammonia flow rates. The Hall-electron concentrations for the first two samples were about $2 \cdot 10^{17}$ and $3 \cdot 10^{16}$ cm^{-3}, respectively. Figures 5.7a-c display the SIMS depth profiles of O, H and Si impurities for those two samples. The sample (A) was prepared with a higher ammonia injector temperature of 600°C, as opposed to 300°C for the others, and a single ammonia purifier at an ammonia flow rate of 4 sccm. This sample exhibited an unintentional electron level of $2 \cdot 10^{18}$ cm^{-3}. The growth condition for sample A was even closer to the N deficient condition than that for sample B. As one can see from Fig.5.7a, the Si impurity levels are lower than 10^{17} atoms/cc (mostly in the 10^{16} range which is close to the lower detection limit of SIMS for Si) for all three samples regardless of the different background doping levels. It is thus reasonable to arrive at the conclu-

160

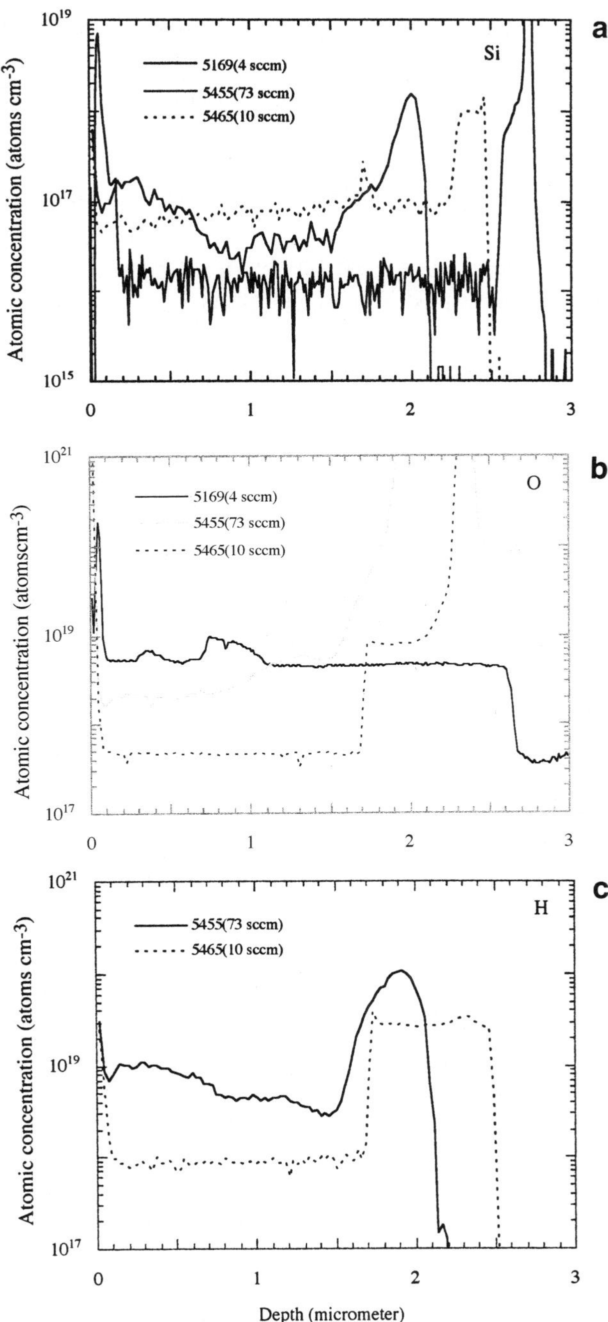

Fig. 5.7. The depth profile of (**a**) Si, (**b**) O, and (**c**) H measured by SIMS for various ammonia fluxes. Samples A, B, and C were grown with 4, 10, and 73 sccm ammonia flow rates, respectively. A single ammonia purifier was used for sample A and double purifiers for the other two samples. Moreover, the injector temperature was 600°C for sample A and 300°C for others. After [5.26]

sion that the Si impurity is not mainly responsible for the autodoping of these undoped GaN samples.

Given the lower ammonia flow rate (4 sccm) and making the reasonable assumption that the source of O is ammonia, the large O concentration in sample A can be attributed to the single ammonia purifying stage. The background O level from the system itself was checked by carrying out SIMS measurements for an undoped GaAs layer grown in the same system. The O concentration in this sample was about $6 \cdot 10^{16}$ atoms/cc and was assumed to be the instrumental detection limit in the GaAs matrix. Thus, it was concluded that the O contamination in the GaN samples is mainly due to ammonia.

Assuming that the O source is ammonia, a GaN sample was prepared with high ammonia flow rate to determine the role of O. Figure 5.7b, in addition, depicts the O concentration for sample C which was grown with a high (73 sccm) ammonia flow rate. The O concentration for this sample was higher than that of sample B, as expected due to the higher ammonia flow rate. The background doping level for this sample was about $3 \cdot 10^{16}$ cm^{-3} which is about one order of magnitude lower than that of sample B. Consequently, the nitrogen vacancy argument is a more plausible explanation of autodoping in GaN barring any active compensation processes which may be in effect during nitrogen rich growth conditions.

Figure 5.7c shows the depth profile of H atoms for samples B and C. The overall H content in sample C is higher than that in sample B, which is consistent with the ammonia flow rate employed for these two films, the former having been grown with a larger ammonia flow rate. It is predicted from a first-principles calculation that H incorporation in n-type bulk GaN is much lower than that in p-type films [5.27] due to the higher formation energy. There are two experimental reports about deuterium incorporation after hydrogenation for both n- and p-GaN [5.28, 29]. The H concentrations in the n- and p-type GaN films were almost the same. Although a detailed study of the incorporation of H in n- and p-type GaN films is necessary, one can suggest that the H concentration in these GaN films is determined by the amount of ammonia regardless of the carrier type, unless there are additional factors which affect the impurity incorporation involved, such as a high density of defects. It should be pointed out that hydrogen incorporation may also be enhanced by extended defects which renders the picture rather complicated to model.

Although different from growing films, support for the donor nature of the nitrogen vacancy has also been extended from high-energy electron irradiation experiments [5.30]. Measurement of the mobility and fitting it to the theory, and the electron concentration as a function of temperature before and after the sample was subjected to irradiation, showed that the acceptor and donor concentrations increased by the same amount after irradiation whereas the mobility decreased accordingly. A consistent picture emerging from this investigation is that nitrogen Frenkel pairs are formed in which the N vacancies are donor-like and the N interstitials are acceptor-like.

5.3.3 Optical Signature of Defects in GaN

PhotoLuminescence (PL) spectroscopy has also been applied to help unravel defects in GaN. PL spectra for two samples grown at 800°C, one undoped (S1), and the other Si-doped (S2), are similar (Fig. 5.8). The ammonia flow rate for S1 is 6 sccm and that for S2 is 16 sccm. Hall measurements indicate that the unintentional electron concentration of the undoped sample is $1.6 \cdot 10^{17}$ cm^{-3}, and the intentional electron concentration of the Si-doped sample is $4.1 \cdot 10^{17}$ cm^{-3}. The PL spectra for S1 and S2 reveal a sharp peak P1 at 362 nm for each of the samples, which could be ascribed to nitrogen vacancies at an energy level very close to the conduction-band edge, E_{CMIN}. That the sharp peak of the Si-doped sample occurs at about the same energy as that of the undoped sample is consistent with the argument again that the donor state in the undoped sample, which is likely to be induced by nitrogen vacancies, behaves essentially in the same way as shallow Si dopant atoms. Utilizing Hall and SIMS measurements, one can argue that V_N may be responsible for the donor level in undoped samples, particularly the earlier varieties. In addition to vacancies, antisites, though their formation may not be favored, and other defect centers [5.25, 31, 32] must be considered.

The yellow-green emission peak P2, occurring at about 550 nm, may be associated with a deep acceptor-like state, about 2.25 eV below the conduction-band edge, E_{CMIN}. The details of this particular emission will be

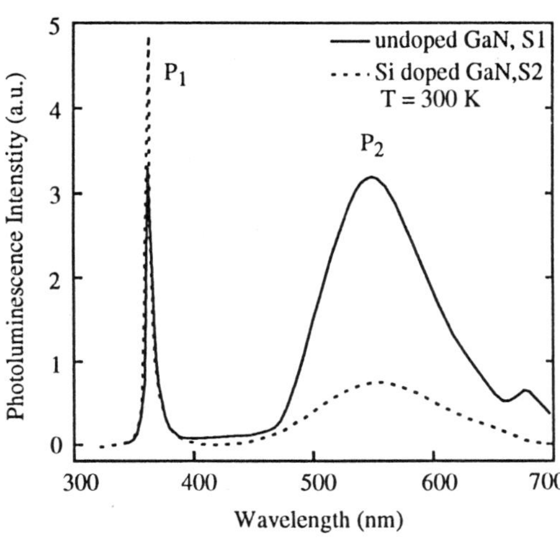

Fig. 5.8. Room-temperature PL spectra of undoped and Si-doped GaN

discussed in Sect.10.3.4a. Suffice it to say here, this emission is likely due to transitions from the conduction band to deep acceptor-like states. The deep states have recently been suggested to be of a **Ga-vacancy nearest-donor complex** nature [5.33], consistent with the calculations suggesting their likely presence [5.14, 25]. A Ga vacancy would be a triply charged state, and it would be doubly charged after forming a complex with a nearby donor. The binding energy of the complex would be reduced. The nearest donor on the nitrogen site has been suggested to be the substitutional oxygen as a sample perceived not to contain discernible quantities of oxygen does not seem to exhibit the yellow emission. Moreover, the lack of the yellow emission after electron-beam irradiation, which would cause native defects to be formed, supports an impurity participation in the complex. Other likely deep levels such as antisites are not thermodynamically favored according to *Neugebaur* and *Van de Walle* [5.15] and can not be present is appreciable quantities. From Fig.5.8, however, it is apparent that the intensities of both P1 and P2 decrease with an increase in the ammonia flow rate. It can be argued that the intensity of P1 decreases because an increase of the ammonia flow rate causes a decrease of the nitrogen vacancies. The variation of the intensity of peak P2 with the ammonia flow rate is depicted in Fig.5.9. When the ammonia flow rate exceeds about 25 sccm/s, the peak P2 practically disappears. This is consistent with the Ga-vacancy nearest-neighbor donor complex argument, as increased ammonia would tend to inhibit the incorporation of residual oxygen on the nitrogen sites.

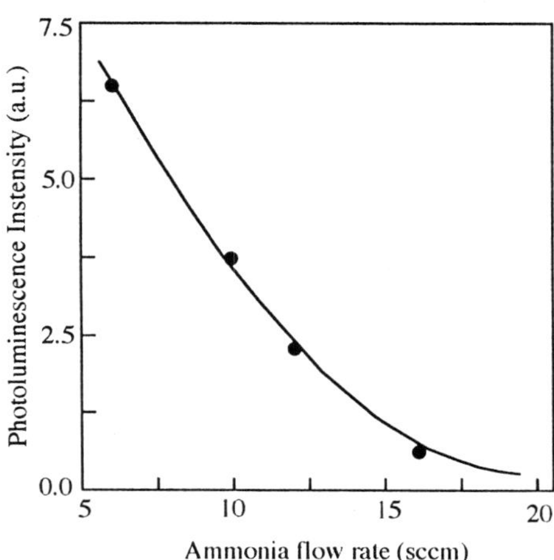

Fig. 5.9. Variation of the room-temperature PL intensity of the observed peak P2 with ammonia flow rate during growth

5.4 Intentional Doping

As is the case in any semiconductor technology, the control of the electrical properties of GaN and related materials remains the foremost obstacle in the progress of device development. Unintentionally doped GaN has, in all cases, been observed to be n-type with the best samples still showing a compensated electron concentration of about $5 \cdot 10^{16}$ cm^{-3}. With the exception of oxygen, no impurity has been found present in a sufficient quantity to account for the carriers, so researchers have attributed the background to native defects, which are widely thought to be nitrogen vacancies, and/or oxygen. All efforts, until recently, to obtain p-type doping have resulted in heavily compensated, highly resistive films. The development of a p-type doping above the 10^{18} cm^{-3} level for GaN remains a primary challenge for researchers. The task is even more challenging for AlGaN and InGaN despite the recent progress.

5.4.1 n-Type Doping with Silicon, Germanium and Selenium

With a successful reduction of the unintentional n-type background, doping GaN with Si has very well been established in both the vapor phase and vacuum deposition techniques. While the current technology is sufficient for light emitters, the background impurities and native defects must further be reduced for certain varieties of field-effect transistors and detectors.

Controlled n-type conductivity in binary and ternary alloys of GaN, AlN and InN is generally achieved by Si doping in both vacuum deposition and metalorganic chemical-vapor deposition techniques [5.34]. Si substitutes a Ga atom in the lattice and provides a loosely **bound** electron. In the dilution limit the ionization energy of the Si level in GaN is about 27 meV and decreases with an increasing doping level due to screening, as discussed in Chap. 7. PL measurements yielded a value of 22 meV [5.35]. The solubility of Si in GaN is high, and on the order of 10^{20} cm^{-3}. Therefore, Si is suitable for group-III N doping and is most frequently used. Electron concentrations vs. silane flow rates in an OMVPE reactor for GaN and AlGaN are exhibited in Fig. 5.10.

Nakamura et al. [5.36] have reported on the properties of Si- and Ge-doped GaN grown by MOVPE. The observed carrier concentrations of Si-doped GaN were in the range $10^{17} \div 2 \cdot 10^{19}$ cm^{-3}, while Ge doping has produced material with electron concentrations of $7 \cdot 10^{16} \div 10^{19}$ cm^{-3}. A linear variation of the electron concentration as a function of both the SiH$_4$ and GeH$_4$ flow rates was observed across the entire experimental range (Fig.

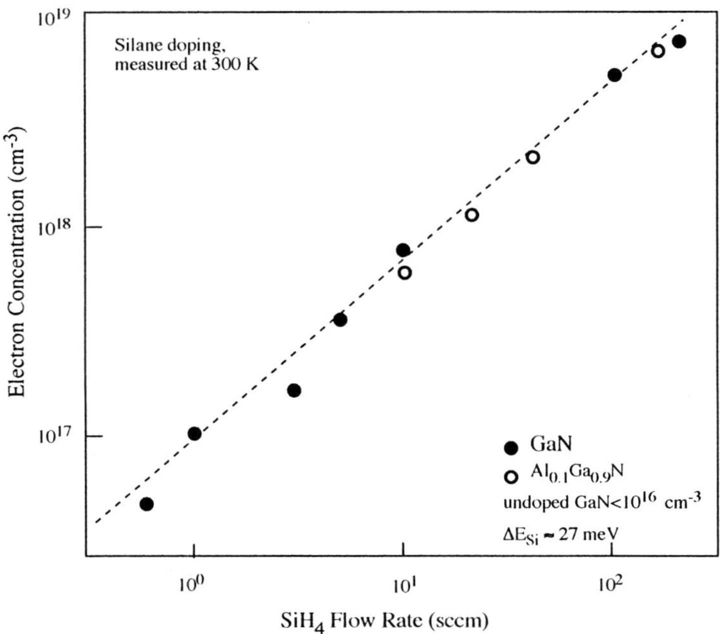

Fig. 5.10. The electron concentration vs. silane flow rate in an OMVPE reactor for GaN (*solid circles*), and AlGaN (*open circles*). Courtesy of I. Akasaki, Meijo University

5.11). Ge incorporation is roughly an order of magnitude less efficient than Si, as judged by the larger GeH$_4$ flow rates required to obtain similar electron concentrations. Other and much safer sources, such as monomethylsilane, have successfully been employed to obtain n-type GaN but their uses are not yet very common. *Goldenberg* et al. [5.37] observed higher n-type GaN conductivities as the NH$_3$ flow was increased during the growth. Although the site selection process, Si substituting for Ga, is very likely enhanced with increased ammonia flow rate, they postulated that H passivation of acceptors in their material leads to the improved electrical characteristics [5.38-40].

Electron concentrations up to $6 \cdot 10^{19}$ cm^{-3} have been achieved in Se-doped GaN MOCVD films [5.41]. The electron concentration was proportional to the H$_2$Se flow rate employed as a dopant source. The room-temperature electron mobilities ranged from 10 to 150 cm^2/V·s. The high compensation ratio of ≈ 0.4 was nearly constant over the concentration range of $10^{18} \div 10^{19}$ cm^{-3} and became even higher at higher electron concentrations. The compensating acceptor was attributed to the Ga vacancy.

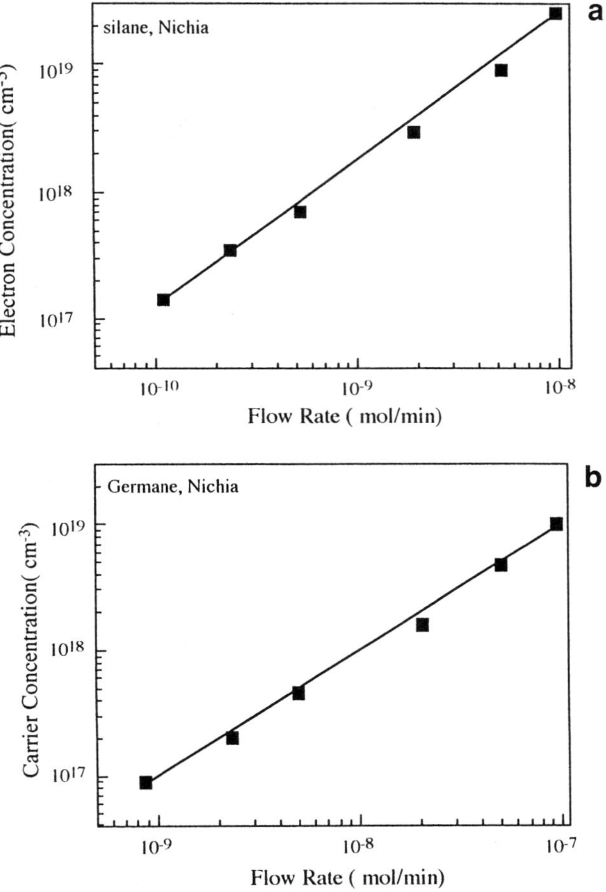

Fig. 5.11. Incorporation rates as a function of gas flow in MOVPE grown GaN, (**a**) Si and (**b**) Ge. Both dopants are well behaved in that the active donor concentrations linearly vary with silane and germane flow rates. However, H_4-Ge requires a factor of 10 higher flow to obtain the same doping level. After [5.33]

5.4.2 p-Type Doping

There has been much effort aimed at doping GaN and its ternaries p-type by introducing group-II and groug-IV elements. Many potential dopants have been incorporated into GaN. Some impurities have been observed to effectively compensate electrons in GaN, leading to highly resistive material. Only recently post-growth electron beam irradiation and thermal annealing which convert Mg-compensated GaN into p-type material, have been developed. No other impurity has been as successful as Mg in rendering GaN p-type. Optical measurements have revealed that the acceptor levels in GaN are several hundred meV above the valence-band edge.

a) Doping with Mg

Due to lack of p-type GaN until 1989, attempts at garnering the potential of GaN for emitters relied on metal insulator n-type (m-i-n) GaN blue-light emitting diodes (LEDs), fabricated at what was then the RCA Laboratories, with doped n-GaN and highly resistive i-GaN due to Zn doping. However, a good p-n junction is required for achieving commercial LEDs, and any injection Laser Diode (LD), as will be detailed in Chaps. 11 and 12. The lack of p-GaN delayed the development of LEDs considerably. p-type GaN layers were eventually achieved in 1989 by Mg doping of MetalOrganic Chemical Vapor Deposition (MOCVD) films which initially required a Low-Energy Electron Beam Irradiation (LEEBI) treatment for the activation of Mg by acceptors [5.38].

Amano et al. [5.42] and *Nakamura* et al. [5.43] achieved p-type nitrides by using atmospheric MOCVD with in-situ Mg doping. *Amano* et al. achieved the initial breakthrough by converting compensated Mg-doped GaN to conductive p-type material by low-energy electron beam irradiation. *Nakamura* et al., again employing LEEBI, achieved GaN with $p = 3 \cdot 10^{18}$ cm^{-3} and a resistivity of 0.2 $\Omega \cdot cm$. Soon thereafter came the observation that thermal annealing at 700°C under an N_2 ambient converts GaN to p-type equally well. The process was observed to be reversible with the GaN reverting to insulating compensated material when annealed under NH_3. Hydrogen was thus identified as the critical compensating agent. Layers of p-type GaN were achieved with no post-growth anneal required using MBE employing activated nitrogen or ammonia as the nitrogen source [5.40]. In the chemical vapor-phase epitaxy technique [5.44], $CC_{p2}Mg$ is utilized for the source of Mg. The Mg chemical concentrations for two different growth temperatures and Mg sources vs. the flow rate of Mg sources are plotted in Fig. 5.12. The upper curve is for $CC_{p2}Mg$ at a growth temperature of 1020°C. The bottom curve is for $MC_{p2}Mg$ at a growth temperature of 950°C. As the data suggest, $CC_{p2}Mg$ is much more effective and less volatile in that it can be incorporated in GaN at high temperatures where the quality of GaN films is superior to those grown at lower temperatures. To transfer to other reactors, other pertinent parameters such as TMG and ammonia flow rates must be known, too. Nevertheless, the point is that Mg concentrations in p-type GaN films are extremely high and approach 1% of the host species. Note that when substitutionally incorporated, only about 1% of the Mg atoms are ionized in GaN. The hole concentration vs. flow rate is not shown because the data are not reliable due to large error bars, the likely source of which will be discussed in Chap. 7.

Theoretical efforts have been attempted to provide a viable explanation for the relative success of the Mg acceptor in GaN, while other group-II metals continue to compensate GaN, even after a LEEBI treatment or an-

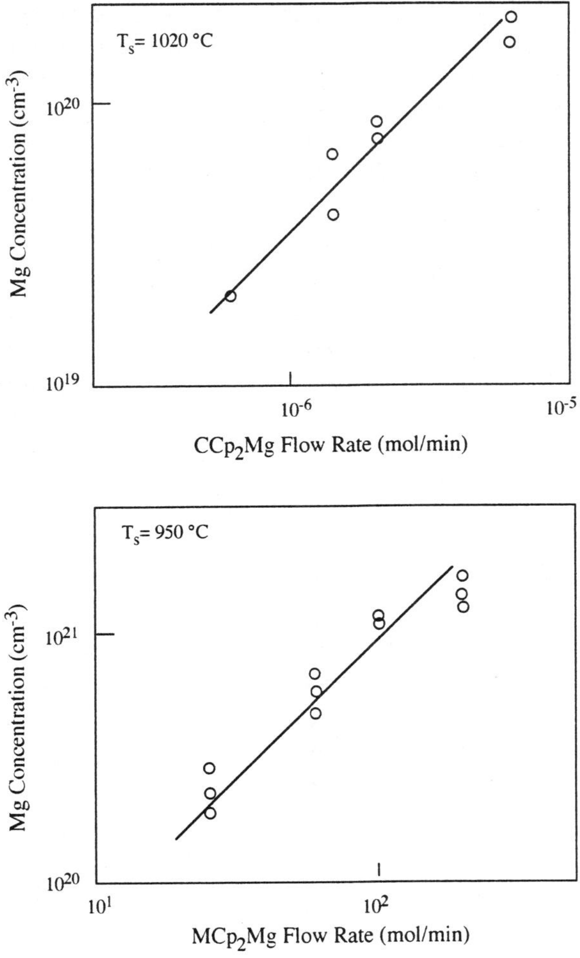

Fig. 5.12. Mg chemical concentrations for two different growth temperatures and Mg sources vs. the flow rate of the Mg source. The *upper curve* is for CC_{p2}Mg at a growth temperature of 1020° C. The *bottom curve* is for MC_{p2}Mg at a growth temperature of 950° C. After [5.44]

nealing. GaN differs considerably from other conventional group-III-V semiconductors. For example, due to the strong covalent character, the charge density in GaAs exhibits its maximum near the Bond Center (BC). There is no such local charge maximum in GaN; rather, there is an increase in the charge density from Ga towards the more electronegative N in GaN, which causes the charge density around N to be high and nearly spherically symmetric. In view of this, group-III-V nitrides are more ionic than other group-III-V semiconductors and resemble the band structure of group-II-VI semiconductors. Specifically, the presence of a large ionic gap in the valence-band density of states causes their lower valence bands to lie deeper beneath the valence-band edge. The resulting energy resonance causes the Ga 3d electrons to strongly hybridize with both the upper and lower valence-band s- and p-levels [5.45]. Such a hybridization is predicted to have a profound influence on the properties of GaN including such quantities as the bandgap, lattice constant, acceptor levels and valence-band heterojunction offsets [5.45]. It is known in the cases of ZnS and ZnSe that potential acceptors such as Cu, whose d electrons are resonant with the Lower Valence Band (LVB), are repelled by the d-hybridized upper valence band in an abnormally deep level, while impurities without resonances form shallow acceptors [5.46]. Mg has no d electrons, as is the case for Be which will be discussed below (Sect.5.4.2d), and turns out to be sufficiently shallow for RT p-type doping of GaN. On the other hand, Zn, Cd, and Hg, which all have d electrons, arguably form deep levels in GaN as evidenced by the resultant high-resistiviy films which at least do not contradict the above argument.

(I) *Role of Hydrogen and Defects in Mg-doped GaN*

The need for post-growth annealing to activate Mg in MOCVD-grown samples and reverting them to high-resistivity compensated material after ammonia annealing have both received a good deal of attention. *Van Vechten* et al. [5.47] suggested that hydrogen passivates Mg, which is also supported by the calculations of *Neugebauer* and *Van de Walle* [5.27]. Moreover, they demonstrated that hydrogen is beneficial to p-type doping by Mg when compared to the hydrogen-free case, since the hydrogen passivates Mg during growth and thus inhibits the formation of native donors self compensating the acceptors. For MOCVD-grown films, it has been argued that H is in its positive charge state, the H^+ proton, and passivates the Mg acceptors which are in their negative charge state during growth; this would prevent the compensating donor-like defects from forming. During post-growth annealing, H^+ is driven out, which results in p-type GaN due to the negatively charged Mg acceptors. Although it has been mentioned frequently that Mg and H form a complex in Mg-doped GaN films, the exact mechanism of formation and the release of H upon a post-growth treatment have not been sufficiently elucidated. In addition, predictions of the position of the H atom from first-principles calculation are not consistent. For example, in the *Neugebauer* et al [5.48] calculations, the **anti-bonding site** (i.e., a hydrogen sitting next to a N atom, about 1 Å away from the N site, at the anti-bonding position of the Mg-N bond) is lower in energy than the **Bond-Center (BC) site**. This follows from the argument that the BC position requires an outward motion of the Mg and N atoms, which is very energy consuming in GaN, as GaN is a very hard material. On the other hand, *Okamoto* et al [5.49] argued in favor of the BC location for H between substitu-

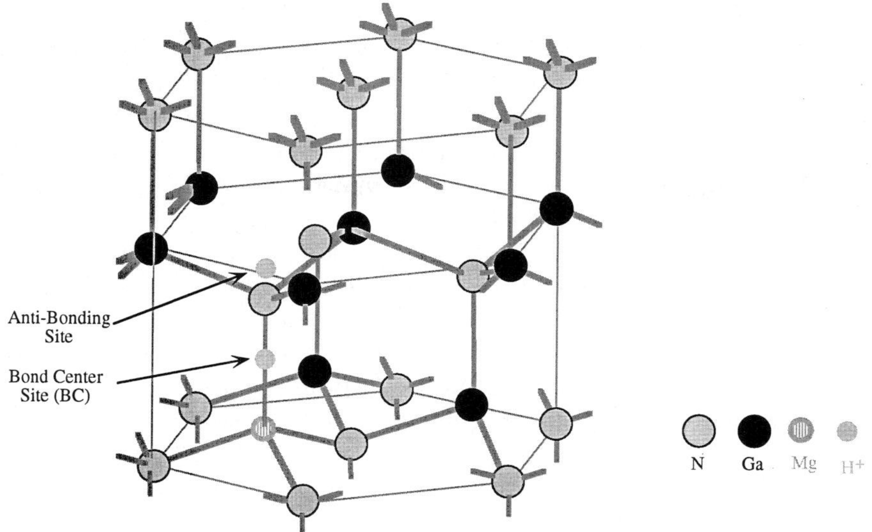

Fig. 5.13. Proposed Bond-Center (BC) and anti-bonding site incorporation of H in GaN and its passivation of Mg during growth (the Mg atom is directly below the H atom)

tional Mg and N nearest neighbors. Unlike the MOCVD-grown layers, Mg doped GaN layers grown by Reactive Molecular Beam Epitaxy (RMBE) with ammonia as the nitrogen source exhibits p-type conductivity without any post-growth treatment [5.50]. In general, H takes the BC position in all other semiconductors. Figure 5.13 displays the passivating H at both the BC and the anti-bonding sites.

Useful is the determination of the **Local Vibrational Modes** (LVMs) of the Mg-H complex in GaN which provides not only the confirmation of hydrogen in GaN, but also give significant information on the structure of the complex. At the present time the stretch frequency is, in fact, the only reliably-established physical parameter available from experiment for the Mg-H complex in GaN. To support their spectroscopic identification, *Götz* et al. [5.51] performed Fourier-transform infrared-absorption spectroscopy on three Mg-doped GaN layers grown by MOCVD: the first sample was as-grown and electrically semi-insulating; the second one was subjected to a thermal anneal and displayed p-type conductivity; and the third sample was exposed to monatomic deuterium at 600°C for 2 h, which increased the resistivity of the material. The as-grown sample displayed a LVM at 3125 cm^{-1} which is in very good agreement with 3360 cm^{-1} predicted by *Neugebauer* et al. [5.48]. After thermal activation of Mg, the intensity of this absorption line is reduced. After deuteration, a new absorption line appeared at 2321 cm^{-1}, which disappeared after a thermal activation treatment. The isotopic shift clearly establishes the presence of hydrogen in the complex.

The controversy regarding the location of H at a Mg-Ga in GaN can be resolved by a Rutherford back-scattering experiment with a good-quality crystal and moderately heavy doping [5.53]. One would direct a well collimated beam of ions down one of the channel axes and observe those that bounce back from a host atom that is not in its perfect lattice site. If H were bond-centered then Mg would protrude out into the channel, which was calculated to be energetically unfavorable by *Neugebauer* et al. [5.48]. If H were in the interstitial space (anti-bonding site), then the incident ions would strike neither Mg, Ga, or N, but rather H; if there were any detectable recoil ions, which would be doubtful, they would easily be distinguished from those that struck the other atoms. Additional support for the interstitial, anti-bonding site, can be discerned from the following phenomenological argument [5.53]. The H^+ proton should act very much like an idealized positive test charge as it does not have to have its wave function orthogonalized to any of the electronic states. The very massive proton is also effectively a point charge, because the energy to confine its wave function to a small volume is less than that to confine the electron to the same volume. Another positive test charge is the positron e^+. A positron's wave function does not have to be orthogonalized to those of the electrons, but the energy to localize it in a small volume is the same as for an electron. In Si, Ge, GaAs, GaP, InSb and most likely in GaN and other semiconductors e^+ travels between interstitial spaces in a perfect crystal, where e^- has the least wave-function density. This is a natural consequence of maintaining the mass and switching the sign of the charge. However, e^+ also needs sufficient volume for its wave function so that the localization energy is not too large, and it finds this volume in the interstitial space.

The analogy between e^+ and H^+ suggests that H^+ will be in the interstitial site surrounded by N ions. However, because H^+ does not need all of the interstitial space, it might find a lower energy site with a much smaller volume. One might think that the bond site would be a logical candidate. There is a local peak in the electron density at the bond site and it does not result from the Coulomb potential of the ions. If it did result from the Coulomb potential, then obviously H^+ would be repelled rather than attracted. The extra electron density in the bond site results from the constructive interference between the atomic wave function centered on the two atoms participating in the bond; this would be a quantum mechanical effect. Thus, H^+ might reside in a low-energy state. However, if it does, it will then certainly perturb all the bonds in the area rather strongly, which is not energetically favorable over the interstitial anti-bonding site [5.48]. Unlike the strongly covalent semiconductors where H^+ is attracted to the bond center, the ionic nature of GaN would tend to expel H^+ into the interstitial site. Thus, it is the interstitial site that will offer the lowest energy [5.53].

The compensating nature of H is also reflected in other group-III-V semiconductors [5.52]. Owing to the incorporation of a relatively large concentration of shallow-acceptor impurities in comparison to the concentration of the native donor-like defects in GaN, one finds that the Mg acceptor concentration is accomplished primarily through the presence of Mg-H complexes. This compensation is lifted (i) when samples are irradiated with an electron beam of $5 \div 15$ keV incident energy for several hours, or (ii) when samples are thermally annealed at constant temperature for half an hour. Also, first-principles calculations demonstrate that the same amount of both Mg and H incorporated into the GaN films when they are grown under an H-ambient growth condition, such as that of MOCVD [5.27]. Furthermore, the calculations predict that more Mg can be incorporated into GaN film when there are more H atoms present.

(II) *Mg in RMBE-grown GaN*

As-grown p-GaN films produced by RMBE have shown p-type conductivity even though these films were grown in a H-containing environment [5.50]. Note that both RMBE and MOCVD growth environments contain copious amounts of H and that MOCVD-grown Mg doped GaN films [5.54] reveal p-type conductivity only after thermal annealing. Actually, the amount of H in a p-GaN film under discussion (Fig.5.14) is much lower than the amount of Mg. This indicates that there is a good deal of Mg atoms which do not form complexes with H in the film if one-to-one Mg-H complex formation occurs. Figure 5.14 exhibits the depth profile of H for the two Mg-doped samples. The Mg concentration for these two samples were about $5 \cdot 10^{18}$ and $3 \cdot 10^{19}$ atoms/cc, respectively, and remain nearly constant through the films. As one can see, the H level in RMBE-grown p-GaN is not negligible. It should be noticed that the H concentration in these two as-grown films are almost the same although there is a distinct difference in the Mg concentration. It was reported [5.26] that thermal annealing or the RTA process did not change the electrical characteristics of these films. But the amount of H decreased by about a factor of three after an RTA treatment for 2 min at 900°C (curve 2 of Fig.5.14). Thus, the release of H atoms from as-grown GaN:Mg did not improve the conductivity of the already conducting RMBE-grown films. Although, a further study is essential to delineate the nature of as-grown p-type conductivity in RMBE-grown GaN:Mg films, the following two explanations may be justified: The first has to do with the in-situ annealing effect during the cooling procedure after film growth. When the film growth is completed, the ammonia flux is normally shut off right away. Consequently, the film is vacuum-annealed, albeit for a short time, in the cooling process during which H may be released, which might not be the case for MOCVD-grown films. It has been

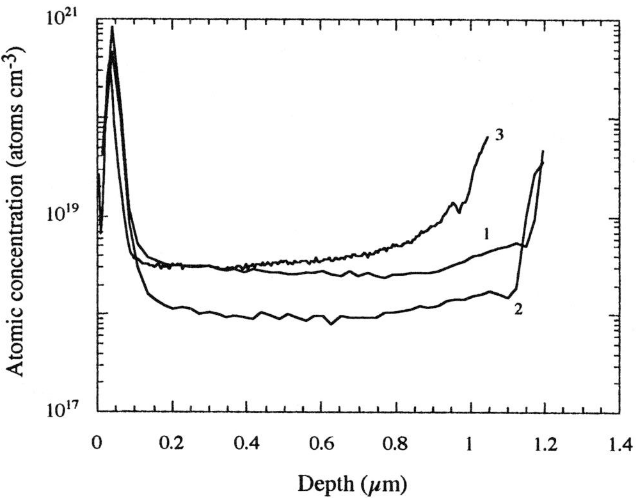

Fig. 5.14. The H profile measured by SIMS for two p-GaN films grown with different Mg fluxes. An ammonia flow rate of 37 sccm was used for both of the samples. (*1*) The H profile for the as-grown sample with low Mg flux. (*2*) The H profile for the same sample shown in curve 1 but after RTA treatment at 900 $_C$ for 2 min. (*3*) The H profile for the as-grown sample with high Mg flux. After [5.26]

reported that annealing in an ammonia atmosphere may passivate a conductive (after a post-growth treatment) p-GaN film, and the conductivity can be restored by annealing the same sample in a nitrogen atmosphere [5.55]. It is, however, not clear if the activation and deactivation processes for already conductive (after a post-growth treatment) p-GaN films are the same as those of highly-resistive as-grown GaN:Mg films.

In an attempt to gain more insight, two Mg-doped GaN films were grown on c-plane sapphire substrates with different Mg fluxes. The flow rate of ammonia was changed in several steps throughout each growth to observe the effect of the ammonia flow rate on the incorporation of Mg into the film. The depth profile of H and Mg atoms for these two samples were analyzed by SIMS. Figures 5.15a and b exhibit the depth profile of H and Mg for these two samples. The numbers for each growth segment indicate the applicable ammonia flow rate. No clear steps are recognized in Fig. 5.15a, whereas two clear steps in H and Mg profiles are seen in Fig. 5.15b when the ammonia flow rate was changed between 10 and 37 sccm. It is noticeable that the Mg and H concentrations change at the same position. In addition, the Mg concentration is more than one order of magnitude higher than the H concentration, as mentioned above. It is tempting to conclude that some of the Mg atoms incorporated into the film are passivated by H, while the majority of the Mg atoms incorporate without passivation by the accompanying H.

174

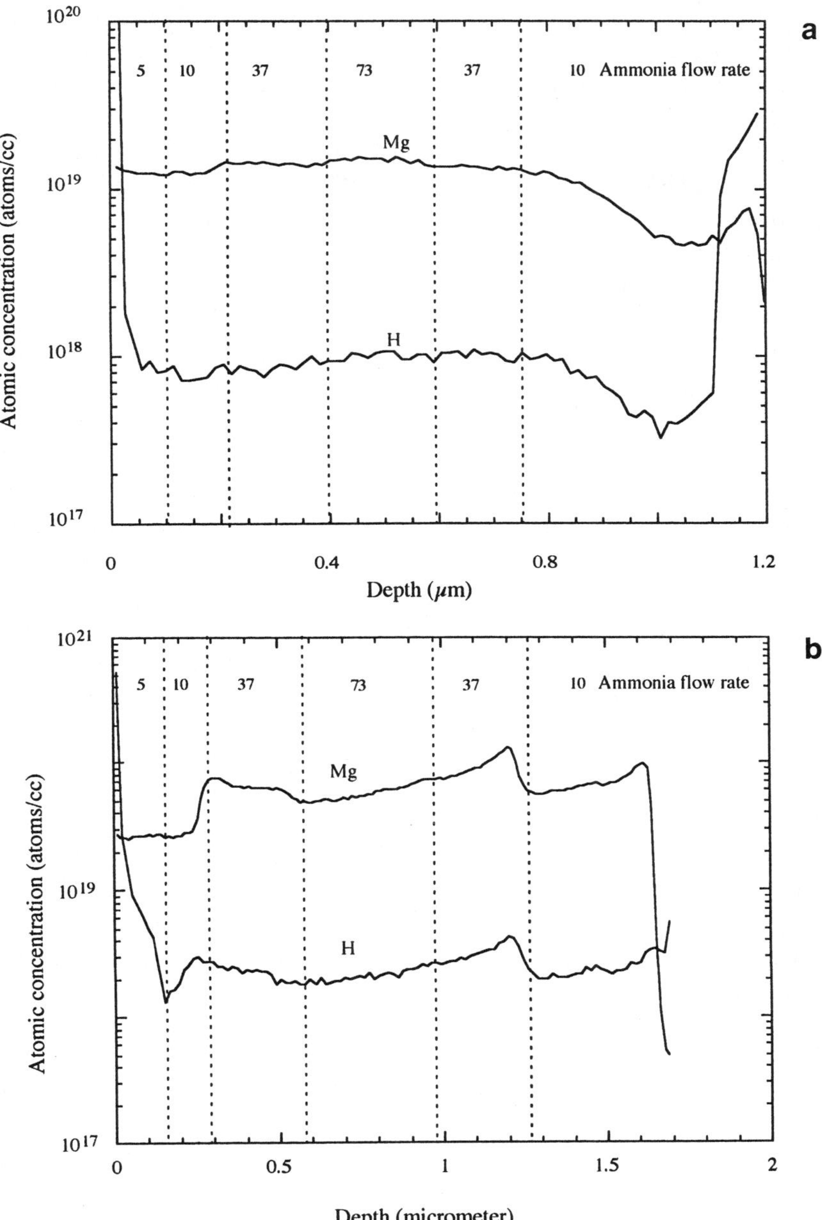

Fig. 5.15a,b. The Mg and H profiles measured by SIMS for two Mg doped samples grown with different Mg fluxes. The ammonia flow rate was changed in several steps throughout each growth. The numbers in each segment represent the ammonia flow rate. (**a**) low Mg flux, (**b**) high Mg flux. After [5.26]

175

The ratio of passivated Mg over the total Mg concentration may or may not be constant, which means that the H profile may or may not follow that of Mg. In fact, the H profile does not match the Mg profile when the same experiment was carried out on a 6H-SiC substrate (Fig. 5.16). At this point, it is not clear whether there is a difference in growth characteristics in terms of H and Mg incorporation when SiC substrates are used. In fact, the H concentration measured is generally about one order of magnitude higher than that in the previous two samples, which might be due to the growth characteristics related to SiC substrates, for example, diffusion of impurities from the SiC surface to the growing film. The H and O concentrations are about one order of magnitude higher near the film/substrate interface as compared to the rest of the film. This might be related to the outdiffusion of impurities from the SiC substrate because the location where the concentration starts to increase approximately matches the point where the Si signal (of SiC) begins to increase drastically. In any case, Fig.5.16 supports the latter idea because there is no such trend in the H profile (Fig. 5.15b) while steps in the Mg concentration are still seen. It appears that "more H during the growth" enhances the incorporation of Mg, as was expected by *Neugebauer* et al. [5.27], when we consider only the sample shown in Fig.5.15b. However, the following argument having to do with the site selection can explain the overall trend of Mg incorporation with the

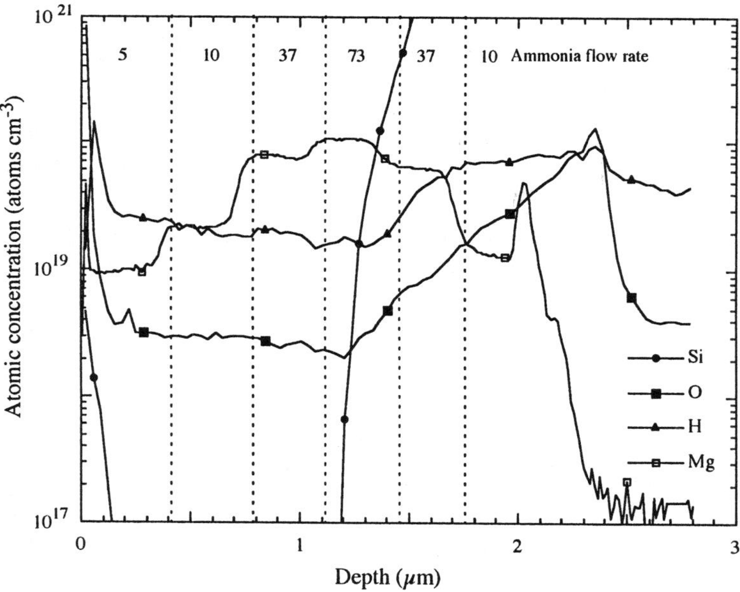

Depth (μm)

Fig. 5.16. The SIMS profile for Si, O, H and Mg for the sample grown on 6H-SiC with the same conditions used for samples in Fig. 5.15. The steps in Mg concentration are clearly seen for each segment. After [5.26]

176

changing ammonia flow rate. If films are grown with a larger amount of active nitrogen species, the concentration of Ga vacancies in the film will be increased and, consequently, it will be more likely for Mg atoms to incorporate into those sites provided that there is sufficient Mg. In fact, "more H" also means "more active nitrogen species" in the nitrogen environment, which is consistent with the expected behavior of Mg.

One can surmise that the absence of steps in Fig.5.15a may be due to the Mg flux not being sufficient to saturate the available Ga vacancy sites; this is consistent with the growth conditions, particularly the 5 sccm ammonia flow rate. The same argument can be applied to the 73 sccm flow-rate portion in Fig.5.15b. However, this does not exclude the presence of Mg atoms at sites other than the Ga sites, and the loss of a reasonable crystalline quality due to a very high Mg concentration. It has not been determined whether Mg atoms still incorporate substitutionally at Ga sites when the concentration is relatively high, while a first-principles calculation predicts that the Ga substitutional site is energetically most stable without considering the effect of concentration [5.48, 56].

Based on the above observations, one can assume that better p-GaN can be grown within a certain limit of Mg concentration for a given amount of active nitrogen species (in this case, ammonia flow rate). It still has to be verified if the upper limit within which better-quality p-GaN can be grown is proportional to the amount of active nitrogen species. Actually, p-GaN films, with a much enhanced quality over the previous ones, were grown with a 73-sccm ammonia flow rate using RMBE (this is about two times higher than the flow rate used for the samples in [5.50]). Moreover, the injector temperature for those samples was 600°C, possibly leading to less active nitrogen species on the growing surface when compared to the lower-injector-temperature case for a given ammonia flow rate. The best sample has a doping level, mobility and resistivity of about $8 \cdot 10^{17}$ cm^{-3}, 26 cm^2/V·s and 0.3 Ω·cm, respectively. The thermal activation energy obtained from the slope of the Arrhenius plot of the hole concentration is about 170 meV. As outlined in Chap.7, the activation energy so determined is really dependent on the level of the acceptor concentration through $E_A = E_{A0} - \beta_A N_A^{1/3}$. This can not be independently substantiated from the hole mobility and its temperature dependence since the p-type mobility has not been calculated reliably and the sample quality is generally very poor.

There are reports of multiple acceptor levels observed in the Deep Level Transient Spectroscopy (DLTS) investigations of *Götz* et al [5.57]. Astonishingly, the Mg concentration in this film is about $8 \cdot 10^{18}$ cm^{-3} which means about 10% activation ratio, and remains unexplained at this point. The enhanced quality is not surprising when this sample is compared to the two samples shown in [Ref.5.50, Fig.5.3]. The Mg concentration in the latter sample (this sample was grown with a 37 sccm ammonia flow rate and a high injector temperature which gave more than two times less active ammonia at the growing surface) was about $5 \cdot 10^{18}$ cm^{-3} and the doping level was $3 \cdot 10^7$ cm^{-3} with a mobility of 3.7 cm^2/V·s. Even after assuming the same Mg concentration, it is not difficult to infer that higher ammonia flow rates (or active nitrogen species) can produce better p-GaN with other conditions being the same. The same experiment was repeated for Si-doped and undoped GaN films to see the effect of the ammonia flow rate on the incorporation of Si and O. No particular steps in the Si and O profiles were seen. The Si concentration was only about $2 \cdot 10^{17}$ cm^{-3} for the Si-doped sample and this level of Si concentration might be low enough to produce the same effect seen in Fig. 5.15a.

Attempts to obtain higher p-doping, above 10^{18} cm^{-3}, by increasing the Mg flux have rarely been successful. As for RMBE and under nitrogen-rich conditions in which the effect of donor-like defects is minimal, to some extent the hole concentration and hole mobility increases. Figure 5.17 displays the conductivity vs. the ammonia mole fraction. Also shown is any effect annealing may have on the conductivity. Overlooking the scatter in the data, there appears to be no noticeable effect of Rapid Thermal Annealing (RTA) and steady-state furnace annealing in a nitrogen atmosphere. In parallel though, increased amounts of ammonia beyond a certain point may also be responsible for paving the way for complexes and defects which limit the hole concentration. In addition, formation of complexes may also limit the availability of Mg to be usable as a dopant.

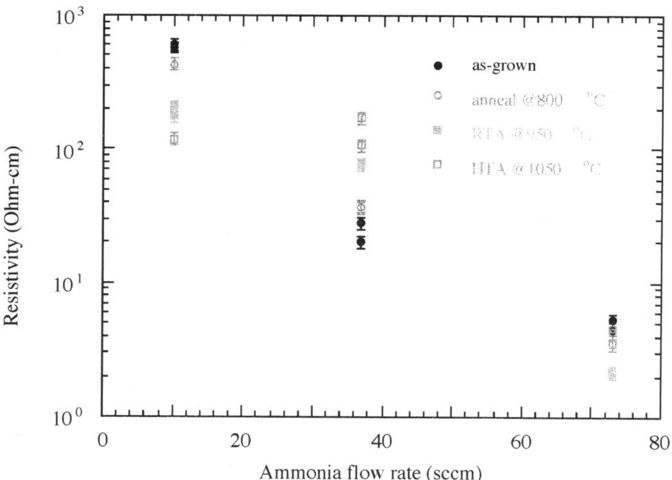

Fig. 5.17. Conductivity of Mg-doped GaN layers grown with RMBE vs. ammonia flow rate. Any effect of the rapid thermal and steady-state thermal annealing is shown

178

5.4.3 Optical Manifestation of Group-II Dopant-Induced Defects in GaN

As is the case of n-type, whether intentionally doped or unintentionally doped, the optical transitions help to elucidate the behavior of p-type dopants. Wide-bandgap semiconductors are notorious for developing defect states when doped with p-type impurities, which generally impedes the achievement of p-type conduction. The case of a well behaved Mg-doped GaN layer is represented in Figs. 10.17 and 18. In this section, optical transitions associated with defects induced while attempting to incorporate group-II elements, namely Mg and Be in GaN will be considered.

The PL spectra of a few representative Mg-doped GaN samples S3, S4 and S5 grown by RMBE are illustrated in Fig. 5.18. The curve associated with sample S3 (growth conditions: Ga cell temperature: 975°C, Mg cell temperature: 290°C, ammonia flow rate: 35 sccm/s, and substrate temperature: 800°C) shows that the Mg-related acceptor states A_M lie about 150 meV above the valence-band maximum, E_{VMAX}. The datum point corresponding to the peak P3 at 380 nm represents the transition from the conduction-band edge to possibly the acceptor level A_M, although it is much shallower than the well-behaved acceptor states in GaN. It is plausible that the p-character of Mg-doped GaN is aided significantly by an increased ammonia flow rate, and shifting the growth from what is most likely to be the gallium-rich condition toward the nitrogen-rich condition. This would

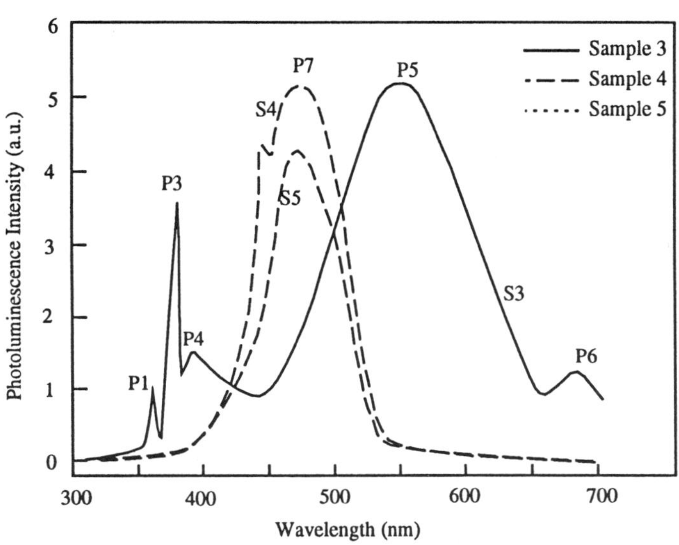

Fig. 5.18. The PL spectra of a few representative Mg doped GaN samples S3, S4, and S5 grown by RMBE where S4 and S5 have been annealed at 1250 and 750°C, respectively

lead to a reduction of both V_N and what little Ga_N is present in p-GaN; it would also ameliorate Mg incorporation on Ga sites. It should be mentioned that GaN is not expected to occur in appreciable quantities as its formation energy is large. However, the picture may change somewhat in defective material. The origin of the transition P5 at 560 nm (sample 3 of Fig.5.18) is more complicated and controversial to assign. As mentioned above in conjunction with the n-type samples, it is probably related to a Ga-vacancy nearest-donor complex [5.33].

The suppositions here are intended to give the reader a cursory view of the defect states involved. Detailed investigations, such as the local coordination, are imperative before deterministic assignments as to the origins of these peaks can be made.

The PL spectrum of p-GaN (curve for sample 3 in Fig.5.18) also exhibits the peaks P6 and P4 at 1.85 and 3.18 eV, respectively. The 3.18 eV transition is possible only if an acceptor-type state lies at about 240 meV above E_{VMAX} and is consistent with a level observed in Mg-doped samples, wherein the transition is attributed to transitions from the conduction band or the donor state to the acceptor level. The origin of the peak P6 is not yet known as it does not always occur in all samples. It is, however, fair to say that it is related to defects formed while attempting to render the semiconductor p-type, which was unsuccessful. Self-compensation of acceptor dopant atoms by spontaneously generated compensating defects is one of the inherent obstacles in obtaining p-type GaN. The self-compensation stems from the natural tendency of the semiconductor to minimize its free energy so that the Fermi level E_F of the semiconductor is as close to the intrinsic Fermi level E_i as possible.

The PL spectrum of the Mg-doped sample S4 annealed at 1250°C is significantly different from that of S5, which was annealed at 750°C (Fig. 5.18). The spectrum associated with S4 lacks the transition peaks P1, P3, P4, P5 and P6, indicating that the sample S4 is somewhat free from many of the native defects of sample 3 discernible by PL. It shows, however, a peak P7 at about 480 nm. Attempts to increase the p-doping beyond $4 \cdot 10^{17}$ cm^{-3} by subjecting the sample to RTA were not successful. The PL spectra of annealed and unannealed samples (curves 4, 5 and 6 of Fig.5.18) indicate that the peak of the PL spectrum at 480 nm remains unchanged even at increasing RTA temperatures, although the PL intensity increases with increasing RTA temperature. This is consistent with the observation of *Zolper* et al. [5.58]. This 480-nm peak is most likely due to transitions from the conduction-band states to deep acceptor-like states caused by Mg impurities. It should be pointed out that much effort, albeit without satisfactory progress, has been expended to correlate the intensity of the peak that commonly appears at about 420 nm at room temperature in Mg-doped GaN with the acceptor concentration. Not all the Mg-doped samples exhibit the peaks

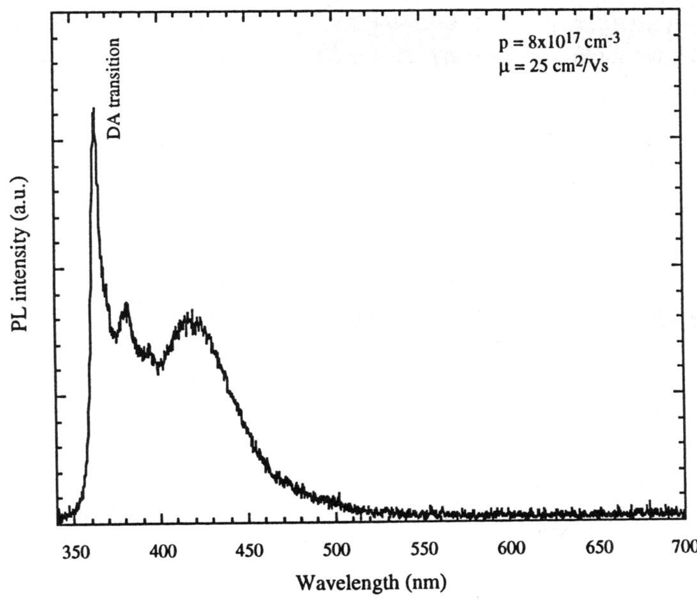

Fig. 5.19. Room-temperature PL spectra for the p-GaN sample which shows a doping level of $8 \cdot 10^{17}$ cm^{-3} with a mobility of about 15 cm^2/V·s Hall mobility. After [5.26]

occurring at 420 nm or 480 nm, as depicted in Fig.5.19, where the 380 nm peak is attributed to the Mg acceptors.

The PL spectra of some representative GaN samples doped with Mg or Be, and yet showing n-type characteristics by Hall measurements, are depicted in Fig.5.20. As may be noted from this figure, all of these samples indicate a strong 362 nm transition. Additional transitions at 380 nm point to the presence of acceptor-like states AM (AB for Be acceptors) near the

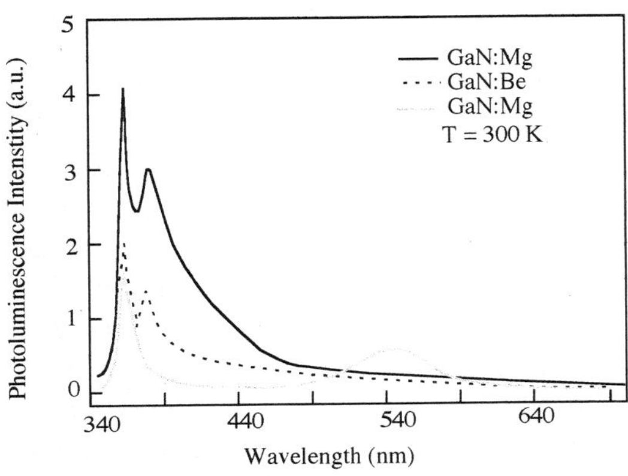

Fig. 5.20. PL spectra of some representative GaN samples doped with Mg or Be and yet showing n-type characteristics by Hall measurements

valence-band maximum. However, the samples have n-type characteristics probably because any acceptor-like states present are more than compensated by donor-like states such as V_N.

a) Doping with Beryllium

Earlier work on Be doping by means of CVD was unsuccessful and led to the incorporation of Be as deep centers [5.59, 60]. Recent predictions based on ab-initio calculations [5.61] point out that Be on Ga sites would form a shallow acceptor with a thermal ionization energy of 60 meV, and a double donor, if incorporated, as interstitial. The error in this method of calculation could be ± 100 meV. Until the experimental investigation of *Salvador* et al [5.62], it was believed that Be doping leads to compensated material and yields high-resistivity GaN. Optical measurements in early samples, dating back to the 1970s, yielded a broad emission at an energy of about 2.2 eV. It should be noted that this is at about the same energy as the notorious yellow peak found in GaN with native defects (Chap. 10). In a recent experiment with RMBE-grown films [5.62], the 380 nm emission peak is the dominant one for low Be-doping levels. In samples with higher doping, $420 \div 430$ nm emission is observed and is attributed to non-effective mass-like acceptor/defect states. Moreover, hot-probe measurements indicate p-type conductivity with p-n junctions attained. The Hall measurements have not yet indicated p-type material because of the low hole concentration and disparity between the higher electron mobility and the lower hole mobility (Sect. 7.2). Astonishingly though, Be-O co-doping in cubic GaN grown on GaN has been reported to result in a high Be activation in excess of 10^{20} cm^{-3} with hole mobilities in the 100 cm^2/V·s range [5.63]. One can conclude that extreme care must be exercised in determining the acceptor binding energies by photoluminescence measurements.

b) Doping with Mercury

Several groups have investigated Hg doping of GaN, but none reported on electrical measurements [5.40]. *Pankove* and *Hutchby* [5.64] performed optical measurements and observed 2.43 eV emission in Hg-doped samples. This is lower than the 2.9 eV emission reported by *Ejder* and *Grimmeiss* [5.65]. On the basis of measured emission, *Ejder* and *Grimmeiss* estimated that the Hg-acceptor level lies about 410 meV above the valence band. We should note that dopant incorporation in nitrides is a rather complex process. Many metals promote defect creation within the gap, Such defects are efficient optical centers and can easily be mistaken for acceptors due to metal incorporation. To this end, available data appear to suggest that unsuccessful attempts end up generating defects. Successful p-type doping, as confirmed by Hall measurements, can be achieved when the unintentional doping level is low, for example, at or below 10^{17} cm^{-3}.

c) Doping with Carbon

Carbon in both GaN and AlN is amphoteric in nature [5.66]. Its salient features of carbon in both materials are similar: Carbon on a Ga site, C_{cation}, is predicted to be an effective mass donor while carbon on a N site, C_{anion}, is an effective mass acceptor. The ionization energies of both donor and acceptor have been predicted to be equal to 0.2 in GaN and 0.4 eV in AlN. Incorporation of C on a nitrogen site is preferable since the C_N-formation energy is lower under both Ga- and-N rich conditions of growth. Carbon on a cation site, C_{cation}, can also assume a metastable DX-like configuration $C^*_{Ga\ or\ Al}$ which requires a broken bond between C and one neighbor. In this configuration, both the host N and carbon atoms are significantly displaced. $C^*_{Ga\ or\ Al}$ introduces a singlet at approximately 0.4 eV and 0.3 eV above the valence bands of GaN and AlN, respectively, and a singlet occupied by one electron at about 0.3 and 1.0 eV below the conduction bands of GaN and AlN, respectively. Investigation of the formation of the C^+_{cation}–C^-_{anion} nearest ion pairs brought to light that the binding energy is substantial and lower than the sum of the formation energies for C^+_{cation} and C^-_{anion}. Therefore, the tendency toward self-compensation is expected to be large.

There has been one report on hole concentrations up to levels of about $3 \cdot 10^{17}$ cm^{-3} with mobilities of about 10^2 cm^2/V·s in CCl$_4$-doped GaN grown by MOMBE on GaAs substrates. Unless the partial cubic nature of the resultant films is responsible for this seemingly high mobility, the results are otherwise considered controversial as the mobility reported is characteristic of n-type mobilities [5.67]. Annealing at 800°C did not increase the hole concentration which was construed that the hydrogen passivation of acceptors is not significant. On the other hand, highly resistive CH$_x$-doped GaN films resulted from plasma-assisted MOCVD. SIMS analyses indicated large amounts of C and H ($\approx 10^{19}$ to 10^{20} cm^{-3}) in the samples. After annealing under a nitrogen atmosphere, the films remained highly resistive and suggest that the C-H complex is thermally stable [5.68]. Since carbon is present as a contaminant in most growth reactors (graphite parts [5.69]) and in metalorganic sources in MOCVD processes, its unintentional role in the compensation and doping of GaN layers might nevertheless be significant. It should be pointed out that any carbon contamination is most likely contributed by sources other than solid sources such as graphite parts, as the vapor pressure of graphite at growth temperatures is very low. Of course, this picture is only applicable when the reaction pathways with those solid sources are rendered inconsequential.

d) Doping with Zinc

Attempts to achieve p-type conductivity by Zn doping of MOCVD grown films were unsuccessful. Highly resistive films with resistivities up to 10^9 $\Omega \cdot$cm were grown [5.70, 71]. It has well been established [5.70] that among others, Zn causes a dominant level that is about 0.5 eV deep both in GaN and InGaN with a low InN mole fraction. In fact, this deep level was exploited in the earlier versions of InGaN LEDs marketed by Nichia Chemical Ltd. in an effort to keep the InN mole fraction in the emission layer small while achieving 450 nm emission that is required by the display society. The nature of Zn centers has received some theoretical attention [5.72]. As many as four different Zn centers with energies in the forbidden gap, corresponding to emission between 1.8 and 2.9 eV, have been observed experimentally [5.73] and predicted theoretically [5.72]. While the 2.9 eV peak is attributed to substitutional Zn on Ga sites, the other three peaks have been assigned by experimentalists to various charged states of Zn on N sites, namely Zn_{N^-}, $Zn_{N^{2-}}$, $Zn_{N^{3-}}$ with Zn atoms binding to as many as three electrons having binding energies of 0.65, 1.02 and 1.43 eV. Since Ga and Zn are both somewhat comparable in size, it is not too difficult to imagine Zn in Ga sites. However, Zn in N sites require large formation energies and are unlikely to form. The thermal activation energy of Zn in Ga sites has been estimated to be 0.33 eV [5.73].

e) Doping with Calcium

Suggestions have been made that Ca may form a shallow acceptor level in the GaN bandgap [5.74]. Ca p-type doping of GaN was achieved by ion implantation of Ca^+ ions or by a co-implantation of Ca^+ and P^+ followed by a rapid thermal annealing at temperatures $\geq 1100\,^\circ$C [5.75]. The ionization energy of 169 meV was found by temperature-dependence measurements of the hole concentration in a sheet. Ca acceptors, like those of Mg, can be passivated by atomic hydrogen at low temperatures ($250\,^\circ$C); and they can be reactivated by thermal annealing at $\leq 500\,^\circ$C for 1 hour.

f) Doping with Rare Earths

Strong emission sources in the range 1.3 to 1.54 μm are in some demand for optical communications on the basis of silica fibers. For these applications, wide-bandgap semiconductors present an advantage in terms of room-temperature stability compared with systems based on narrow-bandgap semiconductors. To this end, GaN and AlN layers have been doped with Er. AlN doped with Er presents a strong PL at 1.54 μm. Incorporation of levels in the range of $3 \cdot 10^{17} \div 2 \cdot 10^{21}$ were achieved in MOMBE systems with a solid Er source [5.76].

5.4.4 Ion Implantation and Diffusion

Ion implantation and diffusion are industrial doping methods for Si. The reports on III-N doping by ion implantation are somewhat contradictory so far. As compared to classical semiconductors, group-III N semiconductors have larger binding energies of the constituent atoms and smaller interatomic distances. Consequently, substitutional impurity incorporation by ion implantation is intrinsically more difficult. On the other hand, the temperatures needed for annealing of the implanted lattice defects are limited to the decomposition temperatures of group-III N compounds, which sometimes are not high enough to anneal out the damage created.

Similarly, it should come as no surprise that experimental reports on diffusion in group-III N semiconductors are limited. Diffusion of Mg at 800°C for 80 h into an unintentionally doped n-type GaN layer in a sealed nitrogen ampoule resulted in p-type material [5.77]. A hole concentration of $2 \cdot 10^{16}$ cm^{-3} and a mobility of 12 cm^2/V·s at room temperature were measured. Lower diffusion temperatures were not successful in converting the sample to p-type. On the other hand, higher diffusion temperatures caused the film to decompose and evaporate. High-energy implants ($80 \div 100$ keV) increased strain and defects in GaN [5.78]. The damage caused by such high-energy implants could not be annealed out after 30 min at 800°C; no conversion to p-type was detected. However, films implanted with Mg ions at lower energies ($40 \div 60$ keV) recovered after annealing at 800°C and retained their original lattice parameters, as determined by X-ray diffraction. Hot-probe tests showed conversion to p-type.

n- and p-type conductivities were accomplished by ion implantation of Si$^+$ (200 keV), Mg$^+$ (180 keV) and Mg$^+$ (180 keV) + P$^+$ (250 keV) [5.78]. For these experiments, undoped layers with background electron concentrations of $1 \div 4 \cdot 10^{16}$ cm^{-3} were employed as well as a post-implant thermal anneal procedure in the range $700 \div 1100$°C for 10 seconds. To prevent or otherwise limit nitrogen loss, another GaN film was firmly pressed against the surface of the implanted sample during annealing. Mg$^+$ implantation alone did not produce any doping effect. Mg$^+$/P$^+$ co-implantation led to a conversion from n- to p-type conductivity after annealing at $1050 \div 1100$°C with an activation percentage of about 62%. The effect of co-implantation here is to increase the vacancy concentration. Si$^+$ implantation resulted in a sharp increase in the n-type conductivity after annealing at $1050 \div 1100$°C with an activation percentage of $\approx 93\%$.

Results somewhat contradictory to the above reports on GaN were obtained by *Wilson* et al [5.79] who observed that no diffusion of implanted Mg and other impurities (Be, C, Zn, Si, Se, and Ge) was discernible even after an anneal at 800°C for 10 min. Only S showed a marked diffusion at temperatures higher than 600°C. Impinging energies of the ions were as follows: H: 40 keV, Li: 100 keV, Be: 100 keV, C: 260 keV, F: 100 keV, Na: 100 keV, Mg: 100 keV, Si: 150 keV, S: 200 keV, Zn: 300 keV, Ge: 500 keV, Se: 500 keV [5.80]. The lack of diffusion in the implanted samples could be attributed to the fact that the energetic ions are capable of producing defects which, in turn, can trap implanted impurities causing them to cluster. The annealing temperatures may not be high enough to anneal out such defects.

Spurred by the predictions that Ca might be a shallow acceptor [5.74], implantation of Ca has been attempted. Implanting Ca^+ at 180 keV and a co-implantation of Ca^+ with P^+ at 130 keV followed by rapid thermal annealing at $T \geq 1100\,°C$ produced p-type doping with an ionization energy of 169 meV for a Ca acceptor level, which is similar to that of Mg [5.81].

In some device-fabrication schemes, selective-area conversion of the sample to high resistivity is very desirable. This serves to eliminate the unwanted current paths in devices such as field-effect transistors. To this end, proton implantation has been undertaken and high-resistance samples have been obtained by implanting H^+ or He^+ [5.82], and N^+ or F^+ [5.83, 84]. As endemic to any aspect of nitride-semiconductor research and development, many more investigations are needed for improved results and reproducibility. We should, however, be prepared for disappointment if ion implantation does not turn out to be universally relevant in nitride development due to a large bond strength and the high N-vapor pressure in this material. On the one hand, one would like to have a strong bond strength for stability, and on the other hand, one would like to have a semiconductor with small bond strength to allow the device engineer to modify its characteristics.

5.5 Defect Analysis by Deep-Level Transient Spectroscopy

To gain some insight into the electrical character of point defects, a popular method that is traditionally applied is the Deep-Level Transient Spectroscopy (DLTS) [5.85, 86]. In this method, a p-n junction or a Schottky barrier structure is biased into depletion which causes the defects above the Fermi level (for n-type samples) to release their electrons. A charging pulse, which reduces the reverse bias across the junction, is applied to fill some of the previously emptied traps increasing the junction capacitance. Upon the removal of the charging pulse, the electrons trapped at the deep level(s) would be freed at a rate which depends on the energy level of the trap. This is also related to the rate at which the capacitance would return to its equilibrium value. In DLTS, the change in capacitance in an appropriate time window is plotted as a function of sample temperature from which the activation energy of the trap is determined. Alternatively, the time rate of change in

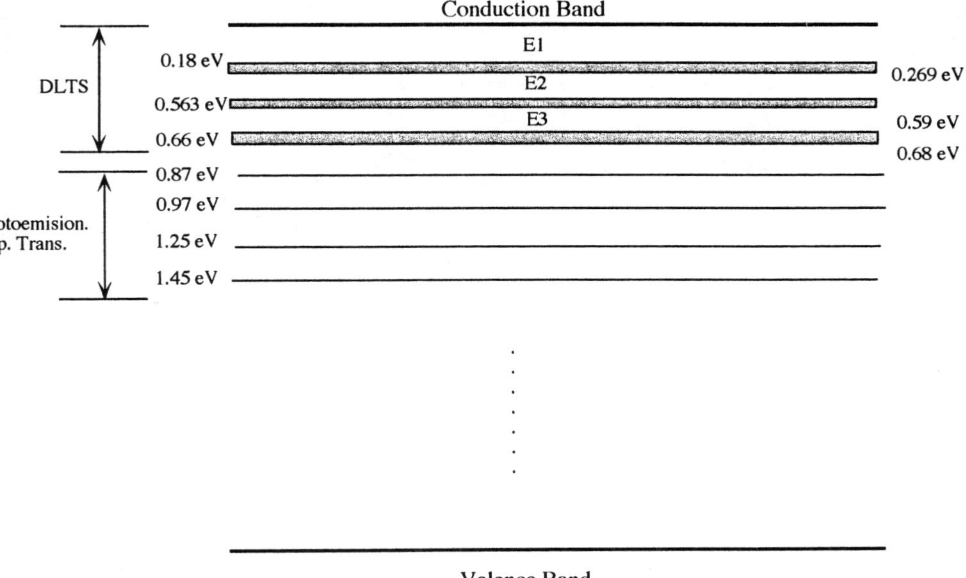

Fig. 5.21. Various deep levels within the gap energy of GaN, as observed by DLTS and photo emission transient spectroscopy.

the capacitance as the traps are emptied can be plotted versus time from which pertinent information about the nature of the centers can be deduced. This technique is called the Isothermal Capacitance Transient Spectroscopy (ICTS) [5.86] which has also been employed for analyzing nitrides.

In wide-bandgap semiconductors, the very deep states may not efficiently release their electrons, particularly at low temperatures rendering the traditional DLTS impractical. To circumvent this problem, optical excitation in conjunction with capacitance transients can be employed. Similar to thermal excitation of electrons, when the electrons are optically excited into the conduction band in a reverse biased junction, the charge in the depletion region increases, following a resonant optical excitation that causes the capacitance to increase [5.87]. Each time the photon energy resonates with a deep level, there is a marked increase in the capacitance. However, the defect state energies measured by this technique, photoemission spectroscopy, can be different from those deduced by DLTS as thermal processes are not involved in the former. Nevertheless, very useful information can be gleaned since DLTS can not probe very deep states.

GaN samples prepared by MOCVD on sapphire substrates have been investigated by DLTS by several groups [5.88-90]. Defect states termed as E_1, E_2, and E_3 have been observed by various investigators with E1 and E2 appearing in ICTS [5.88]. Energies of the aforementioned defects have

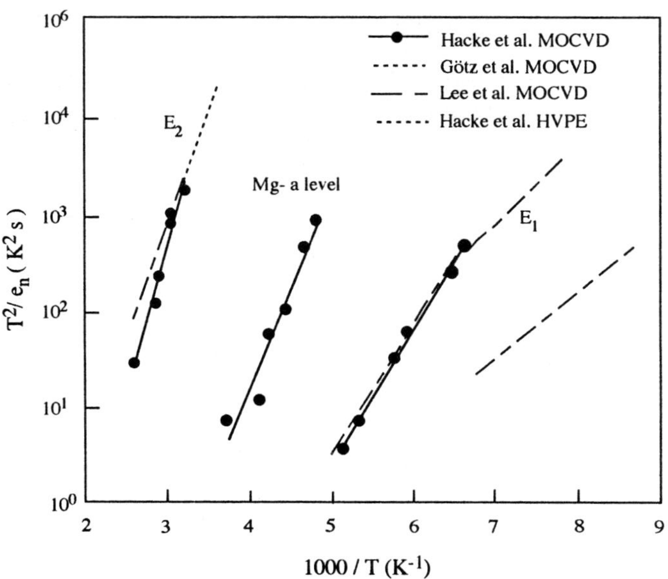

Fig. 5.22. Deep levels observed in GaN by DLTS. After [5.88]

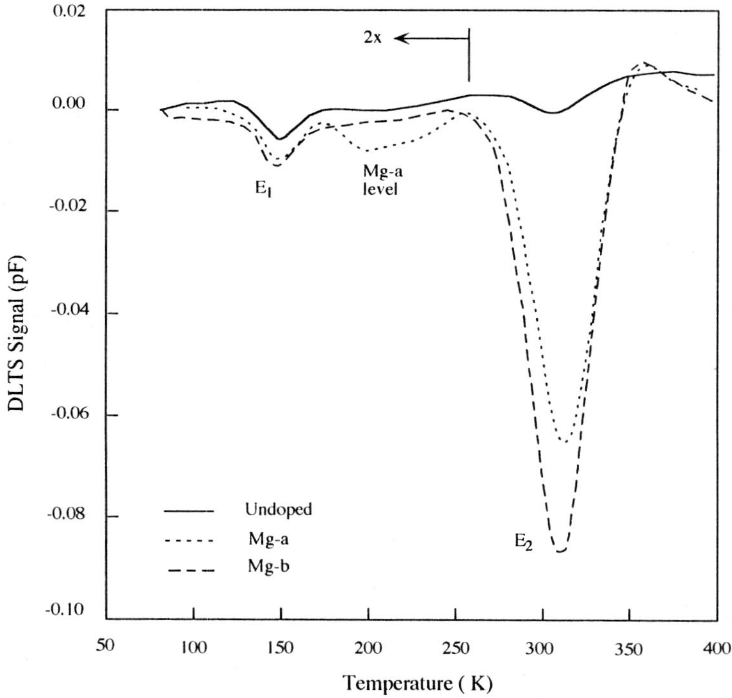

Fig. 5.23. DLTS spectra of undoped and Mg doped GaN layers. After [5.88]

188

been determined to be in the range of, in order, $0.264 \div 0.269$ eV, $0.563 \div 0.59$ eV, and $0.662 \div 0.686$ eV. The concentrations of the same defects, again in the order listed, are in the ranges of $2.6 \cdot 10^{15} \div 6.6 \cdot 10^{16}$, $3 \cdot 10^{13} \div 1.3 \cdot 10^{14}$, and $1.6 \cdot 10^{14} \div 3.8 \cdot 10^{14}$ cm^{-3}. When GaN layers were grown on sputtered/thick ZnO, the concentrations of the E_1, and E_3 defects are reduced to $1.4 \cdot 10^{15}$ and $5.1 \cdot 10^{13}$ cm^{-3}, respectively. The observed trap levels by DLTS and photo-emission transient capacitance methods with the associated spread in their measured energies are schematically shown in Fig.5.21. Deep levels found in GaN by DLTS and by various groups are shown in Fig.5.22 as compiled by *Hacke* et al [5.88]. The shallow level to the right of the E_1 level appears in films grown by the TMG source used. A notorious problem with wide bandgap nitrides is that attempts to accomplish p-type doping often results in creation of structural and electrically compensating defects. Minimization of these defects is imperative and determines whether p-type doping can be successfully attained. Shown in Fig.5.23 are the DLTS spectra obtained in films with and without Mg doping. As can be clearly seen, the E_2 line increases remarkably with addition of Mg. Similar effects are seen for the other defects, albeit not to the same extent. The photo emission transient capacitance technique resulted in the observation

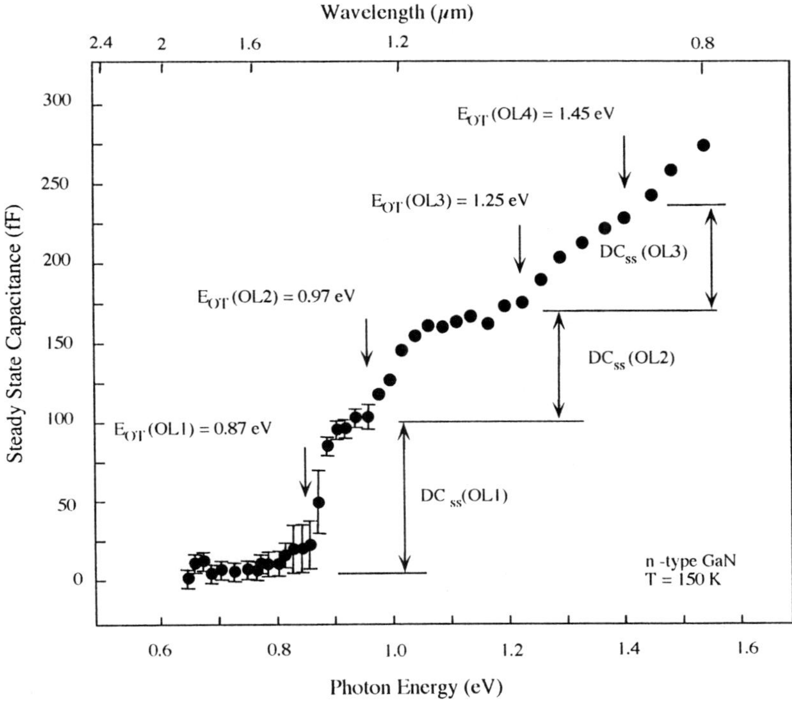

Fig. 5.24. Capacitance vs. optical excitation energy in GaN Schottky diodes. After [5.87]

of much deeper defects as shown in Fig.5.24. Caution must be exercised as optical and thermal activation energies of defects vary substantially.

Deep centers in Si-doped n-GaN layers grown by reactive molecular beam epitaxy have also been studied recently by Deep-Level Transient Spectroscopy (DLTS) along with their dependence on growth conditions. Five deep-level electronic defects were observed with the activation energies $E_1 = 0.234 \pm 0.006$ eV, $E_2 = 0.578 \pm 0.006$ eV, $E_3 = 0.657 \pm 0.031$, $E_4 = 0.961 \pm 0.026$, and $E_5 = 0.240 \pm 0.012$ eV. Among these, the levels labeled E_1, E_2, and E_3 are interpreted as corresponding to deep levels previously reported in n-GaN grown by both hydride vapor-phase epitaxy and metal-organic chemical vapor deposition [5.91]. Si-doped GaN samples grown on Si-doped n^+ GaN contact layers at 800°C show a dominant trap C_1 with an activation energy of $E_T = 0.44$ eV and a capture cross-section of $\sigma_T = 1.3 \cdot 10^{-15}$ cm^{-2}, while samples grown at 750°C on undoped semi-insulating GaN buffer layers exhibit the prominent traps D_1 and E_1 with $E_T = 0.20$ eV and $\sigma_T = 8.4 \cdot 10^{-7}$ cm^2, and $E_T = 0.21$ eV and $\sigma_T = 1.6 \cdot 10^{-14}$ cm^2, respectively. The trap E_1 is believed to be related to a N-vacancy defect since the Arrhenius signature of E_1 is very similar to the previously reported Trap E which has been produced by 1-MeV electron irradiation in GaN grown by both metal-organic chemical vapor deposition and hydide vapor-phase eputaxy [5.92].

5.6 Summary

A simple conclusion that can be drawn is that defects and p-type doping are forever intertwined, and the nature of active p-type impurity incorporation is not sufficiently understood. Site selection and defect creation to minimize the free energy will require more effort than that expended so far. The question of whether acceptors in GaN are effective mass like impurities still lingers despite the predictions that the thermal activation energy for Be is only 60 meV and that for Zn is a large 330 meV [5.93]. Experiments are lacking in that Hall measurements were not possible in p-type Be doped GaN and that Zn doped GaN often has high resistivity. The predicted thermal activation energy for Mg is 230 meV [5.93] which is in the range of experimentally reported figures. Acceptor-like impurity incorporation in nitride semiconductors is one of the most pivotal and outstanding issues, and more reports are certain to follow.

6. Metal Contacts to GaN

It is imperative that a semiconductor device be connected to the outside world with no adverse change to its current-voltage characteristics and no additional voltage drop. This can be accomplished only through low-resistance ohmic contacts on the semiconductor. An ideal contact is one where, when combined with the semiconductor, there are no barriers to the carrier flow in either the positive or negative directions. Ideally, this occurs when the semiconductor and the metal work functions are about the same, and there are no appreciable interface states which tend to pin the Fermi level. Since one can not just dial up ideal work functions for the semiconductor-metal system under consideration, particularly, when the work function of the semiconductor varies with doping, it is usually not possible to find just the right combination. In fact, for large-bandgap semiconductors such as GaN, a metal with a large-enough work function to form an ohmic contact to p-type GaN does not exist.

As will be clear from the analysis to follow, exacerbating the situation is the large effective mass of the carriers, particularly holes in wide-bandgap semiconductors. Consequently, other options must be explored. Traditionally, these solutions center around increasing the surface doping level and affecting the semiconductor surface through chemical interaction with the metal in a way to render it conducive for current conduction without rectification. Ideally, a metal that is either a donor or an acceptor for n- and p-type contacts, respectively, would be very much in demand. A case in point is Al on Si which is also a p-type dopant. Another case is the AuGe on GaAs where Ge is also an n-type dopant. Additional requirements for ohmic contacts are that contacts must be stable both thermally and chemically. The need for stability can not be overstated in devices intended for high-power and high-temperature operation either by design or necessity when the junction temperatures could be very high. For example, in LEDs the power loss at the contacts reduces the wall-plug efficiency and increases the junction temperature. This potentially degrades the operating lifetime. In lasers which require high current levels to operate, particularly in the early stages of development, ohmic contacts may make the difference between a successful attempt and a failed one.

Not being able to count on barrier-free contacts for p-type GaN, an understanding of current conduction in metal-semiconductor systems germane to wide-bandgap semiconductors is warranted. Ironically, in the event that the contact to p-type GaN is not ohmic, a forward biased p-n junction would inevitably transform the metal to a p-GaN contact to a reverse biased Schottky barrier, further exacerbating an already difficult problem. In the absence of defects and high surface doping, only those carriers that have sufficient kinetic energy to surmount the barrier, which is described by thermionic emission and field emission, would contribute to the current flow and power dissipation. In the presence of defects, and in the event of high surface doping which results in thin depletion layers, defect-assisted tunneling, field-aided tunneling, and direct tunneling must also be considered.

6.1 A Primer for Semiconductor-Metal Contacts

When a metal and a semiconductor with no surface states are brought in contact and equilibrium is maintained, their Fermi levels will align. If the Fermi levels of the metal and semiconductor were the same before contact, then there would be no change in the band structure after contact. Since the Fermi level in the semiconductor and thus the work function is carrier-concentration dependent, matching the work functions is difficult even though attempting to do so is better than not.

Let us consider the case of an n-type semiconductor and a metal with a work function that is larger than that of the semiconductor. The requirement that the Fermi levels be aligned after contact, brought about by the charge motion from the higher-toward the lower-energy side, creates a depletion region in the semiconductor and a barrier at the interface. The barrier height ϕ_B is simply the difference between the metal work function ϕ_m and the electron affinity in the n-type semiconductor. The resulting electric field at the metal-semiconductor interface causes an image-force lowering $\Delta\phi$ of this barrier (Fig.6.1). In addition to the equilibrium case, Fig.6.1 also illustrates the picture for the forward- and reverse-bias cases. The former reduces the barrier for carriers from the semiconductor side to the metal, and the latter increases it. The barrier to electron flow from the metal to the semiconductor remains almost unchanged except through a change in $\Delta\phi$. This image-force lowering term increases with increasing electric field due to a reverse bias, and decreases with forward bias. In addition, the position of the peak in the barrier, x_m, moves closer to the metal-semiconductor interface with increasing electric field due to the reverse bias. The picture for a metal p-type semiconductor pair is similar to that for the n-type case (Fig.6.1) and is illustrated in Fig.6.2 for the equilibrium and

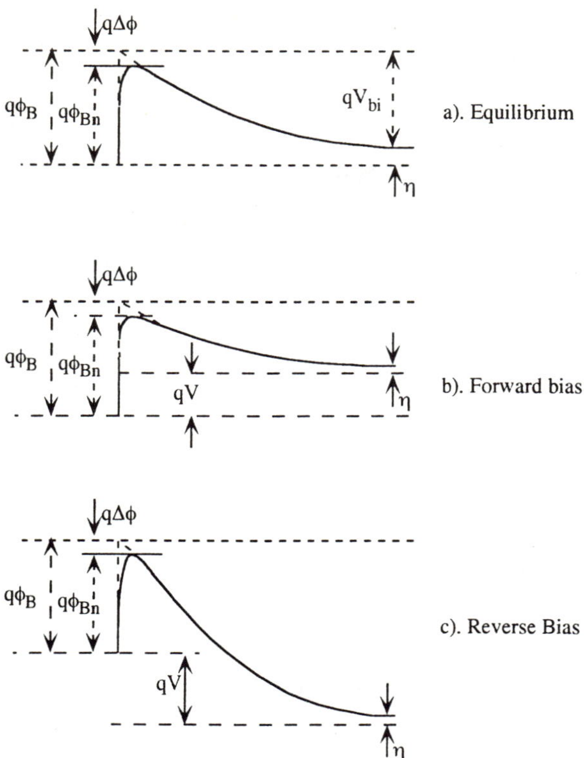

Fig. 6.1a–c. Ideal metal n-type semiconductor contacts under equilibrium, forward and reverse bias. Also shown is the image force lowering of the barrier

reverse-bias cases. The latter is chosen as it depicts precisely what takes place in a forward-biased p-n junction where the contact to the p-type semiconductor is not ohmic. Dependencies of the barrier lowering and the position of the potential maximum on the applied bias presented for the n-type case apply here, too. As indicated in the figure, the current conduction can be due to defect-assisted tunneling current, thermionic field emission and thermionic emission currents. Ideal direct tunneling is unlikely considering the effective mass of the holes, the low hole concentration and a barrier height which is most likely high. In fact, current conduction in GaN-based p-n junctions is generally ill understood and may involve band-tail states, a topic of discussion in Chaps. 9 and 11.

The case where the fortuitous matching of the metal-semiconductor pair occurs is depicted in Fig. 6.3 with an ohmic-contact behavior that results automatically. The condition here is that the metal work function is equal to or slightly smaller than that of the semiconductor. Unlike the case of Fig. 6.1, in which the metal work function is smaller than that of the semiconductor, charge accumulation rather than depletion, occurs with a

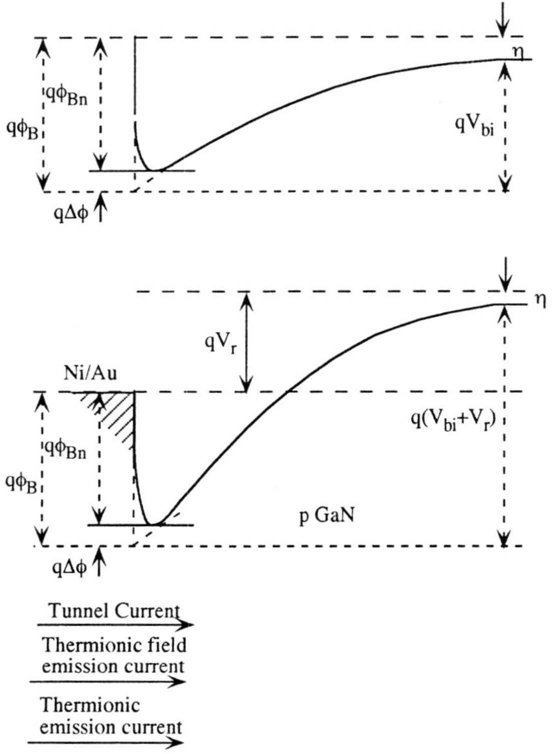

Fig. 6.2. Ideal metal p-type semiconductor contacts under equilibrium and reverse bias. Also shown in the image force lowering of the barrier and the various current conduction mechanisms that may come into play

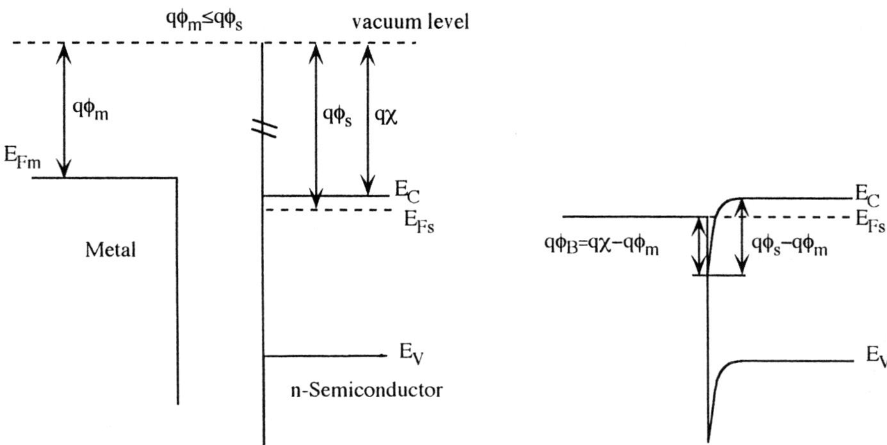

Fig. 6.3. Fortuitous matching of the metal- n- type semiconductor pair with automatic ohmic contact behavior. The results are brought about by assuming that the metal work function is equal to or smaller than that for the semiconductor (*left*: before contact, *right*: after contact)

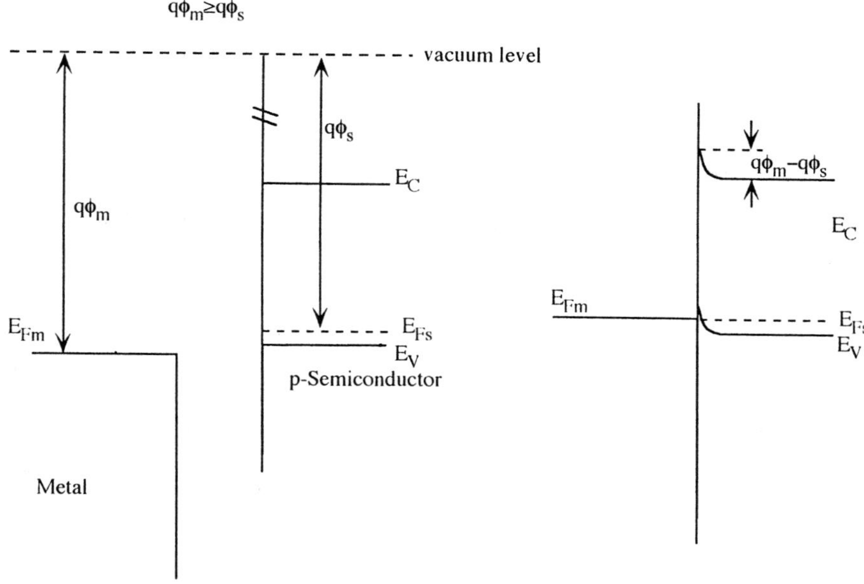

Fig. 6.4. Fortuitous matching of the metal p-type semiconductor pair with automatic ohmic contact behavior brought about by the assumed metal work function being equal to or greater than that for the semiconductor (*left*: before contact, *right*: after contact)

negligible voltage drop. The same is displayed in Fig.6.4 for a p-type semiconductor where the metal work function needs to be equal to or larger than that for the semiconductor. For p-type GaN, this would mean a metal with a work function of about 8 eV; this does not exist. Together with a large hole mass and the difficulty of obtaining high hole concentrations, this paints a very dismal picture regarding ohmic contacts to p-type GaN.

Derivations leading to the image-force lowering $\Delta\phi$, and the position of the maximum of the potential barrier with respect to the interface, x_m, can be found in many texts [6.1, 2], i.e.,

$$\Delta\phi = \sqrt{\frac{qE}{4\pi\epsilon_s}} \quad \text{and} \quad x_m = \sqrt{\frac{q}{16\pi\epsilon_s E}} \tag{6.1}$$

where E is the electric field at the interface, which happens to be its maximum. The other terms have their usual assignments.

6.2 Current Flow in Metal-Semiconductor Junctions

In cases when defects are not involved, there are three mechanisms [6.1, 2] that govern the current flow in a metal-semiconductor system.

(1) **Thermionic Emission** (TE) − for moderately doped semiconductors, $N_d < \approx 10^{17}$ cm^{-3}, the depletion region is relatively wide. It is nearly impossible to tunnel through the barrier unless aided by defects which are considered not to exit in this ideal picture. The electrons, however, can surmount the top of the barrier, which should be small for contacts, by thermionic emission (Fig. 6.5a). For low-doped or high-barrier semiconductors, on the other hand, the vast majority of electrons would be unable to cross in either direction into the semiconductor; and ohmic behavior is not observed.

(2) **Thermionic-Field Emission** (TFE) − for intermediately doped semiconductors, $\approx 10^{17} < N_d < \approx 10^{18}$ [cm^{-3}], the depletion region is not sufficiently thin to allow direct tunneling of carriers that are more or less in equilibrium. But, if the carriers gain a little energy, they may be able to tun-

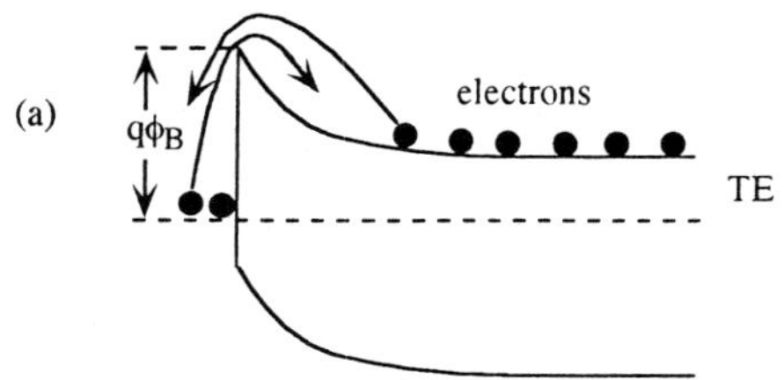

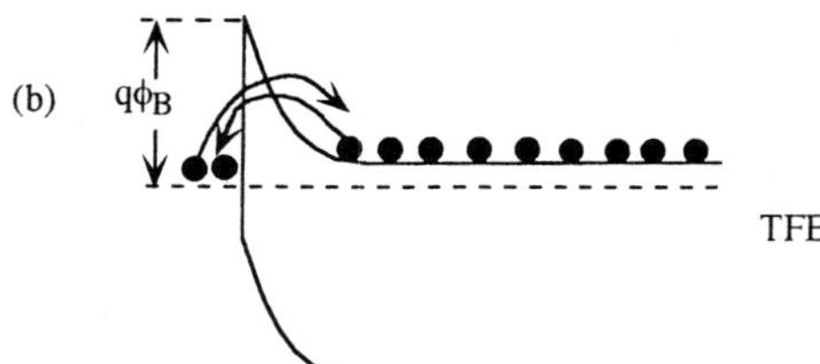

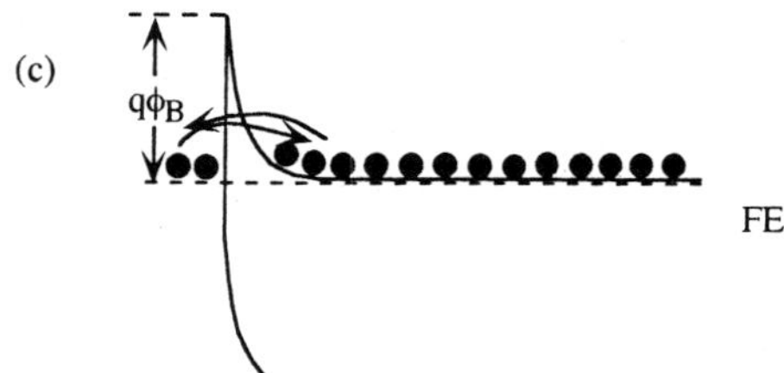

Fig. 6.5. Schematic description of (a) the thermionic emission, (b) thermionic field emission, and (c) tunneling mechanisms in an n-type semiconductor

nel. Consequently, both thermionic emission and tunneling take place (Fig. 6.5b).

(3) **Field Emission** (FE) – for heavily doped semiconductors, $N_d > \approx 10^{18}$ cm^{-3}, the depletion region is narrow, and direct electron tunneling from the metal to the semiconductor is allowed (Fig. 6.5c). In the absence of a good match between the metal and the semiconductor work functions, which is generally the case, this is the best approach to pursue for ohmic contacts.

6.2.1 The Regime Dominated by Thermionic Emission

The traditional current-voltage expression representing thermionic emission is given by

$$J_{te} = J_{te0}\left[\exp\left(\frac{V}{kT}\right) - 1\right]$$

(6.2)

with

$$J_{te0} = A^*T^2 \exp\left(\frac{-q(\phi_B - \Delta\phi)}{kT}\right).$$

(6.3)

where J_{te0} is the saturation value of the current density J_{te}, A^* is the **effective Richardson constant**, ϕ_B is the barrier height, and $\Delta\phi$ is the image-force **barrier lowering**. Equation (6.2) is based on the condition that the series resistance of the circuit is negligibly small.

As the kT term in the exponent indicates, the slope of J_{te0}/T^2 would vary with temperature with a slope of kT in a semi-logarithmic plot. The Richardson constant A is given by

$$A = \frac{4\pi q k^2 m_0}{h^3}$$

(6.4)

which equals to 120 [A$\cdot$cm^{-2}K^{-2}]. The effective Richardson constant is A^* = $A(m_e^*/m_0)$ for n-type and $A^* = A(m_{hh}^*/m_0)$ for p-tape semiconductors under the assumption of single-valley conduction bands such as n-type GaN and single and spherical valence-band conduction. When both heavy and light hole bands are occupied, the effective Richardson constant is given by $A[(m_{hh}^* + m_{lh}^*)/m_0]$. In cubic compound semiconductors, the valence band is degenerate and thus the last expression for the Richardson constant should be used.

Equation (6.2) is a representation of the carrier flux from the semiconductor to the metal, with the barrier being voltage dependent, $\phi_B - V$, and that from the metal to the semiconductor with the barrier fixed at ϕ_B. Since there exists parasitic resistance in the circuit such as semiconductor resistance, the thermionic-emission current expression is modified as

$$J_{te} = J_{teo} \left[\exp\left(\frac{q(V-IR_s)}{kT} \right) - 1 \right] . \tag{6.5}$$

Here, the current I is determined by the product of the current density J and the area of the structure.

6.2.2 Thermionic Field-Emission Regime

In Thermionic Field-Emission (TFE) the tunneling component of the current including the effect of the series resistance can be written as [6.3]

$$J_{tef} = J_{tef0} \left[\exp\left(\frac{q(V-IR_s)}{E_0} \right) - 1 \right] \tag{6.6}$$

where I_{tef0} is the saturation value of the current J_{tef} and expressed by

$$J_{tef0} = \frac{A^*T\,[\pi q E_{00}(\phi_B - V - \eta)]^{1/2}}{k\cosh(E_{00}/kT)} \exp\left[-\frac{q\eta}{kT} - \frac{q(\phi_B - \eta)}{E_0} \right] \tag{6.7}$$

and E_0 is a parameter dependent on the barrier transparency, η is the Fermi level with respect to the conduction-band edge in an n-type semiconductor and with respect to the valence-band edge in a p-type semiconductor. The image-charge lowering can be included by replacing ϕ_B with ϕ_{Bn} ($\phi_{Bn} = \phi_B - \Delta\phi$ with $\Delta\phi$ being the image-charge lowering term). As suggested by *Padovani* and *Stratton* [6.3]

$$E_0 = E_{00} \coth\left(\frac{E_{00}}{kT} \right) \tag{6.8}$$

with

$$E_{00} = \frac{qh}{2} \sqrt{\frac{N_D}{\epsilon_s m^*}} . \tag{6.9}$$

Here, it has been assumed that all the donors are ionized in the region of interest. The effective mass is that of the conduction band when an n-type

198

semiconductor is involved and that of the valence band when a p-type semiconductor is involved. If the doping level is moderately high as in the assumed case, the term η can be neglected.

6.2.3 Direct Tunneling Regime

At low temperatures and high doping concentrations, direct tunneling dominates the current. Referring to Fig.6.1, the density of current flowing from the semiconductor to the metal is proportional to the product of the transmission coefficient, the occupation probability in the semiconductor, f_s, and the unoccupation probability in the metal, $1-f_m$ [6.1][1]

$$J_{s \to m} = \frac{A^*T}{k} \int_0^{q\phi_{Bn}} f_s\, T(\xi)(1 - f_m)\, d\xi \tag{6.10}$$

where $T(\xi)$ is the transmission coefficient and is given by, for low temperatures and/or high doping levels, $T(\xi) = \exp(-q\phi_{Bn}/E_{00})$.

Similarly, the density of current flowing from the metal to the semiconductor is proportional to the product of the transmission coefficient, the unoccupation probability in the semiconductor and the occupation probability in the metal is

$$J_{m \to s} = -\frac{A^*T}{k} \int_0^{q\phi_{Bn}} f_m\, T(\xi)(1 - f_s)\, d\xi \ . \tag{6.11}$$

The total density of current which is simply the sum of the density of current flowing in both directions, can be approximated by

$$J_t = \exp\left(\frac{-q\phi_B}{E_{00}}\right) . \tag{6.12}$$

This expression clearly indicates that the lower the barrier and the higher the doping level, which increases E_{00} through (6.9), the higher the current. This explicitly implies that the resistance is low. The key is then to find a metal with a small barrier which calls for a metal with a work function equal

---

[1] The occupation probability depicts the likelyhood that a state is occupied by an electron, and one minus the occupation probability exhibits that to be free of electrons.

to or smaller than that of the semiconductor for the n-type case. For the p-type case, the same implies that the work function of the metal to be equal to or larger than that of the semiconductor, which is very hard to do for large-bandgap semiconductors. Additionally, the situation is exacerbated by the large hole mass which tends to decrease the E_{00} term, thus to increase the resistance. In fact, the low hole concentration combined with a large hole effective mass precludes direct tunneling in p-type GaN. What may be happening in the presently utilized Ni/Au contacts is that (to be discussed in Sect.6.6), Ni/Au chemically modifies the GaN surface and the current conduction mechanism may be dominated by defect-assisted tunneling which we will call **leakage current** (Sect. 6.2.4). This is nearly impossible to model as the nature of the defects germane to this problem is not known. With improvements in the quality of p-n junctions, one can extract the leakage component from temperature-dependent current measurements with the help of the current-conduction models dicussed in this subsection. This has been done for Schottky barriers in n-type GaN by *Suzue* et al. [6.4].

6.2.4 Leakage Current

In addition to thermionic emission, thermionic field emission and tunneling currents, other currents such as defect-assisted tunneling and others which may have an ohmic nature, can be lumped in the leakage component and can be expressed as

$$I_{lk} = \frac{V - IR_s}{R_t} .\qquad(6.13)$$

R_t is considered to be a fitting parameter which represents defects and in-homogeneities at the metal-semiconductor interface. In semiconductors with less than ideal interfaces, a tunneling barrier E_0 may not be predicted, in which case it should be considered as another fitting parameter. In practice, the terms I_{te0} and I_{tef0} are also considered to be fitting parameters that represent the magnitudes of the contributions to the current from thermionic emission and thermionic field emission, respectively.

6.2.5 The Case of a Forward-Biased p-n Junction

The derivative of the total current which is the sum of all the current components, discussed in Sect.6.2, with respect to voltage forms the inverse of the dynamic resistance that is caused by a metal-semiconductor contact. If the current-voltage characteristic is ohmic, this resistance would be voltage independent. An ideal ohmic contact is one that does not change the current-voltage characteristics appreciably with a resistance smaller than the semiconductor resistance including the spreading resistance. If the ohmic contact is somewhat rectifying as is most likely the case for p-type contacts in GaN unless the hole concentration is very high, this metal semiconductor junction would be under reverse bias when the underlying p-n junction is forward biased, further exacerbating the situation. In such a case, the current flowing through the metal-semiconductor contact is ideally determined by the sum of tunneling, thermionic-emission and leakage currents, and must be equal to the junction current governed generally by diffusion (Fig.6.6). The total voltage across the terminals of a p-n junction is the sum of the diode voltage, the voltage drop across the metal-semiconductor junction, and the voltage across the neutral regions, both p and n type, as well as the n-type ohmic contact. Since the n-type ohmic contact has a relatively low resistivity, and the voltage drop in the neutral regions of the semiconductor is linear with voltage, we will concentrate our attention to the p-type contact related issue. The current continuity dictates that the diode current

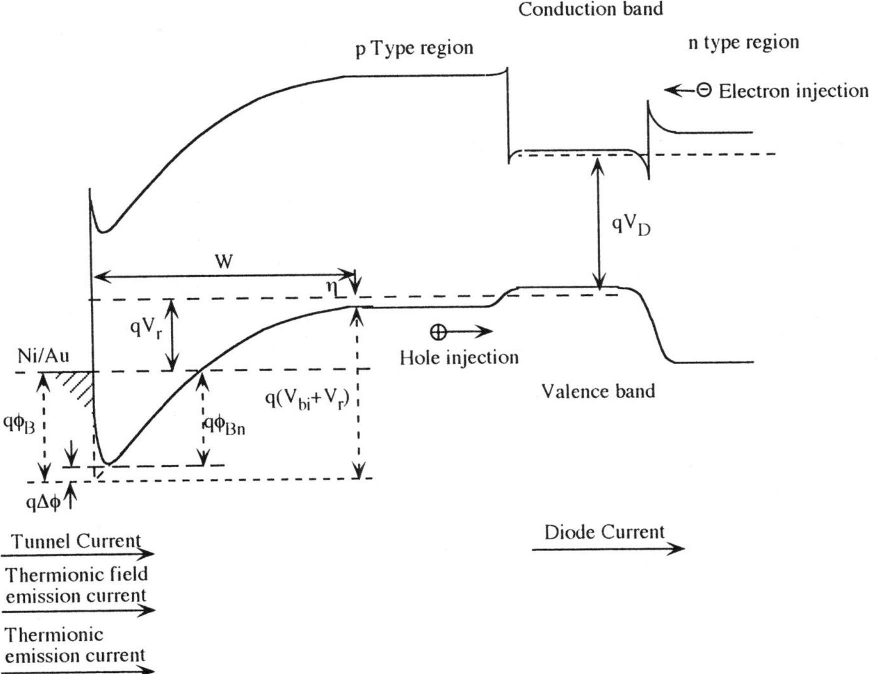

Fig. 6.6. Schematic representation of metal p-type semiconductor and an p-n heterojunction where the contact on the p-type semiconductor is non-ohmic. The structure is biased in such as way as to cause the p-n junction to be forward biased which causes the metal semiconductor junction to be reverse biased

be equal to the metal-semiconductor junction current. This condition determines the partitioning of the total voltage between the metal-semiconductor interface and the junction itself. The individual voltage components can be found if the current-conduction mechanisms are known for both the diode and the p-type contact regions. *Suemune* [6.5] carried out the procedure just outlined for ZnSe lasers assuming that the diode current is governed by diffusion current and the tunneling current is that through the reverse-biased triangular potential barrier formed at the metal-semiconductor interface. The holes that are injected from the metal by tunneling and thermionic emission, drift across the p-type layer before being injected across the p-n junction interface. A similar approach will be outlined here.

The current across the metal-semiconductor interface is the sum of the thermionic emission and thermionic field emission

$$J = J_{te0}\left[1 - \exp\left(\frac{-q(V_{bi} + V_r)}{kT}\right)\right]$$

$$+ J_{tef0}\left[1 - \exp\left(\frac{-q(V_{bi} + V_r)}{E_0}\right)\right] . \tag{6.14}$$

As stated earlier, in an ideal case this current must be equal to the diffusion current. Going through the exercise of setting the diode current and the total metal-semiconductor current equal, finding the voltage partitioning would allow the calculation of the current-voltage characteristic. *Suemune's* calculations for a ZnSe diode clearly show that the current is shifted to higher voltages than would be inferred from the diffusion current only [6.5]. Detailed analysis of this kind for a nitride-based junction must be deferred until after further improvements have been made in the material quality and in the understanding of the current flow in p-n junctions, which, for the sake of this discussion, is assumed to be governed by diffusion.

6.3 Resistance of an Ohmic Contact

An ohmic contact is a metal-semiconductor contact that has a very small contact resistance compared to the bulk or spreading resistance of the semiconductor. It is said that the contact is ohmic when the ratio of the potential drop V across the contact versus the current I flowing through the contact is linear with a constant R_c. Ideally, ohmic contacts do not contribute to the voltage drop across the device and do not alter the current-voltage relationship. Additionally, the contact must remain intact and robust regardless of the environment, and the contact characteristics must not change with storage and dynamic operations. Naturally, not all of these requirements can be met simultaneously but gallant strides should be made to satisfy as many as possible.

6.3.1 Specific Contact Resistivity

Although, the current-voltage (I-V) expression is sufficient, it is customary to deduce the specific resistance near zero bias. Caution should be exercised as the I-V characteristic may not be linear, thus causing a voltage-dependent resistance term. Nevertheless, the specific resistance creates an image of impediments to current flow. It is in this context that we define the specific contact resistivity, in terms of $\Omega \cdot cm^2$. The product of R_c and the area A of the contact is called the **specific contact resistance** ρ_c expressed as

$$\rho_c = \left[\frac{\partial J}{\partial V} \right]_{V=0}^{-1} \quad [\Omega \cdot cm^2] \ . \tag{6.15}$$

For $kT/E_{00} \gg 1$ (*moderate doping concentrations*), the Thermionic Emission (TE) mechanism dominates the current conduction and the specific contact resistance becomes

$$\rho_c = \frac{k}{qA^*T} \exp\left(\frac{q\phi_B}{kT}\right) \ . \tag{6.16}$$

It is clearly dependent on temperature, and at higher temperatures, there is more thermionic-emission current, which results in a smaller ρ_c.

For $kT/E_{00} \approx 1$ (*intermediate doping concentrations*), a mixture of thermionic, thermionic field emission and tunneling mechanisms is observed. The specific contact resistance can be obtained through a derivation of (6.6) with respect to voltage. Setting the voltage to zero and taking the Fermi energy at the conduction-band edge leads to

$$\rho_c = \left[\frac{\partial J_{tef}}{\partial V}\right]^{-1}_{V=0}$$

$$= \frac{k\cosh(E_{00}/kT)}{A^*Tq(\pi E_{00}q\phi_B)^{\frac{1}{2}}}[(1 - \tfrac{1}{2}\pi E_{00})(q\pi E_{00}\phi_B)]$$

$$\times\exp\left(\frac{q\phi_B}{E_{00}\cosh(E_{00}/kT)}\right) \qquad (6.17)$$

or simply

$$\rho_c \propto \exp\left(\frac{q\phi_B}{E_{00}\coth(E_{00}/kT)}\right). \qquad (6.18)$$

For $kT/E_{00} \ll 1$ which is associated with high doping levels, the tunneling current dominates and we have

$$\rho_c \propto \exp\left(\frac{q\phi_B}{E_{00}}\right). \qquad (6.19)$$

In this case ρ_c depends strongly on the doping concentration and the barrier. As the doping concentration is increased further, the depletion width of the Schottky junction decreases. This results in an increase of the tunneling transmission coefficient and a decrease of the resitance.

If a large number of surface states exists on the semiconductor surface, the barrier height is pinned at the semiconductor surface within its energy gap, and is independent of the metal work function. This is the Bardeen limit which contrasts the Schottky limit where the metal-semiconductor contact is assumed ideal and the surface states are ignored [6.6, 7]. In practice, the Fermi levels of most III-V compounds are pinned, and the resultant barriers must be considered. The barrier height depends on the bandgap and the surface-state density of the semiconductor.

6.3.2 Semiconductor Resistance

In addition to the metal-semiconductor resistance, the semiconductor resistance too must be added to the total resistance. The semiconductor resistance, mainly due to the neutral region, may be defined as [6.1]

$$R_s = \frac{1}{A_j} \int_{x_1}^{x_2} \rho(x)\,dx \tag{6.20}$$

where x_1 represents the depletion edge, x_2 denotes the boundary of the epi-taxial-layer, $\rho(x)$ is the resistivity at x, and A_j is the area of the metal-semiconductor junction. The parameter x_1 depends on the depletion width W, which is a function of temperature

$$W = \sqrt{\frac{2\epsilon_s}{qN_{Deff}}\left(V_{bi} - V - \frac{kT}{q}\right)} \tag{6.21}$$

where V_{bi} is the built-in potential given by

$$V_{bi} = \phi_B - \eta = \phi_B - \frac{kT}{q}\ln\left(\frac{N_c}{N_D}\right). \tag{6.22}$$

In GaN based p-n junctions such as those in LEDs and lasers, p-type contact resistance dominates because of a large metal-semiconductor barrier and a large effective mass. In addition, the semiconductor resistance is also large due to a combination of the low hole concentration and the low hole mobility. The non-ohmic behavior caused by a combination of a high metal to a p-semiconductor barrier and a low hole concentration, unless chemical interaction between the metal layers and the semiconductor causes the direct-tunneling-like current to dominate, will give rise to a voltage drop as well as an increased resistance exacerbating the Joule heating and the resultant rise in junction temperature. In addition, sapphire substrates which must be used are semi-insulating, necessitate the use of surface contacts for both n- and p-type regions. Doing so requires current conduction laterally from the n-type contact to the junction area. Due to the considerable distance involved between the metal contact and the junction area, the semiconductor resistance is considerable. It could be lowered if the n-type semiconductor is sufficiently thick. What is exasperating is that highly Si-doped n-type GaN cracks if its thickness is increased beyond about 3 μm due to the residual thermal strain.

6.4 Determination of the Contact Resistivity

The most widely used method for determining the *specific contact resistance* is the method of transfer length or the **Transmission Line Model (TLM)**, first introduced by *Shockley* [6.8], and later refined by others [6.9]. In this particular approach, a linear array of contacts is fabricated with various spacing between them. The pattern used and the resistance versus the gap spacing l (l_1, l_2, ...) are depicted in Fig. 6.7. The total resistance is given by

$$R_T = 2R_C + l\,\frac{R_{sh}}{W_c} \tag{6.23}$$

where the first term represents twice the contact resistance R_C, while the second term is due to the semiconductor. The term R_{sh} denotes the sheet resistance of the semiconductor layer. The contact resistance is related to the transfer length

$$R_C = \frac{R_{shc}L_t}{W_c} \tag{6.24}$$

where R_{shc} and L_t represent the sheet resistance in the contact region and the transfer length which is half the distance between the intercept with the horizontal axis and the origin in Fig. 6.7. Assuming that the sheet resistance under the contact is not modified, the transfer length can be determined and utilized to calculate the specific contact resistance through

$$\rho_c = R_{shc}L_t^2 \,. \tag{6.25}$$

The sheet resistance in the contact region is obviously not the same as that in the gap, which necessitates the more involved treatment found in [6.10]. The TLM method is not accurate for specific contact resistivities near and below $10^{-7}\ \Omega\cdot\text{cm}^2$, in which case the Kelvin probe measurements can be employed.

In the four-terminal (Kelvin) resistor method [6.11], the test pattern consists of four metal pads on an insulator. Two are connected to a semiconductor bar by means of large-area contacts. The other two touch the semiconductor at the contact opening. One pair passes current while the other measures the voltage drop. In contrast to the TLM case, the resistance of the line outside the contact area does not contribute to the contact resistance of this test structure; this allows a more precise measurement. For a

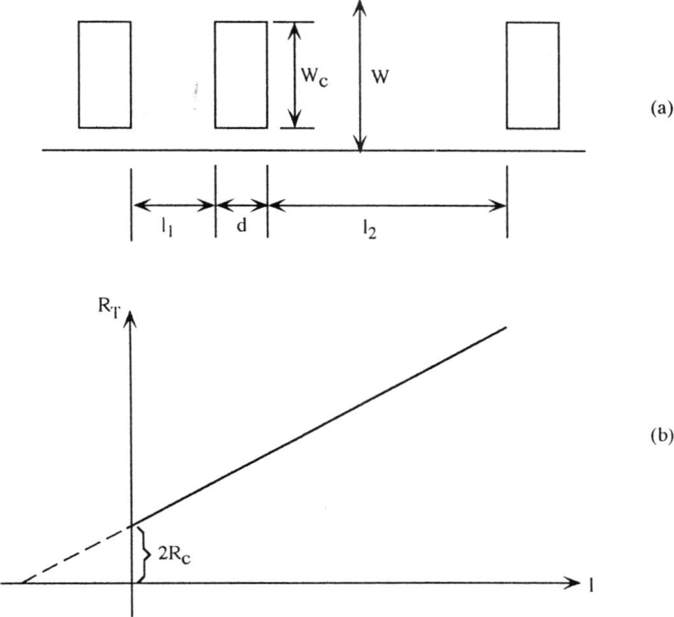

Fig. 6.7. A transmission line pattern (**a**) commonly used to deduce the specific contact resistivity and the variation of the resistance with respect to the gap distance (**b**)

very small contact resistance, this method is more accurate than the TLM method.

The curious reader is referred to *Scorzoni* and *Finetti* [6.12] and *Shen* et al. [6.13] who treated the contact resistivity and its measurement in detail. In addition to defining the relation between contact resistance and contact resistivity, the researchers thoroughly described different types of approaches for improving the measurement techniques. They also noted the mixed use of the terminology *contact resistivity, specific contact resistivity, specific contact resistance, specific interface resistance, specific resistance*, and *contact resistance* for the same term. Here, the term **specific contact resistivity** has been adopted.

6.5 Ohmic Contacts to GaN

Early studies suggested that, although an ohmic contact is possible for n-GaN using Al and Au metallization, the specific contact resistivity was relatively high, about $10^{-4} \div 10^{-3}$ $\Omega \cdot cm^2$ [6.14]. To form an ohmic contact, metals were deposited by evaporation and then patterned by photolithography and lift-off techniques. From current-voltage measurements it was determined that the *as-deposited* Al contacts were ohmic. However, the as-deposited Au contacts were rectifying and became ohmic after annealing at 575°C, paving the way to the argument that work functions of GaN and Al are very close to one another and that the surface of GaN is defect free. Using Ti/Al to contact n-type GaN, the specific contact resistivity was lowered to about $8 \cdot 10^{-6}$ $\Omega \cdot cm^2$ by annealing at 900°C for 30 seconds [6.15]. Au (and later Au/Ni) and Ti/Al were utilized as p- and n-type contacts, respectively, in LED structures [6.16, 17]. Although no contact resistance was reported, the LED operating voltages of 4V at 20 mA forward bias clearly demonstrates that the contact resistance was reasonably low.

6.5.1 Non-Alloyed Ohmic Contacts

Lin et al. [6.15] developed the novel scheme of non-alloyed ohmic contacts on GaN employing a 10 period InN/GaN (10 nm/10 nm) **Short-Period Superlattice** (SPS), as shown in Fig.6.8. The GaN and InN films were doped to about $5 \cdot 10^{18}$ cm^{-3}. The background electron concentration in InN films may be higher though. TLM measurements revealed a specific contact resistivity as low as $6 \cdot 10^{-5}$ $\Omega \cdot cm^2$. The current as a function of the applied gate voltage (Fig.6.9) revealed strong rectifying features with only an InN cap layer, but good ohmic characteristics with the short-period superlattice structures. A speculation suggested that a low Schottky-barrier contact on InN together with an increasing tunneling transmission coefficient may be responsible for the low-resistance contacts, as it was demonstrated for InGaAs [6.18]. The tunneling parameter E_{00} is about 0.37 eV for $N_D = 5 \cdot 10^{18}$ cm^{-3}. Because E_{00} is much larger than the thermal energy kT at room temperature, the possibility of tunneling as the dominant mechanism underlying electron transport through a GaN/InN superlattice ohmic contact is very real.

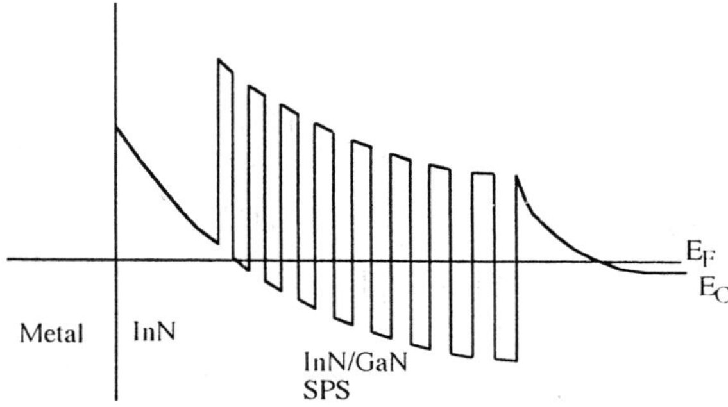

Fig. 6.8. Schematic band structure of a InN/GaN Short-Period Superlattice (SPS) employed to achieve contacts on n-type GaN

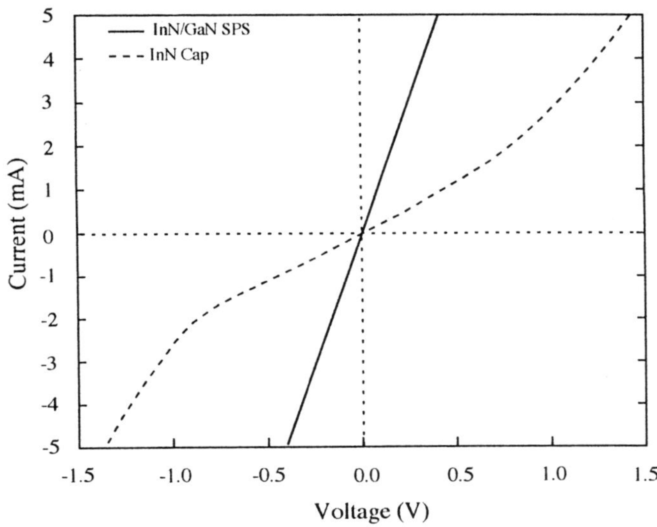

Fig. 6.9. Current-voltage characteristics of a metal-GaN system with an InN surface layer (*dashed line*) and of that with InN/GaN SPS surface layer (*solid line*)

6.5.2 Alloyed Ohmic Contacts

During the past several years several attempts have been made to obtain low-resistance ohmic contacts on GaN [6.14, 15, 19, 20]. Initial attempts of *Foresi* and *Moustakas* [6.14], wherein Au or Al was used after annealing at 575°C, led to contacts with a resistivity of $\rho_c \approx 10^{-3}\ \Omega\cdot\mathrm{cm}^2$. Contacts on n-GaN were later improved significantly by *Lin* et al. [6.15] who employed a Ti/Al bilayer deposited via conventional electron-beam evaporation onto a GaN substrate followed by thermally annealing at 900°C for 30 s in an N_2 ambient. This Ti/Al metallization yielded $\rho_c = 8\cdot10^{-6}\ \Omega\cdot\mathrm{cm}^2$, but suffered from Ga out-diffusion and the subsequent reaction with Al which rendered the surface metal to be discontinuous and have a high resistivity. To minimize the high-resistivity problem, *Wu* et al. [6.19] added a second set of Ti/Al stack, following the annealing step which requires realignment. The ohmic contact resistivity was lowered to $\rho_c \approx 3\cdot10^{-6}\ \Omega\cdot\mathrm{cm}^2$.

6.5.3 Multi-Layer Ohmic Contacts

Building on the earlier work of *Lin* et al. [6.15], a multi-layer ohmic contact to n-GaN based on Ti/Al metallization has been designed by *Fan* et al. [6.21]. The GaN films utilized were grown by MBE and doped with Si to a

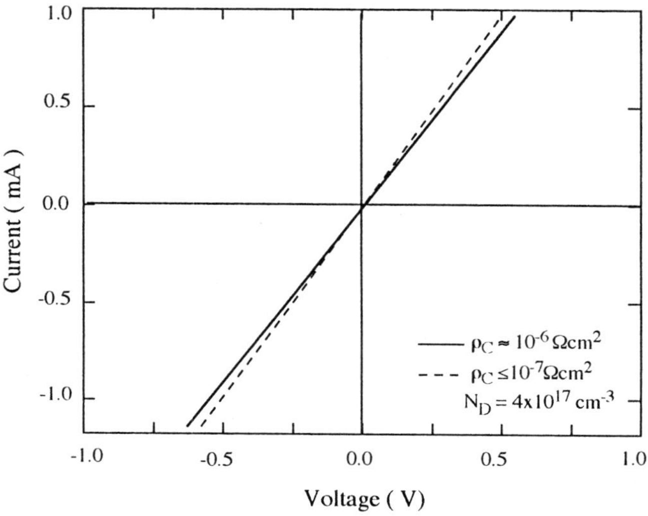

Fig. 6.10. Current-voltage characteristics for non-alloyed (*solid line*) and alloyed (*dashed line*) metal-n-GaN contacts. The GaN surface was subjected to a Reactive Ion Etching (RIE) which is believed to cause a damage-induced increase of the electron concentration on the surface

level of $(1 \div 5) \cdot 10^{17}$ cm^{-3}. The GaN surface was etched first with Cl$_2$ for 20 s, a BCl$_3$ etch for another 20 s, and thereafter a composite-metal layer of Ti/Al/Ni/Au (150 Å/2200 Å/400 Å/500 Å) was deposited. Two measurements of the specific resistivity were carried out, before and after RTA treatments. Measurements of the electrical current through the metal contacts to n-GaN as a function of voltage were performed before (*solid line*) and after (*broken line*) alloying (Fig. 6.10). These measurements indicate that the variation of the current with voltage is nearly linear even at sufficiently large current levels (100 mA) both in alloyed and non-alloyed ohmic contacts. Only annealing reduced the barrier height slightly after RTA, thus resulting in a small increase in current.

Contact resistances were derived from the plot of measured resistance vs. gap spacing by TLM. The total resistance R_T between the two contacts was determined at room temperature employing a four-point probe arrangement. [The contact resistivity ρ_c was derived from a plot of R_T vs. gap length in (6.23)]. The method of least squares was utilized to fit a straight line to the experimental data (Fig. 6.11). The specific contact resistivity depends both on the annealing and the doping concentration of the semiconductor sample. The contact resistivity ρ_c was about $(9 \div 12) \cdot 10^{-8}$ Ω cm^2 for doping levels between $4 \cdot 10^{17}$ and $2 \cdot 10^{17}$ cm^{-3}. For a doping level of 4×10^{17} cm^{-3}, alloying reduced the specific contact resistivity by about a factor of 40.

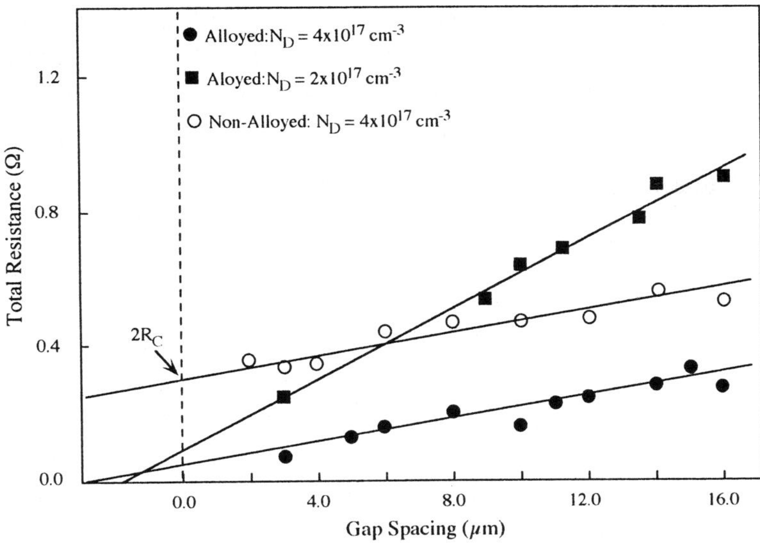

Fig. 6.11. Total resistance versus the gap spacing for alloyed and non-alloyed contacts on GaN shown in Fig. 6.10

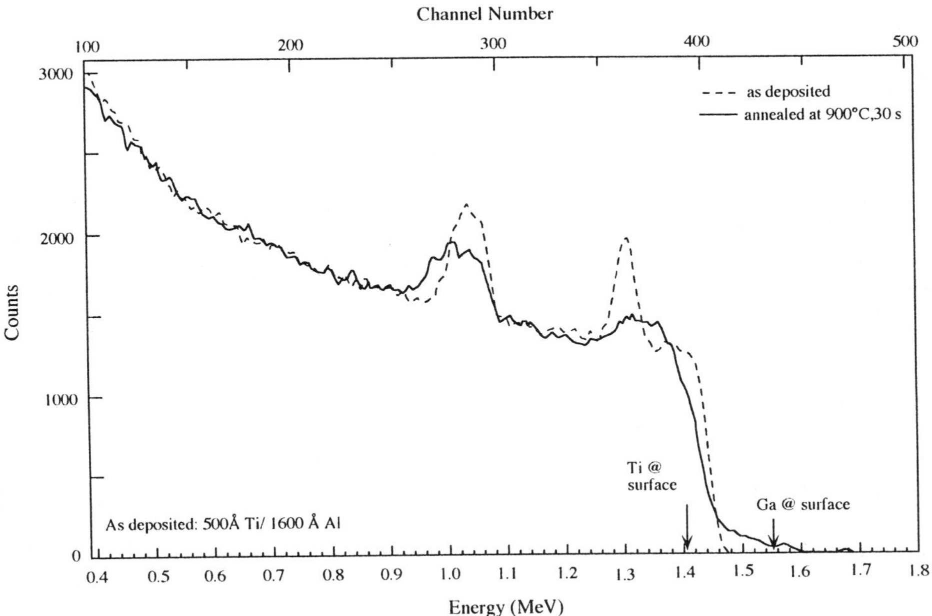

Fig. 6.12. Rutherford Back Scattering (RBS) profile of TiAl contacts on GaN before and after annealing. After [6.22]

The calculation of the contact resistivity was based on the assumption that the semiconductor sheet resistance underneath the contacts remains unchanged, see (6.25), which is not true for alloyed contacts. As mentioned previously, a metal with a low enough work function ϕ_m does not exist for good ohmic contacts. The inevitably large barriers diminish the possibility of thermionic-emission-governed ohmic contacts to GaN. The alternative mechanism is naturally some form of tunneling which may take place if GaN is so heavily doped to cause a very thin depletion region. Tunneling is possible if, due to annealing, for example, at 900 °C for 30 s, Al and Ti along with Ni undergo substantial interaction with each other and GaN. A cursory look would imply that Ti receives N from GaN forming a metallic layer while the lack of N in GaN provides the desired benefit of an increased electron concentration through N vacancy formation. Al acts to passivate the surface and also possibly reacts with Ti forming TiAl. This very possibility was investigated in contacts formed by TiAl, and TiAlNiAu before and after annealing by Rutherford Back Scattering (RBS) by *Duxstad* et al. [6.22]. Annealed and unannealed TiAl contacts on GaN, examined by RBS, indicate some Ti and GaN on the surface (Fig.6.12), which leads to the conclusion that the picture is more complicated. Similar experiments with the multilayered TiAlNiAu (500 Å/2000 Å/700 Å/600 Å) contacts revealed the presence of Ti, Al, and Ni on the surface after annealing (Fig.6.13).

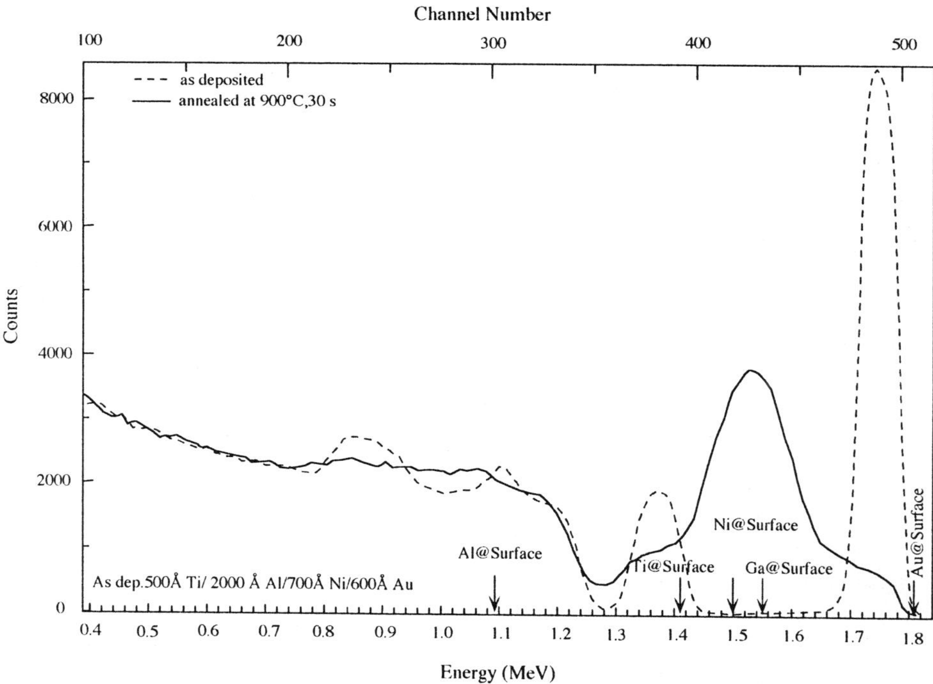

Fig. 6.13. Rutherford Back Scattering (RBS) profile of TiAlNiAu contacts on GaN before and after annealing. After [6.22]

212

6.6 Structural Analysis

To investigate the nature of ohmic-contact formation and the role of Ti, a more in-depth investigation by means of diffraction, high-resolution TEM and Energy Dispersive X-ray (EDX) spectra were undertaken at different metal sublayers to investigate the metal inter-diffusion after thermal annealing [6.23]. The formation of this cubic TiN layer results in a large excess of N vacancies in the GaN close to the interface. This is considered to be the reason for the low resistance of the contact. While no significant inter-diffusion was detected by EDX between Ti and Al in the as-deposited samples, metal inter-diffusion and alloying were observed after alloying. The Ti, Ni and Au metals diffuse into Al and formed alloys with a density higher than that of pure Al. For the Ti/Al contacts, a polycrystalline composite layer which consists of large $Al_{3+x}Ti$ (x < 1) and Al grains with the c-plane or {0001} almost parallel to that of GaN was observed. For the Ti/Al/Ni/Au composite, annealing leads to the formation of at least three polycrystalline layers. The top layer consists mostly of an Al-Ni alloy with some grains of an Al-Au alloy. The middle layer is formed mostly by an Al-Au alloy with some amount of Al-Ni grains. This is consistent with the high diffusivity of Au and shows a high interaction between Al and other species (Au, Ni, Ti). The bottom layer contains mostly Al-Ti and Ti-N alloys, but some amount of other species are also present in the layer.

A high-resolution electron microscopy image of the interfacial area near the GaN surface of Ti/Al/Ni/Au contacts after an RTA anneal is depicted in Fig.6.14. The interfacial cubic TiN which is clearly seen is purported to be responsible for the low ohmic-contact resistance. The interfacial region between the metal-composite layers and the GaN is very complex and also polycrystalline (Figs.6.15a, b). A thin TiN layer was detected by electron diffraction at the Ti-GaN interface even for as-deposited samples. The thickness of this TiN layer increases after annealing of the sam-

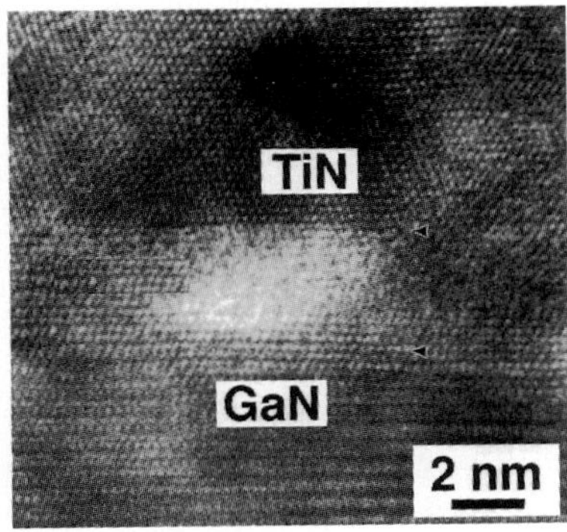

Fig. 6.14. High-resolution electron-microscopy image of the interfacial area near the GaN surface of the Ti/Al/Ni/Au contact in a cross-section after the RTA anneal. After [6.23]

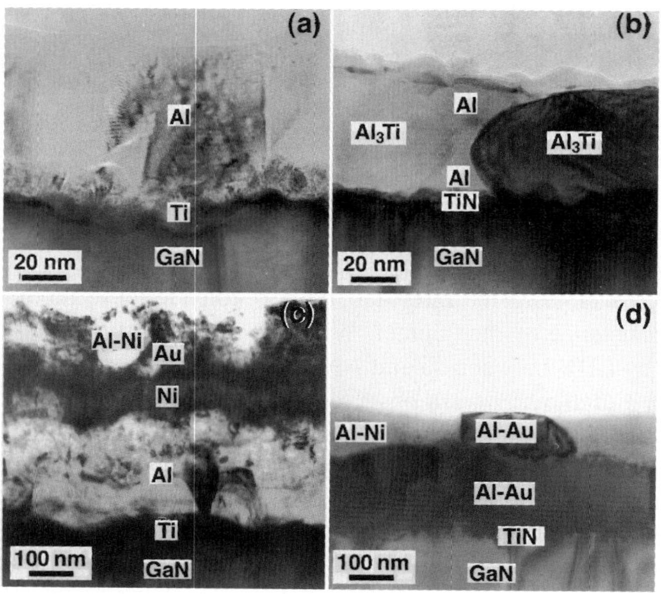

Fig. 6.15a-d. High-resolution electron-microscopy images of the interfacial area near the GaN surface for Ti/Al (**a, c**) and Ti/Al/Ni/Au (**b, d**) contacts in a cross-section before (**a, b**) and after (**c, d**) RTA anneal. After [6.23]

ples and is larger in the case of a Ti/Al/Ni/Au alloy (Fig.6.15b). High-resolution electron micrographs (Fig.6.15) illustrate the interface regions of Ti/Al (*a*) and Ti/Al/Ni/Au (*b*) samples after thermal annealing at 900°C for 30 s. Both HREM and the electron diffraction, not shown, establish the presence of textured polycrystalline TiN on the GaN surface. The orientation relationship between the cubic TiN layer and GaN was determined from the electron diffraction pattern as: {111}TiN//{0001}GaN, [110]TiN //[110]GaN, [112]TiN//[100]GaN. TiN grains are in twinned orientations (Figs.6.15c and d). The mismatch in lattice parameters of TiN and GaN is about 5.6÷6.0%. This leads to the formation of relatively small grains of about 5÷20 nm in lateral size, which are slightly misoriented to each other and with respect to the GaN. Besides Ti, EDX investigations indicate the presence of Al and Ga in the interfacial area that might suggest the formation of Al, Al-Ti-N or Al-Ti-Ga-N phases, in addition to TiN.

The reduction in the resistivity of Ti/Al and Ti/Al/Ni/Au contacts after annealing by two orders of magnitude might be ascribed to the specific structure of the composite layer formed after annealing. It prevents out-diffusion of Ti to the Al layer and, thus, increases the probability of an interaction between Ti and GaN. Indeed, the lower resistivity of the Ti/Al/Ni/Au contact is consistent with the formation of a larger amount of the TiN phase on the GaN surface. Au diffuses through the Ni layer and also forms an alloy with Al. This Au-Al layer might prevent further diffusion of

Ni toward the GaN surface. On the other hand, formation of the Au-Al alloy might prevent out-diffusion of Ti into the Al layer to form an Al-Ti alloy. The role of Reactive Ion Etching (RIE) in the reduction of the contact resistivity has several aspects among which is the roughening of the GaN surface due to preferential etching of dislocated material and, hence, increases the contact area. Moreover, RIE causes radiation damage near the GaN surface and, hence, increases the point-defect density in a thin GaN region near the surface. As a result, it promotes inter-diffusion of Ti into GaN even during the metal deposition.

6.7 Observations

While ohmic contacts to n-type GaN are satisfactory, the field is awaiting the much needed breakthroughs with regard to AlGaN, and particularly to p-type GaN and AlGaN. As the AlN molar fraction is increased, the n doping level that can be obtained is reduced and the metal-semiconductor barrier height is increased exacerbating the situation. As for p-type GaN and AlGaN, the large hole masses, large barriers with metals, and low acceptor concentrations call for a non-traditional approach which must certainly involve chemical modification of the surface layer. It should also be mentioned that the large bandgap discontinuities between the heterogeneous materials used in a heterojunction device increases the resistance to current flow; this also is a nonlinear process. The heterojunctions must therefore be graded to avoid barriers at interfaces. For both n-type and to be developed p-type contacts, reliability data are nascent. Initial data appear to indicate that the Ti/Al based contacts are reasonably robust at temperatures as high as 400°C. But, detailed data are needed before any lasting conclusions can be drawn.

7. Determination of Impurity and Carrier Concentrations

When impurities such as donors and acceptors are introduced into a semiconductor, they produce levels within the energy gap. The energy of a level with respect to the edge of the conduction band in the case of donors, and the valence band in the case of acceptors is called the **ionization energy**. The simplest calculation of an impurity energy level is based on the hydrogenic model.

7.1 Impurity Binding Energy

The ionization energy of a hydrogen atom is given by

$$E_H = \frac{m_0 q^4}{8\epsilon_0^2 h^2} = 13.6 \text{ eV} . \tag{7.1}$$

The ionization energy for a donor atom or an acceptor atom can simply be found by replacing the mass m_0 with the conduction-band conductivity effective mass and the valence-band conductivity effective mass, respectively, and substituting the free-space dielectric constant ϵ_0 with the dielectric constant of the semiconductor. Doing so leads to the convenient equations for the donor and acceptor binding energies as

$$E_d = \left(\frac{\epsilon_0}{\epsilon_s}\right)^2 \frac{m_n^*}{m_0} 13.6 \text{ eV} \quad \text{and} \quad E_a = \left(\frac{\epsilon_0}{\epsilon_s}\right)^2 \frac{m_p^*}{m_0} 13.6 \text{ eV} . \tag{7.2}$$

If we use a relative dielectric constant of 10.4 and an effective conduction band mass of 0.22, the hydrogenic donor binding energy in GaN would be 28 meV. On the other hand, if we adopt an effective hole mass of 0.8, the acceptor binding energy would be about 100 meV which is about one half of the commonly accepted value. The effect of screening on the donor binding energy, which is generally determined from the dependence of the electron concentration on temperature, is to lower it. The value so determined is the screened value and is related to that in a dilute semiconductor by E_D = $E_D^0 - \alpha N_D^{1/3}$, where E_D^0 is the donor binding energy for extremely small donor concentrations (dilute semiconductor), and N_D is the total donor concentration.

7.2 Conductivity Type: Hot Probe and Hall Measurements

It has been difficult to obtain p-type GaN, and when obtained, it is troublesome to analyze it because of the low hole concentrations to the point that an unequivocal determination of the polarity of the conduction is often not possible. For example, in many cases, the hot-probe conductivity measurements would indicate the sample to be p-type while the Hall measurement would suggest n-type conductivity. The mobility thus determined would, of course, be for electrons, but the polarity found by the hot-probe measurements is indicative of p-type material. This inevitably contributes to the controversy regarding the Hall mobility and the hole concentration reported in the literature. In reality, conclusions by both measurement schemes are correct and the discrepancy lies in the low hole concentration coupled with low hole mobility in those samples.

In a p-type GaN, all the donors are ionized. When a hot probe is applied, additional acceptors are ionized and generate excess holes under the hot tip. Excess holes set up a diffusion process with holes moving away from the tip. Since ionized acceptors are negatively charged, the semiconductor under the hot probe will be negatively charged while that under the cold one will be positive. Consequently, a positive electrostatic voltage with respect to the hot probe is formed and can be measured by a high-impedance electrometer. If, on the other hand, the sample is n-type, all the acceptors will be ionized and the hot probe will cause more donors to be ionized under it. The voltage at the hot probe with respect to the cold one is positive.

To determine the hole concentration and the Hall mobility, Hall measurements must be performed. Here a conducting bar is placed in a magnetic field which is normal to the epitaxial layer, and the Hall voltage develops normal to the current flow. In the plane of the epitaxial layer, it is used in conjunction with the conductivity measurements which can simultaneously be carried out to determine the polarity of the sample as well as the mobility and the net free-carrier concentration. The Lorenz force which causes the Hall voltage to develop, triggers electrons as well as holes to accumulate on the same side of the structure (Fig. 7.1). In a semiconductor with mixed conduction, meaning hole and electron concentration that are not drastically

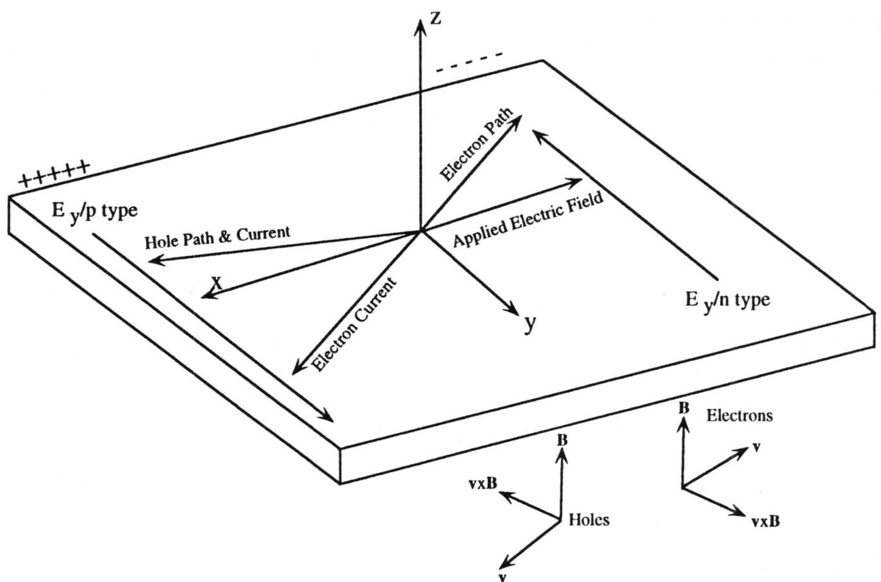

Fig. 7.1. Schematic representation of the Hall-measurement configuration

different from one another, the polarity of the Hall voltage may not correlate with the polarity of the semiconductor:

If the magnetic field is in the z direction, and the current flow is in the x direction, for the magnetic-field **B**, the electric-field **E**, and the velocity **v** we can write

$$\mathbf{B} = B\hat{\mathbf{z}} , \quad \mathbf{E} = E_x\hat{\mathbf{x}} + E_y\hat{\mathbf{y}} , \quad \mathbf{v} = v_x\hat{\mathbf{x}} + v_y\hat{\mathbf{y}} . \tag{7.3}$$

The Lorenz force affecting the carriers is

$$\mathbf{F} = q\mathbf{E} + q(\mathbf{v} \times \mathbf{B}) . \tag{7.4}$$

The nature of the geometry is such that the z component of the source vanishes.

The electron and hole current densities are simply the products of the force on them, the carrier concentration and the Hall mobility, recognizing that the force F on holes and electrons is the same, we have

$$\mathbf{J}_n = -n\mu_n\mathbf{F} \quad \text{and} \quad \mathbf{J}_p = p\mu_p\mathbf{F} \tag{7.5}$$

and the x component of the current

$$J_x = qE_x(n\mu_n + p\mu_p) . \tag{7.6}$$

Upon substitution for the Lorenz force, the y component of the current which is zero, as the open circuit conditions prevail, is

$$J_y = q\mu_n n(E_y - v_{xn} B) + q\mu_p (E_y - v_{xp} B) = 0 \tag{7.7}$$

where v_{xn} and v_{xp} are the velocities of the electrons and holes in the x direction, and E_y is the y-component of the elctric field.

Solving for the y component of the electric field, while utilizing the x component of the current, and multiplying it with the width of the sample in the y direction, we obtain for the Hall voltage V_y

$$V_y = \frac{I_x B}{t} \frac{\mu_p^2 p - \mu_n^2 n}{(\mu_n n + \mu_p p)^2} , \tag{7.8}$$

where I_x is the current in the x direction (the product of the current density and the vertical area), and t is the thickness of the epitaxial layer.

The Hall-voltage V_y must be larger than zero, or

$$\mu_p^2 p - \mu_n^2 n \geq 0 \quad \text{or} \quad p \geq n\mu_n^2 / \mu_p^2 , \tag{7.9}$$

in order for the Hall measurement to indicate that the polarity is p-type.

If we assume values of 300 and 10 for the electron and hole mobilities, respectively, the equilibrium hole concentration must be 900 times that of the equilibrium electron concentration before Hall measurements would give the correct polarity.

7.3 Density of States and Carrier Concentration

The electron concentration can be found from the integral of the product of the density of states and the Fermi-Dirac distribution function over the entire conduction band

$$n = \frac{(2^{1/2} m_n^*)^{3/2}}{\pi^2 \hbar^3} \int_{E_C}^{\infty} \frac{(E - E_C)^{1/2} dE}{1 + \exp[(E - E_F)/kT]} \tag{7.10}$$

where m_n^* is the conduction-band density-of-states effective mass. For a spherical energy surface, it is the same as the effective mass. Equation (7.10) reduces to

$$n = N_C 2\pi^{-1/2} F_{1/2}\left(-\frac{E_C - E_F}{kT}\right) \qquad (7.11)$$

where N_C is the total density of states in the conduction band and given by

$$N_C = 2\left(\frac{2\pi m_n^* kT}{h^2}\right)^{3/2}$$

$$= 2.5 \cdot 10^{19} \left(\frac{m_n^*}{m_0}\right)^{3/2} \left(\frac{T\,[K]}{300}\right)^{3/2} \text{cm}^{-3} \qquad (7.12)$$

and

$$F_{1/2}(x_0) = \int_0^\infty dx\, \frac{x^{1/2}}{1 + \exp(x - x_0)} \qquad (7.13)$$

which for $E_C - E_F \geq 3kT$ reduces to

$$F_{1/2}(x_0) = \pi^{1/2} e^{x_0/2} . \qquad (7.14)$$

Then,

$$n = N_C\, e^{-(E_C - E_F)/kT} . \qquad (7.15)$$

A similar treatment for holes leads to

$$p = N_V 2\pi^{-1/2} F_{1/2}\left(\frac{E_V - E_F}{kT}\right) . \qquad (7.16)$$

Again when the Fermi level is above the valence band by several kT values, the Fermi-Dirac distribution can be replaced by the Boltzmann distribution leading to

$$p = N_V\, e^{(E_V - E_F)/kT} \qquad (7.17)$$

where N_V is the total density of states in the valence band and is given by

$$N_V = 2\left(\frac{2\pi m_p^* kT}{h^2}\right)^{3/2} = 2.5 \cdot 10^{19} \left(\frac{m_p^*}{m_0}\right)^{3/2} \left(\frac{T\,[K]}{300}\right)^{3/2} \text{cm}^{-3} . \qquad (7.18)$$

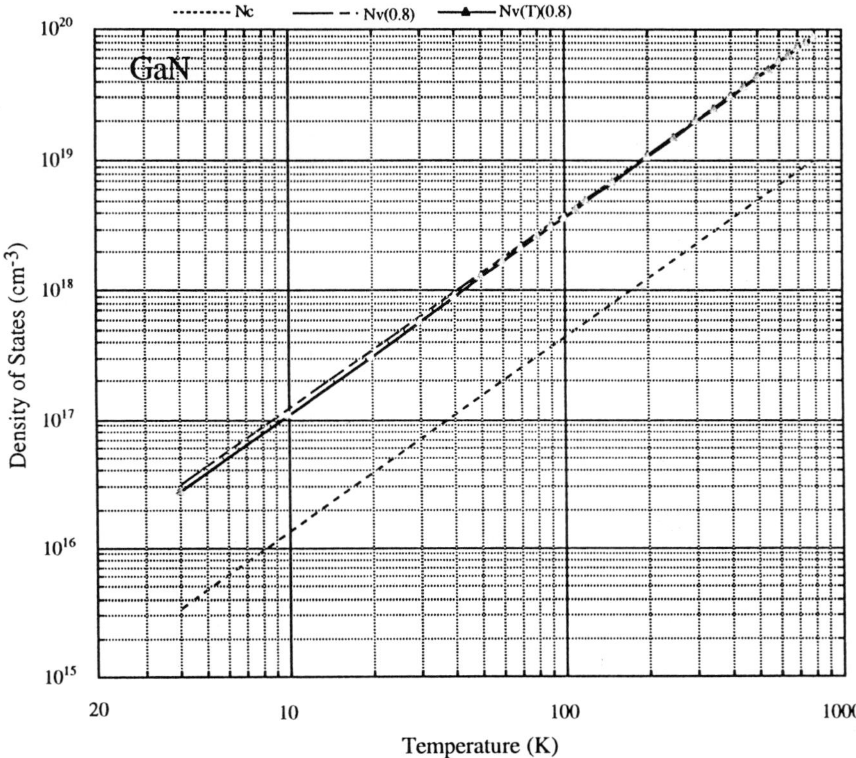

Fig. 7.2. Temperature dependence of the conduction- and valence-band densities of states assuming 0.22 and 0.8 for the electron and hole effective masses, respectively, and parabolic valence and conduction bands. In addition, the valence-band density-of-states curve with HH and LH occupations, separated by 6 meV, and an LH effective mass of 0.2 having a Boltzmann distribution is shown

The temperature dependence of the conduction-band and valence-band densities of states in GaN are depicted in Fig. 7.2 for electron and hole effective masses of 0.22 and 0.8, respectively. Here, the assumption has been made that the conduction and valence bands are parabolic, an assumption which is good for the conduction band, but not so good for the valence band. The valence band is, strictly speaking, not parabolic and there is coupling between the heavy hole (hh) and light hole (lh) states. Also exhibited in Fig. 7.2 is the valence-band density of states calculated with an energy separation of 6 meV and the Boltzmann distribution. Despite the somewhat inaccurate assumption, the data shown in Fig. 7.2 are very valuable for a quick analysis.

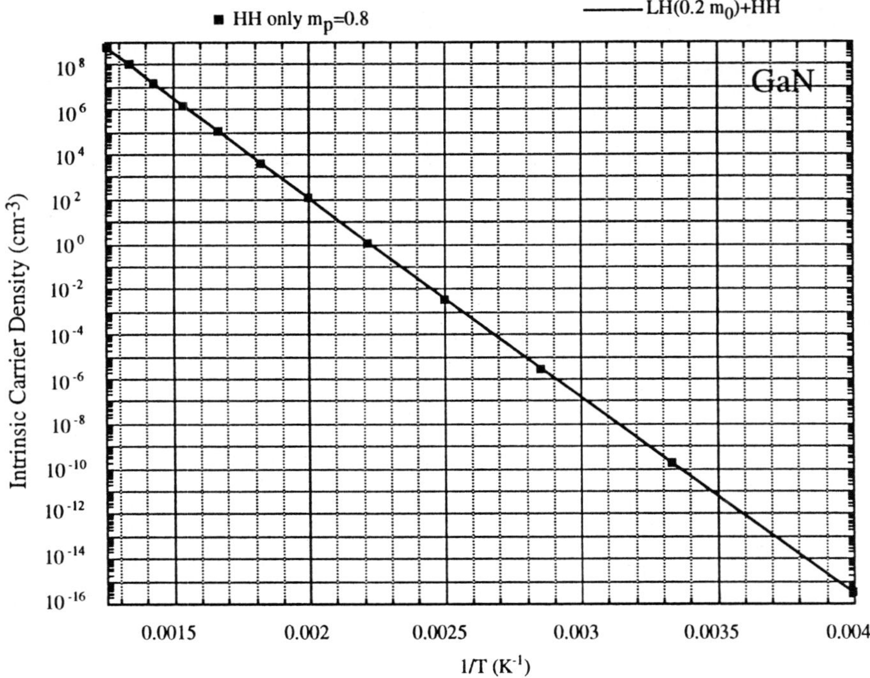

Fig. 7.3. The intrinsic carrier concentration as a function of temperature with 0.22 and 0.8 for the electron and hole effective masses, respectively

Under equilibrium, the intrinsic electron concentration is related to the conduction- and valence-band densities of states by

$$n_i^2 \;=\; np = N_C\,N_V\,e^{-E_g/kT} \tag{7.19}$$

which can be expressed in a more convenient fashion by

$$n_i \;=\; 2.5\cdot 10^{19}\left[\frac{m_n^* m_p^*}{m_0^{\,2}}\right]^{3/4}\left[\frac{T\,[K]}{300}\right]^{3/2} e^{-E_g/2kT}\quad cm^{-3}\;. \tag{7.20}$$

The temperature dependence of the intrinsic carrier concentration in GaN is displayed in Fig. 7.3, again assuming 0.22 and 0.8 for the electron- and valence-band effective masses, respectively.

7.4 Electron and Hole Concentrations

Another expression that must be used in unison with (7.19) is the **charge neutrality equation** which dictates that the positive and negative charges be equal:

$$p + N_D^+ = n + N_A^- \qquad (7.21)$$

where the left-hand side represents the positive charges with p and N_D^+ depicting the hole and ionized donor concentrations, respectively, and vice versa.

If the compensated semiconductor is n-type, the equilibrium electron concentration from the charge balance and the electron-hole product can be expressed as

$$2n_0 = \left[N_D^+ - N_A\right] + \left[\left[N_D^+ - N_A\right]^2 + 4n_i^2\right]^{1/2} . \qquad (7.22)$$

Equation (7.22) takes into consideration that all the acceptors in an n-type semiconductor are ionized. If, on the other hand, the semiconductor is p-type, the equilibrium hole concentration is given by

$$2p_0 = \left[N_A^- - N_D\right] + \left[\left[N_A^- - N_D\right]^2 + 4n_i^2)\right]^{1/2} , \qquad (7.23)$$

with an analogous interpretation of the donors in a p-type semiconductor.

In wide-bandgap semiconductors such GaN the intrinsic carrier concentration n_i is extremely small ($10^{-7}\,\text{cm}^{-3}$ at room temperature). Consequently, when the semiconductor is even moderately n- or p-doped, the opposite free-carrier concentration would be extremely small and can be neglected in the charge-neutrality expression.

For its difficulty and importance, let us consider the case of p-type GaN. The ionized-acceptor concentration is related to the total acceptor concentration through the Fermi distribution as follows (we refer to [7.1] for details)

$$\frac{N_A^-}{N_A} = \frac{1}{1 + g_a \exp[(E_A - E_F)/kT]} \qquad (7.24)$$

with g_a being the **acceptor degeneracy factor**.

In an n-type semiconductor the corresponding expression relating ionized donors to the total donor concentration is

$$\frac{N_D^+}{N_D} = 1 - \frac{1}{1 + (1/g_d)\exp[(E_D - E_F)/kT]} \,. \qquad (7.25)$$

with g_d being the **donor degeneracy factor**.

In compound semiconductors, the donors can get electrons with either spin, and thus the ground state degeneracy for donors is 2. The ground-state degeneracy for acceptors is 4 because each acceptor can accept one hole with either spin, and the impurity level is doubly degenerate. In GaN, the acceptor ground-level degeneracy of 4 gives the best fit to the data. This would imply that either the level is doubly degenerate or also the spin-orbit split valence band is involved.

In the case of multiple donors, the charge neutrality expression can be rewritten as

$$n + n_A^- = p + \sum_i N_{Di}\left[1 - \frac{1}{1 + (1/g_d)\exp[(E_{Di} - E_F)/kT]}\right] \,. \qquad (7.26)$$

This expression will be employed to describe the temperature dependence of the electron concentration observed in some samples where inclusion of two donor levels has been reported to enhance the fit.

7.5 Temperature Dependence of the Hole Concentration

In wide-bandgap materials such as GaN, the minority free-carrier concentration can be neglected because the intrinsic carrier concentration is small and further simplifies the analysis.

Substituting the Fermi distribution into the relation between the ionized and total concentration of acceptors, we get

$$\frac{N_A}{N_A^-} = 1 + g_a \exp\left(\frac{E_A}{kT}\right)\exp\left(\frac{-E_V}{kT}\right) \,. \qquad (7.27)$$

Further manipulation and substitutions lead to

$$\frac{N_A - N_A^-}{N_A^-} = \frac{g_a p}{N_V} \exp\left(\frac{E_A - E_V}{kT}\right), \tag{7.28}$$

$$\frac{N_A - N_D - p}{N_D + p} = \frac{g_a p}{N_V} \exp\left(\frac{E_A - E_V}{kT}\right) \tag{7.29}$$

and

$$\frac{p(N_D + p)}{N_A - N_D - p} = \frac{N_V}{g_a} \exp\left(\frac{E_V - E_A}{kT}\right). \tag{7.30}$$

Hole concentrations measured as a function of temperature and calculated according to (7.30) can be used in unison with donor and acceptor concentrations, and the acceptor binding energy as fitting parameters to determine the quantities contained in (7.30). It should be mentioned that there is no unique solution to this fitting which questions the confidence in the values of the parameters extracted. Note that for p-type semiconductors, all the donors can be assumed ionized. Unfortunately, additional information can not be garnered from the Hall mobility and its dependence on temperature, as the Hall data are not reliable due to non-ohmic contacts and poor mobilities. In addition, the severe non-parabolicity of the valence band and the coupling of the various valence bands necessitate numerical approaches for mobility calculations, which are not yet at a point of development to be of much use. In spite of this, the results of such a fitting for a degeneracy of 4 are shown in Fig.7.4 where an activation energy of 170 meV, acceptor and donor concentrations of $1.6 \cdot 10^{20}$ cm^{-3} and $8 \cdot 10^{18}$ cm^{-3}, 5 % of the acceptor concentration has been used, respectively. Again, the reader is cautioned that the p-type GaN has not yet been well established. It is therefore more difficult to analyze the problem by the lack of good ohmic contacts even at room temperature, let alone at low temperatures. Measurements must be done at temperatures higher than room temperature.

A brute-force approach to determine E_A is one where the semiconductor is assumed to be uncompensated, and the hole concentration is small compared to the acceptor concentration. This transforms (7.30) into the approximate expression

$$p = \sqrt{\frac{N_A N_V}{g_a} \exp\left(\frac{E_V - E_A}{2kT}\right)}. \tag{7.31}$$

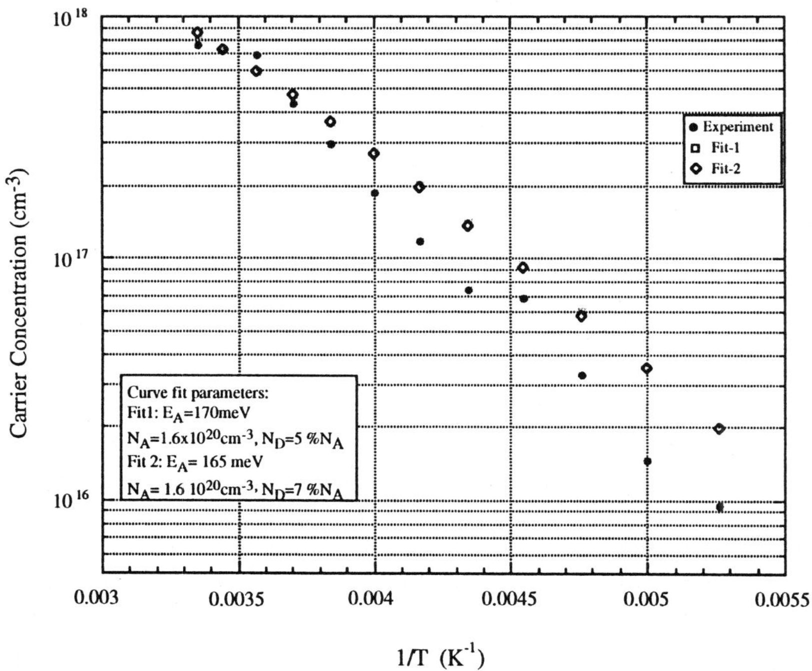

Fig. 7.4. The measured and calculated hole concentrations as a function of temperature with the donor and acceptor concentrations and the acceptor binding energy as fitting parameters

The slope of a plot of the hole concentration vs. temperature dependence in logarithmic scales is half the acceptor activation energy neglecting the temperature dependence of the density of states. This "approach" is used extensively.

Knowledge of the acceptor and donor concentrations, and the acceptor binding energy with the appropriate degeneracy is needed to calculate the hole concentration. Experimentally though the hole concentration is the one that can be measured. If the measurement is done for a range of temperatures, one can deduce the acceptor concentration, the acceptor binding energy, the degeneracy factor and the compensation ratio. The problem, however, is that a unique solution may not be obtained. If the dependence of the hole mobility on the ionized impurities and the temperature is known and utilized in conjunction with the hole-concentration data, as discussed in Chap.8 for electrons, the fitting parameters may be defined with more confidence. Additional confidence can be achieved if the measurements have been conducted for many samples with fixed acceptor binding energy and degeneracy factor.

7.6 Temperature Dependence of the Electron Concentration

Similar to the case of p-type semiconductors, for an n-type wide-bandgap semiconductor with a negligible hole concentration, we have

$$\frac{n(N_D + n)}{N_D - N_A - n} = \frac{N_C}{g_d}\exp\left(\frac{E_D - E_C}{kT}\right) \qquad (7.32)$$

where g_d represents the donor degeneracy factor.

The electron concentration measured as a function of temperature and that calculated from (7.32) can be used in conjunction with donor and acceptor concentrations and the donor binding energy as fitting parameters to determine the aforementioned parameters. Note that for n-type semiconductors, all the acceptors can be assumed ionized. The results of such a fitting exercise is depicted in Fig.7.5. The donor binding energy so determined is the screened binding energy which is related to the binding energy in a dilute semiconductor through $E_D = E_D^0 - \alpha N_D^{1/3}$. Since there are many fitting parameters with possible non-unique solutions, the accuracy of the

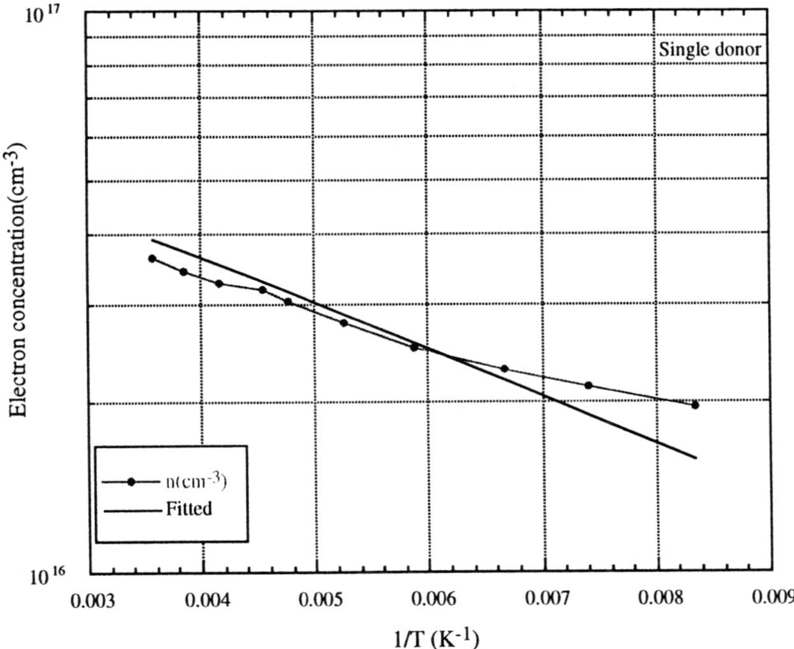

Fig. 7.5. The measured and calculated electron concentrations as a function of temperature with the donor and acceptor concentrations and the donor binding energy as fitting parameters

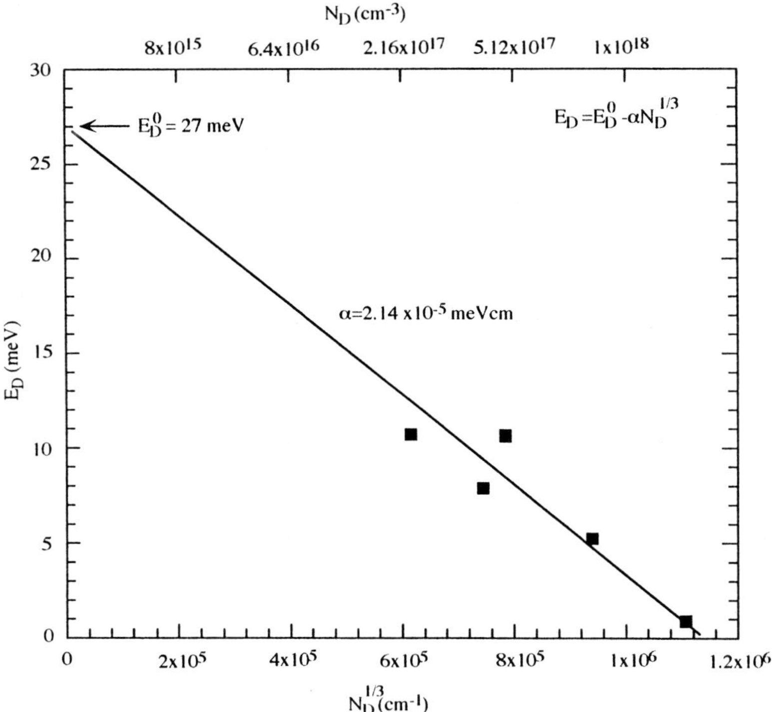

Fig. 7.6. Donor binding energy as determined for a series of samples with varying donor concentrations. Measured, screened, and donor binding energies yield with a dilute value of 27 meV and α of $2.14 \cdot 10^{-5}$ meV·cm

fitting can be improved by utilizing the temperature dependence of the Hall mobility together with an appropriate theory (Chap. 8). Shown in Fig. 7.6 is the donor binding energy, as determined for a series of samples with varying donor concentrations. Clearly, the screened donor binding energy follows the expression for a dilute semiconductor with a value of 27 meV and α of $2.14 \cdot 10^{-5}$ meV·cm.

As in the case of p-type semiconductors and in the spirit of simplicity, the semiconductor can be assumed to be uncompensated, the electron concentration small compared to the donor concentration, and (7.32) leads to the approximate expression

$$ n = \sqrt{\frac{N_D N_C}{g_d} \exp\left(\frac{E_D - E_C}{2kT}\right)} . \tag{7.33} $$

The slope of the electron concentration vs. temperature dependence in a logarithmic scale is half the donor activation energy, neglecting the temper-

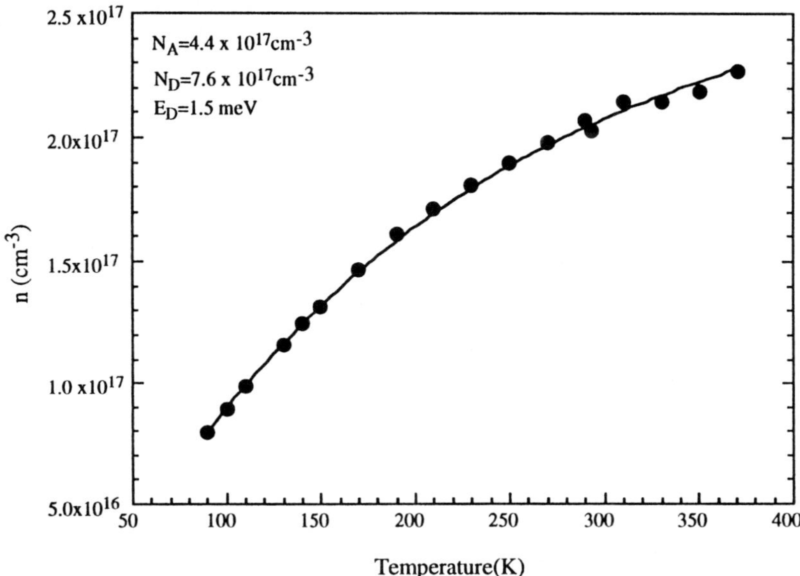

Fig. 7.7. Temperature dependence of the electron concentration, for which the measured data can be fitted with a single-donor model

ature dependence of the density of states. Despite the simplifying assumption, (7.33) is commonly employed to determine the donor binding energy.

For some samples, the single-donor model does not fit the temperature dependence of the electron concentration measured. Invoking the notion of a shallow donor and a deeper donor with the aid of (7.26), the data can be fitted better. This is the **two-donor model** employed by *Götz* et al. [7.2] and *Look* et al. [7.3] in their respective efforts. It should be pointed out that a two-donor model is not needed in all samples for a satisfactory fit. Figure 7.7 exhibits a case where a single-donor model does an adequate job for fitting the data. On the other hand, Fig.7.8 shows temperature-dependent electron concentrations wherein a two-donor model must be invoked for a good fit. Note the shallow and deeper donors and their concentrations. Having reported on the two-donor model, it is encumbered on us to consider alternatives that may be responsible, as has been done by *Look* et al [7.3]. They supposed the presence of a highly conductive interface layer.

7.7 Multiple Occupancy of the Valence Bands

The effective masses used in the treatment so far are actually the density-of-states effective masses. For spherical constant-energy surfaces such as those in GaN, the density-of-states effective masses are the same as the car-

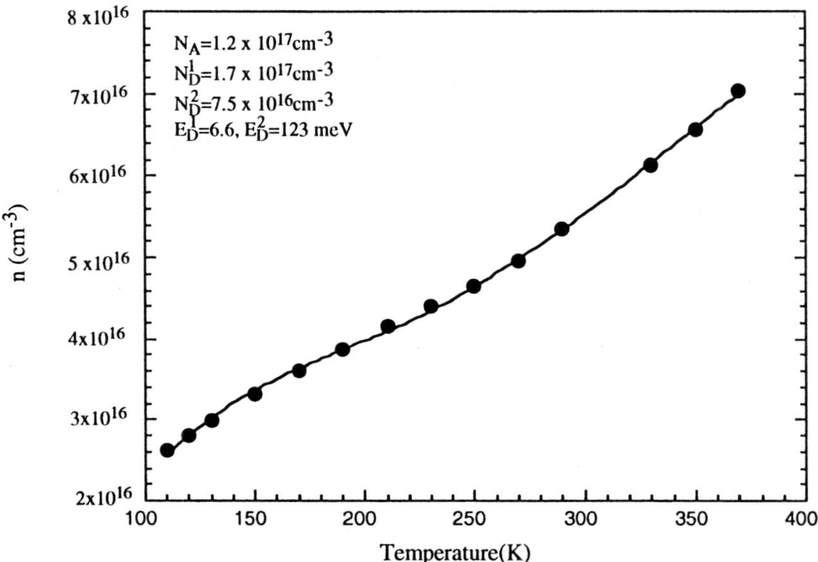

Fig. 7.8. Temperature dependence of the electron concentration, for with the measured data have been fitted with a model invoking the presence of a shallow donor and a deeper donor

rier effective masses provided that only one band is occupied. This holds true for the conduction band. In the valence band, however, any degeneracy and other multiple occupancy would require modifications.

In traditional compound semiconductors such as GaAs, the valence band is degenerate and thus both the heavy and light hole states are occupied in the following manner.

$$p_0 = p_{hh} + p_{lh} = (N_{vhh} + N_{vlh})\exp\left(\frac{E_F - E_V}{kT}\right) \tag{7.34}$$

where N_{vhh} and N_{vlh} indicate the total densities of states for the heavy and light hole bands, respectively. Utilizing the density-of-states expressions, we obtain

$$p_0 = 2\left(\frac{2\pi kT}{h^2}\right)^{3/2}\left[(m_{hh}^*)^{3/2} + (m_{lh}^*)^{3/2}\right]\exp\left(\frac{E_V - E_F}{kT}\right). \tag{7.35}$$

This can be rewritten as

$$p_0 = 2\left(\frac{2\pi kT}{h^2}\right)^{3/2} (m^*_{pdens})^{3/2} \exp\left(\frac{E_V - E_F}{kt}\right),$$
(7.36)

with

$$m^*_{pdens} = \left[(m^*_{hh})^{3/2} + (m^*_{lh})^{3/2}\right]^{2/3}.$$
(7.37)

The valence band of GaN is not degenerate but contains the heavy- and light-hole (spin-orbit splitting) bands that are only about $5 \div 10$ meV apart (Chap. 3). Taking the non-degenerate heavy- and light-hole bands into account, the hole concentration can be expressed as

$$p_0 = 2\left(\frac{2\pi kT}{n^2}\right)^{3/2} \left[(m^*_{hh})^{3/2} + (m^*_{lh})^{3/2} e^{-\delta/kT}\right] \exp\left(\frac{E_V - E_F}{kT}\right)$$
(7.38)

which leads to the density-of-states effective hole mass

$$m^*_{pdens} = \left[(m^*_{hh})^{3/2} + (m^*_{lh})^{3/2} e^{-\delta/kT}\right]^{2/3}$$
(7.39)

where δ is about $5 \div 10$ meV. Considering that the heavy-hole mass is considerably larger that the light-hole mass, the correction due to the light-hole band occupancy would be small. (The heavy-hole and light-hole masses are estimated to be about 0.8 and 0.2, respectively).

In a semiconductor with an anisotropic effective mass as in Si, the density-of-states effective mass is given by

$$(m^*_{cdens})^{3/2} = (m^*_x m^*_y m^*_z)^{1/2}.$$
(7.40)

where m^*_x, m^*_y, m^*_z represent effective masses in the x, y and z directions. For completeness, the conduction mass for an Si-like semiconductor can be written as

$$\frac{1}{m^*_c} = \frac{1}{3}\left(\frac{1}{m^*_x} + \frac{1}{m^*_y} + \frac{1}{m^*_z}\right),$$
(7.41)

because of the 6 conduction minima in each of the 6 $\langle 100 \rangle$-like directions, any anisotropy in the conduction is averaged out.

7A. Appendix: Fermi Integral

The Fermi integrals $F_n(\eta)$ employed in the calculation of the **Fermi-Dirac distribution function** are briefly summarized and appropriate approximations are outlined. The general Fermi integral can be expressed as

$$F_n(\eta) \;=\; \frac{2}{2^{1/2}} \int_0^\infty \frac{x^n\,dx}{1 + \exp(x-\eta)} \;. \tag{7A.1}$$

The Fermi integral of the 1/2 kind can be written as

$$F_{1/2}(\eta) \;=\; \frac{2}{2^{1/2}} \int_0^\infty \frac{x^{1/2}\,dx}{1 + \exp(x-\eta)} \;. \tag{7A.2}$$

To simplify calculations this expression can be approximated by

$$F_{1/2}(\eta) \;=\; e^\eta \qquad \text{for} \quad \eta \ll -1 \tag{7A.3}$$

and

$$F_{1/2}(\eta) \;=\; \frac{4\eta^{3/2}}{3\pi^{1/2}} \quad \text{for} \quad \eta \gg 1 \;. \tag{7A.4}$$

For other values of η, $F_{1/2}(\eta)$ can be obtained from Fig. 7A.1 below.

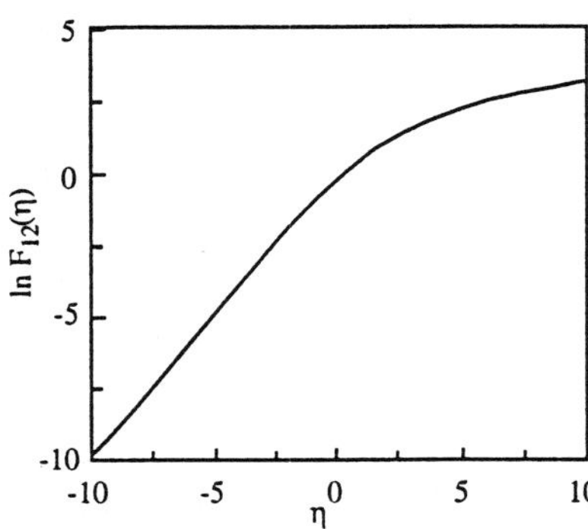

Fig. 7A.1. Fermi integral

8. Carrier Transport

Current conduction, and thus the resistance of a semiconductor material and device, is determined by the ease with which the carriers can traverse through the structure. As the carriers travel through a semiconductor, they undergo a variety interactions with the host material [8.1-3]. In a perfect static crystal, carriers would be accelerated indefinitely by the applied electric field, consistent with the band structure of the crystal. However, the semiconductor crystal contains defects, intentionally added impurities, and even at very low temperatures the semiconductor is in constant motion and far from being static. As free carriers traverse through a semiconductor, they encounter various events referred to as **scattering**, the most effective of which are by charged impurities and/or centers, and by lattice vibrations. The former manisfests itself as deflections of free carries by the long-range Coulomb potential of the charged centers. This can be thought of as a local perturbation of the band edge, which affects the electron motion. The latter is caused by the interaction of a moving charge with lattice vibrations, contraction and dilation, and can liberally be described as follows: As the atoms moves closer to and farther away from one another, the corresponding undulations on the band edge causes scattering, as will be described in Sects. 8.1-4. An additional scattering mechanism is that due to charged dislocations which can be partially screened at high doping levels. Impurity scattering is eleastic or near eleastic, and conserves energy. However, phonon scattering is inelastic and changes the energy and momentum states. In the scattering process, energy can be gained by phonon absorption or lost by phonon emission.

There are many types of lattice vibrations such as acoustic and optical, the former having a longer wavelength whereas the latter is associated with shorter wavelengths. In covalent semiconductors, acoustic and non-polar optical phonon scattering and impurity scattering dominate the carrier motion and thus the mobility. In polar semiconductors such as GaAs and GaN, the Longitudinal Optical (LO) polar optical phonon scattering is the dominant scattering mechanism associated with lattice vibrations. The ionic nature of the bonds in these semiconductors is such that as the neighboring atoms move away from one another, an electric polarization results, which causes an electric field to form. This field interacts with a moving charged particle, which is termed the **polar optical phonon scattering**, and dominates the mobility at high temperatures. Moreover, the lack of center symmetry in compound semiconductors, particularly wide-bandgap nitrides, causes them to be piezoelectric in which phonons scatter electrons.

In alloys such as InGaN and AlGaN, an additional scattering mechanism comes into the picture. The effect of the alloy can be thought of as a local perturbation of the band edge caused by the random distribution of the group-III element, which scatters a moving particle.

The conductivity of a semiconductor is determined not only by the number of available free carriers but also the freedom with which those carriers can move about within the crystal. This freedom is known as **carrier mobility** and depends on whether the transport is in the conduction band, as in the n-type material, or the valence band as in the p-type material, and the host material itself. Carrier mobility is a function of lattice temperature, the electric field, the doping concentration, and the material quality of the semiconductor. It is a universal figure of merit used not only for testing the material quality but also as an indicator of the series resistance in devices such as LEDs and lasers. In some cases, the carrier motion can be within an impurity band or within a band caused by a high concentration of defects. Since the mechanisms affecting the mobility, as mentioned above, depend on the temperature, the carrier mobility is a strong function of temperature. Untold benefits can accrue from the measurement of carrier mobility and free-carrier concentration versus lattice temperature. By making use of them in unison, one can determine more dependably the binding energies of impurities as well as the levels of compensation.

Often times, simplified expressions are derived to avoid complicated algebra. In the relaxation-time approximation, which is applicable when the scattering events are either elastic or nearly elastic such as impurity scattering, the mobility can be expressed as

$$\mu = \frac{q\langle\tau\rangle}{m^*} .$$

(8.1)

The mean-free time τ can be calculated for various scattering events. If the electrons are degenerate, the overall mobility can then be found from Matthiessen's rule [8.1-3] as

$$\mu_{total} = \left(\sum_i \frac{1}{\mu_i}\right)^{-1}$$

(8.2)

which is about 10% accurate. However, when phonon scattering which is very inelastic, is involved, the relaxation-time approximation is not appropriate despite its great appeal. Even though the accuracy is compromised, the relaxation-time approximation has nevertheless been applied to optical phonon scattering as well. Let us now formulate the major scattering events.

8.1 Ionized Impurity Scattering

The ionized impurity scattering is dominant at low temperatures because, as the thermal velocity of the carriers goes down, the effect of long-range Coulombic interactions on their motion is increased. Scattering by ionized impurities has been treated in great detail in many semiconductor texts. The mobility limited by ionized impurities is given by [8.4, 5]

$$\mu_I = \frac{128\,(2\pi)^{1/2}\,\epsilon_s{}^2\,(kT)^{3/2}}{q^3\,(m^*)^{1/2}\,(n+2N_A)}\left[\ln(1+b) - \frac{b}{1+b}\right]^{-1} \tag{8.3}$$

where

$$b = \frac{24m^*\epsilon_s\,(kT)^2}{q^2\,\hbar^2\,n'} \tag{8.4}$$

and

$$n' = n + \frac{(N_D - N_A - n)(n + N_A)}{N_D}. \tag{8.5}$$

This expression, though very widely used, breaks down for degenerate semiconductors. For GaN, it would lead to erroneous results for electron concentrations in excess of 10^{18} cm^{-3}. Since the intrinsic carrier concentration and the density of states drop precipitously, degeneracy can occur readily at low temperatures even with moderate doping levels. A formalism to deal with degenerate semiconductors along with an interpolation method for analytical solutions has been reported in [8.6].

8.2 Polar-Optical Phonon Scattering

When a scattering event is inelastic such as polar scattering, analytical expressions for the mobility limited by that event are not appropriate. The LO-phonon energy in GaN is much larger than the electron energy at room temperature, which precludes the use of analytical descriptions and necessitates numerical solutions of the Boltzmann equation [8.1]. Arguing that the polar optical phonon energy in GaN is high (91.2meV) and would favor elastic scattering, *Gelmont* et al. [8.7] presented an analytical expression for polar optical phonon scattering. Fortunately, utilizing a variational method, *Ehrenreich* [8.8] developed a pseudo-analytical expression which includes a numerically calculated function to preserve the accuracy which had otherwise been lost in the analytical method [8.9]. In fact, this function, G(z), can be modified to account for the screening effect as well and the polar-optical phonon-limited mobility becomes

$$
\mu_{pop} = 0.199 \left(\frac{T}{300}\right)^{1/2} \left(\frac{q}{\epsilon_c{}^*}\right)^2 \left(\frac{m}{m^*}\right)^{3/2}
$$

$$
\times (10^{22}\,M)(10^{23}\,v_a)(10^{-13}\,\omega_{LO})(e^z - 1)G(z) \tag{8.6}
$$

where

$$
\epsilon_c{}^* = \sqrt{\frac{M\omega_{LO}{}^2 v_a}{4\pi\epsilon_0}\left(\frac{\epsilon_0}{\epsilon_\infty} - \frac{\epsilon_0}{\epsilon}\right)} \tag{8.7}
$$

is **Callen's effective ionic charge**. Here $\epsilon_\infty = 5.47\epsilon_0$ is the optical dielectric constant, and $\epsilon = 10.4\epsilon_0$ the low-frequency dielectric constant of GaN for an E field parallel to the c axis. The low-frequency dielectric constant of the GaN semiconductor becomes $\epsilon = 9.5\epsilon_0$ for the E field perpendicular to the c axis. ϵ_0 denotes the dielectric constant of free space. The low-frequency dielectric constant can be related to the optical dielectric constant by

$$
\epsilon = \epsilon_\infty \left(\frac{\omega_{LO}}{\omega_{TO}}\right)^2 . \tag{8.8}
$$

The longitudinal ω_{LO} and transverse ω_{TO} phonon frequencies are 744 and 533 cm^{-1}, respectively.

M denotes the **reduced mass** of the nearest-neighbor atoms in grams and is given by

$$\frac{1}{M} = \frac{1}{M_1} + \frac{1}{M_2} \, , \tag{8.9}$$

where M_1 and M_2 designate the masses of the nearest neighbors. The value of M for Wurtzite GaN is $1.936 \cdot 10^{-23}$ g. The term v_a is the volume of the unit cell which can be calculated from

$$v_a = \frac{3^3 \ a^2 c}{4} \tag{8.10}$$

yielding the value of $2.283 \cdot 10^{-23}$ cm^3. Alternatively,

$$v_a = \frac{\text{mol.weight}}{N_a \rho} = \frac{69.72 + 14.0067}{6.1 \times 6.023 \cdot 10^{23}} = 2.28 \cdot 10^{-23} \text{ cm}^{-1} \, , \tag{8.11}$$

where ρ is the mass density.

The LO phonon angular zone-center frequency for GaN is given by

$$\omega_{LO} = 1.309 \cdot 10^{11} \theta_{LO} \, [K]$$

Table 8.1. The function G(z) which appears in the polar-optical scattering-limited mobility

z	G(z)	z	G(z)
0.0	1.0	2.5	1.041
0.2	0.8957	3.0	1.194
0.4	0.8102	3.5	1.353
0.6	0.7524	4.0	1.495
0.7	0.7340	4.5	1.621
0.8	0.7219	5.0	1.733
1.0	0.7146	6.0	1.919
1.2	0.7263	7.0	2.065
1.4	0.7528	8.0	2.188
1.6	0.7909	9.0	2.2996
1.8	0.8378	10.0	2.394
2.0	0.8911	11.0	2.487
2.2	0.9490		

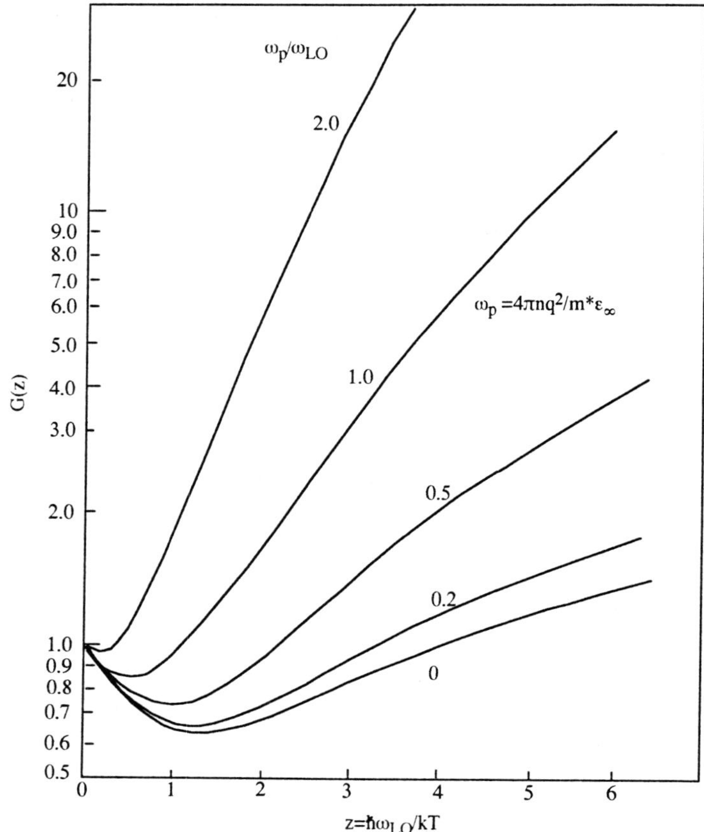

Fig. 8.1. Ehrenreich's relation $G(z = \hbar\omega_{LO}/kT = \theta_{LO}/T)$ as a function of temperature for various electron concentrations

and has a value of $1.367 \cdot 10^{14}$ s^{-1} for wurtzite GaN. The term θ_{LO} [K] is the equivalent LO phonon temperature and has a value of 1044 K for wurtzite GaN.

In (8.6), the term G(z) is a slowly varying function of z (reduced LO phonon energy), which, in turn, is given by

$$z = \frac{\hbar\omega_{LO}}{kT} = \frac{\theta_{LO}}{T}.$$

(8.12)

Neglecting screening effects, *Hammar* and *Magnusson* [8.10] took advantage of an accurate iterative method for the solution of the Boltzmann Transport Equation (**BTE**) to determine the numerical values for this function (Table 8.1). *Ehrenreich* [8.8], on the other hand, included the screening effects and determined G(z) through a variational method (Fig. 8.1).

The temperature dependence of the polar optical phonon scattering is not simple, except at very high temperatures which are above the measurement temperatures commonly employed. At such high temperatures, the dependence is $T^{1/2}$ when one neglects the temperature dependence of the $G(z)$ term, but this is not a good approach.

8.3 Piezoelectric Scattering

The mobility limited by piezoelectric scattering is given by

$$\mu_{pz} = \frac{16(2\pi)^{1/2}\rho s^2 \hbar^2 q}{3(qh_{pz}/\epsilon_s)^2 (m^*)^{3/2} (kT)^{1/2}} \, , \tag{8.13}$$

where $\rho s^2 = C_1$ is the longitudinal elastic constant ($2.65 \cdot 10^{11}$ N/m^2), with s being the sound velocity ($6.59 \cdot 10^3$ m/s) and ρ being the mass density ($6.1 \cdot 10^3$ kg/cm^3); h_{pz} (0.5 C/m^2) is the piezoelectric coefficient, and C stands for the Coulomb unit of charge.

8.4 Acoustic Phonon Scattering

For completeness, another scattering mechanism important in non-polar semiconductors, namely **acoustic phonon scattering**, will be discussed briefly. The mobility limited by this mechanism can be written as [8.11]

$$\mu_{ac} = \frac{2(2\pi)^{1/2}\rho s^2 \hbar^4 q}{3E_{ds}^2 (m^*)^{5/2} (kT)^{3/2}} \, , \tag{8.14}$$

where $\rho s^2 = C_1$ and $E_{ds} = 9.2$ eV are the longitudinal elastic constant and the deformation potential (shift of the band edge per unit dilation in eV), respectively. The elastic constant is equal to the product of the material density and the second power of the sound velocity.

With the aid of the expressions for impurity and lattice scattering, the electron mobility can be calculated for various concentrations of ionized impurities. Figure 8.2 displays the results of such an exercise for a compensation ratio of 20 % and for donor concentrations of $2 \cdot 10^{16}$, $1 \cdot 10^{17}$, $5 \cdot 10^{17}$,

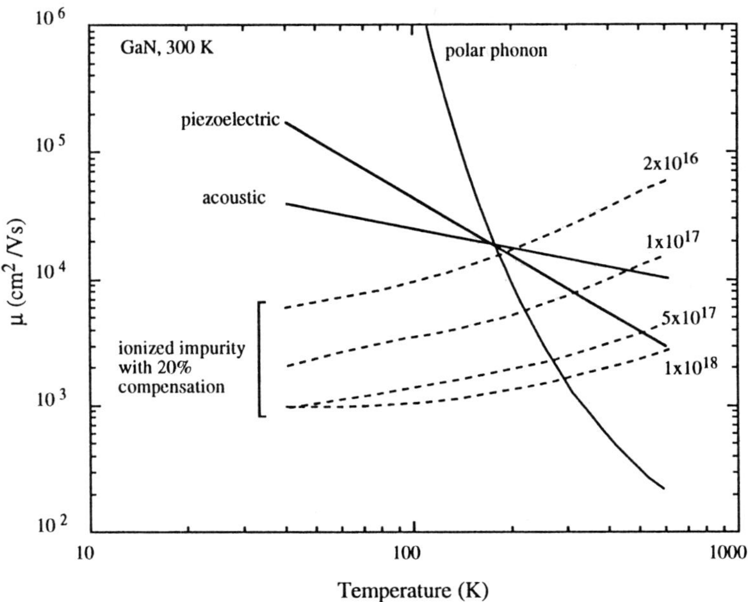

Fig. 8.2. Electron mobility limited by polar optical phonon, piezoelectric, acoustic and ionized-impurity scattering. The ionized-impurity scattering is shown for several concentrations ($2 \cdot 10^{16}$, $1 \cdot 10^{17}$, $5 \cdot 10^{17}$, $1 \cdot 10^{18}$ cm^{-3}), which cover the range characteristic of the available samples

and $1 \cdot 10^{18}$ cm^{-3}. The expression for the ionized impurity scattering is not very accurate for degenerate semiconductors. This is apparent by the lag in the reduction of the impurity mobility at low temperatures for a highly doped sample. As would be expected, the mobility at high temperatures, such as the room temperature, is limited by polar optical phonon scattering. The net mobility for a 20% compensation and donor concentrations of $2 \cdot 10^{16}$, $1 \cdot 10^{17}$, $5 \cdot 10^{17}$, and $1 \cdot 10^{18}$ cm^{-3} is exhibited in Fig. 8.3. Note that the low-temperature mobility at the highest donor level does not degrade as much as it should because the impurity-scattering expression is not that accurate for degenerate semiconductors. The room-temperature electron mobility in GaN for 0, 10, 30, 50, 70 and 90% compensation ratios as a function of $N_D - N_A$ is depicted in Fig. 8.4.

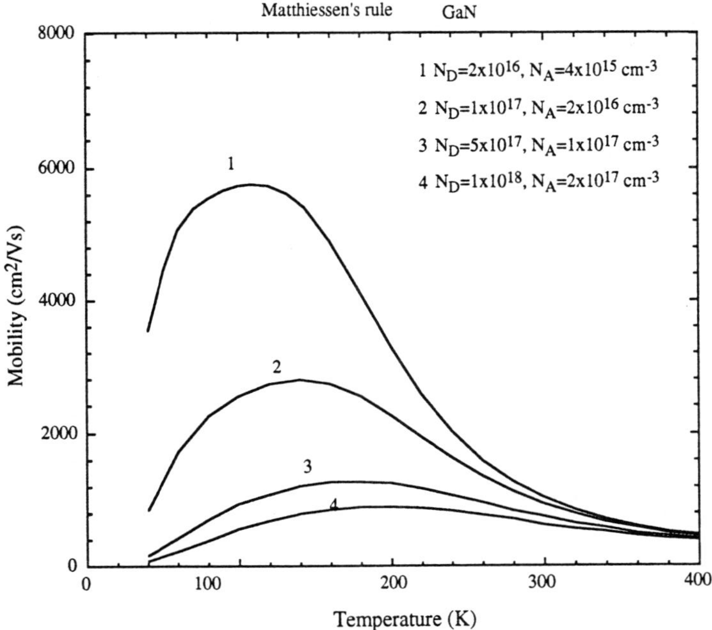

Fig. 8.3. The net mobility for a 20% compensation, donor concentrations of $2\cdot10^{16}$, $1\cdot10^{17}$, $5\cdot10^{17}$, $1\cdot10^{18}$ cm^{-3}. Note that the low-temperature mobility at the highest donor level does not degrade as much as it should because the impurity-scattering expression is not that accurate for degenerate semiconductors

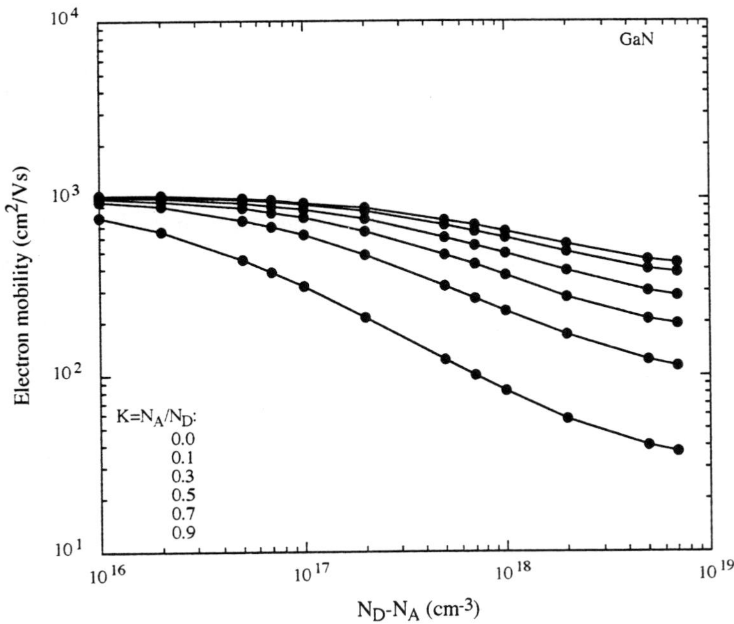

Fig. 8.4. The room-temperature electron mobility in GaN for 0, 10, 30, 50, 70, 90% compensation ratios as a function of $N_D - N_A$

8.5 Alloy Scattering

The random distribution of the constituents present in alloys such as ternaries and quaternaries like AlGaN and InGaN, causes a fluctuation in the potential, which scatters the carriers. While it has been treated by many researchers, the approach recently reported by *Look* et al. [8.12] will be applied here. If V_A and V_B represent the potential at each binary site (i.e., GaN for A and InN for B in InGaN), the average potential is

$$V = (1 - x)V_A + xV_B \, , \tag{8.15}$$

where x represents the molar fraction of InN in the lattice. The potential discontinuities experienced by electrons at the A and B sites are $V - V_A$ and $V - V_B$. The scattering potential at the site A and the position r_A can be approximated by

$$\Delta V \approx v_a (V - V_A) \delta(r - r_A) \tag{8.16}$$

with $v_a \approx 2/N_A$, N_A being the atomic concentration in GaN (namely $8.76 \cdot 10^{22}$ cm^{-3}), it yields $v_a = 2.283 \cdot 10^{-23}$ cm^3, as indicated.

Neglecting screening, the mobility limited by alloy scattering for parabolic and s-type wave functions can be expressed as

$$\mu_{al} = \frac{2^{3/2} \pi^{1/2}}{3} \frac{q \hbar^4}{v_a (1-x)x(V_A - V_B)^2 (m^*)^{5/2} (kT)^{1/2}} \, . \tag{8.17}$$

Note that the $(1-x)x$ term is maximized for $x = 0.5$ which represents the composition for the smallest mobility, all else being equal. In addition, this mobility is inverse-quadratically proportional to the alloy potential $(V_A - V_B)$, and is thus affected significantly by the choice of this potential. If reliable experimental mobility data were to exist, this quadratic term could be treated as a fitting parameter. However, as long as no such data are available, it must be calculated. In fact, the conduction-band potential-discontinuity data are used when no other figures exist. In view of the fact that no consensus exits on the alloy-scattering potentials, *Chin* et al. [8.13] utilized *Phillip's* electronegativity theory [8.14] to obtain these potentials and employed a variational method to calculate the electron mobility for InGaN, InAlN and AlGaN. According to *Chin* et al. [8.13], significant bowing in the mobility of InGaN and InAlN versus the mole fraction occurs. Furthermore, the AlGaN mobility exhibits very little bowing with decreasing mobility and increasing AlN composition, owing to very small alloy potentials.

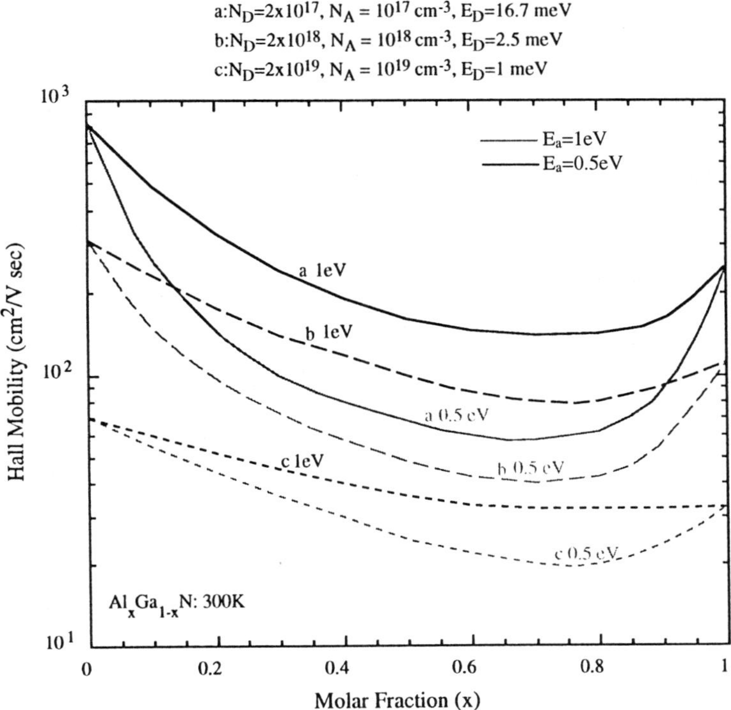

Fig. 8.5. Calculated electron mobility in AlGaN with the following donor, acceptor concentrations: (1) N_D = $2 \cdot 10^{17}$, N_A = $1 \cdot 10^{17}$, (2) N_D = $2 \cdot 10^{18}$, N_A = $1 \cdot 10^{18}$, (3) N_D = $2 \cdot 10^{19}$, N_A = $1 \cdot 10^{19}$ for $(V_A - V_B) = 0.5$ V, $(V_A - V_B) = 1.0$ eV. Courtesy of D.C. Look, Wright Science University, Dayton, OH

Alloy scattering is very dependent on the choice of the alloy potentials, as demonstrated quite clearly by (8.17). If larger alloy potentials were chosen for AlGaN, significant bowing and a reduction of the mobility would occur. In fact, Fig. 8.5 makes this point quite well for AlGaN with significant bowing when alloy-potential values of 0.5 V and 1 V are used for $(V_A - V_B)$. The set of donor and acceptor concentrations adopted for the calculations are (*1*) N_D = $2 \cdot 10^{17}$ cm^{-3}, N_A = $1 \cdot 10^{17}$ cm^{-3}, (*2*) N_D = $2 \cdot 10^{18}$ cm^{-3}, N_A = $1 \cdot 10^{18}$ cm^{-3}, and (*3*) N_D = $2 \cdot 10^{19}$ cm^{-3}, N_A = $1 \cdot 10^{19}$ cm^{-3}. A similar treatment carried out for InGaN also reveals significant bowing in mobility. Figure 8.6 plots the electron mobility for an uncompensated donor concentration of 10^{16} cm^{-3} at 300 K as a function of the InN molar fraction for alloy potentials in the range of 0 to 1.4 V. The mobility in InGaN for an alloy potential of 1 V and donor concentrations of 10^{16}, $5 \cdot 10^{17}$, $1 \cdot 10^{18}$, and $5 \cdot 10^{18}$ cm^{-3} with a compensation ratio of 20 % is displayed in Fig. 8.7.

In Tables 8.2-4, the parameters needed for the mobility calculations for GaN, InGaN and AlGaN are compiled. The values for the ternary alloys

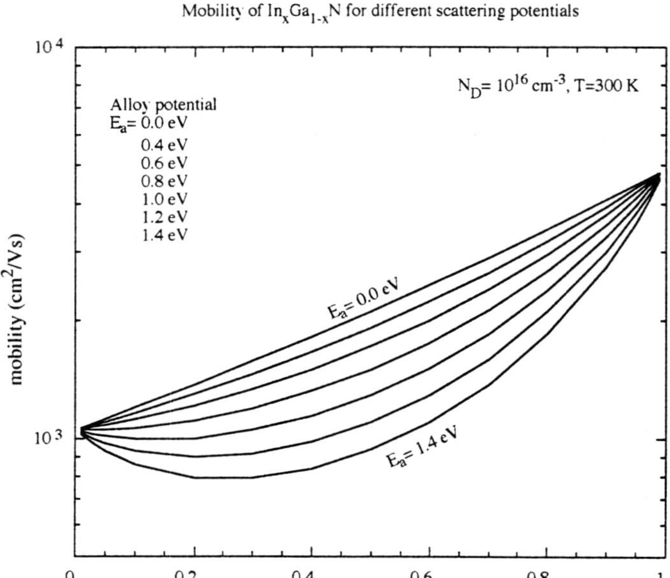

Fig. 8.6. Electron mobility for InGaN as a function of the InN molar fraction for an uncompensated donor concentration of 10^{16} cm^{-3} for alloy potentials in the range from 0 to 1.4 V at 300 K

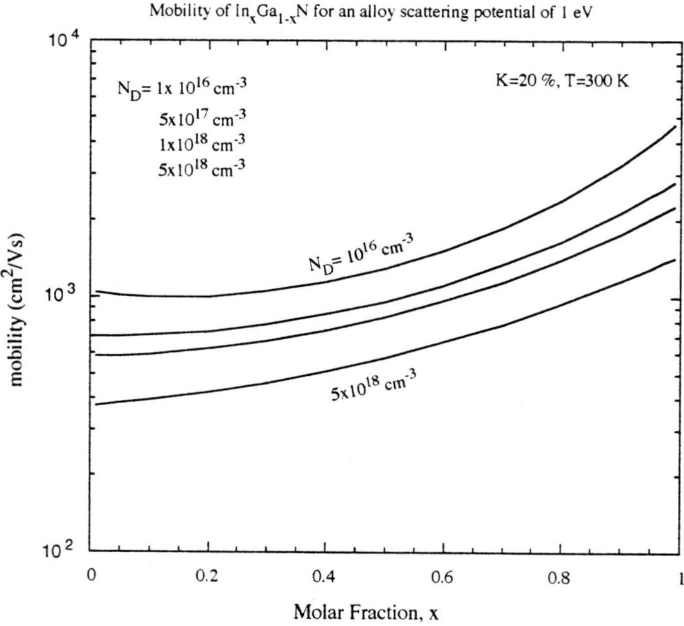

Fig. 8.7. Electron mobility in InGaN-vs.-InN molar fraction for an alloy potential of 1 V, donor concentrations of 10^{16}, $5 \cdot 10^{17}$, $1 \cdot 10^{18}$, $5 \cdot 10^{18}$ cm^{-3} with a compensation ratio of 20%

Table 8.2. Parameters pertinent to mobility calculations in GaN. Source: Complied by *Look* et al. [8.12]

Parameter	Symbol (Unit)	Fitted value	Literature value	Ref.
High-frequency dielectric constant	ϵ_∞ [F/m]		$5.47\epsilon_0$	8.15[a]
Low-frequency dielectric constant	ϵ [F/m]		$10.4\epsilon_0$	8.16
$\epsilon_0/\epsilon_\infty - \epsilon_0/\epsilon$		0.113		8.15,16
Polar phonon Debye temperature	θ_{LO} [K]		1044	8.17
Mass density	ρ [kg/m^3]		$6.10 \cdot 10^3$	8.18
Sound velocity	s [m/s]		$6.59 \cdot 10^3$	8.17[b]
Piezoelectric constant	e_{14} [C/m^3]	$0.5(0.375 \div 0.6)$		8.19
Acoustic deformation potential	E_{ds} [eV]	13.2	9.2	8.19
Effective mass	m^* [kg]		$0.22m_0$	8.16,18

[a] From $\epsilon_\infty = n^2\epsilon_0$ with $n = 2.34$ [8.15]
[b] From $C_1 = \rho s^2$, C_1 being the longitudinal elastic stiffness constant [8.16]

are deduced on the assumption that they can be represented by linear interpolations from the binary end points.

Experimental investigations of the transport in ternary and quaternary layers are minuscule, and alloy potentials are not known. Consequently, further discussion of the matter will be deferred until such time as sufficient progress has been made.

Table 8.3. Parameters compiled and used in the mobility calculations in InGaN

Parameter	Symbol (units)	InN	$In_x Ga_{1-x} N$
High-frequency dielectric constant	ϵ_∞ [F/m]	$8.4\epsilon_0$ [a]	$5.47 + 2.93x$
Low-frequency dielectric constant	ϵ [F/m]	$15.3\epsilon_0$ etimated	$10.4 + 4.9x$
$\epsilon_0/\epsilon_\infty - \epsilon_0/\epsilon$		$\epsilon_0(1/8.4 - 1/10.5)$	
Polar phonon	θ_{LO} [K] or θ_{pO}	1033 or 1038	$1044 - 11x$ or $1044 - 6x$
Debye temperature	[K]		
Mass density	ρ [kg/m^3]	$6.81 \cdot 10^3$	$(6.1 - 0.71) \cdot 10^{-3}$
Sound velocity	s [m/s]	$6.24 \cdot 10^3$	$(6.59 - 0.35) \cdot 10^{-3}$
Piezoelectric constant	e_{14} [C/m^3), h_{pz}	0.5	0.5
Acoustic deformation potential	E_{ds} [eV]	7.1	$9.2 - 2.1x$ 4A.6
Effective mass	m^* [kg]	0.115 or $0.15m_0$	$0.22 - 0.105x$ or $0.22 - 0.07x$

[a] From $\epsilon_\infty = n^2\epsilon_0$ with $n = 2.34$ [8.16]

InN parameters which are not listed above:

$a = 3.548$ Å	Lattice constant (in basal plane)
$c = 5.760$ Å	Lattice constant (in c-direction)
$m^* = 0.115m_0$	Effective mass
ρ	Mass density ($6.8 \cdot 10^3$ kg/m^3)
c_1	Longitudinal elastic constant ($2.65 \cdot 10^{11}$ N/m^2)
M	Reduced mass of an atom ($2.073 \cdot 10^{-23}$ g)
v_a	Volume of the unit cell ($\sqrt{3}a^2 c/4 = 3.140 \cdot 10^{-23}$ cm^3)
ω	Angular frequency of the polar phonon ($1.352 \cdot 10^{14}$ s^{-1})

Table 8.4. Parameters pertinent to mobility calculations in AlGaN, as complied by *Look* et al. [8.17] and from the references stated

Parameter	Symbol [Units]	AlN	$Al_xGa_{1-x}N$	Ref.
High-frequency dielectric constant	ϵ_∞ [F/m]	$4.68\epsilon_0$ [a]	$5.47 \div 0.79x$	8.22
Low-frequency dielectric constant	ϵ [F/m]	$8.5\epsilon_0$	$10.4 \div 1.9x$	8.21
$\epsilon_0/\epsilon_\infty - \epsilon_0/\epsilon$		$\epsilon_0\left[1/4.68 - 1/8.5\right]$		
Polar phonon	θ_{LO} [K]	1150	$1044 + 106x$	
Deby temperature mass density	ρ [kg/m^3]	$3.23 \cdot 10^3$	$(6.1 \div 2.87x) \cdot 10^{-3}$	
Sound velocity[b]	s_v [m/s]	$6.59 \cdot 10^3\sqrt{6.10/3.23}$ = 9.06	$(6.59 - 2.47x) \cdot 10^{-3}$	
Piezoelectric constant	e_{14} [C/m^3], h_{pz}	$0.5\sqrt{23.5/18.5} = 0.56$	$0.5 + 0.66x$	8.21
Acoustic deformation potential	E_{ds} [eV]	9.2 for GaN 9.5	9.2 assumed for GaN $9.2 + 0.3x$	8.20
Effective mass	m^* [kg]	$0.48m_0$	$0.22 + 0.26x$	

[a] From $\epsilon_\infty = n^2\epsilon_0$ with n = 2.34 [8.15]
[b] $C_1 = \rho s^2$ with C_1 being the longitudinal elastic constant [8.17]

8.6 The Hall Factor

In the relaxation-time approximation B (magnetic field) is small so that $\omega_c \tau \ll 1$, and $\omega_c = qB/m^*$ is the cyclotron frequency, the Hall factor r_H is given by

$$r_H = \frac{\langle \tau^2 \rangle}{\langle \tau \rangle^2} \ . \tag{8.18}$$

The Hall factor depends on the scattering mechanism, and τ generally assumes the form $\tau \propto E^s$. Averaging is done over energy which means that the distribution must be known, and thus the Boltzmann distribution comes into play. One can calculate the square of the average and the average of the squared mean-free time, or the relaxation time, to arrive at [8.2, 26]

$$r_H = \frac{\Gamma(5/2+2s)\,\Gamma(5/2)}{[\Gamma(5/2+s)]^2} \tag{8.19}$$

with

$$\Gamma(s) = \int_0^\infty x^{s-1}e^{-x}dx \ , \quad \Gamma(1/2) = \pi^{1/2} \ . \tag{8.20}$$

The most salient appearance of the Hall factor is that it is a prefactor relating the Hall mobility to the drift mobility. In degenerate semiconductors, the Hall factor is unity. The expression for the electron concentration, in light of the non-unity Hall factor, must be modified as

$$n = \frac{r_H}{qR_H} \ . \tag{8.21}$$

For example, for semiconductors with spherical constant-energy surfaces, s is 3/2 for impurity scattering, $-1/2$ for acoustic phonon scattering, 1/2 for piezoelectric scattering, and $-1/2$ for alloy scattering. The Hall factor is $3\pi/8 = 1.18$ and $315\pi/512 = 1.93$) for non-polar optical and ionized impurity scattering, respectively. As indicated above, the polar optical phonon scattering can not be described with the relaxation-time approximation without loss of vital accuracy; and as such it causes r_H to vary with temperature and must be deduced from numerical calculations. The Hall factor can be extracted from simultaneously using and fitting the measured electron concentration and mobility vs. temperature along with the calcu-

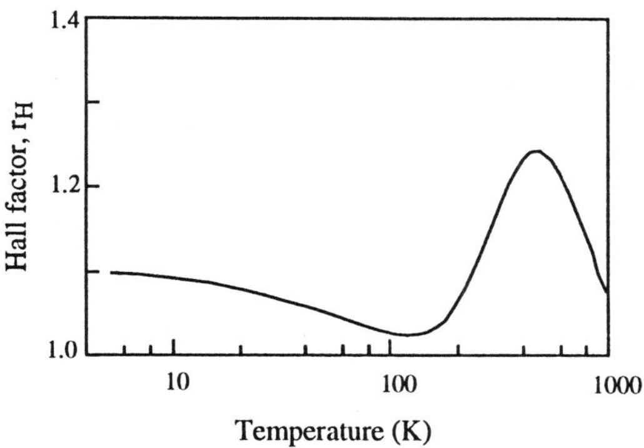

Fig. 8.8. The Hall factor as a function of temperature in pure GaN as calculated by *Rode* [8.28]

lated mobility vs. temperature, as has been done by *Look* et al. [8.27]. It has also been calculated for intrinsic GaN, the result of which is presented in Fig. 8.8.

If $\omega_c \tau > 1$, the Hall factor would tend to unity. It can be measured by Hall measurements conducted at low and high magnetic fields according to

$$r_H = \frac{R_H \text{(low B)}}{R_H \text{(high B)}} \cdot \qquad (8.22)$$

8.7 Other Methods Used for Calculating the Mobility in n-GaN

In polar scattering, an electron occupies the state (E, k) before scattering and $(E \pm \hbar\omega, k')$ after scattering. This information must be used in a numerical solution of the Boltzmann Transport Equation (BTE) that comprises elastic and inelastic scattering processes. *Rode* [8.1] developed an iterative procedure to solve BTE for electrons subjected to small driving forces, and applied it to many III-V and II-VI semiconductors including GaN and ZnO, the latter being considered as a substrate on which to grow GaN. Here, we will limit our discussion to a succinct description of the method to the extent that is relevant to the experimental results that will be discussed later. The calculated electron mobility of impurity-free GaN is shown in Fig. 8.9 for the directions parallel and perpendicular to the c axis. The perpendicular and parallel mobilities are associated with transport perpendicular and parallel to the c axis. In Fig. 8.9, the mobility determined by the iterative method of *Rode* is imbedded in the components of the mobility which are limited by

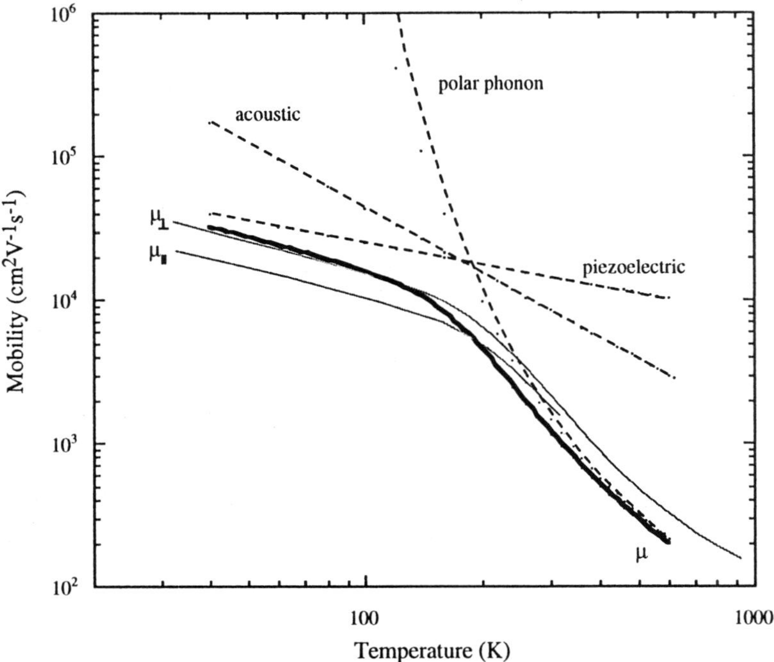

Fig. 8.9. Theoretical electron drift mobility of pure GaN as calculated by *Rode*'s iterative method for a transport transverse to the c axis ($\mu_\perp$) or in the c plane, parallel to the c axis ($\mu_\parallel$) or out of the c plane. The mobility as determined is imbedded in the mobility limited by the well-known scattering processes discussed in the text with the overall mobility being determined by Matthiessen's rule. The FTE result have been taken from [8.1]

the various scattering processes discussed in Sects. 8.1-4. As expected, the electron mobility at room temperature and above is dominated by polar optical (LO phonon) scattering. Superimposed, as a bold line, is the overall mobility calculated with the analytical expressions from Sects. 8.1-4 for impurity-free GaN and Matthiessen's rule. The major disagreement between the iterative method and that relying on the calculations of mobility components occurs at high temperatures. It indicates that the polar optical phonon scattering is the process causing the disagreement.

Besides the above-mentioned approaches, particle-based calculations have also been performed for GaN under low and high fields. Ensemble Monte-Carlo calculations taking into account the details of first-order results for the conduction band within the full Brillouin zone [8.29] have been performed both for zincblende and wurtzite phases of bulk GaN, although all indications are that the Wz form is the technologically important one and the one on which the bulk of the experimental data exists. The band structure throughout the Brillouin zone was determined with the empirical pseudopotential method. The pseudopotential calculations indicate that the two

250

phases of GaN have both similarities (for example, both have direct energy bandgaps at the Γ point, which differ by less than 10 % in magnitude) and dissimilarities. Their conduction bands are particularly different, which lead to differences in their electron transport properties. For example, the conduction-band minimum for ZB GaN is about 1.4 eV below its nearest satellite valley (X-point), and in Wz GaN it is 2 eV below its nearest satellite valley (between the L and M points). Though not important for low-field transport, the calculated electric field vs. velocity calculations peak at $2.57 \cdot 10^7$ cm/s for 112 kV/cm in the case of ZB and at $2.45 \cdot 10^7$ cm/s for 170 kV/cm in the Wz case.

Monte-Carlo simulations of the electron velocity in GaN as a function of the doping concentration, electric field and temperature demonstrated that, when the electron concentration is 10^{17} cm^{-3}, a peak occurs (2.7×10^7 cm/s) in the drift velocity at an electric field of $\approx 1.4 \cdot 10^5$ V/cm [8.30, 31]. When the electron concentration increases to 10^{19} cm^{-3}, the peak value is reduced to $1.9 \cdot 10^7$ cm/s and the electric field is shifted to $1.0 \cdot 10^5$ V/cm which is about twice that of GaAs. These results indicate that, at a temperature of 300 K and a doping concentration of 10^{17} cm^{-3}, the electron mobility for uncompensated films reaches about 900 cm^2/V·s. This compares with some $1400 \div 2000$ predicted by the iterative method of *Rode* and the analytical methods discussed earlier. In fact, the mobility of 900 cm^2/V·s has already been demonstrated for Wz films which still contains a good deal of impurities and defects. Drift velocities calculated for 400, 500, and 750 K suggest that for a sample with an electron concentration of 10^{17} cm^{-3} the peak value of the drift velocity decreases from $2.5 \cdot 10^7$ cm/s at 400 K to $2.0 \cdot 10^7$ cm/s at 750 K. Intervalley transitions seem to play an important role in the lowering of the drift velocity at high temperatures. The calculated temperature dependence of the electron mobility in GaN quantum wells is higher than in the bulk GaN material. Unfortunately, many GaN parameters necessary for Monte-Carlo calculations are not yet precisely known. In view of this uncertainty, it may be unwise to draw any definitive conclusion regarding the calculated effect of compensation on the electron mobility at this point.

8.8 Measured vis. a vis. Calculated Mobilities in GaN

Hall and drift mobilities in n-GaN calculated via a variational solution of the Boltzmann transport equation [8.32] as a function of the carrier concentration are presented in Fig. 8.10 for both 300 K and 77 K and compensation ratios of 0.00, 0.15, 0.30, 0.45, 0.60, 0.75, and 0.90. The drift mobility is

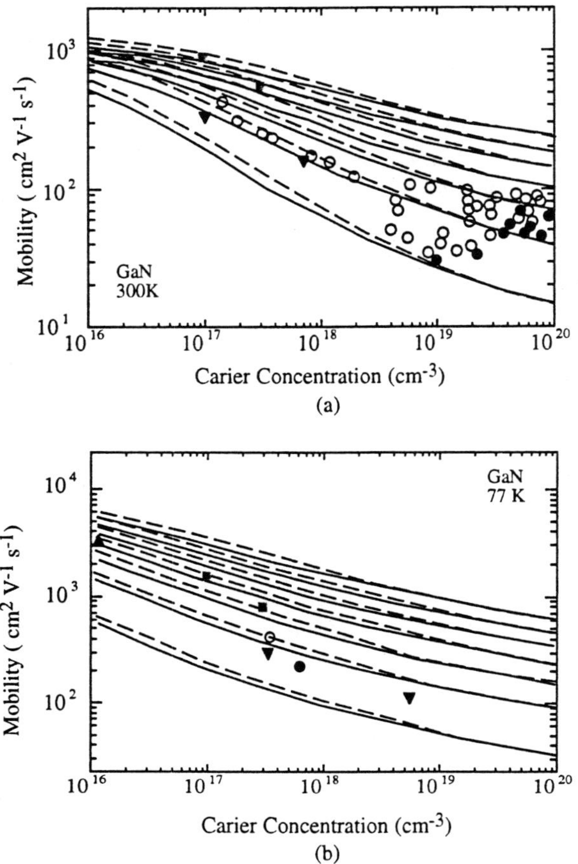

Fig. 8.10a,b. The electron drift (*solid curves*), Hall mobility (*dashed curves*) of GaN as a function of carrier concentration with the compensation ratios 0.00, 0.15, 0.30, 0.45, 0.60, 0.75, 0.90 at (a) 300 K, (b) 77 K, respectively. The experimental data include *solid circles* for Zn-doped samples. The horizontal axis represents the actual electron concentration of the drift mobility, but the Hall concentration for the Hall mobilities [8.32]

the product of the Hall mobility and the Hall factor. The variational method is inherently less accurate than *Rode's* iterative method. Though the effort has been very useful in such an early stage of the development; other likely sources of inaccuracy are due to the use of relaxation approximations in calculating the Hall factor. To reiterate, polar optical phonon scattering is inelastic and is the dominant scattering mechanism at high temperatures; it precludes the application of relaxation approximations. Nevertheless, we present the variation of the electron mobility for n-GaN so calculated as a function of temperature in Fig. 8.11. This figure suggests that at lower doping concentrations of $n \leq 10^{17}\,\mathrm{cm}^{-3}$ the mobility first increases slowly until the temperature increases to about 150 K and then decreases very rapidly

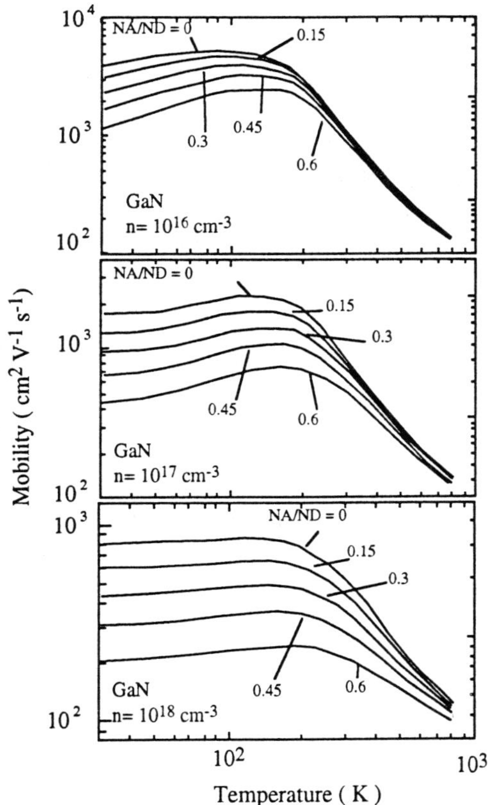

Fig. 8.11. The electron drift mobility in GaN as a function of temperature for carrier concentrations of (a) 10^{16} (b) 10^{17}, (c) 10^{18} cm^{-3} with the compensation ratios 0.00, 0.15, 0.30, 0.45, 0.60 [8.32]

for T $\geq$ 150K. As the carrier concentration increases further to about 10^{18} cm^{-3}, the mobility remains essentially unchanged until the temperature reaches about 150 K, and then decreases rapidly with increase in temperature. This behavior at low temperatures arises from the dominance of piezoelectric scattering (at lower carrier concentrations) and from ionized impurity scattering (at higher carrier concentrations). Above 200 K, the polar-optical phonon contribution is the most important scattering mechanism. In Fig.8.12, the experimental data from one source having compensation ratios between 0.70 and 0.85 are presented as open circles (undoped) and solid squares (Zn doped), all from [8.33]. The Zn-doped case is a special one in that Zn deep defect states can be introduced, which have been exploited for LEDs fabricated in the nineteenseventies and also in the first version of Nichia LEDs introduced in 1993. However, Zn doping can also lead to donor-like defects which most likely are the source of the n-type conductivity (indicated by the solid circles). In addition, the experimental data from other sources are indicated as inverted triangles [8.34], upright triangles [8.35], and solid squares [8.36]. An inspection of this figure indicates

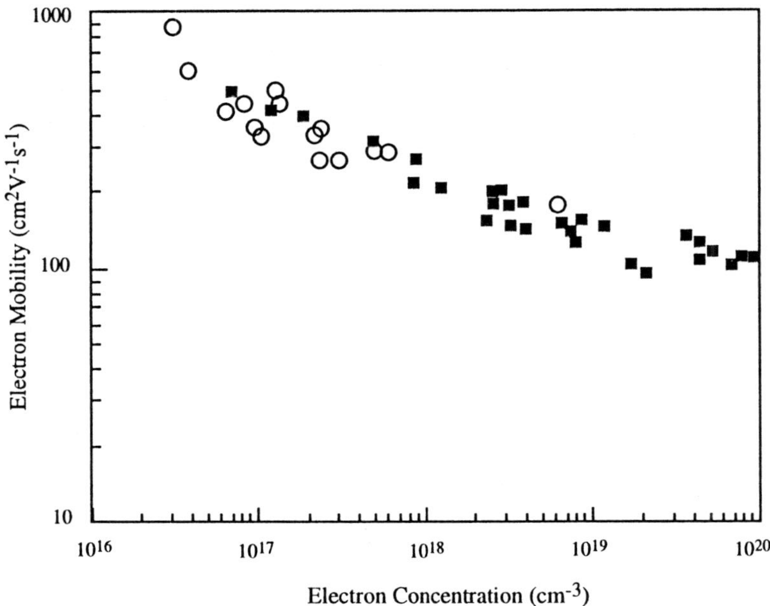

Fig. 8.12. The 300K Hall mobility vs. free-electron concentration for GaN from various groups using both OMVPE and MBE. The *open circles* are for unintentionally doped samples, the *squares* are for samples doped with either Si or Ge [8.40]

that most of the experimental data point to some compensation, and that the theoretical trend is the same as the experimental one.

A bit of mobility evolution in GaN is warranted. Until the advent of low-temperature AlN and GaN buffer layers, the electron mobilities in GaN remained around 100 cm²/Vs [8.37, 38]. With a thin low-temperature AlN layer used as a template for further growth at standard growth temperatures, the electron mobility increased to $350 \div 400$ cm²/Vs at room temperature. With 200-Å thick buffer layer the mobility of the GaN film was reported to also increase to values as high as 900 and 3000 cm²/V·s at room temperature and 70 K, respectively [8.39]. But, the temperature dependence of the mobility reported in [8.39] could not be fitted later by the iterative FTE and the analytical approach discussed above. To asses the distribution of mobilities obtained at various laboratories on top of the aforementioned values, a compilation of the measured electron mobilities as a function of the electron concentration is displayed in Fig. 8.12.

Since the first application of *Rode*'s iterative method to GaN (Fig. 8.9), the method has been revisited with application to GaN again as the material regained interest after a lengthy lull in the nineteeneighties [8.27, 40, 41]. With improved GaN layers and concomitant lowering of the impurity concentration, it has become almost possible to fit the theory to the experiment with parameters utilized in Table 8.2 for GaN. The temperature dependence or the electron mobility can be used in unison with the temperature dependence of the net electron concentration to determine the donor and acceptor concentrations as well, providing that the scattering mechanisms are limited to those discussed earlier. Notwithstanding, the results are much more dependable than just relying on the temperature dependence of the electron concentration despite some fitting involved due to uncertainties in the material properties. By using literature values of the pertinent parameters (Table 8.2) *Look* et al. [8.27] were able to fit the experimental mobility from an MOCVD sample to the calculated ones through *Rode's* iterative method, providing that a rigid downward shift by a factor of 1.4 was employed (Fig. 8.13). Alternatively, a good fit could also be achieved by adopting the par-

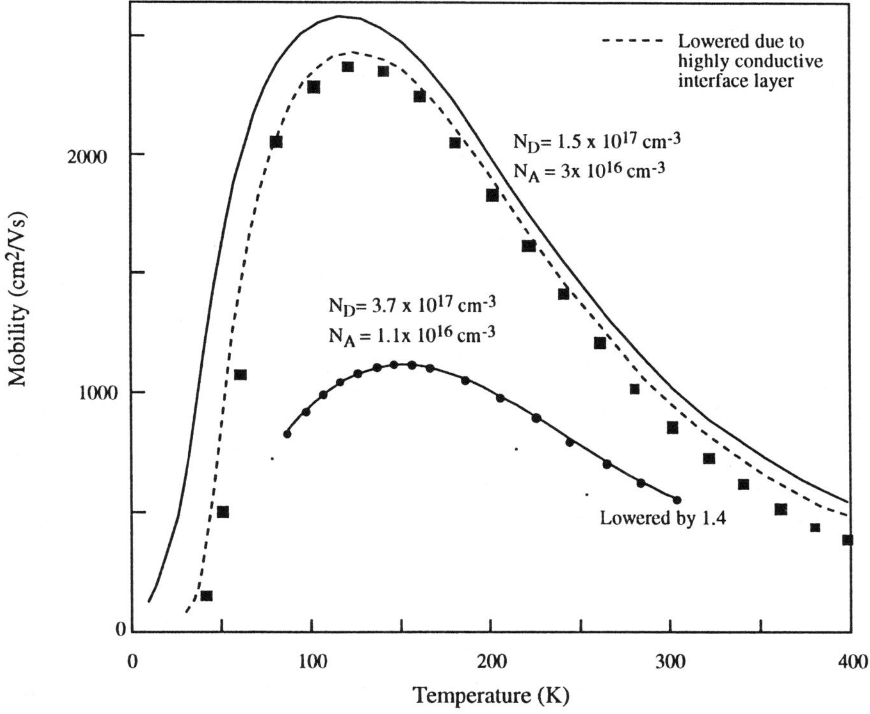

Fig. 8.13. Experimental and calculated electron mobilities for two samples, one grown with HVPE, the other with MOCVD. *Solid squares* and *circles* represent the measured electron mobilities for the 60 μm thick HVPE and MOCVD samples, respectively. The *solid line* on top represents the calculated mobility (*Rode's* iterative method) for the HVPE sample, whereas the *dashed line* gives the calculated mobility for the same sample after a correction to include the effect of a highly doped interface layer. The *solid line* going through the *solid circles* represents the calculated mobility for the MOCVD sample by *Rode's* iterative method with a rigid downward shift amounting to a division by 1.4. The MOCVD sample is after [8.16] while that for the HVPE sample is after [8.42]

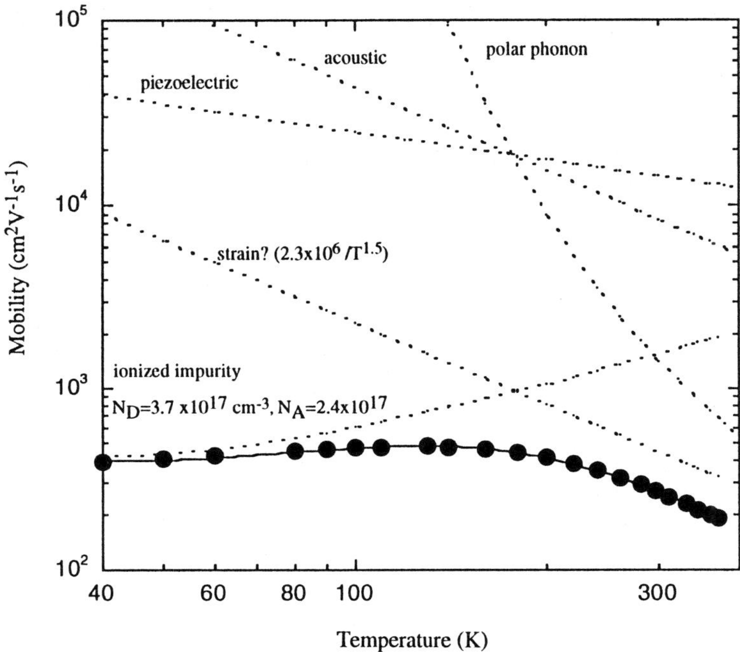

Fig. 8.14. Temperature dependence of the electron mobility in RMBE grown GaN. The experimental results are shown with solid circles. The components of the mobility limited by conventional processes are indicated along with a yet unknown process of the shape $2.39 \cdot 10^6 \, T^{-1.5}$ which when added leads to a good fit to the experiments, as shown by the *solid line* passing through the experimental data

ameters: acoustic deformation potential and $\epsilon_\infty^{-1} - \epsilon^{-1}$ as the fitting parameters. It is clear that samples with higher mobilities are badly needed to have a good fit for the parameters involved. Superimposed on the MOCVD data are those obtained in a 60 μm thick GaN layer prepared by the hydride transport VPE method. The Hall mobility in the HVPE sample [8.42] was characterized between 10 and 400 K and exhibited a room-temperature Hall mobility as high as 900 cm^{2}V^{-1}s^{-1} and a peak mobility of 2300 cm^{2}V$^{-1} \times$ s^{-1} at 110 K. A sample grown by reactive MBE with an ammonia source for the active nitrogen is of interest because the literature values of the parameters displayed in Table 8.2, together with those obtained from the temperature-dependent electron concentration overestimate the mobility. However, good agreement between calculations and experiments is obtained with an additional scattering term. Then, the mobility can be fit over the entire range of measured temperatures (40 ÷ 400 K), as indicated in Fig. 8.14. The additional scattering that must be invoked assumes the form of $2.39 \times 10^6/T^{1.5}$. While an accurate thesis is awaiting further development, one can argue that inhomogeneities may be the cause such as strain and inversion domains with either cations up and anions up. Additional scattering mechanisms such as those caused by charged dislocations may also be of some importance. Moreover, a conduction path through the defective interface layer may also be invoked to account for this inconsistency, as pointed out by *Look* and *Molnar* [8.43]. The donor and acceptor concentrations that are self consistent with the temperature dependence of the electron concentration and mobility are $3.7 \cdot 10^{17}$ cm^{-3} and $2.4 \cdot 10^{17}$ cm^{-3}, respectively. It is imperative that better samples must be obtained so that processes likely to be involved can be interrogated with some being ruled in and some ruled out.

8.9 Transport in 2D n-Type GaN

A 2-Dimensional Electron Gas (2DEG) system is one in which the motion in one direction is ideally eliminated leaving behind only the in-plane motion. This can be accomplished by inducing an electron gas in the interface between two semiconductors with an energy discontinuity and an ideal interface to the extent that the motion perpendicular to the interface is impeded. The carriers donated by donors which are situated in the wider-gap material diffuse to the one with the smaller bandgap, where they are trapped due to the energy barrier on the one side and the charge-induced band bending on the other. Essentially, the carriers and the scattering ionized centers are physically separated from one another, and ionized impurity scattering would be greatly diminished. Additional processes such as screening also takes place and diminishes interactions with the residual charged centers in the smaller-bandgap material where the carriers remain. One would then expect the electron mobility in such a system to be determined by lattice scattering only. To a first extent, this is what happens in a GaAs system with the upper limit for the room-temperature mobility being that of the polar phonon scattering-limited value which has been observed experimentally [8.44]. In a GaN/AlGaN system, measured room-temperature mobilities in a 2DEG system are generally about a factor of two higher than those measured in bulk GaN. Although the exact mechanism is not yet known, one may speculate that one or more of the scattering events is curtailed in the two-dimensional system. If we speculate further, a case can be made that, at least in RMBE samples, the additional scattering attributed to inhomogeneities may be curtailed owing to the supposition that only the nearly forward and backward scattering events may be allowed. Although the topic is in its infancy, room-temperature electron mobilities in GaN/AlGaN modulation-doped structures grown on sapphire, and the 6H-SiC substrates have about 1500 cm^{2}V^{-1}s^{-1}. The figures improved to 5700 and

7500 cm^{2}V^{-1}×s^{-1} at 20 K for samples on sapphire and SiC, respectively [8.45]. In fact, strong Shubnikov-de Haas oscillations in a sample on a 6H-SiC substrate were observed. The mobility in bulk GaN grown in the same reactor was only 385 cm^{2}V^{-1}s^{-1}. With the recent breathtaking developments in AlGaN/GaN modulation-doped field-effect transistors, there will certainly be more future work on transport in reduced-dimensional structures based on nitrides.

8.10 Transport in p-Type GaN and AlGaN

Endemic to wide-bandgap semiconductors is a poor hole mobility, and GaN is no exception. The valence band is non-parabolic and the mobility issue is made more complex by the coupling of various valence-band states. In other words, interband transitions caused by inelastic-collision events require consideration of the full valence band in calculations. As such, no analytical expressions of the type given above for electrons are available, and would not be appropriate. *Rode'*s iterative method, for example, must be applied with the full knowledge of the valence band near the Γ point for low-field transport. At the time of this writing, efforts were under way to tackle this problem from a theoretical point of view. A heavy effective hole mass aided by the difficulty of obtaining high p-type conductivity assures that the problem of transport in p-type nitrides will remain as a major area of research.

Gaskill et al. [8.40] compiled room-temperature experimental data for p-GaN Hall mobilities as a function of the hole concentration (Fig. 8.15). The solid circles correspond to the Mg-doped samples. The Mg-doped layers prepared by MOCVD were activated by an N$_2$ heat treatment or by low-energy electron-beam irradiation, a process which is not needed for the

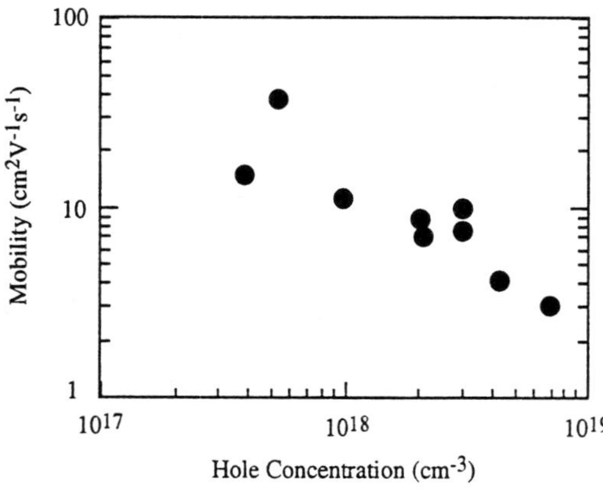

Fig. 8.15. The 300K Hall mobility vs. free-hole concentration for GaN from various groups using both OMVPE, and MBE for the Mg-doped samples [8.40]

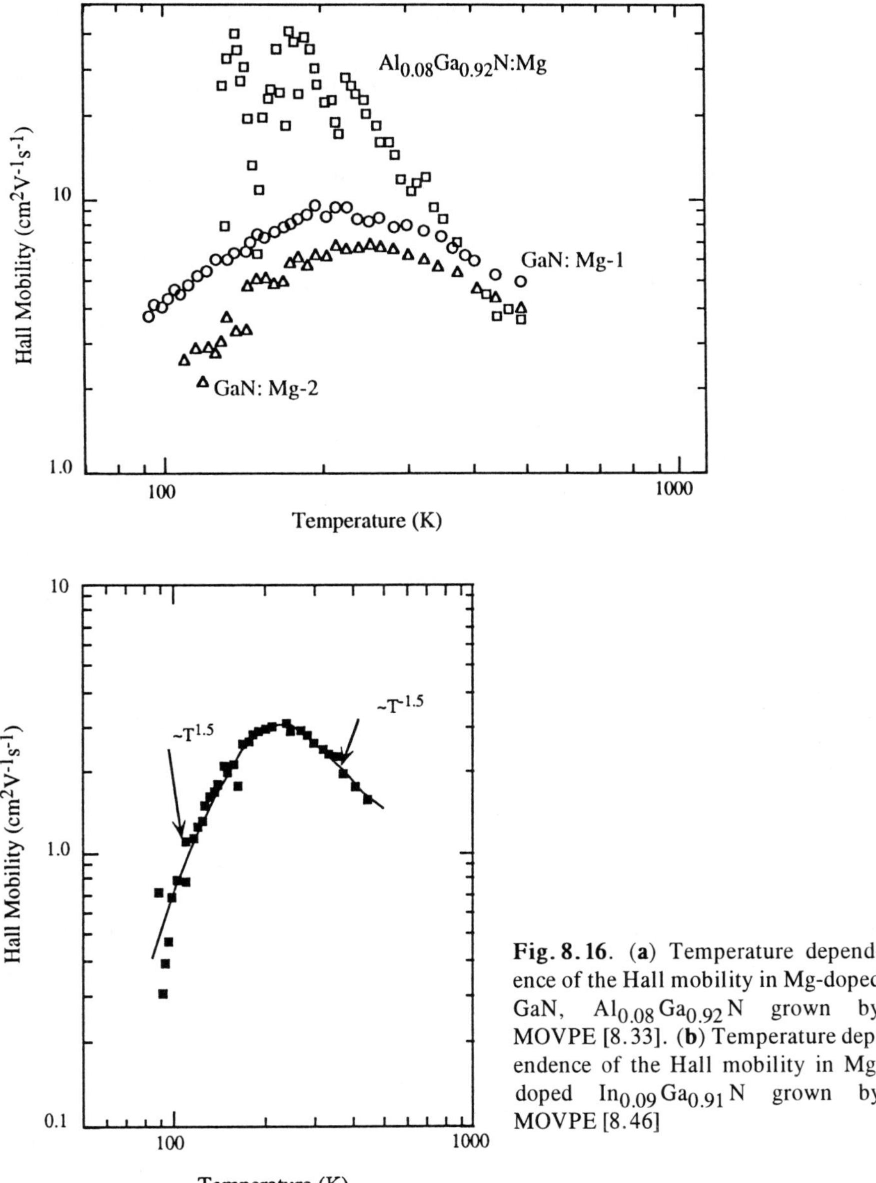

Fig. 8.16. (**a**) Temperature dependence of the Hall mobility in Mg-doped GaN, Al$_{0.08}$Ga$_{0.92}$N grown by MOVPE [8.33]. (**b**) Temperature dependence of the Hall mobility in Mg-doped In$_{0.09}$Ga$_{0.91}$N grown by MOVPE [8.46]

MBE-grown samples. In general, the reliable hole mobilities for p-GaN are in the range of $10 \div 20$ cm^2/V·s. A statistical averaging suggests that the hole mobility generally decreases with increasing carrier concentration. It must be stressed that measurements, and therefore the results, of the p-type GaN must be treated with extreme care. The contacts needed for the measurements are generally not ohmic, and scatter in data is considerable. In addition, attempts to produce p-type material, even with dopants known to produce p-type material, can lead to n-type material with mobilities in the 100 $cm^2/V \cdot s$ range or higher due to donor-like defect generation. Furthermore, though figures above 10^{18} cm^{-3} have been reported, the accuracy of those reports has not been confirmed widely.

Tanaka et al. studied the temperature-dependent p-type conduction in Mg-doped GaN and $Al_{0.08}Ga_{0.92}N$ thin films [8.46], and *Yamasaki* et al. undertook the same task for Mg-doped $Ga_{0.91}In_{0.09}N$ [8.47]. The p-conduction of the samples denoted by GaN:Mg-1, GaN:Mg-2, $Al_{0.08}Ga_{0.92}N$:Mg and $Ga_{0.91}In_{0.09}N$:Mg were interrogated through Hall-effect measurements over a temperature range from 100 to 500 K. The results are presented in Fig. 8.16a for GaN and $Al_{0.08}Ga_{0.92}N$, and in Fig. 8.16b for $Ga_{0.91}In_{0.09}N$. These figures reveal that experimental Hall-mobility data as a function of temperature are quite regular for GaN:Mg-1, GaN:Mg-2, and $Ga_{0.91}In_{0.09} \cdot$ N:Mg samples, but quite irregular for $Al_{0.08}Ga_{0.92}N$ samples. However, the statistical nature of the mobility-vs.-temperature data for the latter sample is not very different from similar curves for others. The peak in the mobility-vs.-temperature curves for both $Al_{0.08}Ga_{0.92}N$:Mg and $Ga_{0.91}In_{0.09} \cdot$ N:Mg occurs at T around 200 K. The same peak for GaN:Mg-1 and GaN:Mg-2 arises around 225 K. Probably due to larger compensation in GaN:Mg-2 than in GaN:Mg-1, the measured mobility in GaN:Mg-2 is slightly lower than that in GaN:Mg-1. The peaks of the curves result from the mobility being affected by phonon scattering at high temperatures and from impurity scattering at low temperatures.

8.11 Carrier Transport in InN

As is the case for GaN, InN too suffers from the lack of a suitable substrate material and, to a much larger extent, high native defect concentrations that really hinder its progress and analysis. Furthermore, the large disparity of the atomic radii of In and N is an additional contributing factor to the difficulty in obtaining InN of good quality. Although not yet confirmed, it is believed that nitrogen vacancies lead to large background electron concentrations in InN, and as such the electron mobilities obtained from various films have very different values [8.48]. By employing a novel deposition technique such as Ultra-High Vacuum (UHV) ECR-assisted Reactive Magnetron Sputtering (RMS) for the film growth, the electron mobility in InN can be as high as 3,000 cm^2/Vs at room temperature [8.49]. A recent study of the electron mobility of InN as a function of growth temperature indi-

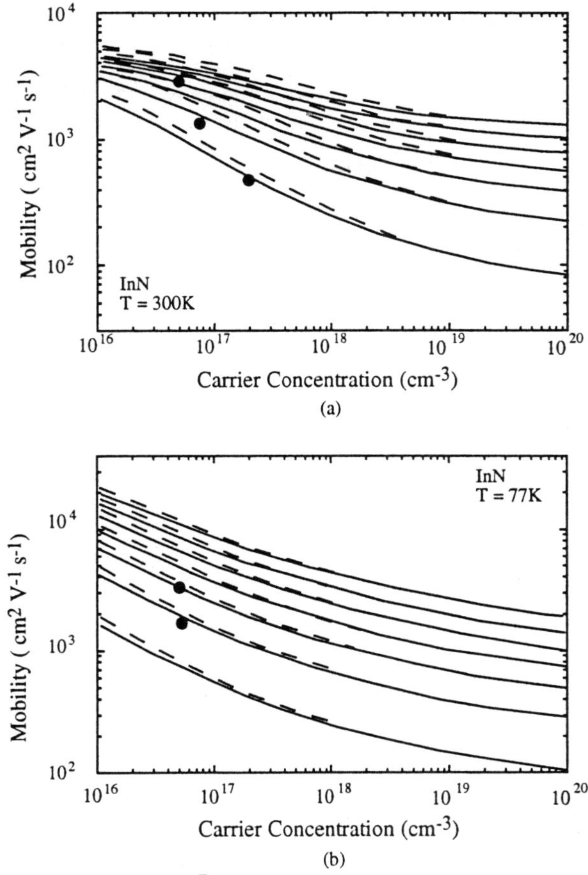

Fig. 8.17. The electron drift (*solid curves*), the Hall mobility (*dashed curves*) of InN as a function of carrier concentration with the compensation ratios 0.00, 0.15, 0.30, 0.45, 0.60, 0.75, 0.90 at (a) 300 K and (b) 77 K, respectively. The horizontal axis represents the actual electron concentration for the drift mobility, but the Hall concentration for the Hall mobilities [8.42]

cates that the mobility of UHV-ECR-RMS-grown InN can be as much as four times the mobility of conventionally grown InN [8.50].

Theoretical modeling, again a variational approach to solving the Boltzmann transport equation, by *Chin* et al. [8.32] carried out for temperatures of 77 and 300 K demonstrated that the carrier-concentration dependence of the Hall and drift mobilities in InN is a significant function of the compensation ratio. These calculated results (Figs. 8.17 and 18) agree well with the little available experimental data for the high-compensation ratios [8.51]: *Chin* et al. suggested that the high densities of deep levels found in the experimental samples may be associated with the antisite defects. Calculated results for the electron drift mobility as a function of temperature,

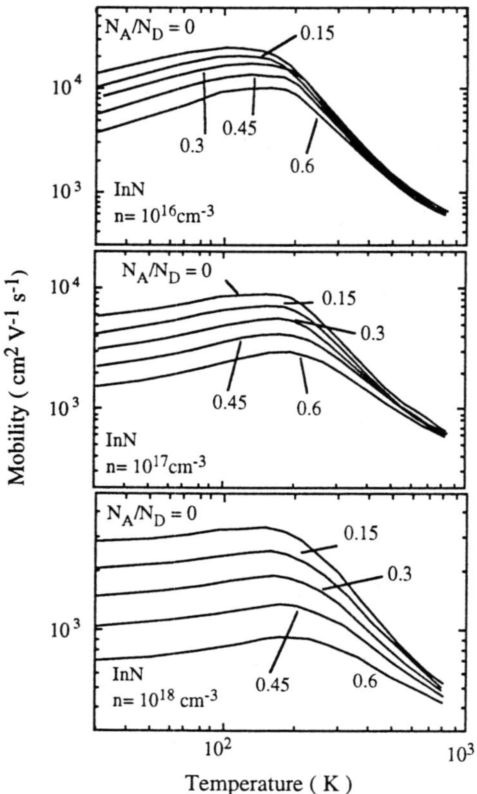

Fig. 8.18. The electron-drift mobility in InN as a function of temperature for carrier concentrations of (a) $5 \cdot 10^{16}$ for the compensation ratios 0.00, 0.60; (b) $8 \cdot 10^{16}$ cm^{-3} for compensation ratios 0.00, 0.30, 0.60, 0.75 [8.42]

compensation ratio, and carrier concentration yield peak electron mobilities of 25000, 12000, and 8000 cm^2/V·s for doping densities of 10^{16}, 10^{17} and 10^{19} cm^{-3}, respectively. Piezoelectric acoustic phonon scattering and ionized impurity scattering are the two dominant scattering mechanisms at temperatures T $\leq$ 200 K, and the polar optical phonon scattering is the most significant scattering mechanism for temperatures T $\geq$ 200 K.

8.12 Carrier Transport in AlN

Because of the perceived insulating nature (owing to the 6.2 eV energy bandgap and the defect nature of the material), the electrical transport properties of AlN have not been studied extensively. With refined growth techniques, AlN with much improved quality has very recently been prepared, it shows both n- and p-type conduction. Consequently, reports on electron Hall mobility have begun to appear in the literature. The reduced effective electron mass for AlN is still relatively unknown and for numerical simulations of the electron mobility, *Chin* et al. [8.32] have estimated it to be 0.48 ± 0.05, as opposed to 0.27 by *Suzuki* and *Uenoyama* [8.51]. Using this and the energy bandgap $E_g = 6.0$ eV (6.2 eV is commonly used), they calculated the polar optical phonon-limited drift mobility as a function of temperature. The mobility was found to decline rapidly at high temperatures, with a value of about 2000 cm^2/Vs at 77 K and dropping to 300 cm^2/Vs at 300 K. Mobility components calculated for each individual scattering mechanism indicate that the scattering due to piezoelectric effects influences the resultant mobility mostly at low temperatures; this is consistent with that for GaN. On the other hand, the scattering due to the acoustic deformation potential becomes dominant in a narrow temperature range between 150 and 200 K (depending on the effective mass). The scattering mechanism due to optical polar phonons is the most dominant one for temperatures above 230 K. *Edwards* et al. [8.53] and *Kawabe* et al. [8.54] carried out some Hall measurements for p-type AlN and produced a very rough estimate of the hole mobility $\mu_p = 14$ cm^2/V·s at 290 K.

8.12.1 Transport in Unintentionally-Doped and High-Resistivity GaN

As briefly alluded to at the beginning of this chapter, semiconductors with high doping levels and/or defect levels can form impurity and/or defect bands within which carrier transport can take place. The transport mechanism for this case will be briefly treated. The conduction through the aforementioned path in GaN has recently been investigated by *Look* et al. [8.55]. Samples intentionally nitrogen-starved to create shallow donor states were produced to interrogate current conduction. The temperature dependencies of ρ, μ_H, and n_H (resistivity, Hall mobility and Hall concentration) for a nitrogen-starved sample are displayed in Fig.8.19 where the dotted line is a fit to the μ-vs.-T data for T = 200 K. Ionized-defect scattering was included through the usual Brooks-Herring formalism, and the Boltzmann transport equation was solved by *Rode's* iterative method. The result was $N_D \approx 6 \cdot 10^{18}$ and $N_A \approx 5 \cdot 10^{18}$, giving $N_I \approx N_D + N_A \approx 1.1 \cdot 10^{19}$ cm^{-3}, N_I being the ionized-defect concentration. Note that μ falls off much more rapidly below 160 K than predicted by theory. This observation suggests that μ is significantly lower in the donor band than in the conduction band. The data indicate that above 140 K the dominant electrical transport mechanism results from conduction by electrons thermally excited from shallow donors into the conduction band. Below 140 K, the dominant mechanism is due to electrons "frozen out" in a band formed by these same shallow

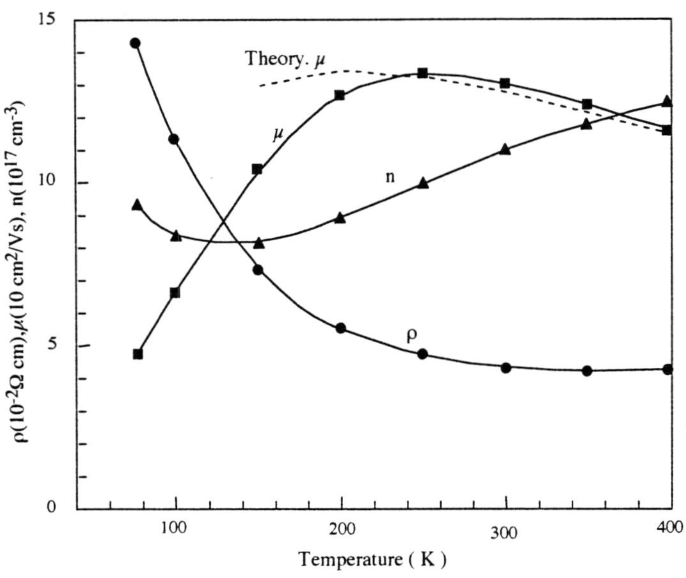

Fig. 8.19. Resistivity ρ, carrier concentration N, and mobility as a function of temperature T for the nitrogen-starved sample. The *solid lines* are added to aid the eye. The theoretical dependence of ρ on T is shown as a *chained line* [8.54]

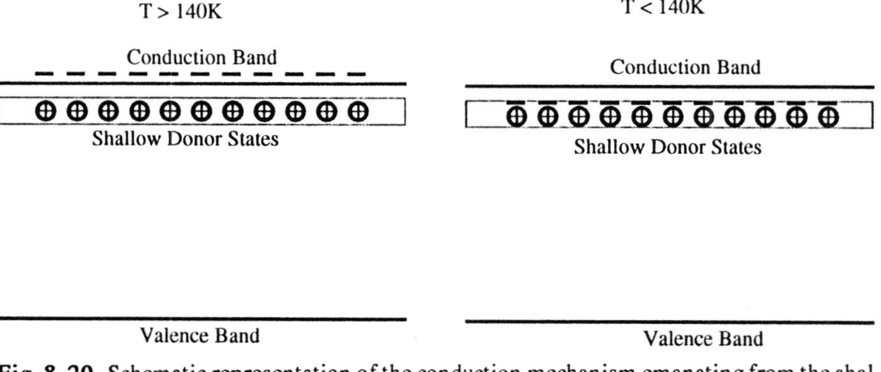

Fig. 8.20. Schematic representation of the conduction mechanism emanating from the shallow states which form a band. At sufficiently high temperatures, the electrons are excited into the conduction band where their motion is governed by the usual scattering mechanisms. At low temperatures, the electrons freeze out in the shallow band where the Hall effect is not measurable

donors, as depicted in Fig. 8.20. This shallow donor band is probably formed from the hydrogenic-type wave functions of electrons loosely bound to N vacancies, as recent experiments seem to suggest [8.42, 55]. An effective mass, $m^* = 0.22m_0$, and a static (low-frequency) dielectric constant ($\epsilon = 10.4$) leads to a Bohr radius of $a_B = 24$ Å and results in a Mott-transition concentration [8.57] $N_{cm} = (0.25/a_B)^3$ of about $1 \cdot 10^{18}$ cm^{-3}. Above this

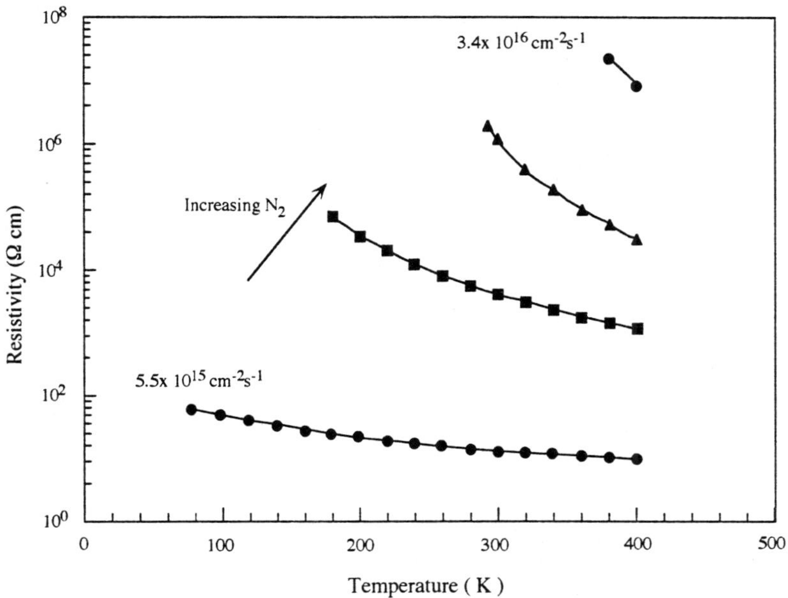

Fig. 8.21. Resistivity ρ vs. temperature T for a series of samples grown by RMBE with increasing flux. The solid lines are guides for the eye [8.54]

concentration the electron motion in the donor band would become free. At a higher concentration ($N_{cb} \approx 5N_{cm} = 5 \cdot 10^{18}$ cm^{-3}) the donor band would merge with the conduction band [8.58].

A sample flooded with nitrogen during growth exhibited very high resistivity and a Hall mobility of $\mu < 0.5$ cm^2/V·s even at 400 K (Fig. 8.21). For conduction-band transport, such a small μ would require $N_I > 5 \cdot 10^{20}$ cm^{-3}, which is not consistent with the sharp excitonic linewidth observed and leads to the conclusion that the higher resistivity and the vanishing Hall mobility cannot be explained by the very large defect concentrations. One consistent picture is that the electrical transport is due to a nearest neighbor multi-phonon hopping process among localized defect centers in semi-insulating samples. Such hopping will, indeed, produce a very small or vanishing Hall coefficient and is in agreement with observation. At high enough temperatures, the energy will not be a limiting factor and Nearest-Neighbor (NN) hopping will dominate.

8.13 Observations

Though tremendous progress has been made in reducing the background impurities and increasing the measured electron mobility, the mobility figures, with the exception of some HVPE material, are still about half of those predicted by various theories. Though one can use compensation to explain the present mobilities, a true picture will likely emerge only after even better samples are available.

The picture regarding the hole mobility is not even at a point where meaningful discussions can be carried on. In addition to p-type dopants being deep, their presence in the growth environment also introduces donorlike, or perhaps even other, defects. There is also some disagreement about the hole effective mass, which recently got even more complicated by suggestions that the effective mass is really much heavier (as high as 2) than that the earlier predictions indicate [8.59].

9. The p-n Junction

As the name suggests, a p-n junction depicts the combination of two semiconductors having n- and p-type conductivities. If the two semiconductors forming the junction are of the same crystal, the term **homojunction** is used to describe the resulting structure. On the other hand, if two different semiconductors with very similar structural, but varying electrical and optical properties are used, the term **heterojunction** is applied. In modern LEDs and lasers, heterojunctions are employed for a variety of purposes which include carrier injection, and carrier and light confinement. In fact, before the advent of heterojunctions many optoelectronic and electronic devices were not possible among which was the CW (Continuous Wave) RT (Room Temperature) laser. Being such an integral part of lasers and LEDs, a concise description of the principles of p-n junctions and their characteristics is warranted. Detailed descriptions of heterojunction properties can be found elsewhere [9.1].

Heterojunctions that are pertinent to the current topic are those between GaN and AlGaN, AlGaN and InGaN, InGaN and GaN, and finally a variety of combination between other less commonly used ternary (InAlN) and quaternary (InGaAlN) alloys. While the bandgaps and the refractive indices of the end binaries are distinctly different, the unfortunate aspect of this family of semiconductors is that their lattice constants differ considerably. The requirement that the bulk of the device structure is free of gross structural defects allows alloys with only a small mole fraction in order not to cause structural defects at the interfaces. However, due to the lack of native substrates in the commercial market many defects are encountered in the bulk of the device structure so much so that the extent of allowable alloy mole fractions is not yet well defined.

9.1 Heterojunctions

When two semiconductors satisfying the above-mentioned description are joined together, the conduction and the valence bands at the junction do not align since the bandgaps of the two semiconductors that compose the heterojunction are not identical. This paves the way for band discontinuities to form in several fashions and the resulting heterojunctions are referred to as **types I , II and broken**. These designations depict the cases wherein **I** the conduction and valence bands of the larger bandgap semiconductor straddle the other, **II** the conduction and valence bands are staggered, and **broken** the valence-band energy of one is above, in energy, the conduction-band energy of the other. Naturally, the sum of discontinuities at the conduction and valence bands in type **I** should add up to the difference in bandgap of the two sides if there is no compositional gradient. An exhaustive review of the band offsets in semiconductor heterojunctions was given by *Yu* et al. [9.2].

In the original picture proposed by *Anderson* [9.3] who considered Ge/GaAs heterojunctions, the electron affinities or the work functions of the semiconductors forming the junction with a common vacuum level were used to determine the partitioning of the difference in bandgap between the conduction and valence bands. It has since been determined that this simple picture, while graphic and educational, does not really apply but hardly anything else has been proposed since. Generally, measurements of many kinds involving Ultraviolett Photoemission Spectroscopy (UPS), X-ray Photoemission Spectroscopy (XPS), C-V and I-V techniques have been used to determine the band discontinuities. In LEDs and lasers, electrons and holes are desired to be in the same layer for efficient recombination. An additional requirement for lasers is that the light confinement requires the active layer to be of a larger-refractive-index material. A type-I band alignment satisfies all these conditions which are formed by the wide-bandgap nitrides to be discussed here.

9.2 Band Discontinuities

Epitaxial crystal-growth techniques such as MBE or OMVPE (MOCVD) of II-VI and III-V heterostructures have proven to be capable of producing abrupt band-edges/discontinuities. Moreover, theoretical calculations indicate that the electronic structure in each layer of a heterojunction becomes very nearly bulk-like, even at a single atomic layer away from the interface, lending credence to the idealized notion of an abrupt band-edge discontinuity. Various types of band alignments are encountered in semiconductor interfaces depending on the relative adjustment of the energy bands with respect to each other. Figure 9.1 depicts two types of possible alignments which occur most commonly in semiconductor heterojunctions. It is worth noting that the device concepts which can be implemented successfully in a

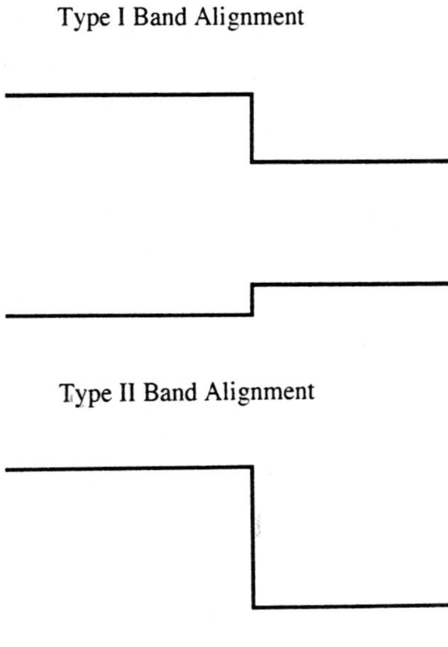

Fig. 9.1. Schematic description of type-I and type-II band alignments

given heterojunction system will depend very strongly on the type of band alignment of the heterojunction, and heterojunction-device performance will often depend critically on the exact values of the band discontinuities. Type-I alignment, in which the band gap of one semiconductor lies completely within the bandgap of the other, is the most useful one for optoelectronic devices. A type-II alignment occurs when the band gaps of the two materials overlap but one does not completely enclose the other (Fig.9.1). Type-II ZnSe/ZnTe heterojunctions have been used to overcome crucial problems related to difficulties in performing p-ohmic contacts for blue ZnSe-based lasers.

Investigations devoted to the determination of band discontinuities in semiconductor heterojunctions yield large discrepancies between measured and calculated values. The origin of the controversy can be related to different factors, among them:

- The technical difficulty and often the indirect nature of the measurements.
- The possible dependence of the band discontinuity on the detailed conditions of the interface preparation.
- The strain and polarity dependencies of the band discontinuities.
- Polarization effects.

The treatment of semiconductor heterojunction discontinuities can be grouped into three categories: The first consists of empirical rules such as electron-affinity rule and the common-anion rule, which give an indication of the type of band alignment and band-discontinuities values. In the second approach, pseudopotential and LCAO theories are utilized to extract band discontinuities from the electronic properties of the bulk semiconductors. The third category involves self-consistent calculations for specific interfaces using a super-cell geometry such as the Linear Muffin-Tin Orbital (LMTO) method [9.4]. It is worth noting that the capabilities of present-day theoretical treatments for band discontinuities are such that consistently reliable predictions for band-discontinuity values in novel semiconductor heterojunctions cannot yet be made. Band offsets must therefore be determined experimentally for each new material system of interest. XPS and UPS are primarily used to determine the band discontinuities by means of electron core-level energies. The energies of the core levels are obtained after an extensive band-structure modeling in order to obtain the valence-band edge. Optical techniques, such as excitation photoluminescence and reflectivity, represent a more accurate tool to determine the band-discontinuity values due to the higher experimental resolution compared to the previous X-ray techniques. Measurements of electrical characteristics such as the capacitance-voltage and current-voltage dependences have also been used with considerable success to determine GaAs/GaAlAs band discontinuities. However, they require an accurate knowledge of the charge density and its distribution throughout the region of the structure that is sampled.

The first detailed experimental investigation of the band discontinuity in the GaN/ AlN heterojunction has been reported by *Martin* et al. [9.5, 6], by taking advantage of the in-situ XPS method. Later, *Martin* [9.7] has extended the investigation to other binary-nitride heterojunctions such as InN/GaN, AlN/InN. All experimental and theoretical estimates of the band discontinuities indicate the occurrence of a type-I alignment between these nitrides materials InN, GaN and AlN.

9.2.1 GaN/AlN Heterostructures

The band-discontinuity values obtained by various experimental methods and the relevant experimental details are listed in Table 9.1. *Martin* et al. [9.6, 7] carried out systematic in-situ XPS studies of both GaN/AlN and AlN/GaN heterostructures with the resultant values obtained for Ga_{2d} and Al_{2p} core levels of both heterostructures; they are nearly the same within experimental errors ($\Delta E_v = 0.67 \pm 0.02\,eV$). *Waldrop* and *Grant* [9.8] reported higher values of the valence-band discontinuity in GaN/AlN grown

Table 9.1. Valence-band discontinuity values given in eV

GaN/AlN	AlN/GaN	Substrate	Ref.
0.8 ± 0.3	0.8 ± 0.3	SiC, Al_2O_3	9.5
0.5 ± 0.5		(polycrystal)	9.9
1.36 ± 0.07		SiC	9.8
0.60 ± 0.24	0.57 ± 0.22	SiC, Al_2O_3	9.6

on SiC ($\Delta E_v = 1.36 \pm 0.07\,eV$) from Ga_{3d} and Al_{2p} core levels measured by XPS in both bulk and heterostructure. The large difference between the values of these two studies is related to the discrepancy in determining the core-level binding energies which depend crucially on how the valence-band edge is fitted in density-of-state calculations.

In a parallel investigation, *Baur* et al. [9.9] estimated the valence band for GaN/AlN by postulating that transition-metal impurity levels in non-adjoining layers act as a common reference level to predict the band alignment in semiconductor heterojunctions. In the nitride system, band discontinuities were extracted using the fact that a characteristic infrared luminescence spectrum for both polycrystalline AlN and GaN is dominated by a zero-phonon line at 1.3 eV of the iron level. It was also observed that by employing photoluminescence excitation spectroscopy the (−/0) acceptor level of iron is located at $E_V + 3.0$ eV for AlN, and at $E_V + 2.5$ eV for GaN. The observed zero-phonon line was assigned to the spin-forbidden internal 3d-3d transition $^4T_1(G) \rightarrow {}^6A_1(S)$ of $Fe_{Al}^{3+}(3d^5)$. Based on this observation the valence-band offset was determined to be $\Delta E_V = 0.5$ eV, and the conduction-band offset to be $\Delta E_C = 2.3$ eV for the GaN/AlN heterostructure. *Albanesi* et al. [9.10] performed self-consistent calculations by means of the LMTO method to estimate the valence-band discontinuity at a zinc-blende AlN/GaN interface, in the [001] direction. The value obtained ($\Delta E_V = 0.85\,eV$) is close to the value found by *Martin* et al. with XPS. The experimental value of ΔE_v measured by *Waldrop* and *Grant* is in accordance with the affinity model prediction [9.8].

9.2.2 GaN/InN and AlN/InN

An in-situ XPS study of *Martin* [9.6] revealed a large asymmetry of the band-discontinuity values depending on the order with which the binary layers are grown, i.e., GaN on InN versus InN on GaN due to a polarity-dependent interface potential and the piezoelectric effect [9.6]. The experimental values have been reported:

$$\text{InN / GaN} = 0.93 \pm 0.25$$
$$\text{GaN / InN} = 0.59 \pm 0.24$$
$$\text{InN / AlN} = 1.71 \pm 0.2$$
$$\text{AlN / InN} = 1.32 \pm 0.14 .$$

It is interesting to note that the mean values obtained for both heterostructures obey the transitivity law when compared to the experimental value for GaN/AlN (Sect.9.2). The piezoelectric effect, to be described below, has been evoked to explain the asymmetry of the band discontinuities. The dangling bonds unique to one polarity of the interface may also induce such an asymmetry. Though this concept accounts for the major features of the observed data, deterministic investigations must be undertaken before a more definitive statement can be made.

Note that the InN/GaN-GaN/InN and InN/AlN-AlN/InN heterojunctions show a significant forward-backward asymmetry, while the AlN/GaN-GaN/AlN heterojunctions give almost identical values. All heterojunctions were in the standard type-I heterojunction alignment. Insight into the asymmetrical nature may be provided by strain-induced piezoelectric fields [9.11]. The lattice constant of GaN is larger than that of AlN by approximately 2.5%, while the lattice constant of InN is larger than that of GaN by

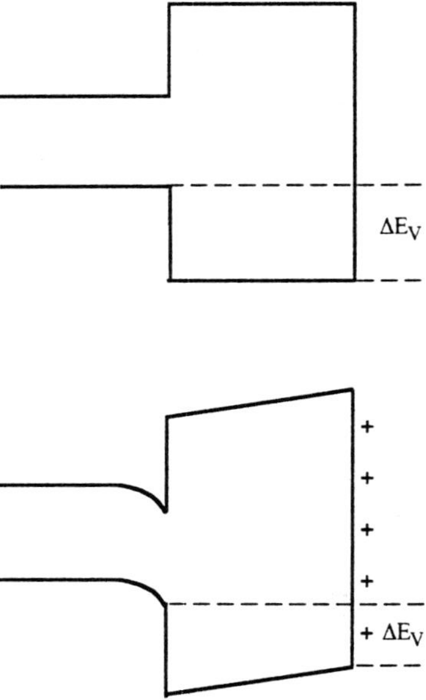

Fig. 9.2. Schematic representation of how piezoelectric effects alter the measured band discontinuities. The screening charge is shown

about 11% [9.12]. The heterojunction under-layers were thick enough to be relaxed to their natural lattice constants at the growth temperature with the residual strain remaining due to thermal mismatch between the substrate and the layers, so the heterojunction thin over-layers used were at least partially strained. The nitrides are piezoelectric materials, so the strain induced static electric fields via the piezoelectric effect play a role. Figure 9.2 illustrates how the strain-induced piezoelectric field changes the apparent valence-band discontinuity. For a Wz material grown in the [0001] direction, the strain-induced electric field is entirely longitudinal. MBE produces, at least on SiC substrates, N-terminated layers, so, in the language of [9.11], each grown layer ends with a B (anion N) face and begins with an A (cation In/Ga/Al) face. The nitrides have negative piezoelectric coefficients just like III-V zinc-blende materials. Furthermore, nitrides have decreasing bandgaps with increasing lattice constants, so that the field directions of [9.11] lead to the result that strain-induced piezoelectric fields always tend to decrease the apparent valence-band discontinuities for nitride materials. The field magnitude has been given by *Bykhovski* et al. [9.13]. Several of the constants appearing in the referenced equation are not well known for InN, GaN, and AlN; the values and sources used in this work are given in Table 9.2.

The calculated electric fields in the strained regions are

$$
\begin{aligned}
\text{InN} \quad &(\text{on GaN}) = 5.5 \cdot 10^8 \text{ V/m} ,\\
\text{GaN} \quad &(\text{on InN}) = 2.1 \cdot 10^9 \text{ V/m} ,\\
\text{GaN} \quad &(\text{on AlN}) = 4.7 \cdot 10^8 \text{ V/m} ,\\
\text{AlN} \quad &(\text{on GaN}) = 6.0 \cdot 10^8 \text{ V/m} ,\\
\text{InN} \quad &(\text{on AlN}) = 5.5 \cdot 10^8 \text{ V/m} ,\\
\text{AlN} \quad &(\text{on InN}) = 2.7 \cdot 10^9 \text{ V/m} .
\end{aligned}
$$

Table 9.2. Values used for calculating strain-induced piezoelectric effects

	InN	GaN	AlN
d_{31} [cm/V]	$-1.1 \cdot 10^{-10}$ [a]	$-17 \cdot 10^{-10}$ [a]	$-2.0 \cdot 10^{-10}$ [b]
ϵ (static)	15.3 [c]	10.0 [c]	8.5 [c]
c_{11} (GPa)	271 [d]	396 [d]	398 [d]
c_{12} (GPa)	124 [d]	144 [d]	140 [d]
c_{13} (GPa)	94 [d]	100 [d]	127 [d]
c_{33} (GPa)	200 [d]	392 [d]	382 [d]
a [Å]	3.548 [d]	3.189 [d]	3.112 [d]

[a] Estimated from the AlN value in the ratio $1/\epsilon$, ϵ being the strain.
[b] [9.24]
[c] [9.25]
[d] [9.26]

An important factor to note is the strong asymmetry for the hetero-junctions containing InN, where InN on the bottom leads to much bigger piezoelectric fields than does InN on the top. Assuming a typical hetero-junction over-layer thickness of 10 Å, these fields lead to voltage changes of $0.47 \div 2.7$ V, values much larger than the observed discrepancies. Under the assumption that the underlying semiconductor does not contain struc-tural defects to begin with, the critical thickness for GaN/AlN pseudo-morphic growth is approximately 30 Å, but the critical thicknesses for GaN/InN and AlN/InN are estimated at a much smaller (6Å), even less than the heterojunction over-layer thickness. In the present-day approaches and because of the lattice mismatch with the substrates, dense networks of thre-ading defects extend from the substrates to the surfaces. Compositional in-homogeneities are present and provide strain-relief mechanisms. Neverthe-less, we proceed with the supposition that heterojunction over-layers are partially strained to an unknown degree. The presence of the strain-induced piezoelectric effect accounts for the observed asymmetries and makes the case for measurements without piezoelectric effects, or at least with the smallest piezoelectric effect possible. After rejecting the cases with smaller valence-band discontinuities (based upon having larger piezoelectric fields) and applying rough corrections for the small strain-induced piezoelectric fields to the cases with larger valence-band discontinuities, the following re-sults

$$
\begin{array}{ll}
\text{InN} \mid \text{GaN} & \Delta E_v = 1.05 \pm 0.25 \text{ eV} \\
\text{GaN} \mid \text{AlN} & \Delta E_v = 0.70 \pm 0.24 \text{ eV} \\
\text{InN} \mid \text{AlN} & \Delta E_v = 1.81 \pm 0.20 \text{ eV}
\end{array}
$$

have been obtained. The interface potential and its polarization dependence may also be a factor [9.14].

These results obey the transitivity[1] well within the experimental ac-curacy. The ratios of conduction-band discontinuities to valence-band discontinuities are roughly 30:70, 75:25, and 60:40, respectively. We brief-ly note that the strain-induced piezoelectric fields lead to a small broaden-ing of the photoemission signals, since the piezoelectric fields create spatial variations of the potential in the near-surface regions where photoelectrons originate. This effect was observed throughout the work of *Martin* [9.7]. Note that the Fermi levels for both GaN and InN were noted just below the conduction-band edges and both materials typically have large n-type back-ground doping when grown by MBE. The Fermi level for AlN was found at roughly 4 eV above the valence-band edge although considerable variation

---

[1] Transitivity means the band alignment relationship of two semiconductors remains the same with respect to each other even when additional heterolayers are inserted between the two.

was seen which was likely due to charging. Having gone through this treatment, one should also point out that asymmetric interfacial defects can also contribute to the observed order dependence, which was briefly considered by *Martin* [9.6, 7].

9.3 Electrostatic Characteristics of p-n Heterojunctions

A detailed treatise of the topic is beyond the scope of this book, and the curious reader is referred to elsewhere [9.1]. A succinct treatment to the extent needed for self sufficiency will be presented. When n- and p-type semiconductors are joined, free carriers near the interface diffuse across the junction until the electric field caused by the depletion charge so created balances the motion. In the process, the Fermi levels on both sides align. To gain an insight, let us first consider the semiconductor component before contact (Fig. 9.3). The terms have their usual meanings. When the semiconductors are brought into contact, the zero steady-current condition necessitates that the Fermi level be constant throughout the semiconductor (Fig. 9.4). In Fig. 9.4a the larger-bandgap semiconductor is chosen as the p-type and in Fig. 9.4b as n-type semiconductor. The total band bending is called the **built-in potential**. This is rather simple to imagine and is illustrated with the bands before contact in Fig. 9.3. After contact, since the Fermi levels have to align, a potential equal to the separation of the Fermi levels must develop across the junction. This potential is comparable to the bandgap when the p and n layers are reasonably doped. When the junction is forward biased, the current increases rapidly as the internal voltage approaches

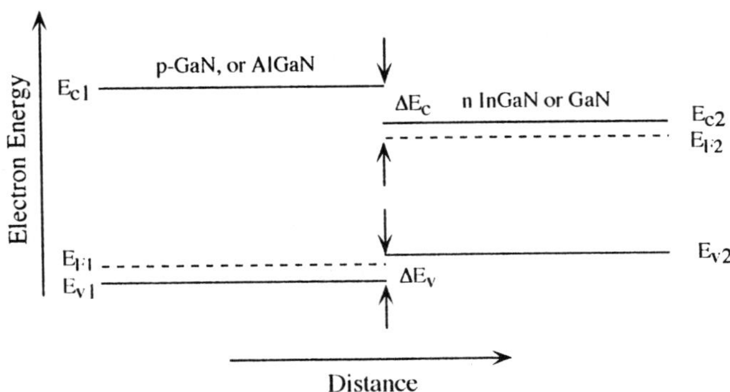

Fig. 9.3. Band diagrams of a larger and a smaller bandgap nitride semiconductor before contact

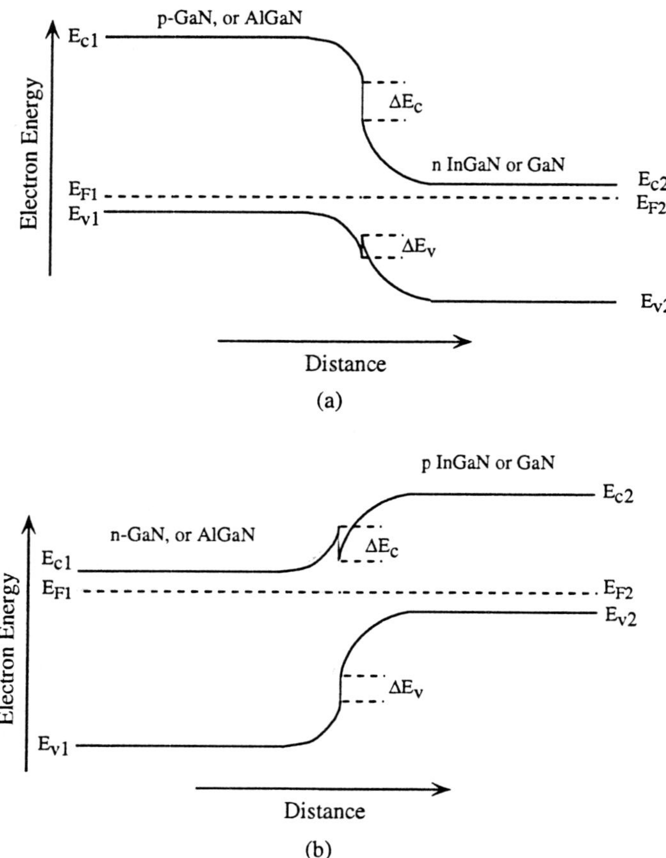

Fig. 9.4. (a) p-n and (b) n-p heterojunction band diagrams after contact at equilibrium

the built-in voltage. It can easily be demonstrated that the built-in voltage, which again is the separation of the Fermi levels, can be written as

$$E_{g1} - (E_{F1} - E_{v1}) + \Delta E_c - (E_{c21} - E_{F2}) - qV_{bi} = 0 \ . \tag{9.1}$$

The Fermi levels can be calculated using expressions of Chap. 7 knowing the electron and hole concentrations in the n- and p-type semiconductors, respectively. If the p- and n-type semiconductors were doped to the onset of degeneracy, meaning the Fermi levels align with the valence- and conduction-band edges in p- and n-type semiconductors, respectively, the built-in voltage would simply be

$$qV_{bi} = E_{g1} + \Delta E_c \ . \tag{9.2}$$

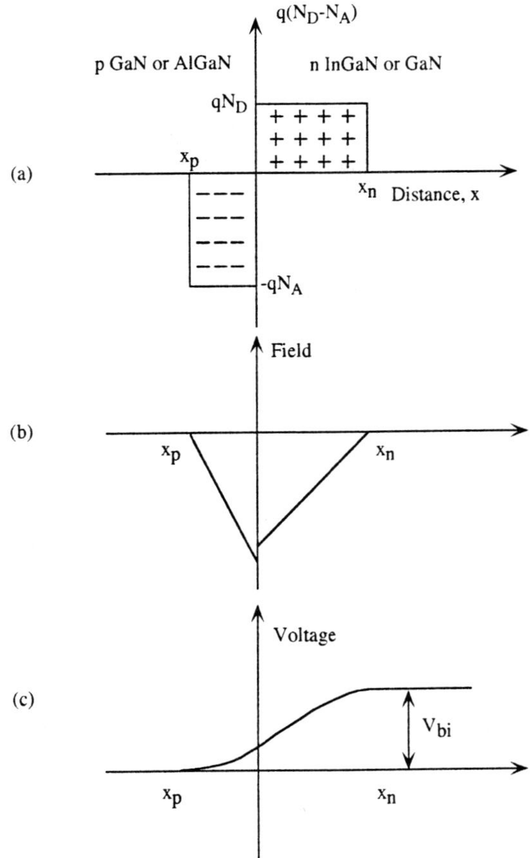

Fig. 9.5. Charge (**a**), field (**b**) and voltage (**c**) profiles of a nitride p/n heterojunction

Shown in Fig. 9.5a is the charge distribution in the space-charge region of the abrupt p-n heterojunction under discussion in the depletion approximation. In other words, free-carrier concentration is zero in the depletion region and the semiconductors are neutral immediately outside of the depletion region. Since the total net charge must be zero, the positive and negative charges on either side of the junction balance out, which is referred to as the **charge balance**. Figures 9.5b and c plot the field and voltage variations as a function of distance. Note the field discontinuity at the heterointerface due to the larger bandgap semiconductor which has the smaller dielectric constant.

Integration of the electric field results in the built-in potential, mentioned above, and that of the Poisson equation for the n and p sides twice, and taking advantage of the built-in voltage. The extent of the depletion region and its dependence on the forward and reverse bias. Poisson's equation can be written as

$$- \frac{\partial^2 V}{\partial x^2} = \frac{\partial E}{\partial x} = \frac{q}{\epsilon}[p(x) - n(x) + N_D^+ - N_A^-] \ . \tag{9.3}$$

In the depletion approximation, and with the additional assumptions that the semiconductor is not compensated and all the dopants are ionized, Poisson's equation is reduced to

$$- \frac{\partial^2 V}{\partial x^2} = \frac{q}{\epsilon_n} N_D \quad \text{for the n side, meaning} \quad 0 < x \leq n_n \tag{9.4}$$

and

$$- \frac{\partial^2 V}{\partial x^2} = - \frac{q}{\epsilon_p} N_A \quad \text{for the p side, meaning} \quad -x_p \leq x < 0 \ , \tag{9.5}$$

where subscripts n and p attached to the dielectric constant depict that the parameter is for the n and p sides, respectively. In a homojunction, they are equal.

Equation (9.3) can be solved with the boundary conditions that the field is zero at the edges of the depletion region, that the normal component of the displacement vector is continuos at the interface between the p and n layers, i.e., $\epsilon_n E_n = \epsilon_p E_p$, and that the voltage at x_n is equal to $V_{bi} - V$, V being the applied bias which is positive for forward bias and negative for reverse bias. Once again, the subscripts n and p designate the n and p sides, respectively. The solution is then

$$x_n^2 = \frac{2 N_A \epsilon_n \epsilon_p (V_{bi} - V)}{q N_D (\epsilon_n N_D + \epsilon_p N_A)} \ . \tag{9.6}$$

Similarly,

$$x_p^2 = \frac{2 N_D \epsilon_n \epsilon_p (V_{bi} - V)}{q N_A (\epsilon_n N_D + \epsilon_p N_A)} \ . \tag{9.7}$$

Knowledge of the extension of the depletion regions paves the way for calculating the junction capacitance, a matter of importance for the modulation bandwidth, the materials characterization, and the maximum electric field.

9.4 Current-Voltage Characteristics of p-n Junctions

In p-n junctions formed in wide-bandgap semiconductors, the diffusion-current component in reverse bias and for small applied voltages in forward bias is small, because of the small minority carrier concentrations. Consequently, the generation-recombination current takes on a strong importance. As the name suggests, the generation-recombination current is caused by capture and release of free carriers by traps. Generally speaking, if these centers are located in the upper half of the bandgap, they capture and emit electrons from and to the conduction band. Conversely, if the centers are in the lower half of the energy gap, they capture and emit holes from and to the valence band. An electron in the conduction band can be captured by a trap if the trap has not already captured an electron. The electron captured by a trap can later be re-emitted back to the conduction band. Similar processes occur involving hole traps and holes. Centers close to the middle of the gap are very efficient in the capture and emission processes, and can capture electrons and holes.

9.4.1 Generation-Recombination Current

The process described above was first developed independently by *Hall* [9.15], *Shockley* and *Read* [9.16], and *Sah* et al. [9.17] and is now commonly referred to as the **SRH generation/recombination process**. Capturing is a bimolecular process in that electrons and available trap states must participate. The rate of electron capture is proportional to the product of electron concentration in the conduction band and the concentration of traps not occupied by electrons. The treatment of p-n homojunctions here follows closely that of *Grove* [9.18].

If N_t is the trap concentration and f is the occupation probability which is nothing more than the Fermi-Dirac distribution function, then $1-f$ is the probability that a center is unoccupied. The occupation probability is given by the Fermi-Dirac statistics as

$$f = \frac{1}{1 + \exp[(E_t - E_F)/kT]} \cdot \tag{9.8}$$

Then, the rate of electron capture can be expressed as

$$r_a = v_{th}\,\sigma_n\,n N_t(1 - f) \tag{9.9}$$

where $v_{th} \sigma_n$ is a proportionality constant, and n is the electron concentration. The term v_{th} is the thermal velocity, and σ_n is the capture cross-section. The thermal velocity can be calculated from the kinetic energy as

$$v_{th} = \sqrt{\frac{3kT}{m_n^*}} \, ,$$

(9.10)

σ_n is a measure of how close the electron must come to the vicinity of the center before capture. Using a parallel argument, the rate of hole capture by a center, the electron dropping into the valence band is given by

$$r_c = v_{th} \sigma_p p N_t f \, .$$

(9.11)

Similarly the rate of electron emission from a trap center into the conduction band can be expressed as

$$r_b = e_n N_t f \, ,$$

(9.12)

where e_n denotes the electron-emission probability. Hole emission from the center to the valence band or electron transfer from the valence band to the center is given by

$$r_d = e_p N_t (1 - f) \, ,$$

(9.13)

where e_p is the hole emission probability. Using the fact that under equilibrium the electron emission and capture rates must be equal, so are the hole emission and capture rates, one can obtain expressions for the emission rates as we have for capture rates. Doing so leads to

$$e_n = v_{th} \sigma_n n_i \exp\left(\frac{E_i - E_F}{kT}\right)$$

(9.14)

and

$$e_p = v_{th} \sigma_p n_i \exp\left(\frac{E_i - E_F}{kT}\right) \, .$$

(9.15)

When the electric field strength is large both the emission and capture probabilities would be modified. This could happen in reverse-biased GaN diodes with large operating voltages. If we assume the effect to be like that due to *Frenkel-Poole* [9.19], then the probability expression would be multiplied by $\exp[(qE/\pi\epsilon)^{1/2}/kT]$, E being the electric field. Under non-equilibrium conditions, generation due to external causes such as injection and optical illumination must be considered. The rate equations for a special case, the steady state, in an n-type semiconductor, can be written as

$$\frac{dn_n}{dt} = G_L - (r_a - r_b) = 0 \, , \qquad (9.16)$$

which is simply the time rate of change of the electron concentration, which is equal to the generation rate minus the capture rate plus the emission rate, which is equal to zero under steady state. Here, we assumed that the electron and hole emission rates do not change other than through changes in the carrier concentrations.

By analogy, we can write for holes

$$\frac{dp_n}{dt} = G_L - (r_c - r_d) = 0 \qquad (9.17)$$

and

$$r_a - r_b = r_c - r_d = U \, . \qquad (9.18)$$

Since the concept of the Fermi level is not meaningful under non-equilibrium conditions, the electron and hole concentrations will depend on the injection level or the external generation rate making f dependent on injection.

The term U is the net generation rate and can be related to the usual parameters as

$$U = \frac{\sigma_n \sigma_p \, v_{th} \, N_t \, (pn - n_i^2)}{\sigma_n \{n + n_i \exp[(E_t - E_i)/kT]\} + \sigma_p \{p + n_i \exp[(E_i - E_t)/kT]\}} \, . \qquad (9.19)$$

If we assume $\sigma_n = \sigma_p = \sigma$,

$$U = \sigma v_{th} N_t \frac{pn - n_i^2}{n + p + 2n_i \cosh[(E_t - E_i)/kT]} \, . \qquad (9.20)$$

We see that the emission rate increases when n+p is at its minimum and the trap center gets closer to the intrinsic level. In short, the centers near the intrinsic level are very effective generation-recombination centers. We should point out that the emission process is enhanced when the center moves close to the respective band, but the recombination at that center is suppressed. In short, the emission and recombination processes together are at their maximum when the center is at the intrinsic level.

If we now consider an n-type semiconductor which means that n is much larger than p, assume a low-level injection and that the traps are not near the conduction band, the latter would make the exponential in (9.20) negligible, the generation rate is then

$$U \approx \sigma_p v_{th} N_t \frac{pn - n_i^2}{n + p} = \sigma_p v_{th} N_t (p_n - p_{n0}) \tag{9.21}$$

where p_n and p_{n0} represent the minority-carrier hole and equilibrium hole concentrations, respectively.

Since $U = (p_n - p_{n0})/\tau_p$, the minority-carrier recombination lifetime is

$$\tau_p = \frac{1}{\sigma_p v_{th} N_t} . \tag{9.22}$$

Similary, for a p-type semiconductor,

$$U = \sigma_n v_{th} N_t (n_p - n_{p0}) \quad \text{and} \quad \tau_n = \frac{1}{\sigma_n v_{th} N_t} . \tag{9.23}$$

9.4.2 Surface Effects

In addition to the generation-recombination processes, the surface-recombination current also plays an important role in wide-bandgap semiconductor, because the diffusion contribution to the current is small, particularly in reverse bias and at low injection levels. Indirect experiments appear to indicate that the surface-recombination velocity is perhaps not very high in GaN. Nevertheless, a treatment is warranted. Besides, processing-induced damage may cause this component to be sizable.

The surface-recombination rate for an n-type semiconductor is given by

$$U_s = \sigma_p v_{th} N_{st} [p_n(0) - p_{n0}] , \tag{9.24}$$

where N_{st} is the center concentration in cm^{-2}, $p_n(0)$ is the average minority-carrier concentration on the surface in question, and p_{n0} is the equilibrium minority carrier concentration. Carrier loss to surface recombination sets a minority carrier profile promoting diffusion to the surface, the process of which can be expressed as

$$D_p \left. \frac{\partial p_n}{\partial n} \right|_{x=0} = \sigma_p \, v_{th} \, N_{st} \, [p_n(0) - p_{n0}] = s_p \, [p_n(0) - p_{n0}] \; , \qquad (9.25)$$

where $s_p = \sigma_p \, v_{th} \, N_{st}$ is the surface-recombination velocity.

This process may be complicated by the presence of ions on the surface which would change the potential distribution, "cause" the space-charge neutrality not to hold: the space charge will be balanced by the ions. For this case the generation-recombination rate (9.19) should be modified as

$$U_s = \frac{\sigma_n \, \sigma_p \, v_{th} \, N_{st} \, (p_s \, n_s - n_i{}^2)}{\sigma_n \, \{n_s + n_i \exp[(E_t - E_i)/kT]\} + \sigma_p \, \{p_s + n_i \exp[(E_i - E_t)/kT]\}} \, , \qquad (9.26)$$

where the subscript refers to the fact that the parameter is associated with the surface. If the centers are located at or near the middle of the gap ($E_t \approx E_i$), which are the most effective ones, and the electron and hole capture cross-sections are equal, U_s becomes

$$U_s = s_p \, \frac{p_{x_d} \, n_{x_d} - n_i{}^2}{n_{x_d} + p_{x_d} + 2n_i} \qquad (9.27)$$

where s_p denotes the surface recombination velocity for a surface without a space charge layer. Again, the carriers must be supplied from the bulk through diffusion, i.e.,

$$D_p \left. \frac{\partial p_n}{\partial x} \right|_{x=x_d} = s_p \, \frac{p_{x_d} \, n_{x_d} - n_i{}^2}{n_{x_d} + p_{x_d} + 2n_i} \qquad (9.28)$$

where x_d is the edge of the space-charge region.

To express the right-hand side in terms of carrier concentrations, we will assume the $n \cdot p$ product to remain constant even though equilibrium does not prevail. In other words,

$$p_{x_d} \, n_{x_d} = p_n(x_d) n_n(x_d) = p_n(x_d) N_D \; , \qquad (9.29)$$

noting that $p_0(x_d) N_D = n_i{}^2$, and we have

$$D_p \left. \frac{\partial p_0}{\partial x} \right|_{x=x_d} = s[p_n(x_d) - p_{n0}] \tag{9.30}$$

with $s \equiv s_p N_D / (n_{x_d} + p_{x_d} + 2n_i)$.

9.4.3 Diode Current Under Reverse Bias

Referring to (9.19, 20), in a reverse-bias junction the free electron and hole concentrations are very small and can be neglected in the denominator. In addition, the electron-hole product $p \cdot n = n_i^2 \exp(qV_r/kT)$ in a reverse-bias junction tends to zero because of the large negative V_r. Then, the net recombination rate reduces to

$$U = -n_i/\tau_e \tag{9.31}$$

where τ_e is an effective recombination lifetime which is given by

$$\tau_e = \frac{\sigma_n n_i \exp[(E_i - E_i)/kT] + \sigma_p n_i \exp[(E_i - E_t)/kT]}{\sigma_n \sigma_p v_{th} N_t}. \tag{9.32}$$

In addition, if we assume the generation-recombination centers to be at or near the intrinsic level, the effective lifetime is reduced to (9.22).

9.4.4 Effect of the Electric Field on the Generation Current

For large electric fields under the reverse-bias condition, the Frenkel-Poole-like effect would modify the net generation rate, assuming an n-type sample where the $p_0 + 2n_i$ term is small compared to n_0, and the net generation rate expression is

$$U \approx \frac{-2n_i^2}{n_0 \tau_p \left[\exp\left(-\sqrt{\frac{qE}{\pi \epsilon} \frac{1}{kT}}\right) \right]} = \frac{-2p_0}{\tau_p \left[\exp\left(-\sqrt{\frac{eE}{\pi \epsilon} \frac{1}{kT}}\right) \right]}. \tag{9.33}$$

The generation current density without the Frenkel-Poole effect is

$$J_{gen} = \int_0^W q|U|dx = \frac{qn_i W}{\tau_e}$$ (9.34)

which reduces for an abrupt junction to

$$j_{gen} = \frac{qn_i}{\tau_e}\sqrt{\frac{2\epsilon_s(V_{bi} + V)}{q}\left(\frac{1}{N_A} + \frac{1}{N_D}\right)}$$ (9.35)

indicating the square-root dependence on the applied voltage V. Here, W denotes the width of the depletion region. It is clear that the higher the trap density, the smaller the effective lifetime, and the higher the generation current.

The generation current modified by the Frenkel-Poole effect is

$$J_{gen} = q\int_0^W |U|dx = q\int_0^w \left| -2p_0 dx / \tau_p \exp\left(-\sqrt{\frac{qE}{\pi\epsilon}\frac{1}{kT}}\right)\right| .$$ (9.36)

For a constant E field in the space-charge region such as that in a pin diode, we have

$$j_{gen} = \frac{-2qWp_0}{\tau_p}\exp\left(\sqrt{\frac{qE}{\pi\epsilon}\frac{1}{kT}}\right) .$$ (9.37)

9.4.5 Diffusion Current

The total diffusion current is the sum of the diffusion currents on the n and p sides caused by both thermal generation and other means within the diffusion length of the depletion region such as the optical excitation, i.e.,

$$J_{diff} = J_{diffn} + J_{diffp}$$

$$= qD_n\frac{n_{p0} + \tau_n G_L}{L_n} + qD_p\frac{p_{n0} + \tau_p G_L}{L_p} ,$$ (9.38)

where G_L is the optical generation rate. In the dark, this expression with $L_{p,n} \equiv (D_{p,n}\tau_{p,n})^{1/2}$ reduces to the more familiar relation

$$J_{diff} = qn_{p0}\sqrt{\frac{D_n}{\tau_n}} + qp_{n0}\sqrt{\frac{D_p}{\tau_p}} \,. \tag{9.39}$$

The diffusion constant for holes and electrons can be estimated from the mobilities, and the hole and electron minority carrier lifetimes would have to be determined.

The total current under reverse bias is

$$J_r = q\left[\frac{n_i^2}{N_A^-}\sqrt{\frac{D_n}{\tau_n}} + \frac{n_i^2}{N_D^+}\sqrt{\frac{D_p}{\tau_p}} \right.$$

$$\left. + \frac{n_i}{\tau_e}\sqrt{\frac{2\epsilon_s(V_{bi}+V)}{q}\left(\frac{1}{N_A} + \frac{1}{N_D}\right)} \right] \,. \tag{9.40}$$

For a heterojunction, this expression should be modified. For a *P*-n heterojunction,[2] the first term in the diffusion-limited current should contain the parameters of the wider-bandgap material. The second term should contain the smaller-bandgap parameters. The generation-recombination current is more complex but can be separated into two parts each dealing with the respective semiconductor. This requires some knowledge of the effective recombination lifetime, in addition to the dielectric constants.

In wide-bandgap materials such as GaN, the n_i^2 term is very small, and thus the reverse current is dominated by the generation-current component when neglecting the surface-recombination current.

If we assume that the surface recombination velocity is large enough to bring the carrier concentration to zero, the surface-recombination velocity for minority holes and electrons is the same, and the region of surface recombination is limited to the depletion depth $W = x_n + x_p$, the surface-recombination current density can be expressed as

$$J_s = \int_0^W q|U_s|\,dx = sq\left(\frac{n_i^2}{N_A^-}x_n + \frac{n_i^2}{N_D^-}x_p\right) \,. \tag{9.41}$$

---

[2] The italic capital letter depicts the wider bandgap semicondutor

The total current, neglecting the diffusion current, is the sum of the generation and surface-recombination currents and is given by

$$J_r = q_n \frac{x_n + x_p}{\tau_e} + sq\left(\frac{n_i^2}{N_A^-}x_n + \frac{n_i^2}{N_D^-}x_p\right) . \tag{9.42}$$

The small intrinsic carrier concentration appears to make the second term negligible but only if the surface recombination velocity ($s = \sigma v_{th} N_{st}$) is small.

9.4.6 Diode Current Under Forward Bias

In a forward biased p-n junction, the minority carriers are injected across the junction and pave the way for the diffusion current. Generation-recombination centers in the depletion region give rise to an additional current component: **generation-recombination current**. The latter is said to cause the diode current to deviate from being ideal. In addition, as in the reverse bias case, surface recombination current too may be noticeable if the junction area is small and the surface recombination velocity is high.

$$J_{diff} = q\left(\left| \frac{n_i^2}{N_A^-}\sqrt{\frac{D_n}{\tau_n}} + \frac{n_i^2}{N_D^+}\sqrt{\frac{D_p}{\tau_p}} \right|\right)\left[\exp\left(\frac{qV}{kt}\right) - 1 \right] . \tag{9.43}$$

This expression can also be modified to give a first-order representation of the current-voltage relationship for a heterojunction. This is accomplished by using the parameters of the p-type layer for the first term in the parentheses and of the n-type layer for the second term.

The forward-biased recombination current, for the case where trapping centers are at the intrinsic level and hole and electron capture cross sections are equal, is again obtained in exactly the same manner as in the reverse bias case employing (9.34). Through a modification of (9.20), the generation rate for the forward biased case can be expressed as

$$U = \sigma v_{th} N_t \frac{n_i^2}{n + p + 2n_i}(e^{qV/kT} - 1) \tag{9.44}$$

where we made use of the quasi-equilibrium expression

$$p \cdot n = n_i^2 e^{qV/kT} .$$

Recognizing that the p-n product is constant throughout the junction, U would be maximized when p+n is minimized, which occurs where n = p. Therefore, at the point where U is maximum,

$$p = n = n_i e^{qV/2kT} \tag{9.45}$$

$$U_{max} = \sigma v_{th} N_t \frac{n_i}{2} \frac{e^{qV/kT} - 1}{e^{qV/2kT} + 1} \; . \tag{9.46}$$

When $V \gg kT/q$, $U_{max} = (n_i/\tau_e)\exp(qV/2kT)$, and $\tau_e = (2\sigma v_{th} N_t)^{-1}$ is the effective lifetime.

The recombination current can be found by integration as

$$J_{rec} = -\frac{qn_i}{\tau_c} W \exp\left(\frac{qV}{2kT}\right)$$

$$= \frac{qn_i}{\tau_e} \sqrt{\frac{2\epsilon_s (V_{bi} - V)}{q}\left(\frac{1}{N_A} + \frac{1}{N_D}\right)} \; \exp\left(\frac{qV}{2kT}\right) . \tag{9.47}$$

The most right side of the equation holds true for an abrupt junction and shows the voltage dependence of the recombination current.

As will be discussed in Chap. 11, the forward current in nitride junctions is not governed by the diffusion process. Tunneling is the dominant process.

9.5 Calculated and Experimental I-V Characteristics of GaN Based p-n Junctions

The calculated forward diffusion and recombination-generation currents in GaN p-n junction at room temperature, using (9.34, 38, 43, 47) are displayed in Fig. 9.6. The parameters assumed are $N_D = 5 \cdot 10^{17}$ cm^{-3} and $N_A = 10^{17}$ cm^{-3} for the n and p sides, respectively. Plots are presented for the effective recombination lifetimes and the minority-carrier lifetimes of 10 ps, 100 ps, 1 ns and 2 ns. Both recombination-generation and diffusion currents increase with decreasing lifetimes. No external effects such as the series resistance have been included. In wide-bandgap semiconductors where the intrinsic-carrier concentration is extremely small, the diffusion

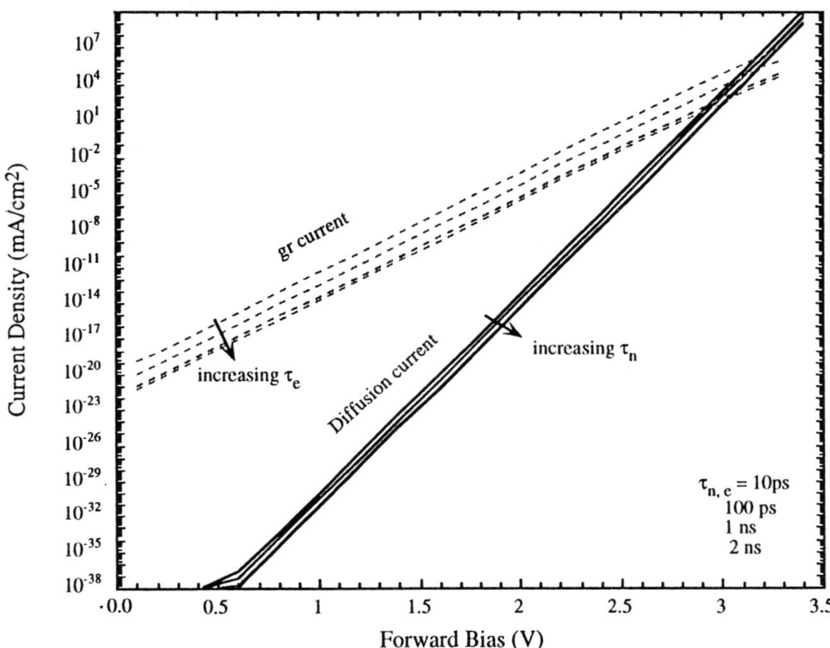

Fig. 9.6. Calculated forward diffusion and recombination-generation currents in GaN p-n junction at room temperature. The assumed parameters are $N_D = 5\cdot10^{17}$ cm^{-3} and $N_A = 10^{17}$ cm^{-3} for the n and p sides, respectively. Plots are presented for the effective recombination lifetimes and minority carrier lifetimes of 10 ps, 100 ps, 1 ns and 2 ns. Both currents increase with decreasing lifetimes. No external effects such as the series resistance are included

does not really become all that noticeable until the forward bias is high. This means that in the absence of tunneling currents, the forward current would be dominated by the 2kT current. In experimental devices, the p-type ohmic contact is problematic in that it is not ohmic, a *poor Schottky barrier*, and the semiconductor itself is very defective. Consequently, the poor Schottky junction may provide the limiting mechanism. Unless the current conduction through this junction is understood and modeled well, it will not be possible to discern the properties of the p-n junction accurately. Having said that, the calculation of the p-n junction current as is done here is warranted as critical information can be extracted.

In the reverse direction, the diffusion current limited by thermal generation of minority carriers is minuscule due to the large bandgap of GaN, as indicated in (9.38). Figure 9.7 depicts the calculated reverse recombination-generation current in GaN p-n junctions at room temperature. The assumed parameters are $N_D = 5\cdot10^{17}$ cm^{-3} and $N_A = 10^{17}$ cm^{-3} for the n and p

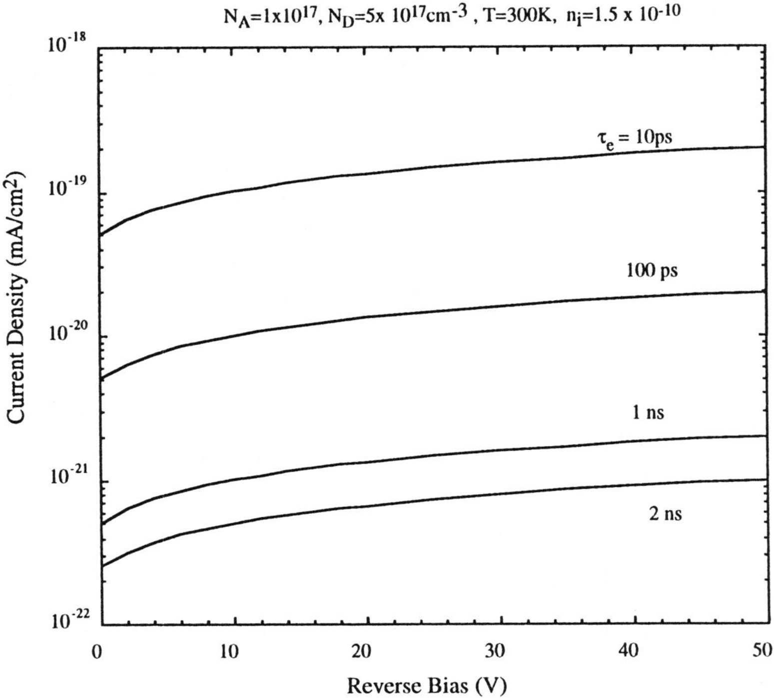

Fig. 9.7. Calculated reverse recombination-generation current in GaN p-n junction at room temperature. The assumed parameters are $N_D = 5 \cdot 10^{17}$ cm^{-3} and $N_A = 10^{17}$ cm^{-3} for the n and p sides, respectively. Plots are presented for effective recombination lifetimes of 10 ps, 100 ps, 1 ns and 2 ns. Both currents increase with decreasing lifetimes

sides, respectively. Plots are presented for the effective recombination lifetimes of 10 ps, 100 ps, 1 ns and 2 ns. Clearly, the current increases with decreasing lifetimes, and the current is extremely small unless the effective recombination lifetime is small.

Owing to the fact that nitride devices are intended for high-power operation and/or operation where high temperatures may be imperative, it is beneficial to consider the generation-recombination current and the diffusion current at high temperatures as well. Choosing 100 ps for the effective lifetime with respect to the generation-recombination current and 1 ns for the minority-carrier lifetime due to the diffusion current, forward-current voltage characteristics for a GaN p-n junction with $N_D = 5 \cdot 10^{17}$ cm $^{-3}$ and $N_A = 10^{17}$ cm^{-3} are displayed in Fig. 9.8 for temperatures of 300, 350, 450, 650, and 750 K; and the same for a reverse bias in Fig. 9.9. The calculations utilized (9.34, 38, 43, 47) and assumed, in the absence of reliable data, that the diffusion length is temperature independent, which is not strictly correct. The reverse-bias current is assumed to be dominated by the recombination-generation current and thus the diffusion current is neglected.

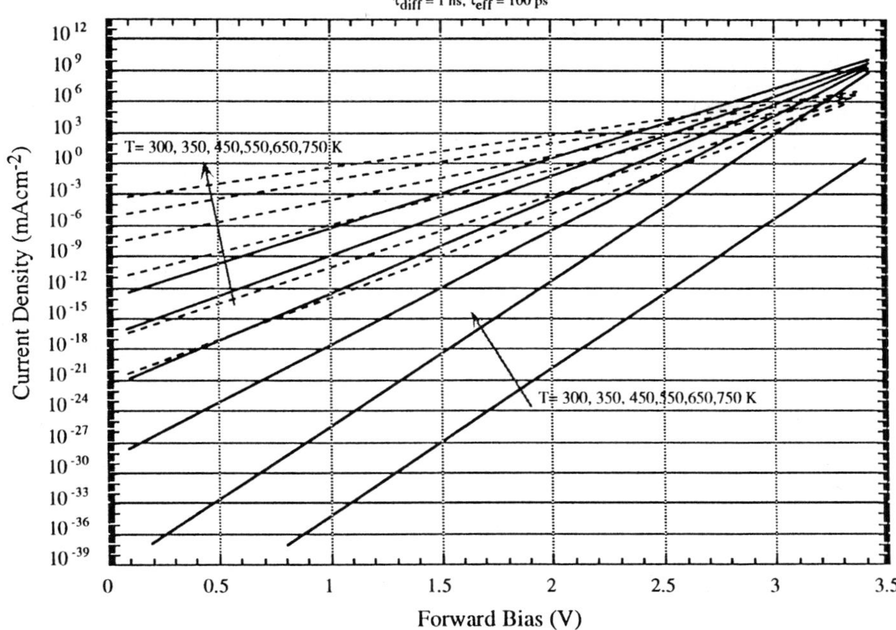

Fig. 9.8. Forward-current voltage characteristics for a GaN p-n junction for temperatures of 300, 350, 450, 650, and 750 K, and for the effective lifetime of 100 ps regarding generation-recombination current, and for the minority carrier lifetime of 1 ns regarding the diffusion current

The forward I-V characteristics of an experimental GaN p-n junction is exhibited in Fig. 9.10 for the given temperatures in the range of $17° \div 250°C$ [9.20]. Superimposed are the slopes of kT diffusion, and 2kT generation-recombination currents at room temperature. Clearly, the ideality factor of the experimental current is larger than 2, indicating current paths other than the aforementioned current components. It is very plausible that the tunneling current through the high-density defects present in GaN takes this path [9.21].

Available data on the current-voltage characteristics of P-n [3] and p-n junctions can not be accounted for by diffusion and generation-recombination current. One such I-V characteristic obtained for a GaN(p)/GaN(n)/GaN(n$^+$) diode is depicted in Fig. 9.11. It demonstrates an exponential dependence of the reverse current on voltage. The exponential voltage dependence of the current, along with the lack of a generation-recombination component is indicative of a tunneling-dominated process. Similar behavior is also prevalent in the forward current of LEDs, as discussed in detail in

---

[3] The capital italic letter denotes the wider-bandgap semiconductor

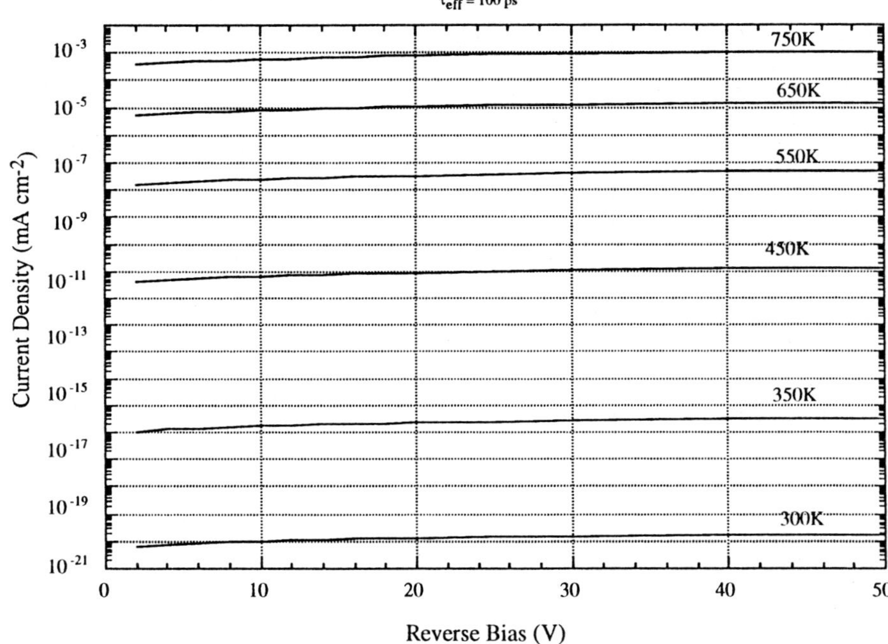

Fig. 9.9. Reverse-bias current-voltage characteristics for a GaN p-n junction for temperatures of 300, 350, 450, 650, and 750 K, and for an effective lifetime of 100 ps. The current is assumed to be dominated by the generation-recombination current, and thus diffusion current is neglected

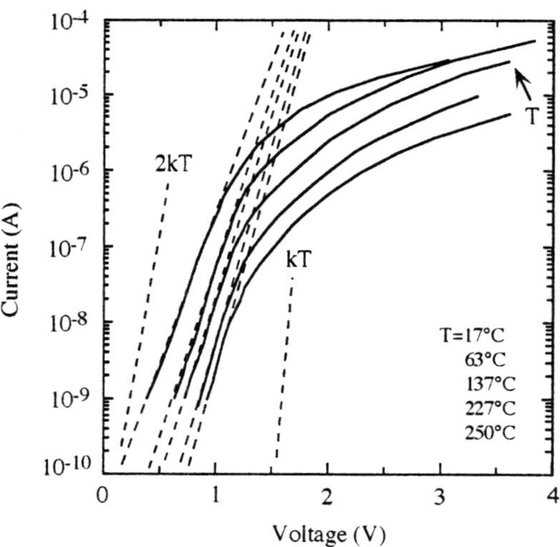

Fig. 9.10. Forward I-V characteristics of an experimental GaN p-n junction for the indicated temperatures in the range of $17° \div 250°$ C. Superimposed are slopes of kT diffusion and 2kT generation-recombination currents at room temperature. The experimental results have been taken from [9.20]

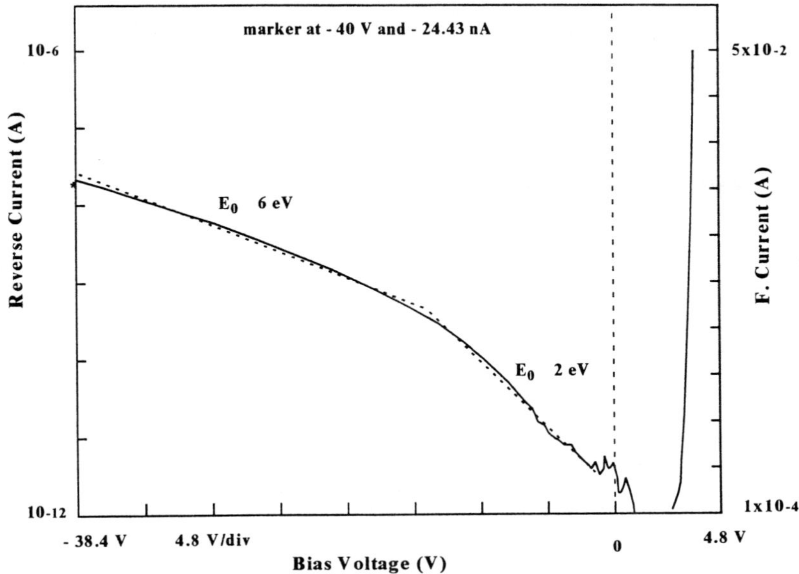

Fig. 9.11. Room-temperature current-voltage characteristic of a GaN(p)/GaN(n)/GaN(n$^+$) diode with a diamter of 200 μm. The reverse current is nearly an exponential function of the voltage with two different ($\approx$5 and 6 eV) energy parameters below about 15 V and above 15 V, respectively

Chap. 11. The reverse I-V characteristics of the aforementioned diode can be represented by the following expression, but with a different E_0 energy parameter for various voltage ranges [9.22, 23]

$$I = I_{10} e^{qV/E_0} \quad ,$$

(9.48)

where V and E_0 are the diode voltage and the energy parameter, respectively, with I_{10} being a prefactor. Preliminary investigations conducted in the author's laboratory indicated that the energy parameter is approximately 2 eV below about 15 V and 6 eV above 15 V. The tunneling current is indicative of a defect involvement and consequently it should be sample dependent as the layers contain many gap states. The Frenkel-Poole current (9.36) would indicate a square-root dependence of the current, in an exponential scale, on the electric field. In addition to this field, the Frenkel-Poole current is temperature activated.

9.6 Concluding Remarks

The lack of band-edge emission in light-emitting diodes and lasers, the lack of kT and even 2kT forward current is all indicative of the current conduction being dominated by others than free carriers in the conduction and valence bands. This behavior is endemic to large-bandgap semiconductor junctions including ZnSe-based LEDs and certainly involve defects either in the form of band-tail states and/or defect-assisted tunneling. Further discussion of the forward current-voltage characteristics can be found in Chap. 11.

10. Optical Processes in Nitride Semiconductors

One of the most amazing properties of semiconductors, particularly direct-bandgap semiconductors, is the light emission, which revolutionized the opto-electronics field. Light emission can be caused through a variety of stimuli among which electroluminescence has seen the most practical application. When an external voltage is applied across a p-n junction, electrons and holes that are injected into the medium recombine. This annihilation results in the emission of a photon whose energy is equal to the difference in the energies of states occupied by electrons and holes prior to recombination. In indirect semiconductors, phonons are generated predominantly, making them inefficient light emitters unless highly localized centers are utilized such as N in GaP.

Another means of light emission, termed **photoluminescence**, is a result of impinging-photon absorption, electron-hole-pair generation, and a photon of different wavelength emission. The incident photons, when absorbed, excite electrons from the valence band into the conduction band through momentum-conserving processes since the photon momentum is negligible. The electrons and holes thermalize to the bottom of their respective bands via phonon emission before recombining across the bandgap and emitting photons of the bandgap energy. While it is not applied to compact devices, this technique is commonly employed in extracting important physical properties and performing materials characterization. Light emission can also be induced by raising the temperature of the semiconductor as well, which is called **thermoluminescence**, again with application to basic studies. A corollary to this, changes in the lattice temperature caused by light absorption, provides a plethora of information about states present in the semiconductor. Photon emission can also be induced by subjecting the semiconductor to electron irradiation and other high-energy particle irradiation. This will not be covered here as it is beyond the scope of this book.

Not all recombination results in the emission of a photon. Recombination process resulting in photon emission is termed **radiative recombination**. This process is called **spontaneous emission**. In other words, electron hole (e-h) pairs are annihilated followed by photon emission. When an electromagnetic field of appropriate frequency is involved in the process, this emission is termed **stimulated emission** such as that found in semiconductor lasers. Naturally, the probability of stimulated emission is proportional to the field strength. That which does not produce photons is termed **non-radiative recombination** in which the energy is exchanged with the lattice as heat through phonons. In a direct-bandgap semiconductor, the latter is enhanced due to defects.

10.1 Absorption and Emission

Among the most important optical processes taking place in semiconductors, which directly influence device operation are the absorption, and emission of photons. Absorption and emission spectroscopies can be employed to extract a plethora of useful information about the semiconductor, particularly a direct-bandgap semiconductor. As such, these techniques are commonly taken advantage of to shed light on the materials properties and gather data which could be used for devices. The fact that a great majority of GaN-based layers are grown on transparent-sapphire substrates paves the way to perform absorption measurements without the need for an often cumbersome substrate removal. Simply saying, the semiconductor is transparent to below the gap radiation while absorbing above the gap radiation. Excitonic absorption is superimposed on band-to-band absorption which makes it convenient to investigate the role of excitons on the device features (Chap. 12).

Let us now turn our attention to the absorption coefficient. If $I(\nu)$ represents the optical intensity at point x in a semiconductor, the spatial rate of change of the intensity at the same point is proportional to the intensity, and is given by

$$\frac{dI}{dx} = -\alpha I \tag{10.1}$$

where α is the absorption coefficient with a unit of inverse cm. In an absorptive medium, the dielectric function is complex and can be expressed as

$$\epsilon = \epsilon' + i\epsilon'' = \epsilon_0 (n + i\kappa)^2 = \epsilon_0 (n^2 - \kappa^2 + i2n\kappa) \tag{10.2}$$

where ϵ' and ϵ'' are the real and imaginary components of the dielectric constant, n and κ, represent the refractive index and the extinction coefficient, respectively.

Since the power of an electromagnetic field propagating along the z direction is proportional to

$$e^{-2i(n + i\kappa)k_0 z} \tag{10.3}$$

where $k_0 = 2\pi/\lambda_0$ is the free-space wave vector, and the real part of the exponent is that associated with the absorption coefficient, we have

$$\alpha = 2\kappa k_0 \tag{10.4a}$$

where

$$\kappa = \frac{\epsilon''}{2\epsilon_0 n} . \tag{10.4b}$$

It is clear that the imaginary part of the dielectric constant is responsible for the loss.

The III-V nitride semiconductors have direct bandgaps, with the exception of the calculated indirect band of cubic AlN. They have large absorption coefficients with near-bandgap values in excess of 10^5 cm^{-1} which bode very well for LEDs, lasers, and detectors.

Einstein [10.1, 2] depicted absorption and stimulated emission per unit electromagnetic energy with an energy between $h\nu$ and $h\nu + \Delta\nu$ due to the transition of an electron from a level 2 to a level 1 by the coefficients B_{21} which is referred to as the transition probability from state 2 to 1. The same for spontaneous emission is depicted as A_{21} which together with B are called the **Einstein's A and B coefficients** [10.3, 4]

In a two-level system with 1 representing the lower and 2 the upper level, the rate of upward and downward transitions for a system at thermal equilibrium at the temperature T, were expressed by *Einstein* [10.1] as

$$R_{21} = [A_{21} + B_{21}\rho(\nu)]N_2 , \tag{10.5}$$

$$R_{12} = B_{12}\rho(\nu)N_1 , \tag{10.6}$$

where N_i $(i = 1, 2)$ represents the number of atoms in their respective states, $\rho(\nu)$ is the photon density in the energy interval $h\nu$ and $h\nu + \Delta\nu$ times $h\nu$, which is given by

$$\rho(\nu) = \frac{8\pi\nu^2 n^3}{c^3} \frac{h\nu}{e^{h\nu/kT} - 1} d\nu . \tag{10.7}$$

$A_{21} N_2$ represents the spontaneous emission process with which the electromagnetic radiation does not participate, and $B_{21} \rho(\nu) N_2$ denotes the stimulated process with the electromagnetic radiation participating. Since the spectral width of the radiation is finite, the unit of the beam intensity is W/m^2 per unit frequency interval.

Energy balance requires that R_{12} and R_{21} be equal, which determines the **spectral distribution**. In other words, the upward transition rate must be equal to the total downward transition rate at thermal equilibrium. Equating the temperature-independent components of this equality leads to

$$A_{21} = \frac{8\pi h\nu^3 n^3}{c^3} B_{21} \qquad (10.8)$$

where c denotes the velocity of light in vacuum, and n is the refractive index.

The term relating the A and B coefficients in (10.8) represents the density of the electromagnetic waves with the frequency between $h\nu$ and $h\Delta\nu$ inside the medium times $h\nu$; in other words, the density of the electromagnetic-wave energy with the frequency between $h\nu$ and $h\Delta\nu$ inside the medium.

Equating the temperature-dependent terms of the same balance equation leads to [10.5]

$$B_{12} = B_{21} . \qquad (10.9)$$

The above treatment can be extended to a semiconductor with the additional conditions that momentum conservation and Pauli's exclusion principle must hold. The population difference between levels 2 and 1 is

$$N_2 - N_1 = V \frac{8\pi k^2 dk}{(2\pi)^3} [f_c(1-f_v) - f_v(1-f_c)] \qquad (10.10)$$

where f_c and f_v denote the electron occupancy factors for the conduction and the valence bands in a semiconductor with excess carriers. Needless to say, $1-f_c$ and $1-f_v$ depict the probability of a corresponding state in the conduction and the valence band, respectively, being empty to satisfy the Pauli's exclusion principle. $(8\pi k^2 dk)(2\pi)^{-3}$ accounts for the density of the electromagnetic waves in k space, and V is the volume.

For a semiconductor with parabolic bands, the upper states correspond to the conduction band and the lower to the valence band. We can write, by assuming one participating valence band only,

$$E_2 = E_c + \frac{\hbar^2 k^2}{2m_n^*} \quad \text{and} \quad E_1 = E_v - \frac{\hbar^2 k^2}{2m_p^*} \tag{10.11}$$

with the aid of the reduced effective mass, $1/m_r^* = 1/m_n^* + 1/m_p^*$ with m_n^* and m_p^* being the electron and hole effective masses, respectively, i.e.,

$$E_2 - E_1 = h\nu = E_g + \frac{\hbar^2 k^2}{2m_r^*} . \tag{10.12}$$

By solving for k and inserting it in the expression for the density of the electromagnetic waves in the k space (10.10), or the density of directly associated states, we have

$$N(h\nu)\,d(h\nu) = \frac{(2m_r^*)^{3/2}}{2\pi^2 \hbar^3}(h\nu - E_g)^{1/2}\,d(h\nu) . \tag{10.13}$$

The absorption coefficient for a given $h\nu$ is proportional to the probability for a transition from the initial state to the final state, and to the density of available electrons in the first state and the density of empty states in the excited state. Eq.(10.13) accounts for the dependence of the absorption coefficient on energy in a direct-bandgap semiconductor. In other words, the absorption coefficient is proportional to the square root of the energy above the gap energy. Below the gap energy and in this ideal picture, the absorption coefficient tends to zero.

In an absorption measurement on a high-purity sample, the probabilities of having an electron in the lower state (valence band) and that for the higher state (conduction band) can be taken as 1 and 0, respectively. Hence, the absorption coefficient reduces to [10.6]

$$\alpha(h\nu) = A^*(h\nu - E_g)^{1/2} \tag{10.14}$$

with

$$A^* \approx \frac{q^2 (2m_r^*)^{3/2}}{nch^2 m_n^*} . \tag{10.15}$$

where c is the velocity of light in vacuum. Once the absorption coefficient vs. photon energy is either calculated or measured, the spontaneous emission spectrum can be calculated from

$$I(h\nu) = (h\nu)r(h\nu) = (h\nu)g(h\nu) \tag{10.16}$$

where r(hν) and g(hν) are the recombination and generation rates which are equal at thermal equilibrium. In a photoluminescence experiment, this assumption can be used provided that the exciting light intensity is very low.

$$I(h\nu) \; = \; \frac{8\pi\nu^2}{c^2} h\nu \alpha(h\nu)[f_c(1-f_v)] \tag{10.17a}$$

where

$$f_c \; = \; \left[1 + \exp\left(\frac{E_c - F_n}{kT}\right)\right]^{-1} \tag{10.17b}$$

and

$$f_v \; = \; \left[1 + \exp\left(\frac{F_p - E_v}{kT}\right)\right]^{-1} , \tag{10.17c}$$

f_c representing the occupation probability of being in the upper (conduction) state and f_v in the lower (valence) state. F_n and F_p are the quasi-Fermi levels for electrons and holes, respectively. Simplifying these probabilities for a non-degenerate semiconductor, which means replacing them with their Boltzmann factors, we get for I(hν)

$$I(h\nu) \; = \; \frac{8\pi n^2 h\nu^3}{c^2} \alpha(h\nu)\exp\left(\frac{F_n - F_p}{kT}\right)\exp\left(\frac{-h\nu}{kT}\right) . \tag{10.18}$$

The spectral emission response is proportional to the product of the absorption coefficient and $\exp(-h\nu/kT)$. On the lower-energy side, the emission spectrum has the spectral dependence of the absorption coefficients and above the bandgap, it roughly declines exponentially.

10.2 Band-to-Band Transitions

The average lifetime of carriers before radiative recombination is called the **radiative lifetime** τ_r. The rate of emission of photons by recombining electrons, n, and holes, p, is a bimolecular process and is giving by

$$R \; = \; Bn \cdot p , \tag{10.19}$$

where B is the radiative recombination probability. R and B have the units of $cm^{-3}s^{-1}$ and $cm^3 s^{-1}$, respectively. For a p-type semiconductor where the excess carrier concentration is much less than the equilibrium hole concentration, radiative recombination lifetime reduces to $\tau_r = (\rho B)^{-1}$. For thermalized electrons and holes, the recombination time depends on the electron and hole energies, which means that it will depend on the photon energy. Consequently, it is customary to define an average lifetime as $\langle \tau_r \rangle$ which depends on the k selection rules as is the case of perfect or nearly perfect semiconductors. In heavily excited semiconductors, this does not hold. In electroluminescent devices, such as in LEDs and lasers, with p-type active regions, when an electron is injected in thermal equilibrium, in terms of electron and hole distributions, we define a lifetime called the **minority carrier radiative lifetime** τ_{rad} to depict the recombination process. This is the time it takes for an extra minority carrier to be annihilated radiatively by a majority hole. In intrinsic and/or near-intrinsic semiconductors with very low electron/hole concentrations, the minority radiative recombination time is rather long as in the case of indirect semiconductors because the probability of these processes is very small. The radiative recombination time can be made smaller with increased doping up to a certain limit as the more the doping is increased the less the above expression is valid. The stimulated-emission lifetime does not follow this rule as the stimulated-emission rate depends also on the photon density.

In addition to the radiative processes, there are non-radiative processes in semiconductors because of imperfections which act as non-radiative centers. We should mention some defects as radiative recombination centers which in a PhotoLuminescence (PL) experiment can shed light on the energy levels of defect states. For a semiconductor containing non-radiative traps, in an experiment such as time-dependent PL, the decay rate can be expressed as

$$\frac{1}{\tau_{total}} = \frac{1}{\tau_{rad}} + \frac{1}{\tau_{nonrad}} . \tag{10.20}$$

In a PL experiment, the emission lineshape is determined by the joint density of states, and the probability of participating states being available for recombination. The former has the form $(h\nu - E_g)^{1/2}$ and the latter $e^{-E/kT} = e^{-h\nu/kT}$. When the semiconductor is excited by the above-bandgap photon radiation, the two lineshapes put together lead to a lineshape of the form

$$(h\nu - E_g)^{1/2} e^{-h\nu/kT} . \tag{10.21}$$

Lower-energy photons do not excite electrons into the conduction band, therefore the PL band-to-band emission will be zero in "linear" experiments.

10.2.1 Excitonic Transitions

When the charged center concentration in semiconductors is sufficiently
low, a free electron and a free hole can be attracted to one another through
Coulombic attraction. An electron can orbit the hole, called **exciton**, much
the same way as in the case of a hydrogen atom. At low temperatures and in
high-quality samples with low donor and acceptor concentrations as well as
a low density of defect centers, the photo-excited carriers with opposite
charge not only are attracted to one another, but also to neutral centers via a
Van der Waals interaction. Since this additional attraction reduces the exciton binding energy, the neutral impurities are efficient in trapping excitons
to form bound excitons as denoted by **D° X** and **A° X** for neutral donor and
acceptor bound excitons. Excitons can also be bound to ionized donors and
acceptors and are termed **D⁺X** and **A⁻ X**. The donor-bound exciton can be
considered analogous to a molecule ion with a binding energy of about $0.1 \div$
0.2 of that of the hydrogen atom. Not to be confused with an ionized donor-
bound exciton is the hole bound to a neutral donor, which is depicted as
D° h and has the same charge state as the ionized donor-bound exciton.
Actually, **D° h⁺** should be reserved for describing a transition. To a first
extent, the one with the larger binding energy should be more stable and is
therefore more likely to occur. But, whether this is so depends on the electron and hole masses. The exact nature and behavior of free excitons in
semiconductors are rather complex. For example, excitons formed in the
continuum state can interact with photons and relax through phonon emission. In the process, exciton dispersion curve splits into upper and lower
branches, as will be discussed shortly.

Using the hydrogenic model, the binding energy, or the ionization energy, of such a system is given by

$$
E_x \; = \; \frac{n_r^* q^4}{2h^2 \epsilon^2 n^2} \tag{10.22}
$$

where the quantities have their usual meanings, and n is an integer with
n = 1 corresponding to the ground state of the exciton. If the disparity of the
electron and hole effective masses is large, and the electron effective mass
is the smaller of the two, the reduced mass can be approximated by the electron effective mass. As can be seen, in large-bandgap semiconductors
which have large effective masses, the exciton binding energy is relatively
large in the range of 20 meV.

10.3 Optical Transitions in GaN

Optical transitions in semiconductors, GaN is no exception, can be grouped into two categories. The **intrinsic transitions** are those that are associated with semiconductors void of impurities and defects. The **extrinsic transitions** have their genesis in impurities and defects. Free excitons and their phonon replicas, if any, and free-to-free transitions represent the intrinsic transitions. The impurity bound excitons, transitions involving impurities such as free-to-bound and bound-to-bound, and defects constitute extrinsic transitions. A collage of the intrinsic and extrinsic transitions are sketched in Fig. 10.1.

10.3.1 Excitonic Transitions in GaN

Excitons are first classified into free and bound excitons. In high-quality samples with low impurity concentrations, the free excitons can also exhibit excited states, in addition to their ground-state transitions. Wurtzite structures are more interesting due to the splitting of the valence band by crystal field and spin-orbit interactions, as will be described now.

a) Free Excitons

Excitons in GaN take on a special meaning in that the valence band is not degenerate due to the crystal field and spin-orbit interactions at the Γ point (Fig. 3.4). The three emerging states are termed Γ_9^v, upper Γ_7^v, and lower Γ_7^v. The related free-exciton transitions from the conduction band to these

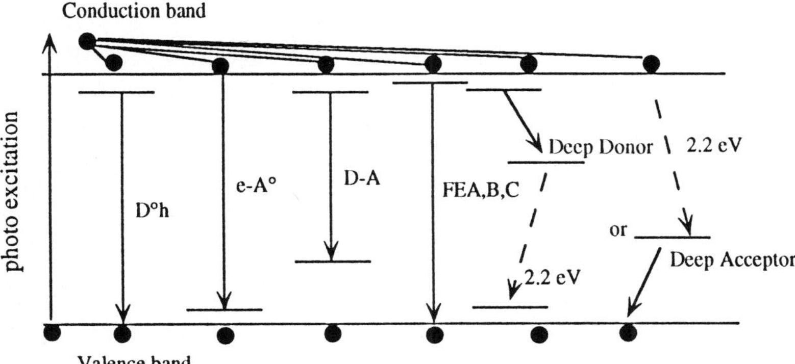

Fig. 10.1. Intrinsic and extrinsic optical transitions that occur in response to an above-bandgap excitation in semiconductor GaN

three valence bands are termed **A, B and C excitons**. In terms of symbols, they are A $\equiv$ $\Gamma_7^c \rightarrow \Gamma_9^v$ (also referred to as the **heavy hole state**), B $\equiv$ $\Gamma_7^c \rightarrow \Gamma_7^v$ the upper one (also referred to as the **light hole state**), and C $\equiv$ $\Gamma_7^c \rightarrow \Gamma_7^v$ lower (also referred to as the **crystal-field split band**). In ideal Wz crystals, i.e. strain free, they have the following symmetries: Excitons associated with all three bands are allowed in the α **polarization** (**E** $\perp$c and **k** $\parallel$c). In the σ **polarization** (**E** $\perp$c and **k** $\perp$c), A and B excitons are observable with the C exciton being very weak. In the π **polarization** (**E** $\parallel$c and **k** $\perp$c), A and C excitons are observable with the B exciton being weak. Here **E** and **k** are the electric field and momentum vectors, and c denotes the c-axis of the crystal. GaN samples on SiC, which are under symmetry-reducing compressive strain, exhibit a strong C exciton in σ polarization. Each of these three excitonic states are expected to have fine structure due exciton polariton longitudinal transverse splitting which is on the order of $1 \div 2$ meV. Until very recently, this splitting has nearly been impossible to observe due to wide spectral widths exhibited by available samples; they are on the order of about 1 meV in the best samples [10.7].

An observation of only the intrinsic transitions and the excited states of the A-and B-band excitons in GaN was reported by *Reynolds* et al. [10.8] and *Smith* et al. [10.9], see Fig. 10.2. This paves the way for the determination of exciton binding energies, exciton Bohr radii, the dielectric constant, and with the aid of the quasi-cubic model, the spin-orbit and crystal field parameters, Δ_{so} and Δ_{cf}. The absolute energy values of these transitions, however, depend on the local strain and thus do not represent a fully relaxed GaN. Ground-state exciton transitions associated with the A- and B-bands, the n = 2 excited state of the A exciton, and the n = 2 and 3 excited state of the B exciton are indicated. The exciton associated with the C-band is seen in the reflection spectrum only (Fig. 10.2). Assuming hydrogen-like excitons, the A and B exciton binding energies can be deduced to be about 20 meV and 22 meV, respectively. The exact value of the exciton binding energy is somewhat controversial in that values as high as 26 meV have been reported for a thick sample grown by HVPE, wherein the ground and excited states of the A exciton were observed, in a thin sample on sapphire.

Absorption and/or reflectance and photoluminescence-excitation measurements can discern higher-order transitions much more readily. The PhotoReflectance (PR) measurements have been refined to take the form of modulated photoreflectance in which the built-in field in the sample is modulated with the aid of modulation-injected carriers by a chopped laser beam. The change in the photoreflectivity is $\Delta R/R$. The differential manifests itself as sharp lines corresponding to transitions in the Brillouin zone. The exact position of an optical transition can be found more accurately by fitting a functional form where the details as applied to GaN can be found in [10.10]. The photoreflectance spectrum of a GaN film grown on sapphire is

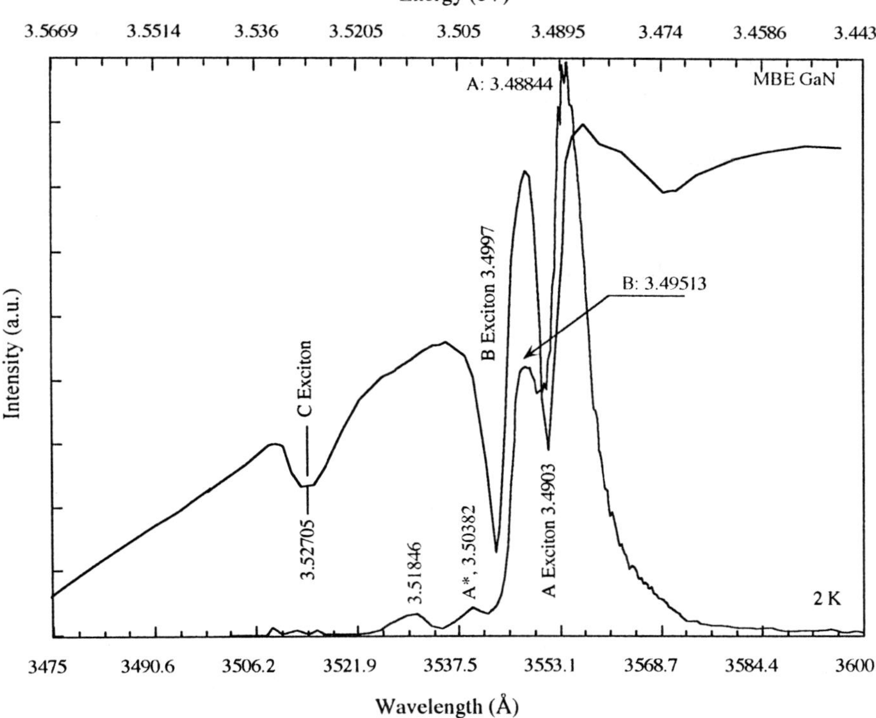

Fig. 10.2. Emission and reflection spectra of a GaN epitaxial layer grown on a sapphire substrate by MBE. After [10.8]

depicted in Fig. 10.3 where the experimental data along with the fitting function are shown. The collective use of the two leads to determine the excitonic transition and some of their excited states, as in the case of the emission spectrum shown in Fig. 10.2. By noting several of the excited states and using the Elliot's theory [10.11] relating the excited states of the excitonic transitions as

$$E_n - E_m = -E_b(1/n^2 - 1/m^2) \qquad (10.23)$$

where n and m are integers, and E_b denotes the binding energy of the exciton in question, one can get the exciton binding energy. From the data of Fig. 10.3, the A- and B-exciton binding energies were determined to be 21 meV while that of the C exciton was deduced to be 23 meV. The A- and B-exciton binding energies are in good agreement with those deduced from the emission measurements (Fig. 10.2).

The large excitonic binding energy manifests itself as the excitonic absorption feature being observable at lattice temperatures as high as 400 K, as illustrated by Fig. 10.4. It is this high exciton binding energy that led to

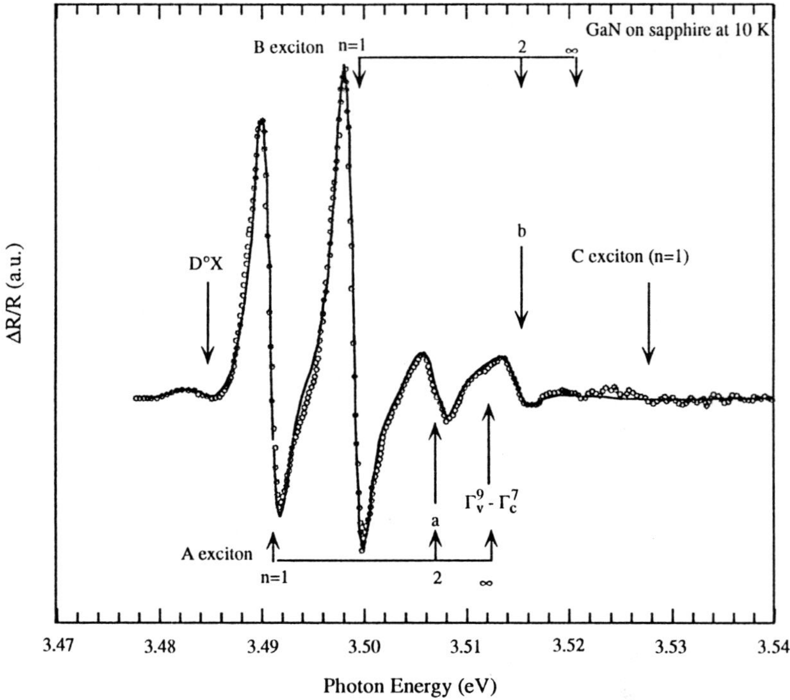

Fig. 10.3. Modulated photoreflectance spectrum obtained at 10 K in a 7.2 μm GaN film grown on sapphire with open circles and the solid line indicating the experimental data and the fit. After [10.10]

many proposals to the effect that excitonic processes may be involved in lasers, a topic which is discussed thoroughly in Chap. 12. Excitons may also play an important role in photodetectors utilizing GaN, the embryonic stage of which warrants further discussion to be deferred.

Assuming that $E_{1,2}$ and $E_{2,3}$ can be represented by the difference in energy between the A and B bandgaps, and B and C exciton bandgaps (further assuming that the binding energy of B and C excitons are equal), then $\Delta_{so} = 17.3$ meV and $\Delta_{cr} = 24.7$ meV result. However, cumulative data suggest that the spin-orbit interaction term is between 15 and 20 meV while the crystal splitting term is between 10 and 25 meV. The quasi-cubic model is not expected to be very accurate for GaN since the three valence-band states are close in energy with mixing likely to occur. The lack of samples with definite linewidths may also contribute to the spread in the data. Using an electron mass of $m_n^* = 0.22 m_0$ and a hole mass of $m_p^* = 0.8 m_0$ (leading to a reduced exciton mass, $m_r^* = 0.172 m_0$), combined with the A-exciton binding energy, dielectric constant of $\epsilon = 10.7$ was obtained. Using the expression

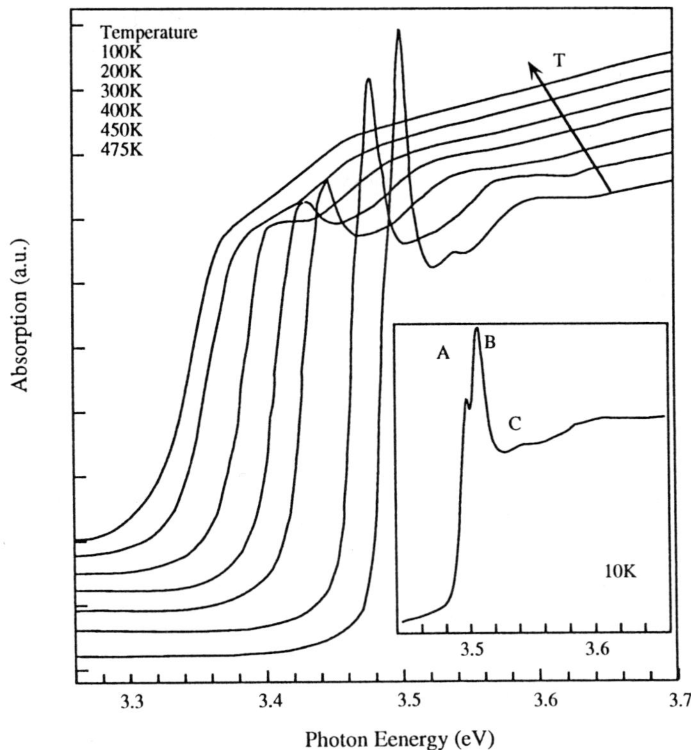

Fig. 10.4. Absorption spectra of GaN epitaxial layer near the fundamental band edge taken at lattice temperatures of 100, 200, 300, 350, 400, and 450 K. Note the presence of excitonic contribution up to about 350 K due to the large (20 meV) exciton binding energy. After [10.12]

$$a_r = \frac{4\pi\epsilon\hbar^2}{m_r^* q^2} \qquad (10.24)$$

the Bohr radius a_r of the A- and B- excitons was determined to be about 31 and 28 Å, respectively.

Excitonic transitions in GaN are very efficient. One school of thought invokes exciton-polariton transitions which have previously been observed in CdS and GaAs. Exciton polariton transitions come about due to the interaction of excitons with photons which gives rise to an **Upper Polariton Branch** (UPB) and a **Lower Polariton Branch** (LPB) [10.13], as depicted in Fig. 10.5. Normally, the k-selection rules for the optical matrix element are such that excitons having exactly the same photon **k** vector are allowed. However, within the frame work of the simplest model and in a pure bulk material, the coupling of the electromagnetic field represented by a photon

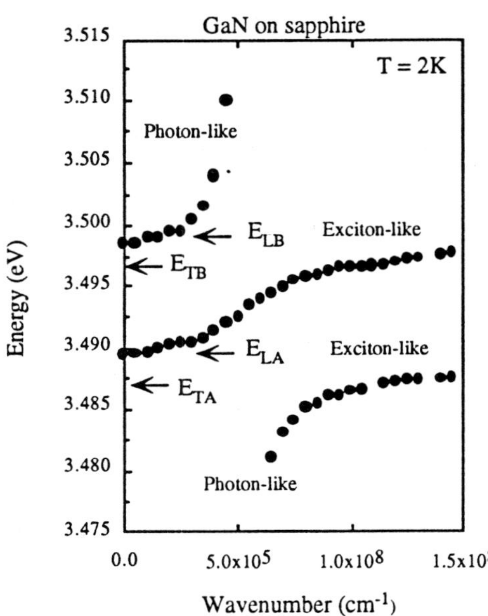

Fig. 10.5. Polariton dispersion relations in the energy region of strong exciton-photon coupling with one resonance, with two oscillators in the realistic case of GaN. After [10.14]

having an energy of $E = hkc/2\pi$ with the exciton $E = E_T + h^2k^2/8\pi^2m$ gives rise to the notion of an **exciton polariton** [10.14], which is an efficient radiator. Since we are concerned with A and B excitons in GaN, Fig. 10.5 is for the two-oscillator case. This coupling relaxes the k-selection rule and transforms the fluorescence mechanism into a transport of the coupled excitations from the bulk towards the surface where part of it is reflected, part of it is transmitted. A complete interpretation of the fluorescence spectra requires a treatment beyond the scope of this book. Suffice it to say that sophisticated models involving imbricated (overlapping) contributions of group velocities, momentum relaxation mechanisms, and penetration depth of the polariton distribution are required.

Under the non-resonant excitation conditions, an electron-hole pair in the continuum of states will relax to lower-energy states within the two polariton branches, most likely through acoustic-phonon emission, down to the knee of the polariton dispersion curve just below the transverse frequency due to the high density of states in this region. Below the transverse frequency, a decrease in both the scattering matrix element and the density of final states combined with a large increase of the group velocity $d\omega/dk$ causes the radiative recombination to dominate over thermal relaxation. In this region, an escape of the photon out of the crystal becomes very efficient, which may be responsible for the efficient excitonic transitions in GaN. The ensuing polariton-relaxation bottleneck leads to polaritons exhibiting a pronounced distribution peak just below the exciton energy. In GaN, peaks below the A and B exciton energies associated with the LPB of A and

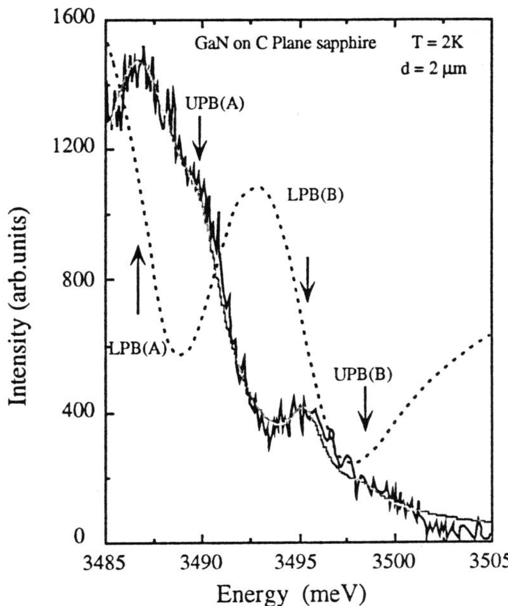

Fig. 10.6. 2K reflectance spectrum (*dashed line*) and the corresponding 2K photoluminescence (*solid line*) spectrum which shows the LPB and UPB contributions to the GaN fluorescence. The smooth solid line is the lineshape fitting of the photoluminescence spectrum by using four Lorentzian functions. After [10.14]

B excitons have been observed at 3486.6 meV and 3495.4 meV, as shown in Fig. 10.6 with the labeling of LPB(A) and LPB(B). Also observable are the higher-energy peaks labeled as UPB(A) and UPB(B) due to a significant population of the upper polariton. *Gil* et al. [10.14] have also deduced the longitudinal-transverse splitting by reflectance lineshape fitting to the data. For each exciton, the longitudinal transverse splitting is given by $\omega_{LT} \approx 2\pi\alpha\omega/\epsilon_b$, which culminates in 2.9 meV and 1.8 meV for the A and B excitons, respectively. This compares well, within the experimental accuracy, with the splittings of 2.4 meV and 1.8 meV between the energies of the dips in the PL bands at 3489.4 meV and 3497.8 meV, and transverse excitons.

Since GaN with a low-background concentration is not yet available, fundamental optical studies have been done in epitaxial layers which unfortunately were strained to varying degrees due the mismatch of the thermal expansion coefficients between the layer and the substrate on which it is deposited [10.15]. To circumvent this complication, thick and presumably unstrained GaN layers on sapphire have been used to determine the energies of intrinsic and extrinsic transitions in GaN. However, even the thick films, tens of micrometers, on sapphire may not be completely strain free. Evidence for the residual strain is provided by curling of these samples when the sapphire substrates are thinned. *Monemar* et al. [10.7], who have examined numerous thin and thick layers on various substrates including homoepitaxial layers on GaN substrates, reported in [10.16], concluded that the

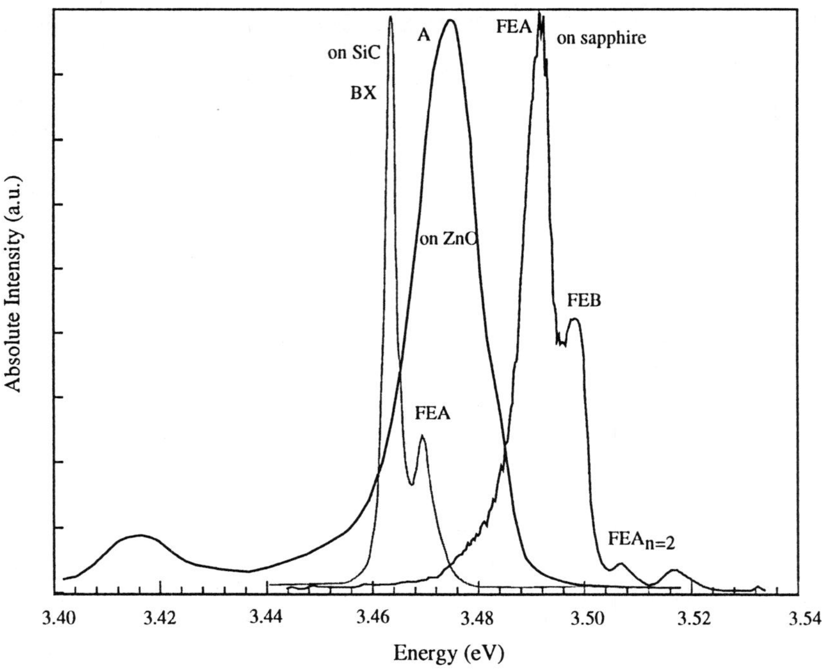

Fig. 10.7. The fuorescence spectra of GaN epitaxial layers grown on c-planes of sapphire, ZnO and SiC:Si. The data on SiC is from [10.10]

A, B, and C exciton lines in GaN relaxed to an accuracy of ±2 meV are 3.478, 3.484, and 3.502 eV, respectively, at 2 K. Extreme caution should be exercised as the samples could still be under strain due to mounting. The improved linewidths in GaN grown by MOCVD on GaN lead to the resolution of three transitions near the donor-bound exciton region. It is tempting to suggest that there are three different donors. The exchange-split component may not apply as one of the lines which must be involved is absent in some samples. It is very likely that local inhomogeneities may be responsible for these observations. Displayed in Fig. 10.7 are the photoluminescence spectra taken in samples grown on sapphire, SiC and ZnO to display the shift in the excitonic energies caused by residual strain. The reader should be cautioned that the spectrum presented for the sapphire substrate case does not represent the sharpest linewidth available. To date, the sharpest linewidth obtained on sapphire is about 1 meV with values as low as 0.25 meV in homoepitaxial layers on GaN [10.17] (Fig. 10.8). The sample on ZnO utilized a substrate prepared by the hydrothermal method which has not generated the best quality. Much better ZnO substrates are beginning to be produced via the vapor-transport method which will undoubtedly have a favorable impact on GaN grown on ZnO. The MBE-grown sample on a GaN substrate depicted in Fig. 10.8 is of interest in that it shows two sharp

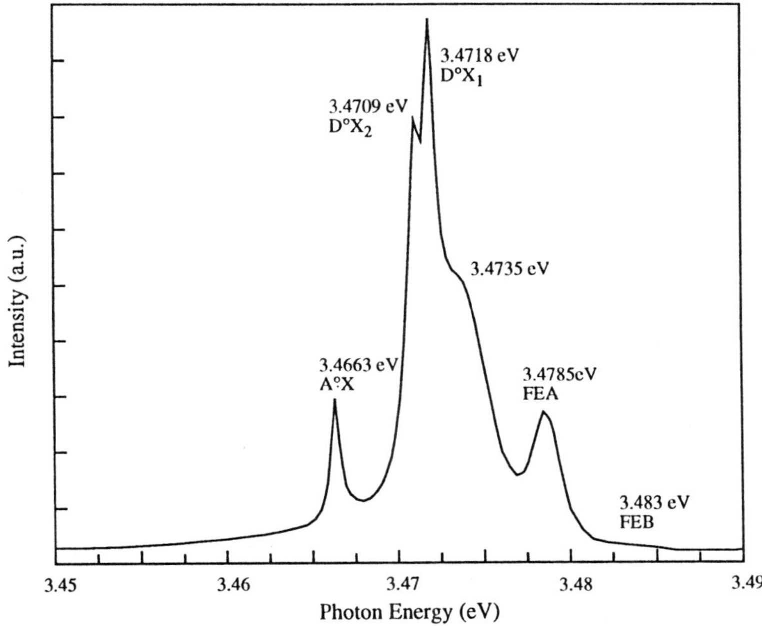

Fig. 10.8. Photoluminescence spectrum of a 0.3 μm thick GaN epitaxial layer grown on a c-xis GaN substrate grown under a high pressure. After [10.17], see also [10.18]

neutral Donor Bound Exciton (DBE) lines, most likely associated with two different donors at 3.4709 and 3.4718 eV. These two levels could also be attributed to spin splitting of the same donor. The sample also exhibits an Acceptor Bound Exciton (ABE) at 3.4663 eV, and the A and B free-exciton transitions at 3.4785 and 3.483 eV, respectively.

As noted above, the transition energies are dependent on the substrate employed due to the expansion-coefficient mismatch. Moreover, these energies vary from one sample to the next, even on the same type of substrate and growth temperature, presumably due to variations in the local structure, such as inhomogeneities. Using the photoreflectance data on various samples, the excitonic transitions can be identified (Fig. 10.9). The data were taken from GaN on sapphire having thicknesses of 7.2 μm and 4.2 μm, and GaN on SiC with a thickness of 3.7 μm.

Figure 10.10 exhibits data similar to those shown in Fig. 10.9 except in terms of strain along the c and a axes. The lattice parameters used, together with optical reflectance measurements to generate the data for Fig. 10.10 were obtained from X-ray scattering using a triple axes, four-crystal X-ray diffractometer [10.18]. The complicating factors in situations such as this is that films could be partially relaxed by creating more defects aided by the fact that the semiconductor contains many defects to begin with. Together with inhomogeneities, this may explain in part the spread in transition ener-

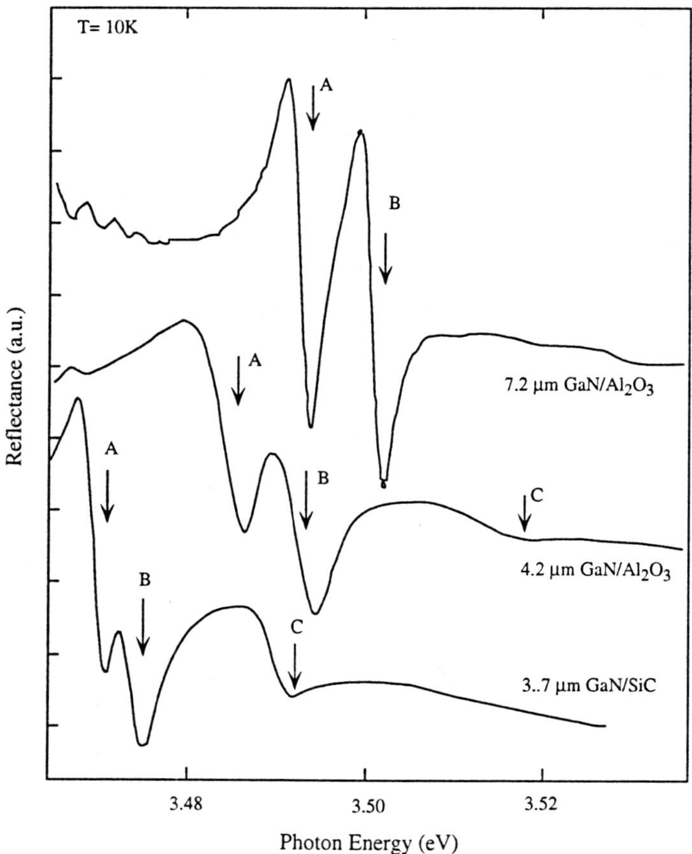

Fig. 10.9. Photoreflectance spectra of excitonic transitions in GaN grown on sapphire, having thicknesses of 7.2 μm and 4.2 μm, and on SiC having a thickness of 3.7 μm. After [10.10]

gies from sample to sample. Further complications arise from partial relaxation during growth which, in general, can be lumped in with inhomogeneities.

Figure 10.11 compiles the exciton energies observed in samples which are under varying degrees of compressive (on sapphire) and tensile (on SiC) strain. Different techniques such as reflection, photoluminescence, and calorimetric absorption have been used to extract excitonic resonances. The hollow symbols represent the data for GaN on ZnO substrates indicating very little residual strain as the transition energies are very close to those of the homoepitaxial layers. Due to the relatively larger thermal expansion coefficient of sapphire, GaN films on that substrate undergo an in-plane compressive strain. Of course, the reverse occurs for the out-of-plane strain. In contrast, the GaN films on SiC are under an in-plane tensile strain [10.7, 19, 20] which concurs with the energy values reported above.

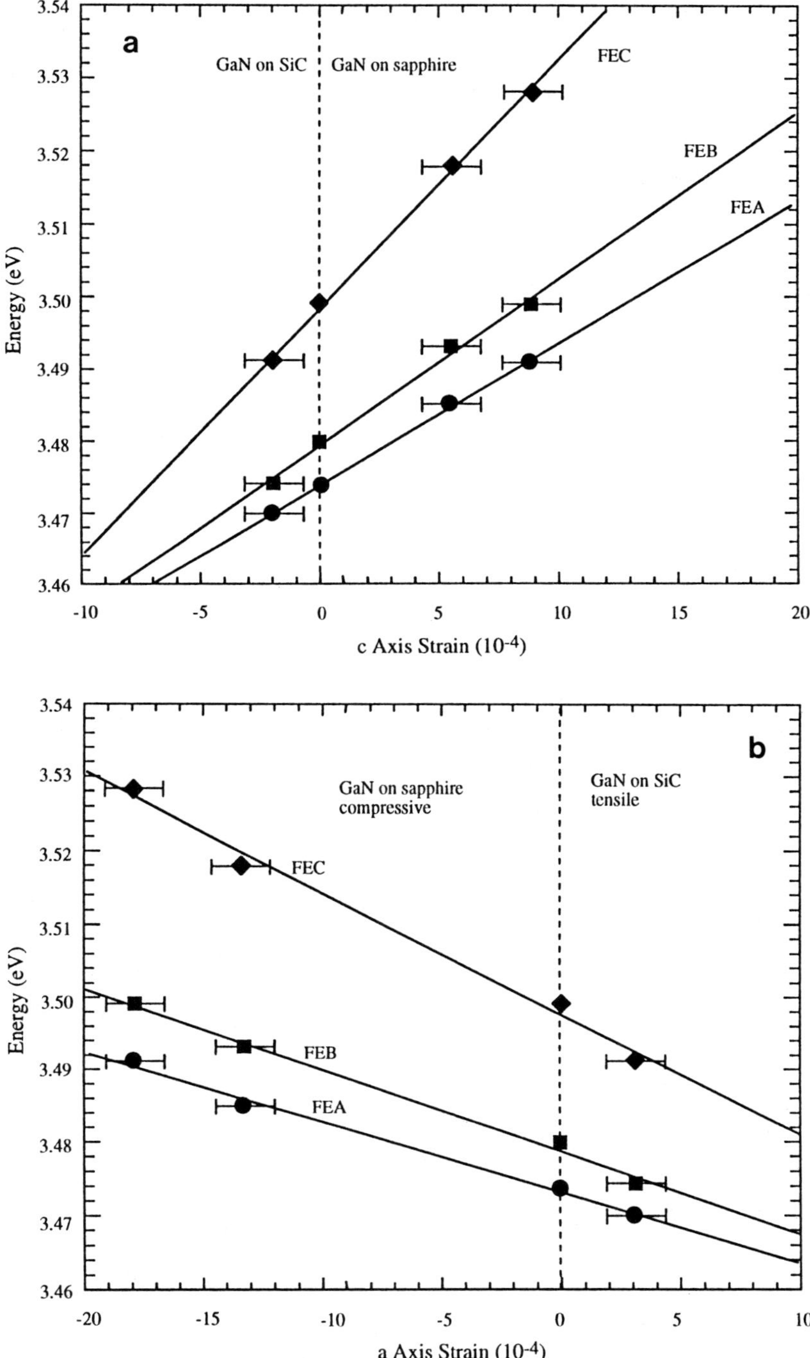

Fig. 10.10. (**a**) Measured excitonic transition energies vs. the c-axis lattice constant. The former and latter were obtained by reflectance and x-ray diffraction measurements. (**b**) The same as that shown in (**a**) except versus the in-plane lattice constant. After [10.10]

313

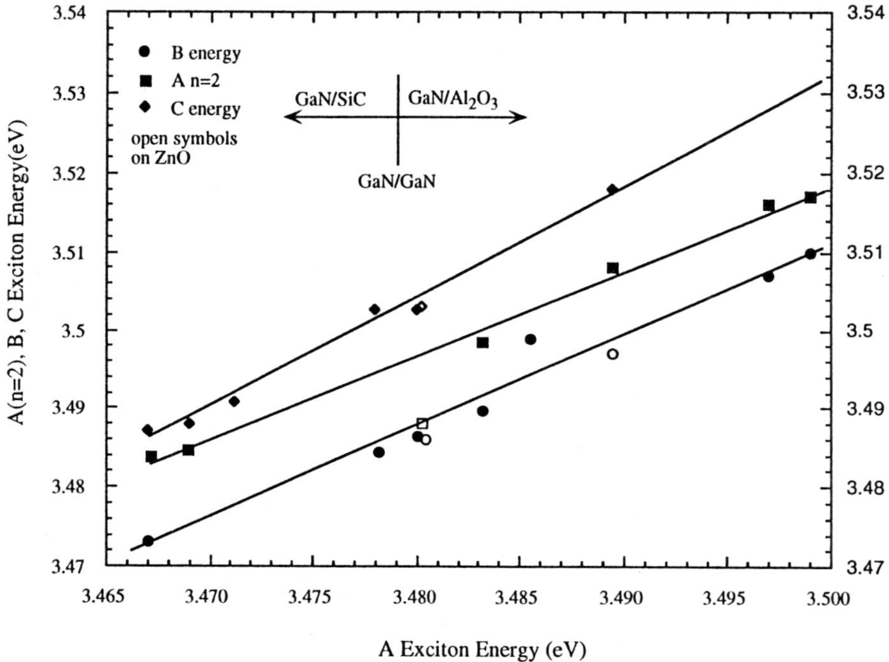

Fig. 10.11. Strain-dependent energy of the first excited state of the A exciton, and B and C excitons with respect to the energy of the A exciton in GaN grown on SiC, GaN, ZnO, and sapphire substrates. The *open symbols* are for samples on ZnO substrates which are preliminary in nature. The data contain those collated by B. Gil, a private communication, *Monemar* [10.20], and *Meyer* [10.28]

Observed shifts in transition energies can be used to get an estimate of the hydrostatic and uniaxial deformation potentials [10.15, 21]. Strain-induced modifications of the exciton energies can be obtained through the use of the strain Hamiltonian for a Wz crystal such as that reported by *Pikus* [10.22], and *Ivchenko* and *Pikus* [10.23]. The energies of the three free excitons are given by

$$
\begin{aligned}
E_A &= E_{A0} + a_z\epsilon_{zz} + a_{xy}(\epsilon_{xx}+\epsilon_{yy}) + b_z\epsilon_{zz} + b_{xy}(\epsilon_{xx}+\epsilon_{yy}) , \\
E_B &= E_{B0} + a_z\epsilon_{zz} + a_{xy}(\epsilon_{xx}+\epsilon_{yy}) + \Delta_+[b_z\epsilon_{zz} + b_{xy}(\epsilon_{xx}+\epsilon_{yy})] , \\
E_C &= E_{C0} + a_z\epsilon_{zz} + a_{xy}(\epsilon_{xx}+\epsilon_{yy}) + \Delta_-[b_z\epsilon_{zz} + b_{xy}(\epsilon_{xx}+\epsilon_{yy})] ,
\end{aligned}
\tag{10.25}
$$

where E_{i0}, and a_{ij} and b_{ij} represent, in order, the strain-free exciton energies, combined hydrostatic deformation potentials, and uniaxial deformation potentials, and ϵ_{ij} is the particular component of the stress tensor. The majority of GaN films were grown on the c plane, which means that the parallel or the in-plane component is in the xy plane, leaving the c direction as the z direction. The strain components are

314

$$\epsilon_{xx} = \epsilon_{yy} = \epsilon_{\parallel} = \frac{a_s - a_o}{a_o} \qquad (10.26a)$$

and

$$\epsilon_{zz} = \epsilon_{\perp} = \frac{c_s - c_o}{c_o} \qquad (10.26b)$$

where a and c represent the in-plane and out-of-plane lattice constants with the subscripts s and o indicating strained and relaxed parameters. Under biaxial strain conditions, the in- and out-of-plane strains are related through the stiffness (elastic) coefficients as

$$\epsilon_{\perp} = \frac{-2C_{13}}{C_{33}} \epsilon_{\parallel} \; . \qquad (10.27)$$

The coefficients Δ_+ and Δ_- in (10.25) account for the valence-band mixing through spin-orbit interaction and are given as [10.24]

$$\Delta_{\pm} = \frac{1}{2} \left\{ 1 \pm \left[1 + 8 \left(\frac{\Delta_3}{\Delta_1 - \Delta_2} \right)^2 \right] \right\} , \qquad (10.28)$$

where Δ_1 is the splitting of the Γ_9 and Γ_7 valence bands due to the crystal field. The symbols Δ_2 and Δ_3 represent the spin-orbit coupling. These splitting parameters are discussed in Chap. 3. By plotting the measured excitonic transition energies as a function of strain, as determined from X-ray diffraction, one can obtain for $\Delta_3/(\Delta_1-\Delta_2)$ a value of about 0.531 [10.19]. Utilizing the experimentally observed excitonic transition energies together with (10.25), one can separate out the uniaxial component of the strain-induced shift. Doing so leads to

$$b_z - \frac{C_{33}}{C_{13}} b_{xy} = 15.2 \text{ eV} \; . \qquad (10.29)$$

A similar approach for the in-plane strain leads to

$$a_z - \frac{C_{33}}{C_{13}} a_{xy} = 37.9 \text{ eV} \; . \qquad (10.30)$$

Utilizing the quasi-cubic approximation, the $C_{11} \approx C_{33}$ assumption [10.25], which is justified by the small strain-induced shift relative to the total excitonic energies, leads to $b_z \approx -2b_{xy}$, and $a_z - a_{xy} \approx 2b_{xy}$. One then

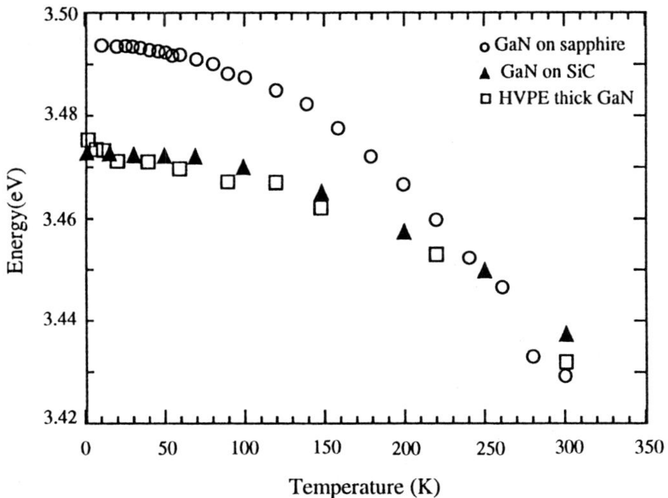

Fig. 10.12. Temperature dependence of the A exciton PL peak energy for GaN samples on sapphire, SiC and thick GaN which is construed as bulk. After [10.19]

obtains $b_z \approx -5.3$ eV and $b_{xy} \approx 2.7$ eV [10.19] by using 106 and 398 GPa for C_{13} and C_{33}, respectively [10.26].

The temperature dependence of the excitonic resonances is also dependent on the particular sample and the local strain. Figure 10.12 exhibits the dependence for three samples, presumably relaxed, under compressive strain and under tensile strain. The reduced deviation with increasing temperature has been attributed to partial relaxation. The temperature dependence of the excitonic resonance can be determined with *Varshni's* empirical relation $E_g(T) = E_0(0) - \alpha T^2/(\beta + T)$ [10.27]. In samples where both A ($\Gamma_7^c \rightarrow \Gamma_9^v$) amd B ($\Gamma_7^c \rightarrow \Gamma_7^v$) (upper band) ground-state excitons as well as their excited states are seen, the experimental data can be fit well with $\alpha = 7.32 \cdot 10^{-4}$ eV/K and $\beta = 700$ K. The fact that the bound-exciton transitions decrease rapidly while the free-exciton transitions are observable even at room temperature is indicative of small localization energies associated with the bound excitons. *Meyer* [10.28] followed the thermally activated dissociation of bound excitons and described the dependence on temperature with two activation energies. The above-bandgap excitation creates free excitons which can localize at impurity sites. As the temperature is elevated, impurities such as donors ionize with an interplay with localization, which causes the bound excitons to decay as the temperature is increased and eventually completely dissociate leaving the free exciton A as the dominant transition. The ratio of the emission intensities at absolute zero and at the temperature T can be formulated as [10.25]

$$\frac{I(0)}{I(T)} = \frac{1}{1 + c_1 \exp(-\Delta E_1/kT) + c_2 \exp(-\Delta E_2/kT)} \, , \qquad (10.31)$$

where c_1 and c_2 are prefactor constants, and ΔE_1 ($=4.5\pm1\text{meV}$) and Δ_2 ($=32\pm2\text{meV}$) are the thermal activation energies. The presence of two activation energies is intriguing in the sense that it suggests that there are two different donors present in the film.

b) Bound Excitons

The A, B and C excitons discussed above represent intrinsic processes as they do not involve pathways requiring extrinsic centers. Available semiconductors contain impurities such as donors and acceptors, and shallow donor- and acceptor-like defects. In theory, excitons could be bound to neutral and ionized donors and acceptors. Not all may be observed in a given semiconductor as some of them may not be stable. In GaN, the neutral donor-bound exciton I_2 often is dominant because of the presence of donors, either due to impurities and/or nitrogen vacancies, and its efficient radiation characteristics. In samples containing acceptors, the acceptor-bound exciton I_1 is observed. Very recently, the ionized donor-bound exciton has also been seen. Figure 10.13 exhibits the part of the spectrum displaying the aforementioned bound excitons [10.29]. The identification was confirmed by the

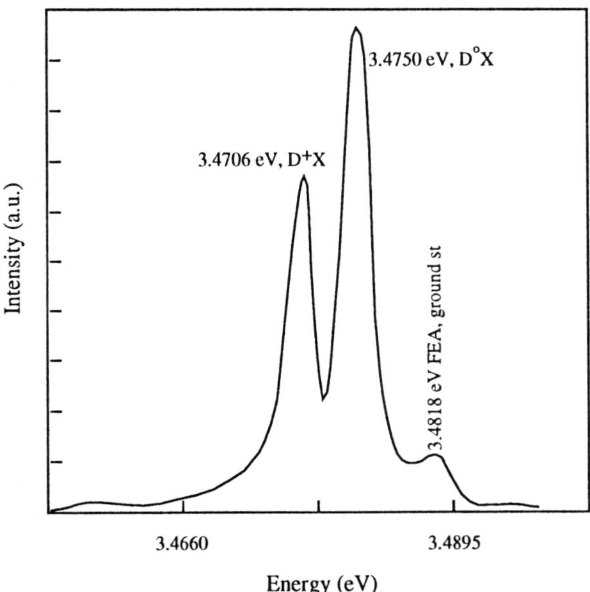

Fig. 10.13. PL spectrum showing the excitons bound to neutral and ionized donors in GaN. After [10.27]

energy ordering of this particular transition with respect to the neutral donor-bound exciton. Additional confirmation was obtained by observing the increased intensity of the peak as the donor concentration was increased, a task accomplished by irradiating the sample with high-energy electrons which create donor-like nitrogen vacancies. The neutral donor-bound exciton I_2 can be seen in Fig. 10.7. Excitons bound to the neutral acceptors I_1 can be found in samples doped with p-type impurities, meaning acceptor sites are present, as will be demonstrated in Sect. 10.3.2. The localization energy of bound excitons can be utilized to extract binding-energy information through *Haynes'* rule [10.30] which has been successfully applied to Si. Doing so for GaN led to the localization energy being about 10% of the binding energy for acceptor-bound excitons and 20% for the donor-bound excitons [10.28]. We must caution that the *Haynes* rule has not been uniformly successful for all semiconductors and can lead to large errors. However, in the absence of any other data, this deduction might be of some value. The localization energy for donor-bound excitons is about $6 \div 7$ meV, whereas that for Mg is about 19 meV. Note that getting the impurity binding energies from the exciton localization energy is at best an indirect measurement. Doing so, for acceptors leads to binding energies of about 120 meV. In fact, there is no other supporting evidence for acceptors of such low binding energies. Other methods such as the Donor Acceptor (DA) transitions and the thermal ionization methods with proper screening should be utilized collectively to get at the binding energies of donors and acceptors [10.28, 31].

c) Exciton Recombination Dynamics

The temporal behavior of excitons is of importance for emitters in that it provides a window on the dynamics of recombination processes. The lifetime of the excitons is generally determined from the decay of the excitonic emission following the removal of the excitation source. The ensuing decay can be described by a single exponential time constant called the **lifetime**. Having two or more decay times can shed some light into the homogeneity of the crystal. Slower decay would be indicative of inhomogeneities in that excitons localized in good portions of the semiconductor would have a longer decay time. The time-dependent intensity is given by

$$I(t) \; = \; I_{01} \, e^{-t/\tau_1} + I_{02} \, e^{-t/\tau_2} \; . \tag{10.32}$$

Recently, upconverted Ti-sapphire lasers with pulse lengths on the order of a few picoseconds have been used as the excitation source. An unusually short exciton lifetime would mean excessive non-radiative recombination processes which are detrimental to light emitters and detectors. Additionally, owing to their large binding energies, the excitons in wide-bandgap semiconductors are purported to participate in the light-emission processes even at room temperature although the extent of this participation is hotly debated. At very high injection rates, it has not yet been established that excitons will dissociate. The issue of excitonic processes in relation to lasing will be discussed in Chap. 12. For the present case, the exciton dynamics will be treated, but only for the low-to-intermediate excitation levels. The overall recombination rate is determined by radiative and non-radiative processes. GaN with its large defect concentration is an interesting case because it is different from other semiconductors with similar defect concentrations where the non-radiative processes dominate. A plausible suggestion is that carriers/excitons are localized in pseudo-individual columns with good local material quality. The radiative lifetime in an excited state can be expressed as [10.32, 33]

$$
\tau_r = \frac{2\pi \epsilon m c^2}{n_r q^2 \omega^2 f} ,
\tag{10.33}
$$

where f is the oscillator strength, $\omega \approx 5.3 \cdot 10^{15}$ Hz for GaN, and other symbols have their usual meanings. The oscillator strength for a free exciton in the effective-mass approximation is given by

$$
f = \frac{E_p}{\pi \hbar \omega v_a / a_r^3}
\tag{10.34}
$$

where E_{p_o} is the Kane matrix element, v_a is the volume of the unit cell, and $a_r = 28$ Å is the exciton Bohr radius. If one uses a matrix element of 18 eV (it may actually be as high as 26 eV), one arrives at about 0.01 for the oscillator strength, which gives rise to exciton lifetimes on the order of tens of nanoseconds. As is always the case, simple expressions are useful for gaining insights and quick ball-park determinations. A detailed investigation of the oscillator strength reveals that it is dependent on strain as well and could vary from sample to sample due to inhomogeneities [10.21]. Figure 10.14 depicts the oscillator strength of A and B free excitons vs. strain through the energy of the A free exciton which is affected by strain. The oscillator strength varies between about 0.1 and 0.5. This would mean that the exciton lifetimes should be around 1 to about 8 ns depending on the strain. Exciton lifetimes much shorter than these numbers would imply faster non-radiative recombination processes. As if to make matters more complicated, *Monemar* et al. [10.7] estimated the exciton lifetime by scaling the value for GaAs using the GaN parameters and arrived at a value of about 200 ps at 2 K which is not too far from the experimentally observed values in thick

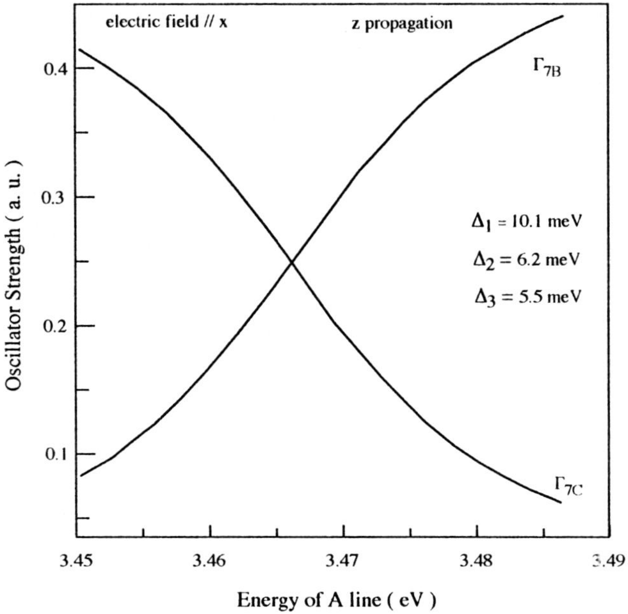

Fig. 10.14. Oscillator strength of the A and B excitons as a function of strain in GaN. After [10.20]

layers grown by HVPE. The radiative recombination lifetime can be expressed with the following simple expression [10.34]

$$\tau_{ex} \propto \frac{1}{n_r^2 E_g^2 v_a \alpha_{ex}}, \qquad (10.35)$$

where n_r, E_g, v_a, and α_{ex} are the refractive index, the bandgap, the unit-cell volume ($2.283 \cdot 10^{-23}$ cm^3, Chap. 8), and the excitonic absorption coefficient which is about $2 \cdot 10^5$ cm^{-1}. Assuming that the exciton lifetime is ≈ 1 ns for GaAs and scaling it to GaN with the aid of (10.35), one arrives at a value of about 200 ps [10.7].

On the experimental side, the temporal behavior at low excitation levels indicates processes, depending on the sample, that are as fast as about 35 ps and 50 ps for free and donor-bound excitons. The decay time increases as the excitation intensity is moderately increased, and decreases with temperature, which is indicative of defect participation. The fast decay is generally attributed to non-radiative processes whose rates are much faster than the radiative recombination. Non-radiative processes such as multiphonon emission and the capture by deep levels, relax electrons rapidly to lower states from which they radiatively recombine with holes or other centers non-radiatively [10.7, 9].

320

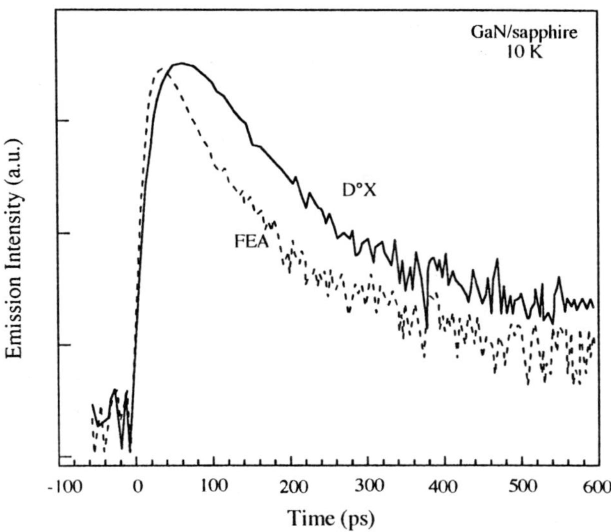

Fig. 10.15. Temporal response of the free exciton A and the donor-bound exciton in GaN on sapphire at T = 10 K. After [10.12]

A typical temporal response of Free Exciton A (FEA) in GaN at 10 K is displayed in Fig. 10.15. The decay in the luminescence intensity can not be described with a single decay time. Rather, a combination of a fast decay and a slow decay describes the process reasonably well. The rapid decay is followed by a slower process with an approximately 300 ps time constant which may be associated with excitons weakly localized in good portions of the crystal, perhaps due to potential fluctuations as a consequence of inhomogeneous strain fields such as those caused by columnar growth. In these regions, non-radiative processes may be suppressed. The defects can be saturated with moderately increased excitation levels leading to slower decay processes. In a good crystal, localization would increase the recombination rate reducing the lifetime. The defective nature of GaN samples is also manifested in that the decay time goes down as the lattice temperature is increased. A thick GaN layer grown by HVPE [10.7] exhibited decay times of about 200 ps at 2 K independent of a moderate excitation intensity suggesting that this is the radiative lifetime. Due to defects in all the available samples, no experimental evidence from photoluminescence decay is available for the room-temperature radiative lifetime of excitons. An estimate derived from the homogenous linewidth sets this lifetime at 2 ns at room temperature [10.7]. Much more research is warranted before conclusive comments can be drawn.

d) High Injection Levels

Carrier dynamics at high injection levels including the excitonic processes are of paramount importance for lasers where the injection levels are on the order of 10^{18} cm^{-3} or higher. At high levels of excitation, the exciton density is high and exciton-exciton interactions would occur. Even processes where two interacting excitons are in their ground states upon recombination can produce an exciton in an excited state. In addition, electron-hole plasmas can form, which are purported to dominate over the excitonic processes. Recently, a large number of reports began to appear which attempt to explain the role of excitons in stimulated emission. The large threshold injection levels needed for lasers reported so far have been attributed to electron-hole plasma formation, as the excitonic lasing process would not require such high injection levels for stimulated emission. Until such time when the defect concentrations in films are reduced, the issue will certainly remain ambigeous. The issues in this respect are very similar in nature to the case of ZnSe involving the argument that excitons take part in the gain process. This has been investigated in detail with no clear conclusion. At very high injection levels, processes such as carrier-carrier interactions, and electron-hole plasma formation are all too complex to analyze. For a detailed treatment of the issue, the reader is referred to Chap. 12.

10.3.2 Free-to-Bound Transitions

At low temperatures the carriers are frozen on the impurities because the thermal energy required for their ionization is no longer available. In a photoluminescence experiment, electrons excited into the conduction band can recombine radiatively with holes frozen on the acceptors at the acceptor energy levels. Thus, the resultant emitted photon energy is the difference between the conduction band and acceptor level or the bandgap minus the acceptor binding energy, $E_g - E_A$. These transitions involving free carriers and bound charges constitute **free-to-bound transitions**. If the impurity concentration is increased, acceptors become closer to one another and their wave functions begin to overlap with an associated broadening of the level which is called the **impurity band**. With continued increase in impurity concentration, the impurity band may widen so much so that it would overlap with the nearest band, the conduction band in the case of donors and the valence band in the case of acceptors. As a result, the carriers are freed and this delocalization is called the **Mott transition** which begins to be observed even before such overlap because the impurity band is half filled owing to spin degeneracy. The higher portion of the photoluminescence spectrum deviates from the simple exponential by spreading out to

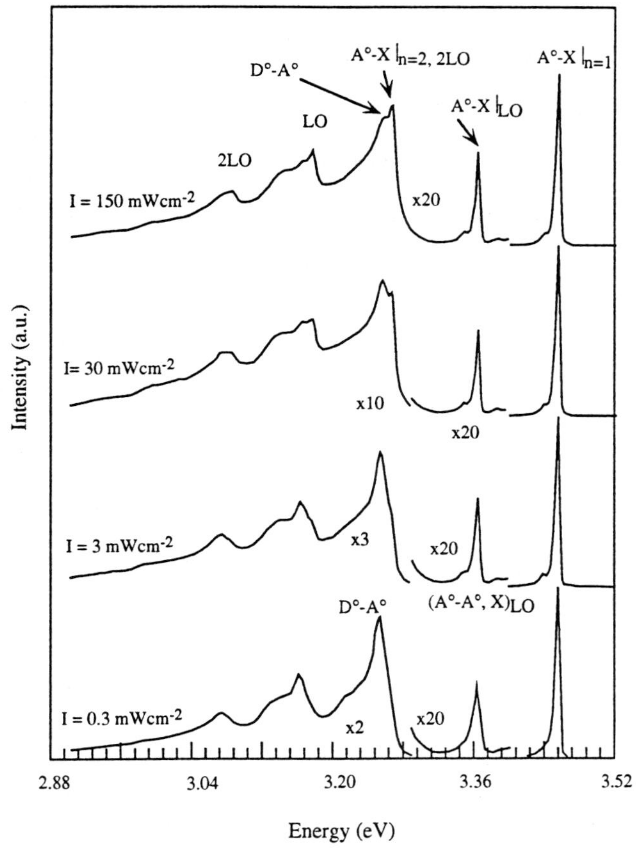

Fig. 10.16. Excitation intensity-dependent PL spectra for a Mg doped GaN sample at 1.7 K. After [10.37]

higher and lower energies, and also changing because of bandgap renormalization (**red shift**). Deep impurities and defects causing radiative decay of electrons to their ground states can be probed with luminescence experiments, and GaN is no exception. The free-to-bound (*electron-acceptor*) and bound to bound (*donor-acceptor*) transitions, the latter to be discussed below, are illustrated in Figs. 10.16 and 17.

10.3.3 Donor-Acceptor Transitions

Though some semiconductors represent the purest materials available, all contain both donors and acceptors, and are known by the term *compensated*. The nomenclature results from acceptors capturing electrons from the donor states. Consequently, a **compensated semiconductor** contains both ionized donors and acceptors. Carriers generated by optical excitation

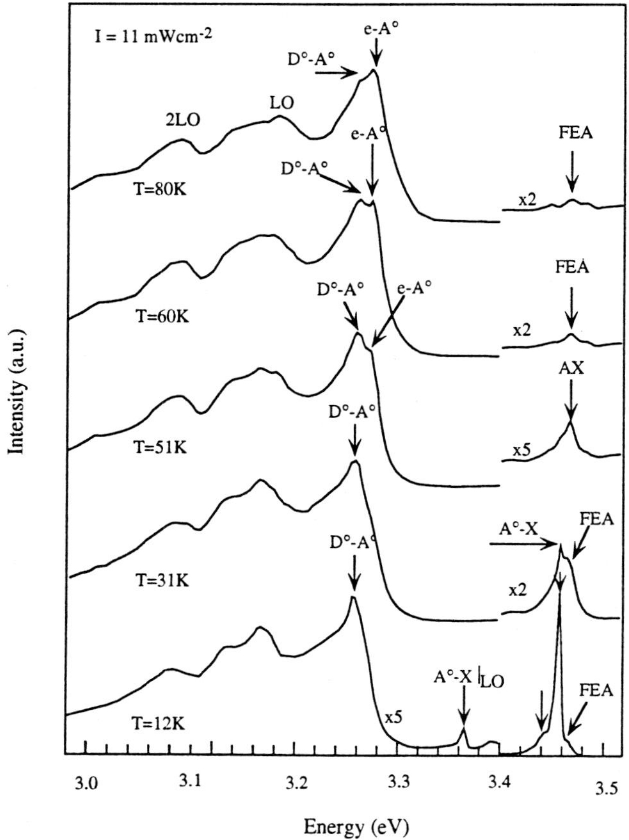

Fig. 10.17. PL spectra of a Mg-doped GaN layer as a function of temperature. After [10.35]

can be trapped at the donor and acceptor sites causing them to be neutral. In returning towards equilibrium, some electrons on the neutral donor sites will recombine with holes on the neutral acceptors, a process termed **Donor Acceptor Pair (DAP) transition**. The DAP transition energy is given by

$$h\nu = E_g - E_D - E_A + \frac{q^2}{\epsilon R}, \tag{10.36}$$

where E_D and E_A are the donor and acceptor binding energies, and the last term on the right-hand side is the Coulomb interaction contribution resulting from the interaction of ionized donors and acceptors. R is the distance between such donors and acceptors, and is assumed much larger than the lattice constant. This is easily satisfied in semiconductors that are not highly doped. In high-quality samples, many DAP transitions can be observed for many values of R. The Coulomb interaction causes the energy of

324

the final state to be lowered. **Pair spectra** have attracted much discussion in the semiconductors GaP [10.35] and GaAs [10.36]. Unfortunately, the term pair spectrum is commonly used very loosely, and is attributed to any donor acceptor transition observed in GaN.

In GaN, free and donor-bound excitons are prevalent as the samples are naturally n-type. As the p doping, i.e. with Mg, increases, the PL spectrum reveals the acceptor-bound exciton (I_1 line). As depicted in Fig. 10.16, with increasing Mg concentration, the neutral Donor Acceptor ($D°$-$A°$) transition begins to appear whose strength shows a nearly monotonic increase at around $3.25 \div 3.26$ eV with its LO phonon replicas relative to the exciton emission at 1.7 K (Fig. 10.16).

In addition to the peak position, further confirmation for this identification is provided by a blue shift at a rate of about $2 \div 3$ meV per decade of intensity as the excitation intensity is increased [10.37]. The ($D°$-$A°$) peak position can not be used unequivocally to determine the acceptor binding energy, as it depends on the average strength of the Coulomb interaction between recombining pairs which, in turn, is a function of the doping level, and the non-radiative recombination rate. At the highest pump powers (Fig. 10.16), however, a relatively new sharp peak at 3.267 eV, about 189 meV below the neutral acceptor-bound exciton ($A°, X$) peak, appears which has been assigned to a "two-hole" replica neutral acceptor-bound exciton ($A°, X$) peak in which the acceptor is left in its first excited e-like state after the exciton recombines. This identification implies a 1s-2s separation of 189 meV for Mg acceptors, which is a relatively large fraction (0.844) of their binding energy. While this value is larger than that for a purely hydrogenic case (0.75) and larger than typical values of semiconductors with large spin-orbit, no quantitative calculations have been presented in the literature for the 2s excited state of acceptors in the limit where the acceptor binding energy is much higher than the spin-orbit splitting of the valence band.

To observe the corresponding conduction-band to acceptor (e-$A°$) transitions, whose energy is a much better measure of the acceptor binding energy, variable-temperature PL measurements were performed [10.37]. These data are depicted in Fig. 10.17 for a moderately-doped sample.

As the temperature is raised, the shallow donors ($E_D \approx 29$ meV) thermally ionize into the conduction band, quenching the ($D°$-$A°$) peak in favor of a higher-energy (e-$A°$) peak that becomes dominant above about 60 K. The (e-$A°$) peak position at 60 K is 3.273 eV, and the thermal energy contribution ($k_B T/2$) is 2.6 meV at this temperature, which, with the help of an exciton binding energy of 20 meV and Fig. 10.18, leads to a Mg-acceptor optical binding energy E_A of about 230 meV. This compares with about 170 meV deduced from transport measurements under assumptions which may be questioned.

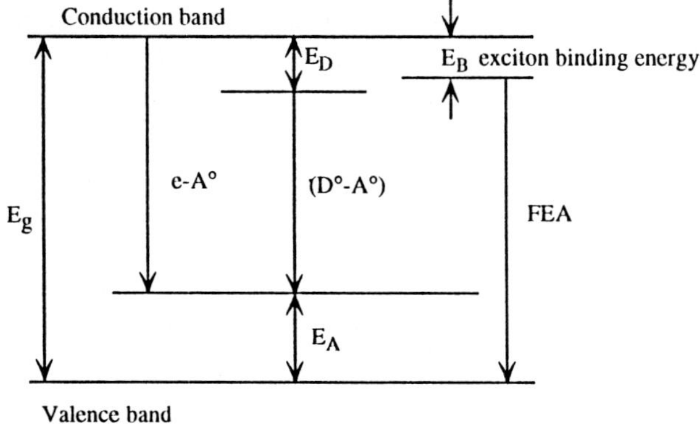

$$E_A (Mg) = FEA + E_B(20\,meV) - (e\text{-}A°)$$

Fig. 10.18. Schematic diagram of e-A°, D°-A°, FEA transitions with respect to the conduction- and valence-band edges which one can use to calculate optical binding energy of the acceptor level in question.

10.3.4 Defect-Related Transitions

A number of theoretical calculations have been performed on the native defects in GaN [10.38-44]. While none of these theories agree with one another fully, many of them suggest that a nitrogen vacancy V_N produces a single donor and is responsible for the n-type behavior in undoped GaN. Furthermore, if it is a simple donor with its s-like deep level in the gap just below the conduction-band edge, the nitrogen vacancy would provide a natural explanation of the 0.2 eV feature in the Tansley-Foley optical absorption data [10.45]. The predominance of donor-bound excitons discussed above and the n-type background in GaN films are consistent with the N vacancy premise [10.46]. Many defects caused by vacancies, antisites and interstitials are possible in GaN [10.44, 45]. Some, if not all, of these defects appear in luminescence experiments where there are two prominent transitions which are discussed below. One transition appears at about 2.2 eV, proverbially referred to as the **yellow luminescence**, and the other at about 2.9 eV, the **blue emission** which is still relied on in some LEDs, will be discussed below. The other details of defects and their interplay with doping are discussed in Chap. 5.

a) Yellow Luminescence

Many wide-bandgap semiconductors, GaN is no exception, suffer from emission near the midgap, which is yellow-green in the case of GaN. The situation is exacerbated by the enhanced response of the human eye to this region of the spectrum as compared to blue. In LEDs, this yellow emission causes a serious deviation from the saturated blue color and inhibits the achievement of all colors when used with other primary colors [10.47]. There are two arguments for the origin of this transition. One suggests transitions from shallow donors to deep acceptors, while the other evokes transitions from deep donor states to shallow acceptor states. Exhaustive studies [10.48-49] culminated in the conclusion that the yellow emission is due to transitions from deep donor-like states to shallow acceptors. In contrast, the pressure dependence of this yellow transition is such that it seems to follow the conduction band; and, after accounting for the fact that the transition should have the character of about half the bandgap of GaN, it seems to support the shallow donor-like state to a deep acceptor-like state transition. *Hoffmann* et al. [10.50] suggested that the yellow band in GaN results from the recombination between a shallow donor and a deep level. They proposed that the deep level may be a double donor, although an acceptor cannot be ruled out. *Reynolds* et al. [10.51] reported on the similarities between the green luminescence band in ZnO and the yellow luminescence band in GaN, and drew the conclusion that the genesis of these bands may be the same. The modulated nature of the broad yellow emission on the high-energy side of the band is depicted in Fig. 10.19 for ZnO. In the figure the sample quality is sufficiently high to allow the observation of the LO phonon

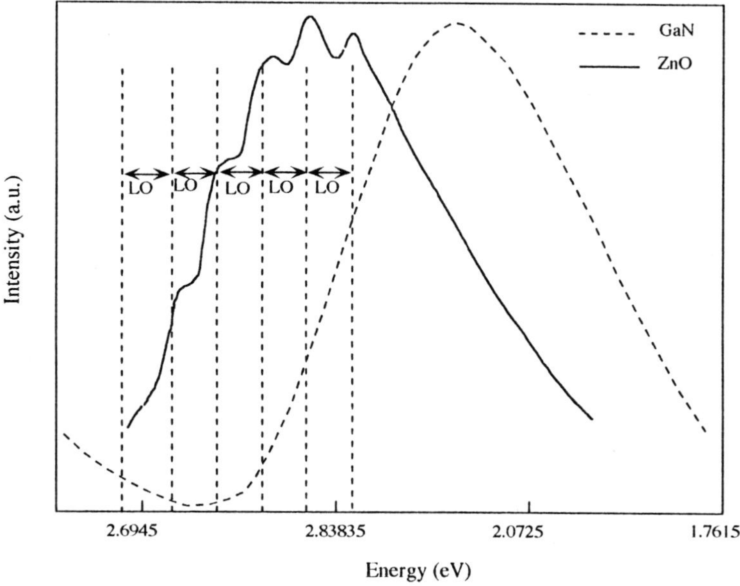

Fig. 10.19. The yellow emission band (*dashed curve*) observed for GaN and the green emission band (*solid curve*) observed for ZnO. The structured peaks occurring on the high-energy side of the green band are separated in energy by that of the longitudinal optical phonon in ZnO. After [10.51]

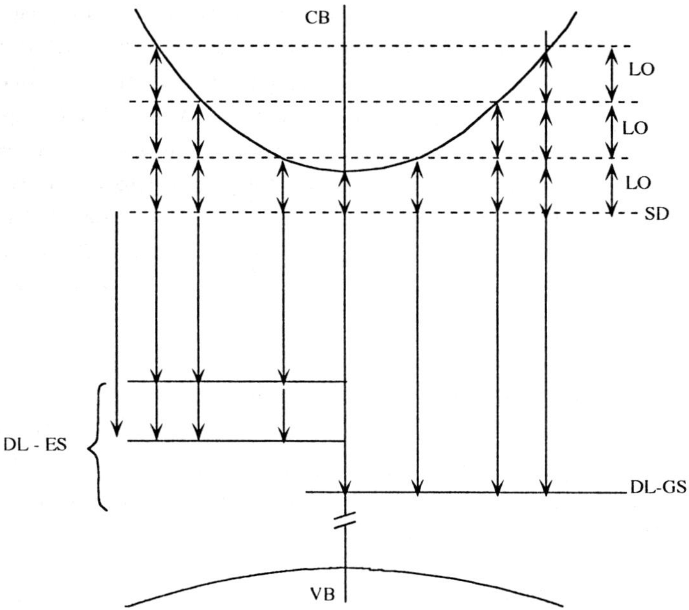

Fig. 10.20. Model to explain the green-emission band in ZnO. Band model in k-space shows the conduction band, the Shallow-Donor (SD) level, the Deep-Level Ground-State (DL-GS), and the Deep-Level Excited States (DL-ES). The Longitudinal-Optical (LO) phonons are designated as LO. The energies of the above states are all related to the valence band. After [10.51]

replicas at intervals of the LO phonon energy and may hold the clue to some of its important aspects. Assuming that the genesis of the green emission in ZnO is the same as that of the yellow emission in GaN, one can suggest that the yellow emission results from the recombination between a shallow donor level and a deep level.

The excitation source for PL measurements creates hot electrons in the conduction band. Peaks in the PL emission band occur whenever the energy of the PL peak coincides with the sum of the energies of the donor level plus an integral multiple of a principal optical-phonon energy (Fig. 10.20). At adjacent energy values, an equilibrium number of electrons will arrive at the donor level and thus take part in the recombination with the deep level. The deep level will also have accompanying excited states due to an interaction with local vibrational modes as well as lattice modes. It would be expected that the dominant transition would occur between the shallow donor and the ground state of the deep level. Transitions will also occur between the shallow donor and the excited states of the deep level, with reduced oscillator strengths. This model agrees with that proposed by *Hoffmann* et al. [10.50] for GaN and has the added advantage that it can explain the broad nature of the emission band. It is noted in Fig. 10.19 that the modulated energy structure does not occur on the low-energy side of the green band. This would be expected since the phonons that are involved in cascading hot electrons from the conduction band to the donor level are not involved with the low-energy emission. This emission is accounted for by the recombination of donor electrons with excited states of the deep level. The deep level may be a complex center whose excited states consist of both local vibrational modes and lattice modes. These excited states are so distributed that they do not produce a resolvable modulated structure on the low-energy side of the green band. This would support the model of *Hoffmann* et al. [10.50].

b) Group-II Element Related Transitions

In wide-bandgap nitrides, addition of what should be p-type dopant impurities leads to many complex phenomena. In GaN, for example, while the effective-mass-like acceptor is about 230 meV from the valence band, Mg-, Be- and Zn-doped GaN exhibits emission at centers which are about 0.5 eV above the valence band when the Mg concentration exceeds a certain level. In optical spectra of Mg-doped samples, two broad emission bands, one at about 3.21 eV, 290 meV below the gap (dominant for T < 150 K) and the other at about 2.95 eV, 550 meV below the gap (dominant for T > 150 K) are observed. A typical CW PL spectrum of p-type GaN layers at 10 K is dominated by a band at about 3.21 eV which nearly disappears for T > 150 K (Fig. 10.21). As the temperature is increased above 150 K, a weak emission band at about 2.95 eV appears. Moreover, the peak position of the lower-energy emission (2.95 eV at 150 K) red shifts considerably as the Mg-doping level is increased. Similar observations have been made for Zn-doped GaN which may erroneously lead one to believe that the binding energy of the acceptor is very deep. Paradoxically, annealing experiments to activate Mg acceptors in MOCVD samples have originally centered around maximizing the 420 nm emission peak as if it would increase the hole concentration in the films although some connection between this peak and hole concentration can not be ruled out at this point. There is some evidence now which suggests that group-II impurities are not stable on N sites and cause defects which may be responsible for the observation of the aforementioned deep defects. In fact, the peak position of this lower-energy emission band can be varied from 430 to about 700 nm at room temperature.

In order to explore the physical origin of the observed emission lines, their dynamical behavior have been studied [10.52]. At low temperatures, PL decay is non-exponential, but can be approximated by a two-exponential decay process. The typical lifetime of the fast component which contributes 90 percent of the PL signal is about 0.6 ns and the slow component is about

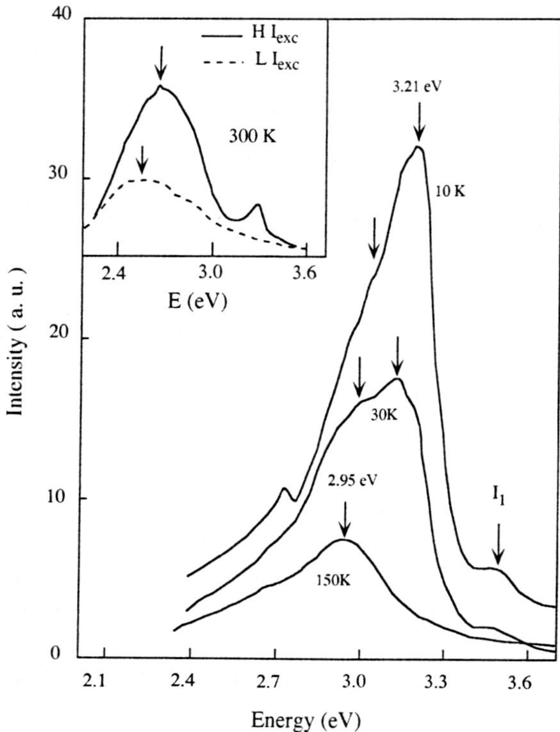

Fig. 10.21. CW PhotoLuminescence (PL) spectra measured at three representative temperatures for Mg-doped p-GaN samples, which exhibit a dominant emission line at about 3.21 eV at low temperatures. Another lower-energy emission line at about 2.95 eV (T = 10K) dominates at higher temperatures (T > 150K). The inset presents the room-temperature PL spectra associated with the lower energy emission band obtained for higher excitation (H I$_{exc}$) and lower excitation intensities (L I$_{exc}$, showing that the spectral peak position shifts toward higher energies as the excitation intensity increases. After [10.53]

5 ns. In the temperature region T < 150 K where the 3.21 eV emission band dominates, the recombination lifetime decreases progressively from 0.6 to 0.3 ns as the temperature increases from 10 to 140 K. This behavior can be accounted for by an increased non-radiative recombination rate at higher temperatures, caused by the non-radiative carrier transfer to the lower-energy recombination channels. This is consistent with the observation of thermal quenching of the 3.21 eV emission line and the subsequent increase in the emission intensity of the lower energy band at 2.95 eV.

In the higher temperature region (T > 150K) where the lower-energy emission band dominates, the fast decay component contributes nearly 95 percent of the PL signal and consequently the decay kinetics of PL are closely described by a single exponential. The temperature dependence of the recombination lifetime of the lower-energy emission band indicates an increase with temperature and reaches 0.3 ns at room temperature. This is

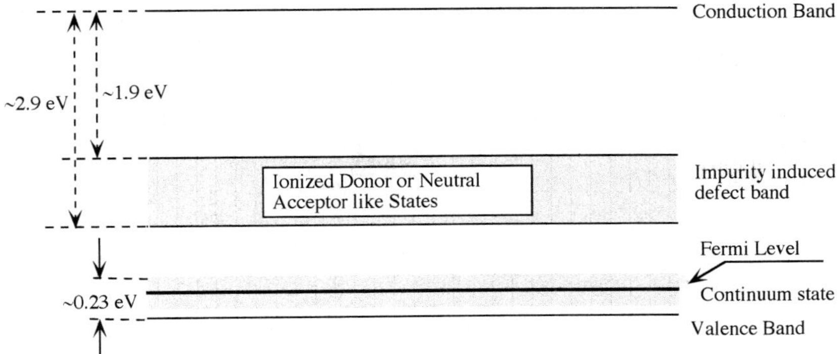

Fig. 10.22. Schematic representation of the proposed acceptor level/band and deep levels caused by group-II impurities

due to the carrier transfer from the 3.21 eV recombination channel, as discussed in the previous subsection. At this point of development or lack thereof, one can suggest that the deep-lying states are caused by the large supply of group-II dopants. To depict this phenomenological picture, Fig. 10.22 exhibits the shallow band associated with the substitutional group-II dopants. A band can be formed when the group-II dopants concentration is high. The deeper band, which is associated with the defects, is caused by large concentrations of group-II dopants.

10.4 Optical Properties of Nitride Heterostructures

To observe a sizable quantization, large electron effective masses in GaN and related materials require the well thickness in the few tens of Å. Although several of the heterostructures reported thus far really fit this criterion, the term quantum well will be interpreted liberally here, and heterostructures approaching this criterion will be treated. We must also mention that the quality of the quantum-well-like structures grown so far is not sufficiently advanced to reach the final resolution of the confined states. Additional complication arises due to piezoelectric effects, the lack of native substrates, and a small critical thickness in the nitride system. Nevertheless, great strides that have been made in a considerably short period of time are compelling enough to begin investigation of such structures.

10.4.1 GaN/AlGaN Heterostructures

GaN/AlGaN heterostructures are by far the most studied and best understood ones among those investigated so far. Figure 10.23 displays the (CW) PL spectra of the GaN/Al$_x$Ga$_{1-x}$N Multiple Quantum Wells (MQW) sample obtained at (a) T = 300 K and (b) T = 10 K [10.53]. For comparison, PL spectra of a GaN epilayer are also shown. In the GaN epilayer, the dominating transition line at T = 10 K is due to the recombination of the ground-state of an A-exciton. in the MQW, the excitonic transition peak position is blue shifted due to the well known effect of quantum confinement of electrons and holes. The blue shift at room temperature (79meV) is what is expected for a MQW structure with a 67% (33%) conduction (valence) band offset. One of the interesting features depicted in Fig. 10.23 is that the blue shift observed at 10 K is 54 meV, which is about 25 meV less than the shift of 79 meV seen at 300 K. One can attribute this difference to the fact that the PL emission in MQW measured at low and room temperatures resulted

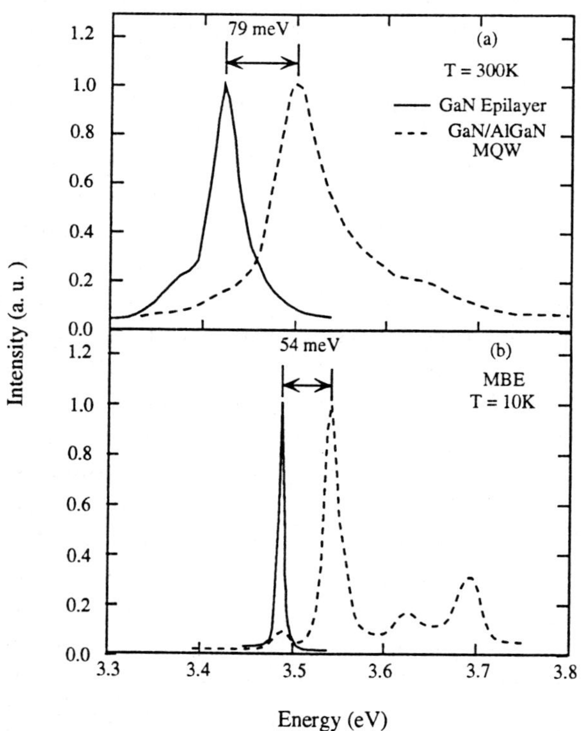

Fig. 10.23. CW PL spectra of a GaN epilayer (*solid dots*) and GaN/Al$_x$Ga$_{1-x}$N MQW measured at (a) T = 300 K and (b) T = 10 K. Note that the blue shift of the excitonic transition in the MQW at T = 10 K (54 meV) is about 25 meV less than that at T = 300 K (79 meV). After [10.54]

from the recombination of localized excitons and free excitons, respectively. The exciton localization at low temperatures may be caused by the interface roughness of MQW and quantum-well-width fluctuations. A closer look indicates transitions at 3.692 eV, 3.625 eV, 3.558 eV, and 3.489 eV which are assigned to no phonon and three phonon replicas of the excitonic transitions in $Al_xGa_{1-x}N$ barriers. This assignment is based on the fact that these transition lines are separated by an equal energy space (67.5 meV). Moreover, the relative emission intensities of these transitions also support our assignment. The excitonic transitions from the $Al_xGa_{1-x}N$ barriers are easily seen here because the AlN mole fraction in the barrier material is relatively low (x $\approx$ 0.07). This makes the energy difference between excitons in the wells and the barriers relatively small. The emission intensities of the excitonic transition at 3.692 eV and its phonon replicas decrease with an increase in temperature. This is due to an increased rate of exciton transfer from the $Al_xGa_{1-x}N$ barriers to the GaN wells.

There are many unsettled issues clouding the attempts to characterize nitride-based heterostructures. What little solace there is has to do with the AlGaN system being the least complicated one. The InGaN-containing structure have the added complication of phase separation of this ternary which complicates matters considerably. Returning to the AlGaN case, there are basically two complications which arise from the non-uniform strain and a large piezoelectric effect. The former results are due to defect-laden buffers on which these structures are grown and a sizable lattice mismatch between GaN and AlGaN. All of these issues are intertwined and feed on one another. Nevertheless, the problem is an important one deserving attention. In this vein, attempts have been made to investigate GaN/AlGaN quantum wells with varying well thicknesses, with and without doping [10.54]. In addition, Si doping has been introduced since it, like other impurities, can fill vacancies and reduce vacancy-driven defects.

Low-temperature (10K) CW PL spectra for two representative GaN/$Al_xGa_{1-x}N$ MQW samples with well thicknesses of L_W = 25 Å and L_W = 50 Å are presented in Fig. 10.24. For comparison, the PL spectrum, not shown, of a GaN epilayer deposited under similar conditions has a dominant transition at 3.485 eV and 10 K, which is the ground state of the A-exciton. In the 25 Å well MQW sample, the 56 meV blue shift in the excitonic feature is attributed to the well-known effect of quantum confinement of electrons and holes which compares to a 25 meV shift that would have been seen due to strain alone. The higher-energy emission peaks are due to an exciton transition and its LO phonon replicas in the AlGaN barrier regions [10.54]. In sharp contrast, the dominant emission in the 50 Å MQW sample is at 3.414 eV, which is 71 meV below the exciton energy.

Additional MQW samples with 20, 40 and 60 Å well thicknesses were examined. They indicated that the features of the 20 Å MQW are similar to

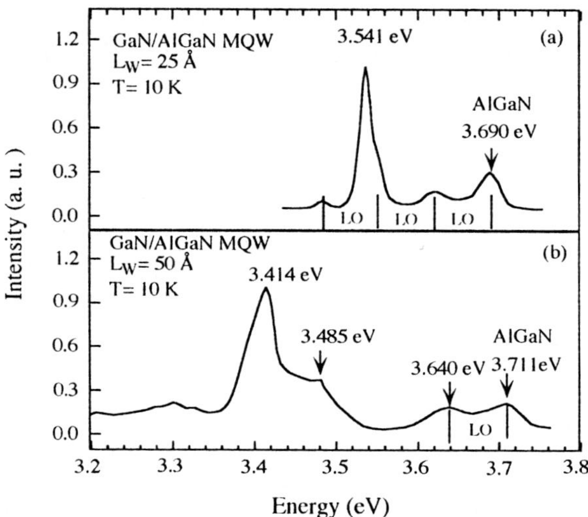

Fig. 10.24. Low-temperature (10K) PL spectra of an nominally undoped (a) GaN/AlGaN MQW sample with the well thickness $L_W = 25$ Å, (b) GaN/AlGaN MQW sample with the well thickness $L_W = 50$ Å, and (c) GaN epilayer grown under identical conditions as the MQW samples. After [10.52]

the 25 Å MQW sample, and the 60 Å one exhibited characteristics similar to the 50 Å case. In order to further identify the origin of the optical transitions, the dynamic behavior of the optical transitions in these MQWs were investigated. The excitonic decay resulting from the well regions of the 20 and 25 Å MQWs is exponential below 150 K. The exciton recombination lifetime in both 20 and 25 Å MQWs increases linearly with temperature up to 60 K. This is consistent with observations in bulk GaN and is also similar to the behavior seen in the GaAs/AlGaAs QWs. On the contrary, because of the coupling between the impurity-related transitions and the intrinsic transition, the decay of the dominant transition in the 50 Å MQW (at 3.414eV and 10K) follows $I(t) = A_1 \exp(-t/\tau_1) + A_2 \exp(-t/\tau_2)$ with the faster decay-time constant τ_1 being almost independent of temperature, and the slower decay-time constant τ_2 decreasing monotonically with temperature. The decreasing amplitudes of A_1 and A_2 with decreasing temperature are indicative of the faster and slower transitions being associated with intrinsic and band-to-impurity (defect) transitions, respectively. The emission intensity of the band-to-impurity (defect) transition is largely quenched at temperatures above 200 K.

The PL characteristics of an undoped 50 Å MQW and a Si doped 60 Å MQW unequivocally illustrate the benefits of Si doping. At 10 K, no blue shift was seen for the 50 Å MQW. The dominant transition at 300 K is the band-to-band (or free-electron to free-hole) recombination which is indica-

334

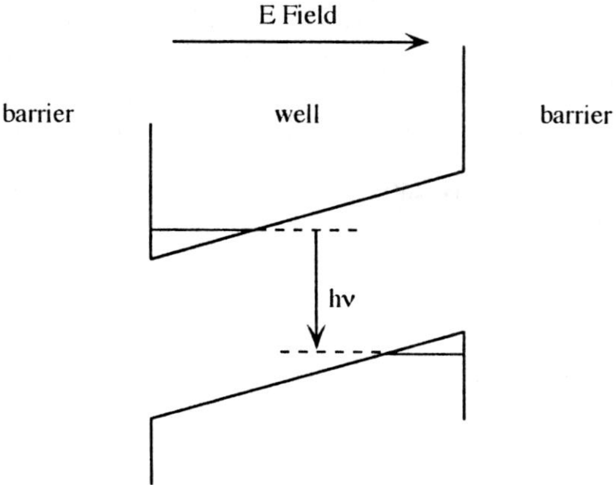

Fig. 10.25. Schematic illustration of the quantum confined Stark effect caused by the electric field induced by the piezoelectric effect

tive of the presence of larger densities of dislocations, which tend to break up the exciton into free carriers in this MQW. In contrast, the room-temperature PL spectrum of a 60 Å well QW doped with Si exhibited a dominant PL emission peak at 3.466 eV with a 45 meV blue shift due to the quantum-confinement effect. That Si doping improves the crystalline quality is consistent with the notion that Si fills point defects. It is further supported by the observations of *Ruvimov* et al. [10.55] who reported a reduction in dislocation density in Si-doped bulk GaN. Supporting the same argument are the observations of *Nakamura* et al. [10.56] who have reduced the threshold current density of InGaN MQW-structure laser diodes from 10 to 3.6 kA/cm^2 by Si doping the active layer along with other improvements.

Despite some progress, characterization of heterostructures is really in an embryonic stage. In addition to the role of defects and impurities which make the identification of transitions difficult, the inherent piezoelectric effect causes an electric field to be induced as a result of strain in the layers [10.57-59]. The piezoelectric effect is due to the residual strain in the films as well as the lattice mismatch between the components of the heterostructure. The strength of the field is dependent on the piezoelectric constants, the elastic stiffness coefficients, and the extent of strain (Chap.3). The field causes band bending and forces the carriers to the opposite side of the well (Fig.10.25). The transition taking place between the electrons and holes would radiate at energies smaller than the bandgap. In fact, the wider the wells the larger the red shift. Forcing the carriers to the opposite ends of the wells causes a spatial separation which makes the transitions less efficient. In short, the larger the well thickness the less efficient the transition. Likewise, the larger the field, the larger is the red shift. The band bending can be assumed linear if the light-generated carrier density is low. If the carrier density is high, the field generated by the spatially separated carriers would be in the opposite direction to the piezoelectric field, reducing its impact. This is termed the **Coulomb screening effect**. In other words, the extent of the red shift would be reduced with increasing excitation power. On the experimental side, the picture is rather complicated as filling of the band-tail states and localized states within the gap also causes a blue shift and makes the analysis complex. If one can assume reproducibility in the production of structures, which is not possible in its true sense, the thickness of the well could be varied which would shed some light on the extent of the piezoelectric effect and filling of the states.

10.4.2 InGaN/GaN and InGaN/InGaN Heterostructures

InGaN is imperative in the operation of optoelectronic emitters that have so far been demonstrated. With its smaller bandgap, as compared to GaN, and robustness against defect propagation, InGaN is firmly established as the desired quantum well or active region for emitters. The inferior quality and/or the high background dopant concentration of InGaN has made it difficult to determine its bandgap, let alone any quantum confinement effect. The analyses such as X-ray diffraction to determine the equilibrium lattice constant are frustrated due to phase separation and inhomogeneous strain. While this may very well be advantageous in combating the adverse effects of lattice and thermal mismatch with the available substrates, it makes it almost impossible to characterize and model the quantum wells formed. The situation is exacerbated by the piezoelectric strain effect involved, which is not well understood either. Nevertheless, it is imperative that what is known, no matter how incomplete, be discussed for the reader to be abreast of the developments.

To restate, the high equilibrium pressure of InGaN and its tendency to phase separate and cause compositional inhomogeneities make it more difficult to analyze heterostructures containing InGaN. In very thin films, however, the severe phase segregation and pooling are reduced to some extent when these thin layers of InGaN are sandwiched between GaN or AlGaN layers such as is the situation in quantum wells. This is evident from the work of *Koike* et al. [10.60] who reported much improved cathodoluminescence in $In_{0.08}Ga_{0.92}N$/GaN multiple quantum wells when compared to bulk InGaN.

Nakamura et al. [10.61] have undertaken a study of $In_xGa_{1-x}N$/$In_yGa_{1-y}N$ heterostructures which, as indicated above, form the active layers of lasers to follow. Structural as well as emission characteristics were analyzed by X-ray diffraction and photoluminescence. The former was employed to determine the periodicity and the latter to measure the emission intensity and energy. The PL emission spectra for two such samples having a periodicity of 60 and 200 Å, as determined by X-ray diffraction, also show the expected satellite peaks. The analysis was aided by a Kronig-Penny formulation with the applicable and available carrier effective masses of 0.2 and 0.8 for electrons and holes in GaN, and 0.11 and 1.6 in InN. We should note that since the publication of these papers, the preponderance of the available data indicates the relative electron effective mass in GaN to be 0.23, and the composition of the well layer was determined to be $In_{0.22}Ga_{0.78}N$. What is technologically relevant here is that the structures of this kind were used to optimize the growth conditions which laid the ground work for high-performance LEDs to be produced. It should be noted that, as part of this particular study, the position of the PL emission peak in bulk InGaN was also investigated with a varying layer thickness resulting in the observation that the peak position is thickness inde-

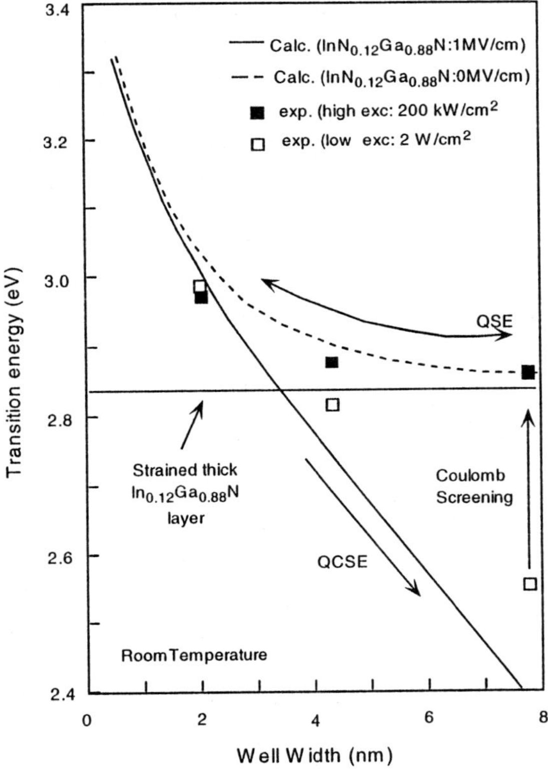

Fig. 10.26. Emission energy in a 10 period $Ga_{0.88}In_{0.12}N/Ga_{0.97}In_{0.03}N$ MQW sample as a function of well width. Solid and open squares represent the measurements at 300 K with high (200 kW/cm^2) and low (2 W/cm^2) excitation intensities, respectively. Solid and dashed lines represent the calculations with a 1 MV/cm piezoelectric-induced electric field and without any field, respectively. The measured PL peak energy of $Ga_{0.88}In_{0.12}N$ bulk layer is also shown. After [10.62]

pendent. This would imply that any In surface segregation is minimized. What is striking is that the structures investigated and optimized are in many ways similar to the active layers used by *Nakamura* and his colleagues in their laser structures.

A more recent investigation of *Amano* et al. [10.62] illustrates the complexity in determining the composition of InGaN. They attempted to determine the bulk properties of InGaN together with InGaN quantum wells with the inclusion of the piezoelectric effect along with the quantum confinement effect. Figure 10.26 shows the room-temperature PL peak energies of the $Ga_{0.88}In_{0.12}N$ strained QWs as a function of well width along with the calculations. The solid and open squares are for PL transition energies under high excitation intensities, produced by a N_2 laser, and low excitation intensities, supplied by a He-Cd laser. The effect of the electric field induced by the piezoelectric effect follows the discussion of Chap.3. Piezoelectric fields up to 1.0 MV/cm are predicted along the [0001] direction for an InN mole fraction of 0.12. The solid line in Fig.10.26 denotes the calculated transition energies assuming a strain-induced electric filed. The transition energy without the electric field is also shown by a dashed line for comparison. It is clear that the red shift increases with increased well thickness. In wells less than 3.5 nm, a blue shift due to the Quantum Size Effect (QSE) is noticeable. Under low excitation, the transition energy of the well layer is

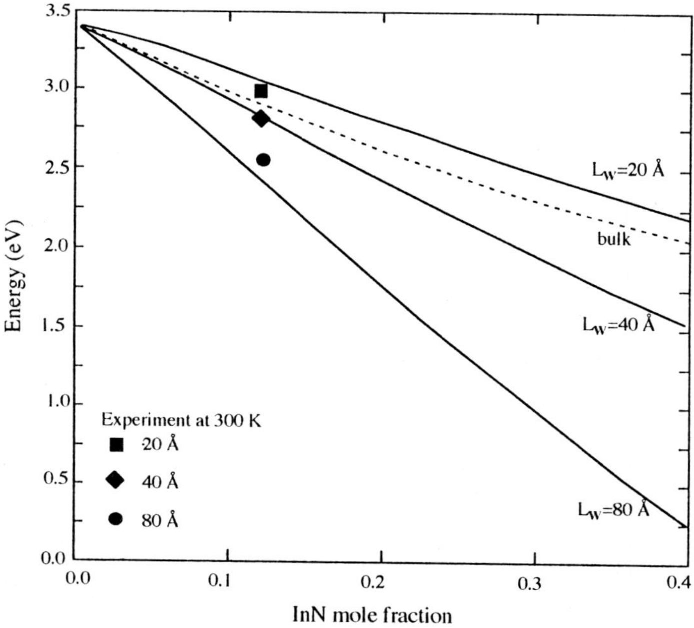

Fig. 10.27. Compositional dependence of the transition energy of strained MQW as a function of the well width. That of bulk GaInN is also shown for comparison. After [10.62]

smaller than that of a thick GaInN single layer, particularly for thicker wells because of the field-induced Quantum Confined Stark Effect (QCSE), illustrated schematically in Fig. 10.25. As the excitation intensity increases, a blue shift occurs as a result of Coulomb screening which compensates somewhat the effect of the piezoelectrically induced electric field. Figure 10.27 summarizes the compositional dependence of the lowest transition energy of strained GaInN as a function of well width. The effective band-gap of MQWs having rather thick wells is expected to show large red shifts due to the QCSE and even a energy smaller than the bulk for the same composition, all else being equal.

In closing, it is imperative that uniformly strained InGaN films, hopefully with lower background concentrations, be prepared to allow such parameters as composition and band discontinuity to be determined in order to establish the composition versus bandgap in this very important ternary. Since bulk layers of high quality can not be prepared, at the time of this writing, the above determination hinges on our ability to measure the effective masses and other applicable parameters.

11. Light-Emitting Diodes

Light Emitting Diodes (LEDs) convert electrical power to generally visible optical power and are simply p-n-junction devices, when biased in the forward direction. They produce light through spontaneous emission whose wavelength is determined by the bandgap of the semiconductor in which the carrier recombination takes place. Unlike the semiconductor laser, generally the junction is not biased to and beyond transparency. Consequently self absorption occurs and photons are emitted in random directions. A modern LED is generally of a double-heterojunction type with the active layer being the only absorbing layer including the substrate. In addition, a plastic dome to increase the light collection cone and to focus the light is employed. Nitride-based LEDs with InGaN-actice regions span the visible spectrum from yellow to violet, as illustrated in Fig. 11.1.

The three types of LEDs are **surface emitters**, **edge emitters** generally intended for fiber-optic communications and **super-radiant** or **super-luminescent** devices which are biased not quite to the point of lasing but are biased enough to provide some gain and narrowing of the spectrum. Anti-reflection coatings or some other measures are taken to ensure that the device does not lase. Surface emitters are divided into those with plastic domes and those with flat surface-mount varieties lacking the dome.

Among the applications of LEDs are displays, indicator lights, signs, traffic lights and lighting (potentially) which requires emission in the visible part of the spectrum, printers and telecommunications. While saturated-color red LEDs can be produced using semiconductors such as GaP, AlGaAs, AlGaInP, the green and blue commercial LEDs with sufficient brightness to be of use for outdoor applications have so far been manufactured with nitride semiconductors. Figure 11.2 exhibits the various ternary and quater-

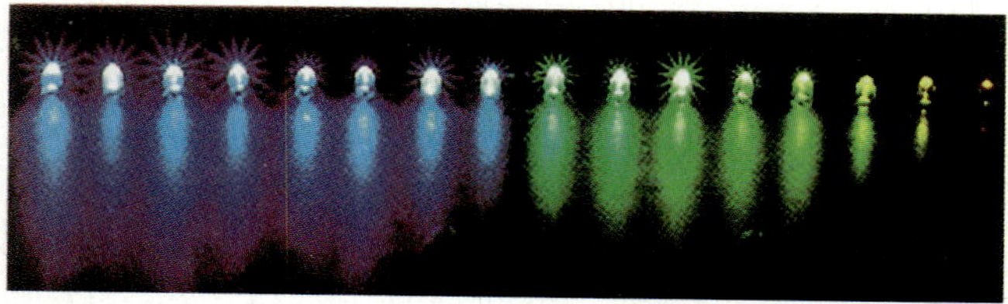

Fig. 11.1. This should be placed at the beginning of the chapter just below or the above the title. InGaN LEDs spanning the spectral range from violet to orange. Courtesy of S. Nakamura, Nichia Chemical Co. Ltd.

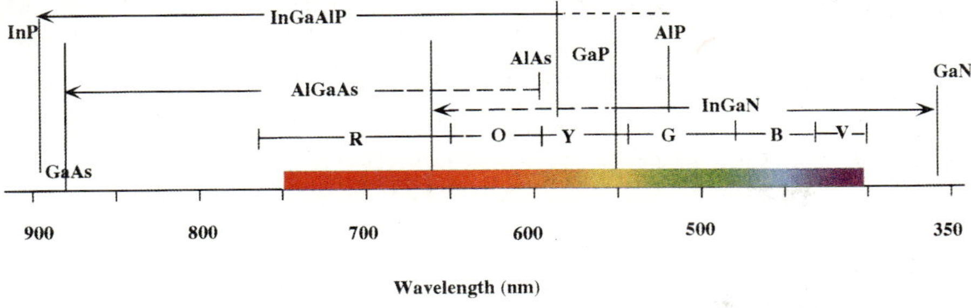

Fig. 11.2. The LED materials and range of wavelength of the emission associated with them. The color band indicates the visible region of the spectrum

nary materials used for LEDs with the wavelength ranges indicated. The color bar corresponds to the visible portion of the spectrum. Though not on the figure, due to an insufficient operating lifetime, it should be mentioned that ZnTeSe/ZeSe double-heterostructure LEDs have been developed through a collaborative effort of the group of J. Schetzina at North Carolina State University and Eagle Picher Co. of Miami, Oklahoma, USA, with a performance comparable to the nitride-based ones. While ZnSe itself has a bandgap of 2.67 eV, it can be increased and reduced, within some technological limitations, by additions of Mg and Cd, respectively [11.1].

Even though the fundamentals of radiative recombination, in general, and LED operation, in particular, in nitride semiconductors are still hotly debated, the basics of LEDs, assuming that the semiconductors of interest are well behaved, will be treated first. This will be followed by the performance of available nitride LEDs and their characteristics. The discussion is completed with succinct treatments of the reliability of nitride-based LEDs, and of organic LEDs which have progressed to the point that indoor applications are being considered.

11.1 Current-Conduction Mechanism in LED-Like Structures

For simplicity, let us assume that a double-heterojunction device is one in which all the carriers recombine in the smaller-bandgap active region. Furthermore, recombination takes place in the bulk, part of which is non-radiative, and at the two non-radiative interfaces. The larger-bandgap AlGaN n and p layers are doped rather heavily so that no field exists in these regions. The treatment here will be developed in a manner similar to that of *Lee* et al. [11.2] and *Wang* [11.3].

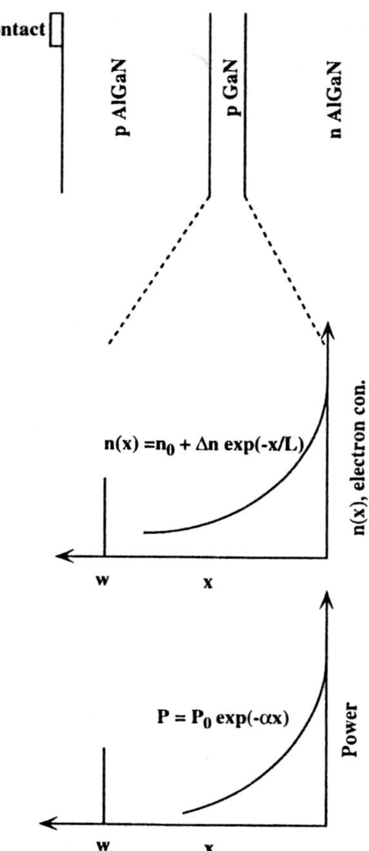

Fig. 11.3. The spatial carrier and light distribution in a double-hetero-structure LED structure

Consider an AlGaN(p)/GaN(p)/AlGaN(n) device which is forward biased. The carrier and light distribution in the active layer are depicted schematically in Fig. 11.3. Since the active layer is p-type, we will be dealing with minority electron carriers. The continuity equation for electrons can be written as

$$D \frac{\partial^2 n}{\partial x^2} - \frac{n - n_0}{\tau} + g = \frac{\partial n}{\partial t} \tag{11.1}$$

where n and n_0 represent the minority-carrier concentration and the equilibrium minority-carrier concentration, respectively. The terms D, g, and τ represent the electron diffusion length, the generation rate, and the carrier lifetime, respectively, and x and t have their usual meaning. If the active layer were n-type, the same equations would apply with the minority electron parameters replaced with the minority hole parameters.

Under steady-state conditions and large injections levels such as is the case for LEDs, the time dependence vanishes, the generation rate and the equilibrium minority carrier concentration can be neglected, and the continuity expression reduces to

$$D\frac{d^2n}{dx^2} - \frac{n}{\tau} = 0 \; . \tag{11.2}$$

This second-order differential equation can be solved with appropriate boundary conditions which can be arrived at by considering the rate of change in the carrier concentration at each side of the p-active layer.

The general solution of the continuity equation is given by

$$n(x) = A\,e^{-x/L} + B\,e^{x/L} \tag{11.3}$$

or in the p-region

$$n(x) = A\sinh\frac{w-x}{L} + B\cosh\frac{w-x}{L} \; . \tag{11.4}$$

Here, L is the diffusion length $L = (D\tau)^{1/2}$, and the constants A and B can be found subject to the boundary conditions as described below.

The rate of change in carrier concentration at $x = 0$ is the difference between the injection rate and the interface recombination rate. The rate of change in carrier concentration at $x = w$ is the difference between the injection rate at $x = w$ and the interface recombination rate at $x = w$:

$$-\frac{dn}{dx}\bigg|_{x=0} = \frac{J_{diff}(0)}{qD} - \frac{v_S\,n(0)}{D} \quad \text{at} \quad x = 0 \tag{11.5}$$

and

$$-\frac{dn}{dx}\bigg|_{x=w} = \frac{J_{diff}(w)}{qD} - \frac{v_S\,n(w)}{D} \quad \text{at} \quad x = w \tag{11.6}$$

where q is the electronic charge, J_{diff} is the diffusion current density, and v_S [cm/s] is the interface recombination velocity. It is assumed that $J_{diff}(w)$ is negligible as is the case when the p-layer is thicker than the diffusion length. One must keep in mind that the rate of change in the minority carrier is always negative.

The solution to the continuity equation subject to the above boundary conditions is

$$n(x) = \frac{J_{\text{Diff}}(x=0)}{q} \sqrt{\frac{\tau}{D}} \left(\frac{\cosh[(w-x)/L] + v_S \sqrt{\tau/D}\,\sinh[(w-x)/L]}{(v_S^2 \tau/D + 1)\sinh(w/L) + [2v_S \sqrt{\tau/D}\,\cosh(w/L)]} \right).$$

(11.7)

Here $J_{\text{diff}}(x=0)$ can be assumed to be the terminal current as the hole injection is negligible, given the very small intrinsic carrier concentration.

The average electron concentration in the active region can then be calculated from the integral

$$n_{\text{ave}} = \frac{1}{w} \int_0^w n(x)\,dx = \frac{J\tau_{\text{eff}}}{qw}.$$

(11.8)

Substitution of the electron concentration (11.7) into (11.8) leads to an effective carrier lifetime which reduces to

$$\frac{1}{\tau_{\text{eff}}} = \frac{1}{\tau} + \frac{2v_S}{w} = \frac{1}{\tau_{\text{rad}}} + \frac{1}{\tau_{\text{nrad}}} + \frac{2v_S}{w}$$

(11.9)

if $w/L < 1$, and $v_S^2 \tau/D \ll 1$. In addition, in the absence of interface recombination, the effective lifetime would reduce to τ.

11.2 Optical Output Power

As seen by the electron-density expression, the electron density and thus the photon density is reduced in the area away from the junction. Consequently, increasing the active thickness does not lead to a continually increasing optical power. In addition, the light generated in the active layer itself is self absorbed in the active layer. Here, it is assumed that the rest of the structure is a larger-bandgap semiconductor which would not be absorbing. The photon-flux density can be approximated by a Gaussian function of the form

$$S(\lambda) = S_0 \exp\left[\frac{-4(\lambda - \lambda_0)^2}{(\Delta\lambda)^2}\right]$$

(11.10)

where S is the number of photons per unit time per unit volume with S_0 representing the same in the center of the spectrum.

At a give point x in the active layer $S_0 = \Delta n(x)/\tau_{rad} = n(x)/\tau_{rad}$, τ_{rad} being the radiative lifetime. Recognizing that the photon energy equals $h\nu = hc/\lambda$, the power is given by

$$P = Ahc \int_0^\infty \frac{S(\lambda)}{\lambda} d\lambda . \tag{11.11}$$

With further manipulation and substitutions connecting the photon density to the carrier density, we obtain

$$P = Ahc \int_0^\infty \frac{d\lambda}{\lambda} \int_0^W \frac{n(x)\,e^{-\alpha(\lambda)\,x}}{\tau_{rad}} dx . \tag{11.12}$$

11.3 Losses and Efficiency

One must now grapple with the fact that the photons generated in the active layer are emitted in all directions with only a fraction of them escaping the device to reach the human eye (Fig. 11.4). Absorption within the LED (η_A in terms of efficiency), critical-angle loss (η_c) and reflections (η_F) (*Frensel loss*) represent the main sources of loss. Absorption of photons emitted

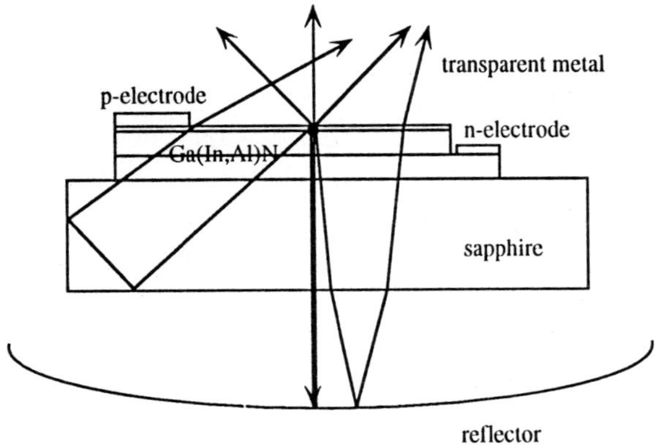

Fig. 11.4. Schematic diagram of an LED intended for as much light extraction as possible with a back reflector and a transparent substrate as that employed in InGaN-based devices

down into the semiconductor structure can be reduced by utilising layers and substrates transparent to the radiant wavelength and coupled with a highly reflective back at the bottom of the substrate. These loss factors are not yet considered in the derivation of (11.12). In GaAs diodes with a GaAs substrate, about 85% of the photons generated are absorbed. If transparent substrates are used, such as GaP, only about 25% is lost. In the case of InGaN diodes, the entire structure, of course, with the exception of the active layer itself, is transparent and absorptive losses are therefore almost eliminated.

As for the reflection at the semiconductor-air interface, when light passes from a medium of refractive index n_2, which is the active layer here, to a medium with the refractive index n_1, being air in this case, a portion of the radiation is reflected at the interface. This loss, which is called the **Frensel loss**, is given in the case of normal incidence by

$$R = \frac{n_2 - n_1}{n_2 + n_1} \,. \tag{11.13}$$

The **Frensel-loss efficiency** [11.4] can be defined as $\eta_F = (1-R)$.

The **critical angle for total reflection** ϑ_c – total reflection taking place above this angle – is determined by **Snell's law**

$$\vartheta_c = \sin^{-1}(n_1/n_2) \,. \tag{11.14}$$

For GaAs and GaP, these angles are $16°$ and $17°$, respectively. For a GaN-air interface, the critical angle is about $21°$, $24°$ and $25°$ at the wavelengths of 365, 450, and 520 nm, respectively. The **critical-loss efficiency** can be expressed as $\eta_c = \sin^2\theta_c$ or $1-\cos^2\theta_c$.

If the efficiency term which is associated with internal losses including interface recombination and self absorption is denoted by η_A, then $\eta_{opt} = \eta_F\,\eta_c\,\eta_A$ would represent the efficiency of the total power extraction.

The optical power at the central wavelength λ_0 can be obtained as

$$P_0 = \frac{Ahc}{\lambda_0\,\tau_{rad}} \int_0^W n(x)\,e^{-\alpha_0 x}\,dx = \frac{Ahc}{q\lambda_0\,\tau_{rad}}J\,\tau_{eff} \,. \tag{11.15}$$

Recognizing that hc/λ_0 represents the photon energy and if the photon energy is E_{ph} in terms of eV, one can define the *internal* **quantum efficiency** as

$$\eta_{int} = \frac{P_0}{IE_{ph}} \,.$$

(11.16)

Utilizing (11.15) for the power, we obtain

$$\eta_{int} = \frac{hc}{q\lambda_0} \frac{I}{\tau_{eff}\tau_{rad}} \frac{1}{IE_{ph}} = \frac{\tau_{eff}}{\tau_{rad}} \,.$$

(11.17)

Multiplying the internal quantum efficiency by the combined loss and efficiency factors, the *external* **quantum efficiency** becomes

$$\eta_{ext} = \eta_{opt} \frac{\tau_{eff}}{\tau_{rad}}$$

(11.18)

which is about 10 % for UV and blue GaN-based diodes. In the case where there are ohmic losses, the term E_{ph} must be replaced by the energy corresponding to the applied voltage qV_{appl}. Then, the external quantum efficiency will assume the form

$$\eta_{ext} = \eta_{opt} \frac{\tau_{eff}}{\tau_{rad}} \frac{E_{ph}}{qV_{appl}} \,.$$

(11.19)

The optical power extracted from the LED is given by

$$P_0 = \eta_{opt} \frac{hc}{q\lambda_0} \frac{I}{\tau_{rad}/\tau_{eff}} \,.$$

(11.20)

For a double-heterojunction LED, where the active layer is the only absorbing layer in the entire structure on a transparent substrate, the internal absorption term η_A, including interface recombination, has been determined to be [11.2]

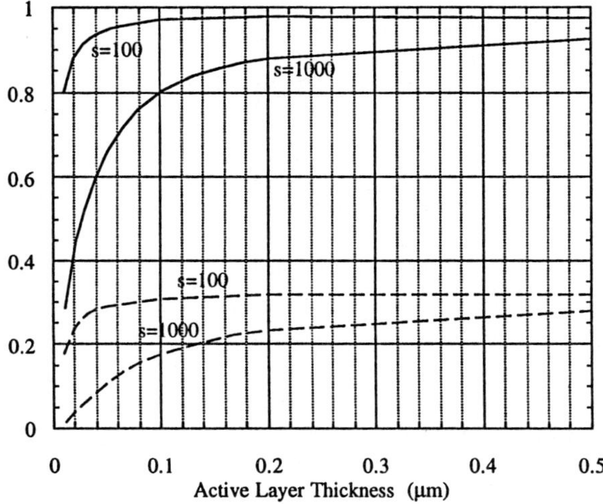

Fig. 11.5. The efficiency reduction term caused by interface recombination and bulk absorption in an otherwise ideal GaN based LED for surface recombination velocities of 100 and 1000 cm/s. The coefficient term relates the output power to the injection current

$$\eta_A = \left\{ \frac{1+V_s}{1+\alpha L}\left[1-\exp\left(\frac{1+\alpha L}{L/w}\right)e^{w/L}\right] - \frac{1-V_s}{1-\alpha L}\left[1-\exp\left(\frac{1-\alpha L}{L/w}\right)e^{-w/L}\right]\right\}$$

$$\times \left\{2\left[(V_s^2+1)\sinh\left(\frac{w}{L}\right) + 2V_s\cosh\left(\frac{w}{L}\right)\right]\right\}^{-1} \tag{11.21}$$

with $V_s = v_s L/D$.

Figure 11.5 exhibits η_A as a function of the active layer thickness for two surface recombination velocities (100 and 1000 cm/s). The other parameters used are for GaN, even though all the LEDs are made of InGaN (center wavelength: 450 nm, electron mobility: 600 cm^2/V·s). The effective carrier lifetime is as indicated (radiative lifetime $= 2 \cdot 10^{-9}$ s, absorption coefficient $\alpha = 10^5$ cm^{-1}, and refractive index: 2.6). Moreover, the coefficient in front of the injection current in (11.15) relating the power to the injection is also plotted. In the absence of available data, what would be plausible was chosen based on the assertion that the GaN surface is reasonably inert. Further consideration was given to the observation that the Schottky-barrier height seems to become higher with an increased work function of the metal. Figure 11.6 displays the same parameters as a function of the surface recombination velocity in the range of 1 to 10,000 for several thicknesses of the active layer ranging from 20 nm to 1000 nm.

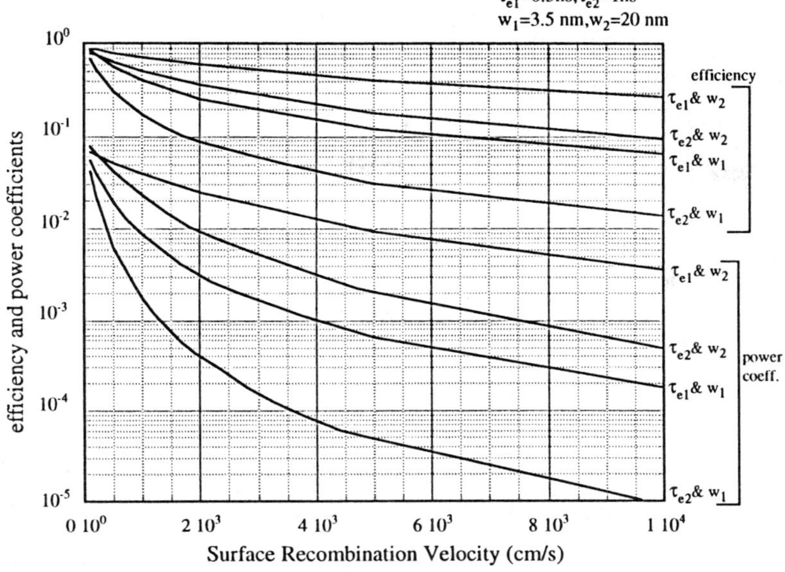

Fig. 11.6. The parameters of Fig. 11.5 as a function of the surface-recombination velocity for active layer thicknesses ranging from 20 to 1000 nm

Having done the analysis, we must recognize that the underlying assumptions made are that the carrier motion in the active layer is driven by conventional diffusion. The InGaN-active layer utilized in an LED is highly clustered and textured, and far from ideal for the diffusion-limited current to be applicable. In fact, these clusters may be responsible for the carrier localization and therefore the enhanced radiative recombination, which may explain the efficient light emission even in the presence of extremely high concentrations of defects. The expressions above are meant to provide the reader with a guide to which parameters are important and what role they play in the device operation. Appropriate carrier lifetimes, when available, can be used in conjunction with the expressions provided here to arrive at characteristics representative of devices available today. In short, to increase light extraction from a p-n junction, the active layer is placed close to the surface, the entire layer structure outside of the active layer is made of a transparent (larger-bandgap) semiconductor. A material of lower refractive index is placed on the top of the active layer. Moreover, a dome of lower refractive index (lens) is placed on top of the device, which increases the collection cone and causes the photons entering it to strike the domed surface at or near a normal angle with an escape certainty of unity (Fig. 11.7). For collecting as much of the light generated as possible, a transparent top ohmic contact coupled with, when applicable, a high-reflectivity back contact is also employed. The dome increases the efficiency by about

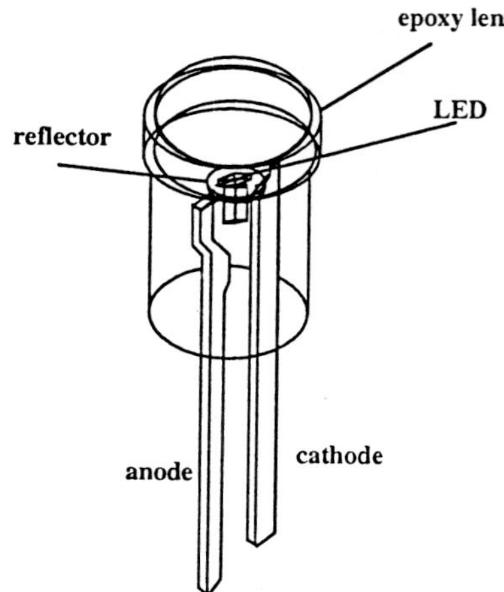

Fig. 11.7. Schematic representation of a domed LED for enhanced light collection as well as focusing

twice the square of the refractive index of the semiconductor. The dome also serves to focus the light concentrating the radiation within the field of view. Various dome approaches are available among which are hemispheres, truncated spheres, and paraboloidal types with narrowing radiation patterns. The cone defined by the total-reflection angle in larger-bandgap semiconductors, such as GaN, is larger and increases the light collection.

11.4 Visible-Light Emitting Diodes

The optical power generated by a light-emitting diode must excite the human eye. This brings into the discussion the color perception of the human eye which has been standardized by the Commission Internationale de ℓ'Eclairage (CIE) [11.5]. This commission produces charts used by the display society to define colors. Detection and measurement of radiant electromagnetic energy is called **radiometry**, which when applied to the visible portion of the spectrum involving the human eye, is termed **photometry**. The nomenclature for the latter delineates itself from the former by adding the adjective luminous to those terms used for the former. For example, energy in the former is called the **luminous energy** in the latter. The former can be converted to the latter, and vice versa, if the perception of color by the human eye is known. In daylight, the human eye is most sensitive to the wavelength of 555 nm with a maximum sensitivity of 683 lm/W. This is called **photopic vision**. In low-light and night situations, the peak sensitivity blue-shifts to 507 nm. This is classified as **scotopic vision**. The maximum sensitivity for the scotopic vision is 1754 lm/W. At the red and blue extremes, the sensitivity of the human eye drops dramatically [11.6-9]. Consequently, light sources near and at these extremes must be very efficient emitters to be practical. The terms employed to describe **LED performance** in terms of photometric terms are as follows:

Brightness: A subjective term used to describe the perception of the human eye, such as very dim, on the one hand, to blinding, on the other. The relationship between brightness and luminance is very nonlinear.

Luminance: The luminous intensity per unit area projected in a certain direction in SI units [candela/m^2].

Luminous efficiency: The power in photometric terms, measured in lumens per Watt divided by the electric power that generates it. To avoid confusion, *luminous efficacy* is used in the display field. In the LED literature one finds *luminous performance* for this term.

Luminous flux: Power of visible light in photometric terms.

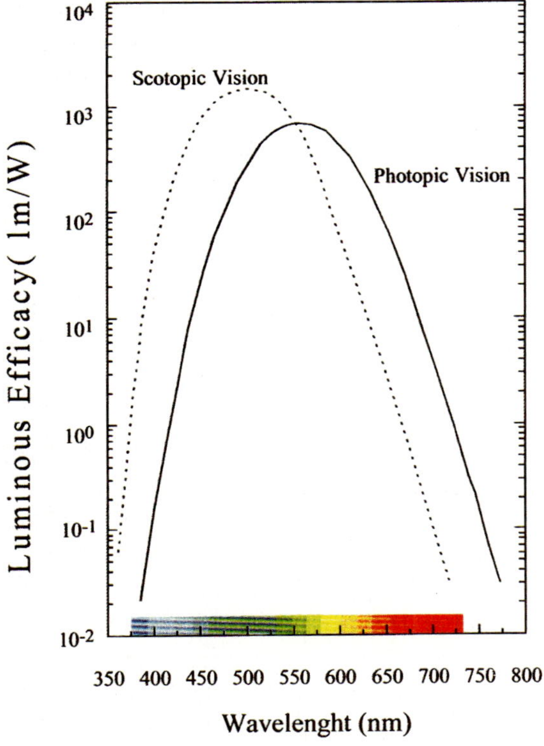

Fig. 11.8. Luminous efficacy curves for the human eye under light and dim conditions. The band indicates the color at a particular wavelength

Luminous intensity: The luminous flux per solid angle emitted from a point. The unit is lumens per steradian, or candelas (cd). This term is dependent on the package and the angle of measurement, and as such is not reliable.

Figure 11.8 presents the **luminous efficacy** W(λ), which represents the effectiveness of the radiant power of a monochromatic light source in stimulating the visual response for daylight vision (*photopic vision*) and night vision (*scotopic vision*). Under daylight situation, the human eye is most sensitive at the wavelength of 555 nm, while at night the peak sensitivity shifts to 507 nm. The specified total luminous flux Φ of a LED is determined for photopic vision, and can be calculated through the relation

$$\Phi = S(\lambda) W(\lambda) d\lambda \tag{11.22}$$

where S(λ) is the spectral power output of the LED. A more relevant criterion to describe the performance of an LED is the **luminous performance** which is the amount of electrical power converted to luminous power. This is to be contrasted to the **conversion efficiency** which is the amount of electrical power converted to radiant power. As seen in Fig. 11.8, the efficacy curve falls drastically at both ends of the visual spectrum. This increases the requirements for the output power and the external quantum efficiency for emitters in the blue and red regions to achieve the same brightness or luminous performance offered by green-light sources and for practical visual displays.

11.5 Nitride LED Performance

Light-emitting diodes have undergone a tremendous advancement in performance and are now used in nearly every aspect of life. The future of many technologies including printing, communications, displays, and sensors depend profoundly on the development of compact, reliable and inexpensive light sources. Currently, conversion efficiencies of commercial LEDs emitting in the red (650 to 660 nm) stand at 16%, and of lasers at 75%. A primary goal of GaN research is to efficiently harness its direct energy bandgap for optical emission. Though the band-edge emission in GaN occurs at about 362 nm, which is in the UV, by appropriately alloying GaN with its cousins AlN and InN, the energy bandgap of the resulting Al(In)GaN can be altered for emission in the range of ultraviolet to yellow or even red. The first GaN LEDs were reported some twenty five years ago by *Pankove* et al. [11.10]. Due to difficulties in doping GaN p-type at that time, these LEDs were MIS LEDs, rather than p-n junction LEDs. The electroluminescence of these LEDs could be varied from blue to yellow depending on the doping of the insulator layer. Unfortunately, the measured efficiencies of these preliminary MIS LEDs were not sufficient to compete with the commercially available LEDs of that time.

The first GaN p-n junction LED was demonstrated by *Amano* et al. [11.11]. The fabricated device consisted of a Mg-doped GaN layer grown on top of an undoped n-type ($n = 2 \cdot 10^{17}\,cm^{-3}$) GaN film with the chemical Mg concentration estimated to be $2 \cdot 10^{20}\,cm^{-3}$. The electroluminescence of the devices was dominated by near-band-edge emission at 375 nm, which was attributed to transitions involving injected electrons and Mg-associated centers in the p-GaN region. Additionally, a small shoulder extending to 420 nm, due to defect levels, was also observed.

One of the timely advancements in the nitride effort has been the exploitation of double heterostructures (DHs) for light-emission devices [11.12-14]. The advantage of DH LEDs over homojunction LEDs is that the entire structure outside of the active region where the light is generated is transparent. This reduces the internal absorption losses. Furthermore, this cladding region serves as an interface for scattering light, thus minimizing the probability of total internal reflections within the device. These two factors together enhance the probability of escape for the light out of the device.

In order to achieve other desired colors, InGaN alloys for emission media are required. While an increased InN mole fraction in GaN red shifts the spectrum, this would be at the expense of introducing additional structural defects unless InGaN is made sufficiently thin since it is not lattice matched to GaN. At least this is the picture in classical semiconductors whose properties are not dominated by inhomogeneities. In homogeneous semiconductors, lattice-mismatched films can be grown up to a certain thickness called the **critical thickness** for a given composition. The larger the composition, the smaller the critical thickness. In view of this, there should be substantial effort devoted to optimization. In this vein, near-band-edge emission was also obtained for LEDs employing Si-doped InGaN quantum wells as the active region in a GaN/InGaN DH LED [11.15]. The In mole-fraction content of the active layer was varied and resulted in a shift of the peak wavelengths of the device's electroluminescent spectra from 411 to 420 nm. Impressively, researchers at Nichia Chemical [11.16] were later successful in reducing the thickness of InGaN emission layers to about 30 Å. With this achievement, InGaN quantum wells with InN mole fractions up to 70 % have been obtained, and light emitting diodes with commercial capabilities are now possible in the blue, green, and amber. A schematic representation of one such Nichia LED is illustrated in Fig. 11.9. The

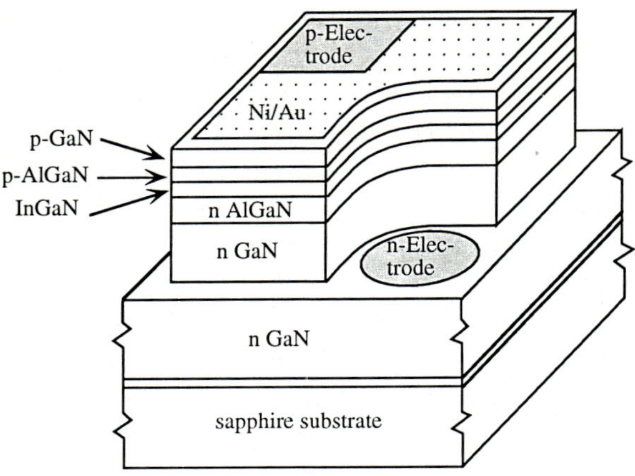

Fig. 11.9. Artist's view of a Nichia LED with transparent large-area contact to the top p-type GaN

light generated in the active InGaN layer traverses without any appreciable absorption as the entire LED chip is transparent to the radiation wavelength except for the self absorption in the thin active layer. Due to the problematic nature of p doping and the low hole mobility, the spreading resistance in the top layer is large. To combat this problem, only a small portion of the p layer, where the wire bond is made, is covered with a thick metallization with the rest being covered with a transparent metal contact, thin in this case (Fig. 11.9).

The blue and blue-green LEDs developed by Nichia Chemical initially relied on the transitions to Zn centers in InGaN (Fig. 11.10). Although it was suggested [11.17] that Zn levels are deep, no direct evidence was provided to verify this assertion while granting that deep levels are introduced by the presence of Zn in the growth environment. It is clear, however, that

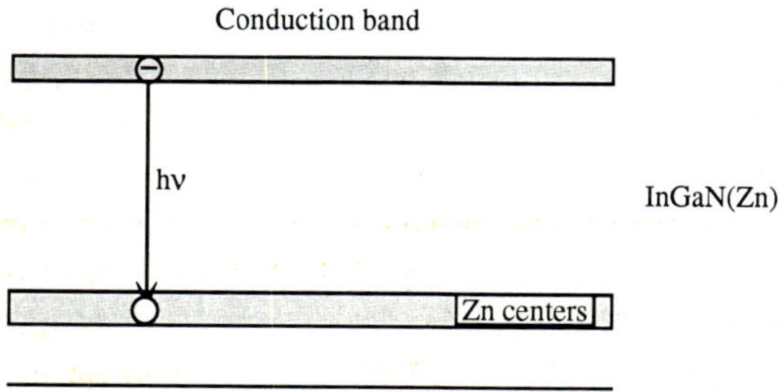

Fig. 11.10. Schematic depicting transitions from states near the conduction band to Zn centers in the earlier versions of commercial nitride LEDs. The Zn centers were also used in the original GaN LEDs fabricated in the 1970's. A schematic representation of optical transitions in Zn-doped and unintentionally doped InGaN LEDs

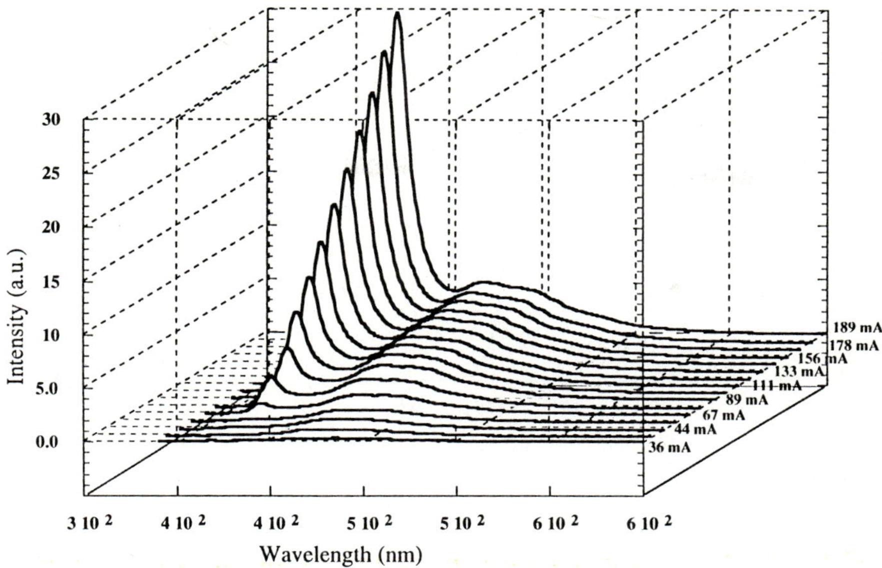

Fig. 11.11. Spectral response of a blue LED utilizing Zn centers with increasing injection level. Note the saturation of the Zn-related transition with an accompanying shift near the band edge as the injection is increased

the presence of Zn causes the film to be of high electrical resistivity. However, Zn centers situated about 500 meV above the valence-band edge of GaN are very efficient centers for recombination. The addition of Zn was originally necessitated by the need to extend the wavelength to the desired values, and the amount of In that could be added was limited while maintaining a good crystalline quality. As depicted in Fig. 11.11, these LEDs had the undesirable characteristics of wide spectral widths and a saturation of the light output with injection current accompanied by the blue shift. Figure 11.12 plots the *electroluminescence spectrum of one such blue LED*, whose radiative transitions are from near the conduction band to the Zn centers, along with Zn-free blue and green LEDs. Also shown is the electroluminescence of so-called **quantum-well devices not containing Zn centers**. The large spectral width spoiled the color saturation with the undesirable outcome that not all the colors could be obtained through color mixing. It should be noted that, in the **quantum-well approach**, the term quantum well has been used very loosely, as in many cases the InGaN layers are not thin enough for carrier confinement and the transitions rely on band-tail states near the conduction and/or valence bands. With the so-called **quantum-well approach**, the In mole fraction can be extended to about 70% which paves the way for excellent violet, blue, green, and yellow/amber InGaN LEDs. The commercial LEDs exhibit power levels of 5 and 3 mW at 20 mA injection current for the wavelengths of 450 and 525 nm, respectively.

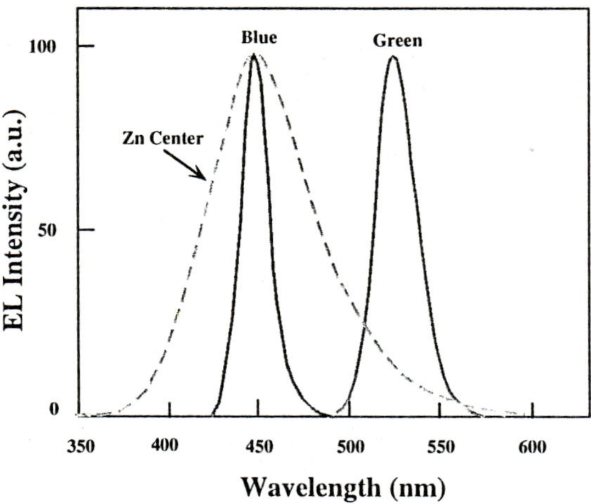

Fig. 11.12. Electroluminescence spectra of the Nichia blue and green LEDs. By way of comparison to the one with Zn centers, the contrast is drawn to the improved spectrum in devices relying on near band-to-band transitions

Elimination of the Zn centers led to the very important consequence that the FWHM of the emission spectra were reduced to 20, 30, and 80 nm for blue, green, and yellow LEDs, respectively (Fig. 11.13). For yellow, the InN molar fraction approaches 70. It should be stressed that an accurate determination of the InN molar fraction is difficult due to inhomogeneities in composition and strain. The In mole fractions used are $15 \div 20$, $40 \div 45$, and about 60 % for the 450, 525, and 590 nm emission, respectively. In Zn-

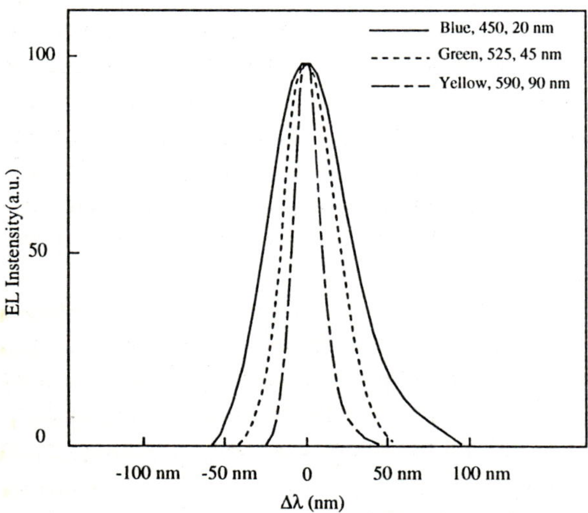

Fig. 11.13. Spectral linewidth of blue, green and yellow InGaN-based LEDs, the so-called quantum well types. Increasing line broadening due to strain and compositional inhomogeneities with increasing InN mole fraction is noticeable

356

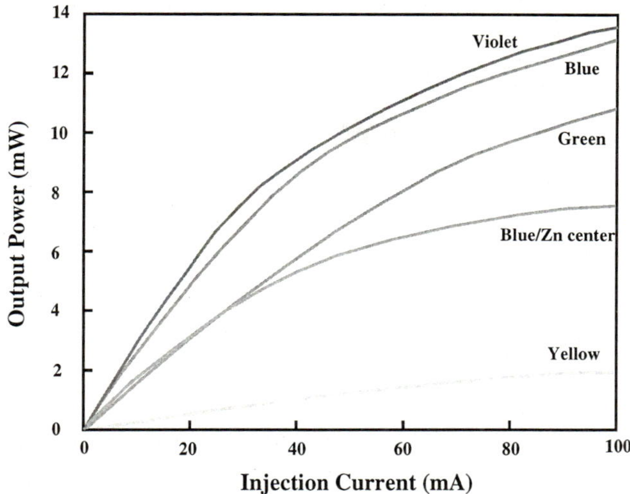

Fig. 11.14. Output power versus the injection current in nitride-based Nichia LEDs up to 100 mA of DC current

free blue LEDs, the linewidth has a temperature-dependent as well as a temperature-independent term. The latter is dominant and is attributable to inhomogeneities in the semiconductor such as compositional and strain variations. The behavior in green LEDs is similar, but to a different extent.

The output power as a function of injection current for the blue, green and yellow LEDs mentioned above is displayed in Fig. 11.14. At the wavelength corresponding to green, the 3-mW power level which corresponds to 12 cd in a 10 degree viewing cone, was obtained with an efficiency of 6.3%. Most recent results for the GaN- and ZnTeSe-material systems are above the 20 cd level. In summary, at an injection current of 20 mA, the blue, green, and yellow LEDs produce 5 mW (efficiency: 9.1%), 3 mW (efficiency: 6.3%), and 1 mW (efficiency: 2.3%), respectively. The UV LEDs operating at 400 nm exhibit efficiencies of 10% at I = 20 mA. At a current injection level of 100 mA, 400 and 450 nm LEDs exhibit power levels of 13 and 12 mW, respectively.

Light-emitting diodes and lasers have been achieved with ZnSe-based heterostructures prior to the recent resurgence of activity in nitrides. These devices, however, have had to deal with a low damage threshold and ensuing lifetime problems. By growing the films on ZnSe substrates, the stacking faults in the active part of the device structure can be reduced giving rise to increased performance and lifetime. Through a collaborative effort between J. Schetzina and Eagle Picher Co., ZnSe/ZnTeSe/ZeSe double-heterostructure LEDs emitting at 512 nm exhibited a spectral half-width of 40 nm and an output power of 1.69 mW at 10 mA. The diode voltage was 2.4 V, an

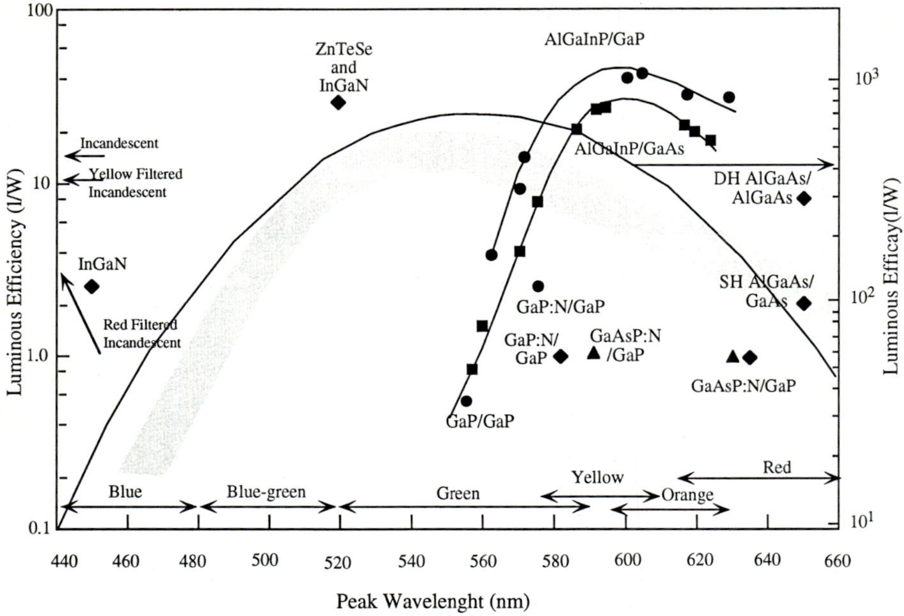

Fig. 11.15. Photometric performance, taking the eye response into account, of visible LEDs fabricated with conventional III-V semiconductors, the nitrides, and ZnSe-based materials. The band indicates the performance of laboratory models of organic LEDs. The efficay of the human eye is indicated as a line. The conventional III-V data have been reproduced by the courtesy of S. Lester wherefrom

external quantum efficiency and the power efficiency (wall-plug efficiency) of 6.9% and 6.5%, respectively, and an operating lifetime of larger than 10,000 h have been attained.

Figure 11.15 exhibits the performance of visible-light emitting diodes in photometric terms [lumens/Watt] which also indicate the wall-plug efficiency. For comparison, the performance of a green CdZnSe-based LED emanating from the joint efforts of J. Schetzina and Eagle Picher Co. is also shown. To gain an insight about the extent of color saturation, the LED output is generally indicated on the chromaticity diagram, as depicted in Fig. 11.16. The outer periphery corresponds to saturated colors. Also shown in dotted lines are the boundaries delineating the well-known colors. The oval near the center indicates various grades of "white light". The line through the white-light region indicates the color diagram for white light with the accompanying color temperature. Moreover, the output of an optically pumped YAG medium doped for yellow emission is marked with data points indicative of various Gd concentrations. The narrower the output spectrum of an LED, the closer its color is to the outer periphery. As the spectrum gets wider, the corresponding color on the chromaticity diagram is pulled toward the center reducing the range of colors that can be obtained

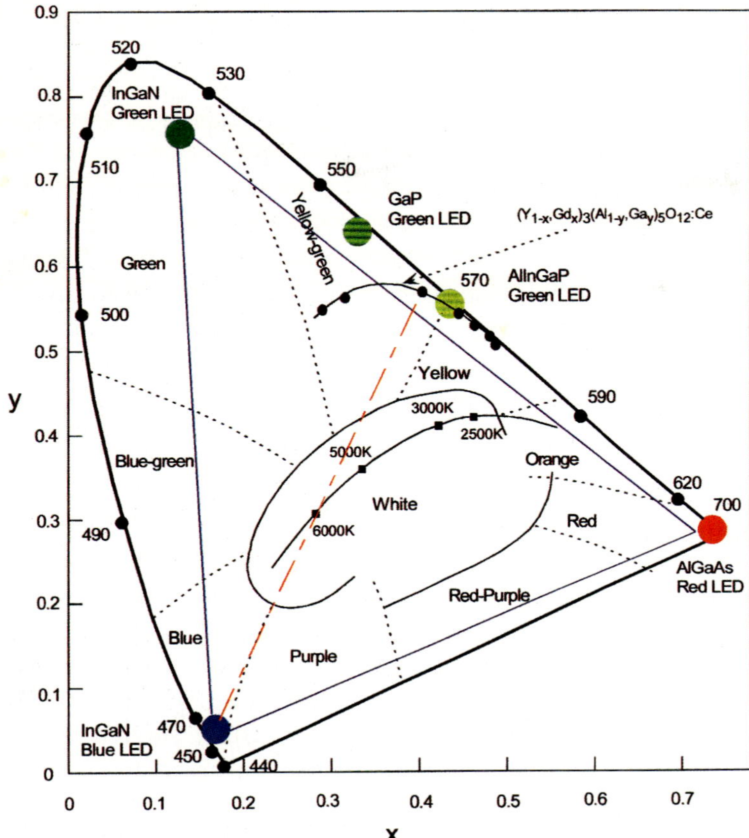

Fig. 11.16. The chromaticity diagram along with available commercial LED performance data. Clearly, blue and green InGaN LEDs constitute the two important legs of the triad, the three primary colors which are needed for full-color displays. Moreover, the output of an optically pumped YAG medium doped for yellow emission is shown with data points indicative of various Gd concentrations. The broken line which connects the blue LED to a one particular composition of the YAG medium indicates the range of warm white colors that can be obtained. Courtesy of S. Nakamura, Nishia Chemical Co., Ltd.

by the color-mixing scheme. The blue and red LEDs available commercially are almost saturated while the same can not yet be said for green ones. The spectral broadening observed for green LEDs is attributed to compositional inhomogeneities which get larger with increasing InN mole fraction. When used in conjunction with the available red and blue LEDs, present InGaN green LEDs provide the means for achieving some $70 \div 80\%$ of all the color possible. As will be discussed in Sect. 11.8, a YAG medium pumped with a blue LED can produce white light. The broken line in Fig. 11.16 which connects the blue LED to a one particular composition of the YAG medium indicates the range of colors that can be obtained by simply adjusting the concentration of the phosphor centers in the YAG medium.

11.6 On the Nature of Light Emission in Nitride-Based LEDs

Although nitride LEDs were introduced as a commercial product in the late 1993, there are still many unanswered questions regarding the optical emission processes responsible for their outstanding operation. One of the questions sending everyone searching for answers is the unusually high efficiency of light emission in the presence of large concentrations of defects in the material as well as a current-conduction mechanism which is inconsistent with band-to-band recombination. It is known that the radiative recombination in wide-bandgap semiconductors such ZnS and CdS deposited on glass is very efficient. The mechanism responsible for this has been attributed to carrier localization and, of course, a parallel here is very enticing to make.

11.6.1 Pressure Dependence of Spectra

In order to take a glimpse at the origin of the strong electroluminescence in nitride-based LEDs, and more importantly to understand the origin of light emission, *Perlin* et al. [11.18] examined the photoluminescence and electroluminescence emission in commercially available blue and green LEDs under hydrostatic pressure after they had been decapsulated. To complement the pressure experiments, *Perlin* et al. [11.19] also embarked on an extensive investigation of Electro-Luminescence (EL) from blue and green LEDs over a broad current and temperature range in an effort to gain some insight into the genesis of radiative transitions. The blue and green LEDs exhibited similar behavior with the green ones accenting the anomalies mentioned. Consequently, the present discussion will be limited to the case of green devices.

The green Nichia LED investigated has an undoped active layer consisting of a 30 Å thick $In_{0.45}Ga_{0.55}N$ layer sandwiched between n-GaN on the bottom and p-$Al_{0.2}Ga_{0.8}N$ on the top layers. The lattice mismatch between the active layer and the barrier materials is about 6% bringing into question whether the structure has a pseudomorphic character or the strain is relaxed by a large concentration of dislocations which is on the order of 10^{10} cm^{-2} [11.20]. Another perplexing aspect of these diodes is the $2.28 \div 2.33$ eV photon energy which is somewhat smaller than the $2.50 \div 2.68$ eV expected from the bandgap of $In_{0.45}Ga_{0.55}N$, although the clustered nature of InGaN makes the reported compositional dependence of its bandgap a suspect [11.21]. *Nakamura* et al. [11.22] suggested that this discrepancy could be caused by tensile strain in the quantum well, which had been induced by differences in the thermal expansion coefficients of the quantum well and barrier materials. Other suggestions involve unidentified and elu-

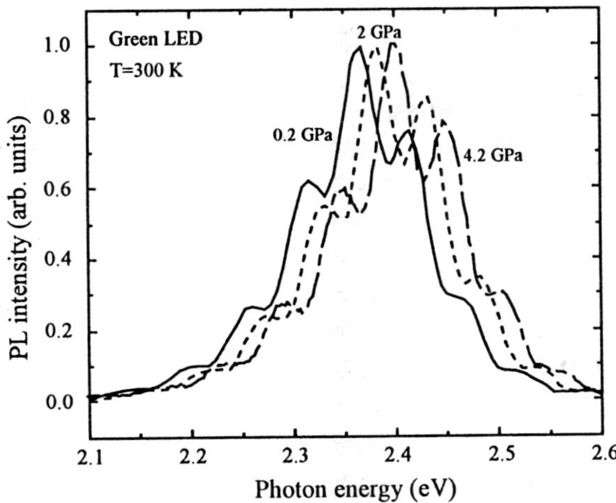

Fig. 11.17. Photoluminescence spectra of a Nichia green LED for the three indicated uniaxial pressures at 300 K obtained after decapsulation. After [11.14]

sive localized states [11.23] or localized excitons [11.24]. In any case, no definitive answer has been given yet as to the nature of the recombination in the so-called **Nichia Single Quantum Well LEDs**.

Figure 11.17 exhibits the photoluminescence spectra of a green Nichia diode as a function of hydrostatic pressure. The observed Fabry-Perot interference fringes are indicative of the fact that the structure is mostly transparent to the particular radiation. The distance between fringe maxima is close to 50 meV ($500\,\mathrm{cm}^{-1}$) and gives a total cavity thickness, sapphire on the one side and air on the other, of approximately 5 mm which agrees well with the total thickness of these GaN/InGaN/AlGaN structures (4.6mm). The presence of interference fringes can cause misidentification of the peak position which is made worse as the fringe periodicity can change with pressure due to the pressure dependence of the refractive index (about 6 meV/GPa). For accuracy, the Gaussian peak profile and an oscillating function representing the interference fringes were fitted to the data [11.18] which paved the way to determine the peak shift in luminescence and electroluminescence with hydrostatic pressure, the latter of which is displayed in Fig. 11.18 for a green LED. The observed linear shifts with pressures of 16 and 12 meV/GPa, compare with 40 (experimental) and 33 (theoretical) for GaN and InN, respectively, and indicate that the transition energy is not of traditional band-edge origin. The contribution of pressure to any confinement due to an increasing effective mass has been determined to be about 2 meV/GPa [11.18]; it is very small. In short, one can conclude that the LED spectra do not follow the band edge.

A pressure coefficient lagging behind the band edge can be expected from deep states. For example, transitions between uncorrelated electrons

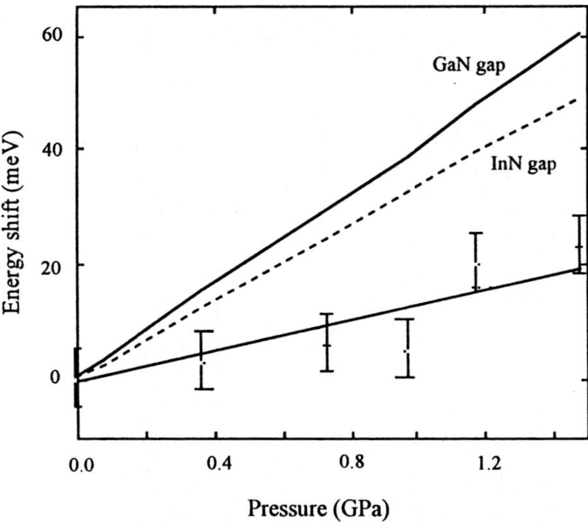

Fig. 11.18. Shift of the ElectroLuminescence (EL) peak position as a function of the hydrostatic pressure for a Nichia green LED. The data along with error bars and the expected shift of the band-edge for GaN (*experimental*) and InN (*theory*). The solid line through the data points is a guide for the eye. After [11.14]

and holes, meaning their wave functions do not overlap, trapped in band-tail states caused by indium clusters/compositional fluctuations lead to pressure coefficients smaller than the band-edge value. Band-tail states can extend deep into the forbidden gap, and deep states have a pressure coefficient that is an average across the entire Brillouin zone. This average coefficient is much lower than that at the Γ-point direct bandgap. Localized excitons have also been postulated as being responsible for the transitions in InGaN LEDs [11.24]. However, as will be demonstrated in Sect. 11.6.2, the observed behavior retains the same trend at high-current-injection levels where excitons would certainly dissociate and would therefore not be responsible. Localization effects such as the postulated quantum dots are so far limited to casual observations of compositional variations [11.25]. Shown in Fig. 11.19 is the compositional variation of InN in an InGaN quantum well manufactured by Nichia Chemical Co. similar to that used in LEDs. Clearly, the molar fraction varies in the growth direction, and in the plane of growth leading to clusters. Moreover, dots of the kind postulated would have pressure coefficients similar to the band edge. In short, the available pressure dependence indicates that the transitions are due to uncorrelated electron-hole pairs localized deep in band tails, which are most likely caused by inhomogeneous InN mole fraction and strain. Ironically, we would not be too illogical if we were to argue that it is precisely the presence of these band tails that are responsible for the extraordinary performance.

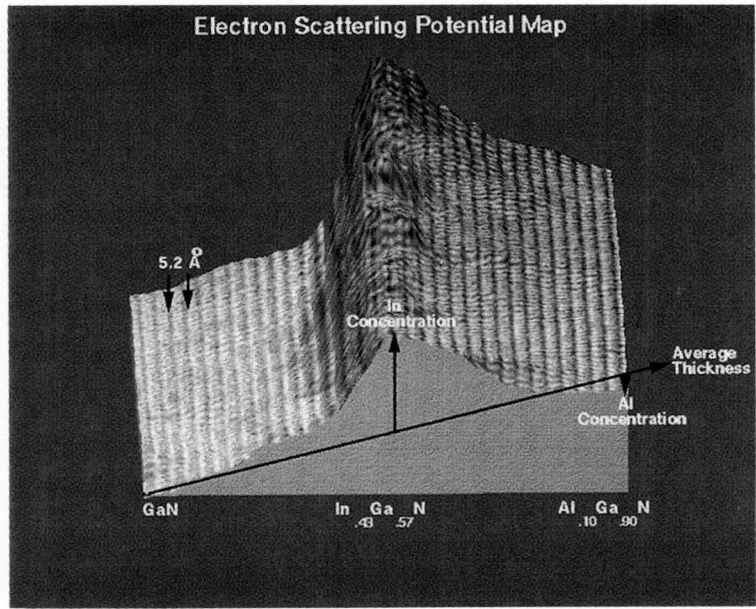

Fig. 11.19. Out-of-plane and in-plane compositional variation of InGaN in a GaN/InGaN well structure similar to that used in the Nichia LEDs. Courtesy of C. Ksielowski, Lawrence Berkeley Nat'l Laboratory

11.6.2 Current and Temperature Dependence of Spectra

Figure 11.20 exhibits the peak position of the EL emission at several temperatures from the green LED discussed above, with a clear shift to higher energies as the injection current exceeds 0.1 mA. For a current level of 0.1 mA (preceding the band-tail filling), the EL emission undergoes a blue shift with increasing temperature. Between 15 and 300 K, this shift is as large as 70 meV for the lowest applied current. As the temperature is increased, the low-current plateau also shifts towards higher energies. Furthermore, the emission bandwidth ($\approx 130\,$meV at 1mA, not shown) remains practically unchanged over the entire temperature range. This blue shift is about two orders of magnitude larger than what we would expect from the filling of the conduction-band states. Consequently, the transition responsible must be due to states with very low density of states compared to the conduction band. This observation is also consistent with the premise that deeper states are the origin of the observed transitions. Deeper states here include band tails as well as other pseudo-continuous states. Interestingly enough, the energy of the emitted photons at the largest applied currents is quite close to the estimated separation between the confined states in a 3-nm thick $In_{0.45}Ga_{0.55}N$ quantum well. Figure 11.21 depicts the same shift at 300 K

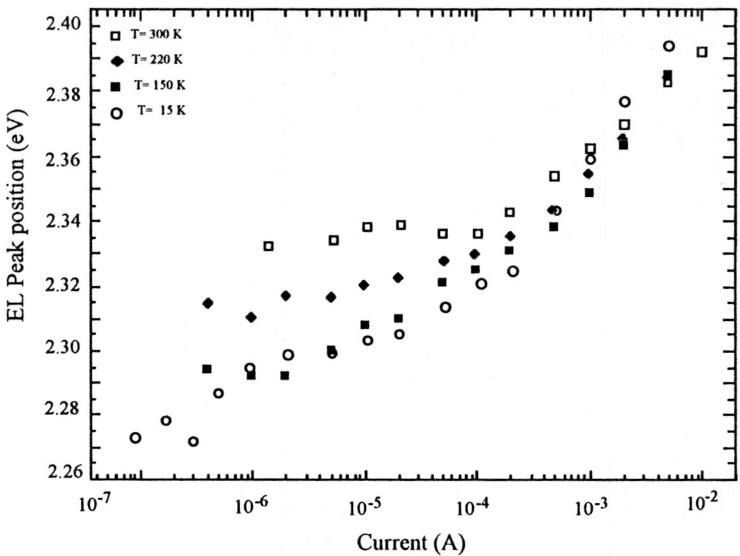

Fig. 11.20. The peak position of the EL emission at several temperatures from a green Nichia LED indicating a clear shift to higher energies as the injection current exceeds 0.1 mA. After [11.15]

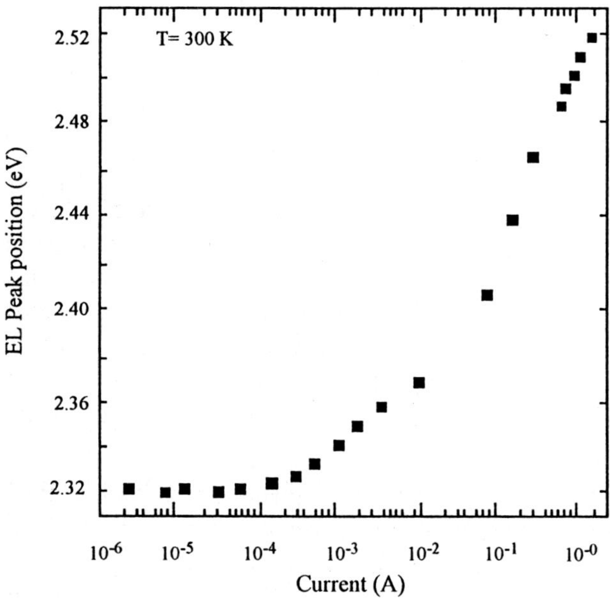

Fig. 11.21. The peak position of the EL emission at several temperatures from the same green LED as in Fig. 11.20, showing the blue shift at 300 K up to a pulsed current level of 2 A accenting very well the blue shift with injection current. After [11.15]

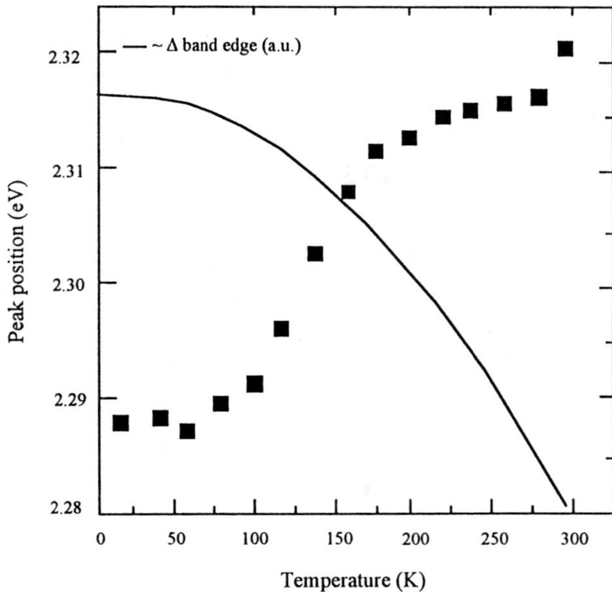

Fig. 11.22. The PL peak position of the same green LED as in Fig. 11.20 measured at different temperatures. The PL peak shifts to higher energies with increasing temperature. This is to be contrasted to the band-edge dependence on temperature, which is depicted by the solid line. This line is intended to show the trend only with attention to be paid to the absolute values. After [11.15]

up to a pulsed current level of 2 A and very well accentuates the blue shift very well with injection current.

Figure 11.22 plots the temperature dependence of the peak energy of the EL emission. The solid line depicts the expected trend of the band edge of GaN shifted rigidly to represent InGaN. The available data, in addition to those displayed in Fig. 11.22, indicate that for a current level of 0.1 mA (preceding the band-tail filling), the EL emission undergoes a blue shift with increasing temperature. Between 15 and 300 K, this shift can be as large as about 70 meV, for the lowest applied current. Furthermore, the emission bandwidth ($\approx 130\,\text{meV}$ at 1mA, not shown) remains practically unchanged over the entire temperature range.

The data presented above suggest that the radiative recombination does not directly involve the conduction and valence bands and, in the case of quantization, quantum-well sub-bands. Though radiative recombination has been suggested to relate to excitons localized in regions containing large InN molar fraction that are caused by fluctuations of the indium contents in the active layer material, the data of *Perlin* support the band-tailing effect. On the contrary, the same data on high-carrier-injection levels do not support the exciton premise. Particularly, the rapid blue shift of the electroluminescence with injection is indicative of a continuous density of states favoring band-tailing effects that are caused by strain and compositional fluctuations.

11.6.3 I-V Characteristics of Nitride LEDs

Data regarding the behavior of optical transitions indicate the origin of emission to be not due to band-to-band transitions in the classical sense but rather to localized deep states. In an electrically pumped system such as LED, the natural question that arises is how the carriers get to the region of recombination. Although sufficient data do not yet exist to answer this question unequivocally, what is known is that the current-voltage characteristics of these devices do not follow the classical diffusion-limited current. Figure 11.23 depicts the room-temperature I-V characteristics of blue and green Nichia LEDs. Also shown are the calculated diffusion and generation-recombination currents. The former has been calculated assuming a minority-carrier lifetime of 1 ns and the latter supposing mid-gap states with a concentration in such a way that the effective lifetime is 100 ps.

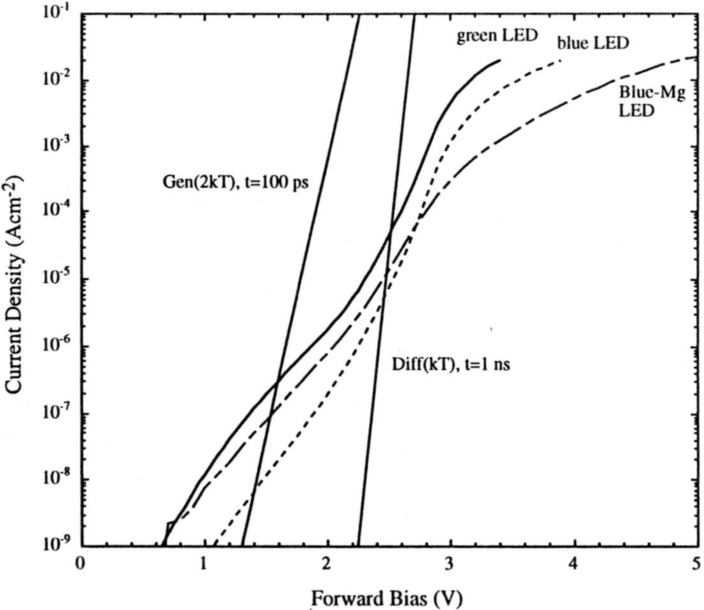

Fig. 11.23. The room-temperature I-V characteristics of blue and green Nichia LEDs along with a Cree GaN:Mg-based LED. Also shown are the calculated diffusion and generation-recombination currents. The former was calculated assuming a minority-carrier lifetime of 1 ns and the latter assuming mid-gap states with a concentration in such a way that the effective lifetime is 100 ps

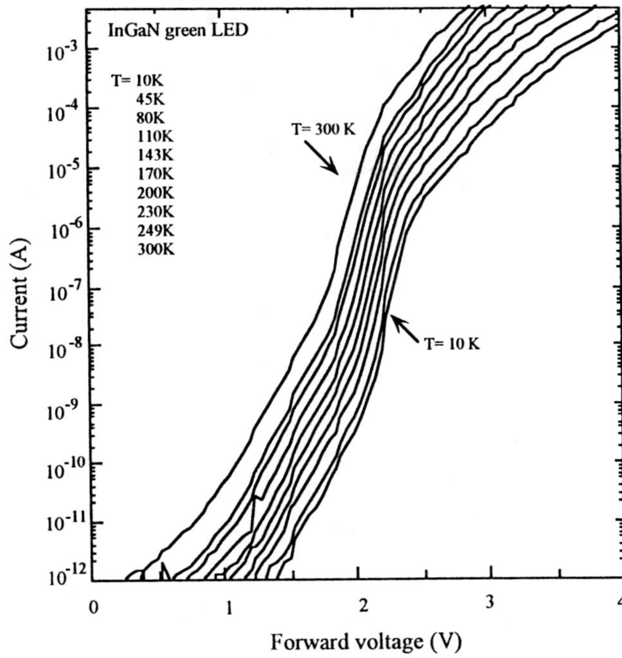

Fig. 11.24. The forward part of the I-V characteristics of the Nichia green InGaN LED between 10 and 300 K indicating hardly any change of the slope. After [11.15]

Figure 11.24 demonstrates that the slope of the forward part of the I-V characteristics of a green InGaN LED is practically temperature independent. If Shockley's thermal diffusion, the kT current, were dominant, the slope would have been inversely proportional to the temperature. This observation, coupled with a lack of 2kT current space charge or generation-recombination current and the voltage dependence of the current resembling tunneling current, is indicative of a tunneling dominated process. Similar behavior is prevalent in blue diodes as well. Inter-band tunneling can be ruled out because of the low doping levels and the large bandgaps involved. Blue LEDs manufactured by Cree Research, Inc., which rely on Mg-induced deep centers in GaN, also indicate the absence of a slope change with temperature (Fig. 11.25). Because of the extremely small intrinsic carrier concentrations in wide-bandgap materials such as Ga-rich InGaN, GaN and AlGaN, the diffusion current would not become noticeable unless large forward biases are applied (refer to Fig. 9.6). The current well below large forward voltages approaching $3 \div 4$ V, neglecting resistive drops, must therefore involve tunneling into and through the gap states [11.26-28].

Three different regions in the forward I-V characteristics of the green LED can be discerned and each of which can be represented by the follow-

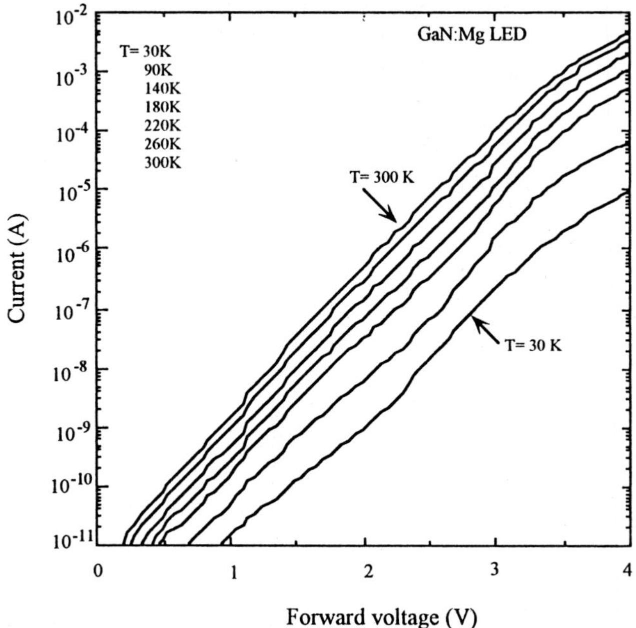

Fig. 11.25. The forward part of the I-V characteristics of a Cree blue GaN:Mg LED between 30 and 300 K indicating hardly any change of the slope. After [11.15]

ing expression, but with a slightly different E_1 energy parameter for the various voltage ranges [11.26]

$$I = I_1 e^{qV_d/E_1} \tag{11.23}$$

where V_d and E_1 are the diode voltage and the energy parameter, respectively, with I_1 being a prefactor. The behavior described by (11.23) is generally ascribed to the tunneling current. For voltages below 2 V, the tunneling current with an energy parameter of about 180 meV supports non-radiative processes as no light emission takes place. For the region of $2V < V_d < 2.3V$, tunneling of carriers result in radiative recombination. By subtracting the current at low voltages from that at medium voltages, one gets the applicable energy parameter which is between 50 meV and 90 meV. This component is tentatively suggested as being associated with the hole tunneling into the InGaN-active layer [11.26]. For voltages larger than 2.3 V, the series resistance of the device dominates and the slope gradually decreases with temperature. Measurements encompassing the dependence of the light-intensity spectrum of the diode current (Fig. 11.26) reveal a considerable blue shift of the sharp short-wavelength edge of the emission band. At room temperature, in addition to the emission-peak shift, a separate short-wavelength peak at about 3.2 eV appears at voltages exceeding ap-

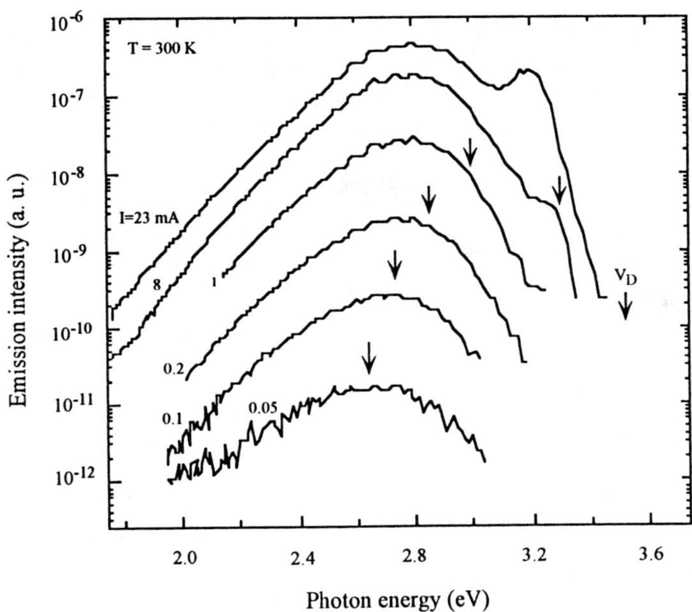

Fig. 11.26. Electroluminescence spectra of an LED at 300 K. Arrows indicate the corresponding diode voltage. After [11.22]

proximately 3.3 V. The latter is obviously associated with interband transitions in the active InGaN region. The large electroluminescence spectral linewidth is most likely induced by the strong lattice coupling of the Zn state both in GaN and in InGaN.

To reiterate, the observation that the slope of the measured current does not follow kT and 2kT slopes (Figs. 11.24 and 25) is indicative of current conduction due to tunneling assisted by states below the bandgap, which is consistent with the band-tail states postulated by *Perlin* et al. [11.18]. It is very plausible that tunneling at voltages considerably smaller than the built-in voltage could occur from the p-type layer into the localized states within the energy gap of the n-type layer, which is InGaN in the case of LEDs and lasers. At voltages approaching the built-in voltage, tunneling of holes could also occur from the valence band of the p-type layer to the narrower-bandgap n layer. This sort of minority-carrier injection is consistent with the large quantum efficiencies observed in LEDs. Moreover, shifting spectra with the injection current suggest photon-assisted tunneling and band filling. The former is due to the electrons and holes tunneling into the depletion region where they radiatively recombine and emit photons at energies near the applied voltage. In the band-filling case, as was discussed in Sect. 11.5.2, the blue shift occurs owing to filling of the empty states by injected minority carriers. The effect of gap states, compositional fluctuations, and imhomogeneities in strain would all be convoluted and prevent the true picture from emerging until such time as high-quality material can be prepared hopefully together with a description of the tunneling current behavior in wide-bandgap nitrides. *Having argued in favor of tunneling current, we should note that the dependence of the light intensity on the diode voltage at room temperature, in the so-called **single quantum-well LEDs**, exhibits the characteristic diffusion-current dependence described by exp(qV/kT), albeit much smaller than that of the tunneling current.*

11.7 LED Degradation

Prior to the debut of GaN, experience with all LED materials had been that acceptable quantum efficiencies, say over 1%, can be obtained only when the defect concentration in the semiconductor is well below about 10^4 cm^{-2}. GaN with defect concentrations about 6 orders of magnitude higher than this was not consistent with the trend of the time and GaN was discounted by many LED manufactures. Even with this large defect concentration, GaN LEDs with a longevity well over 10,000 hours, a minimum required by the display society (CIE), were marketed in early 1994, which took many by surprise. In order to reduce the amount of InN employed, early devices utilized deep Zn centers in the 20 nm thick active layer to shift the wavelength to about 450 nm, which is defined as blue by CIE. These devices were shown to have more gradual and graceful aging than the AlGaAs diodes, particularly in steam tests. Nevertheless, degradation caused by p-metallization was notable. In the second wave of devices, the Zn centers were eliminated and the true blue color was obtained by an increased InN mole fraction in the lattice with an accompanying decrease in the active layer thickness to about $3 \div 4$ nm. These devices exhibit a longevity well over that required by CIE.

Life testing of Nichia LEDs have been undertaken by *Osinski* et al. [11.29], *Barton* et al. [11.30], and *Barton* and *Osinski* [11.31]. The lot contained three types of devices, as the technology evolved with two containing Zn deep centers and wider InGaN layers, while the third set of devices did not utilize Zn but took advantage of the increased InN mole fraction to obtain the 450 nm blue radiation. Some 20 devices were mounted inside a large environmental chamber whose temperature was maintained constant. The light output of each LED was sampled by an optical fiber which was connected to its own photovoltaic detector located outside the chamber. The LED-to-fiber connection was both mechanically stable and light-tight which eliminated intensity variations due to mechanical misalignments and ambient light. The system used a switching device to select a single detector's current which was fed to a meter for automated reading of each LED's output. Measures were taken to ensure that the current through each device remained constant. The test was fully computer controlled with data automatically gathered every 12 hours or at the operator's request.

For the test, eighteen Nichia blue LEDs (numbered $1 \div 18$ for the test) were selected from a new and improved batch purported to have higher Zn concentration and two devices (numbered 19 and 20) from an older batch. Two "improved" devices (labeled A and B) were left untested to serve as controls. The LEDs were subjected to CW operation after pre-test power measurements were taken on all 20 devices. Figure 11.27 exhibits the relative luminous intensity from all 20 devices tested and normalized to their initial readings. The general trend of the 18 LEDs was for the output intensity to increase at a faster rate within the first $50 \div 100$ h and then at a slower rate over the remainder of the test. The output intensity of the two older LEDs containing less Zn increased within the first 50 h and then decreased during the remainder of the test.

After the first 1000 hours, the drive currents of the LEDs were increased to try to accelerate the degradation process in some of the devices under test. The previously tested eighteen devices from the new batch were

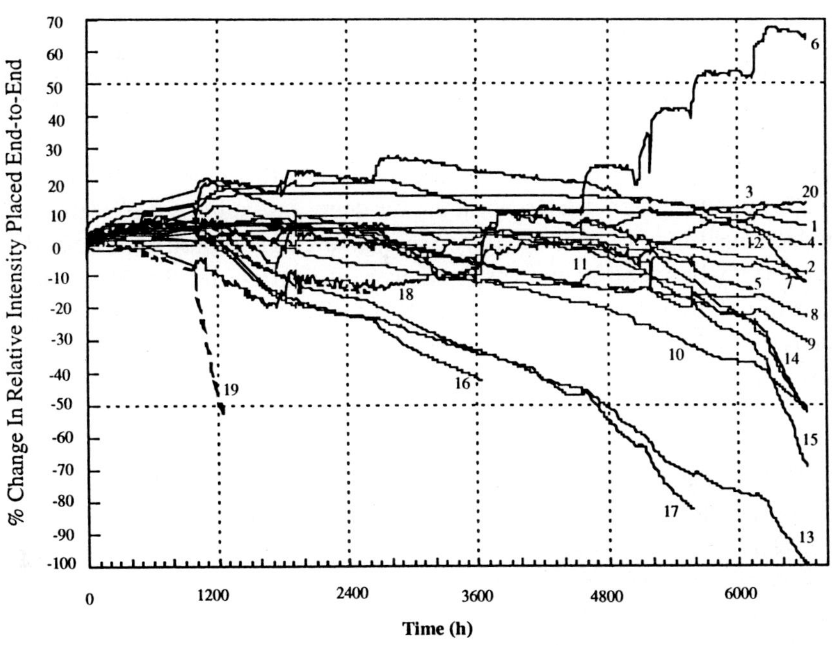

Fig. 11.27. The evolution of the relative luminous intensity from 20 Nichia LEDs tested and normalized to their initial readings. Courtesy of M. Osinski, New Mexico State University

divided into six groups of three. Each group was driven at one of six currents, namely 20, 30, 40, 50, 60, or 70 mA. Of the two older devices, one (#19) was subjected to a high current of 70 mA and the other (#20) remained driven at 20 mA. The maximum current level of 70 mA is close to the condition producing a maximum CW output power from the LEDs with the observation of the onset of a thermal rollover at 80 mA with slightly decreased output.

The relative intensity of one of the older generation LEDs (#19) dropped to about half of its initial value after approximately 1200 h. In this case, the high current (70 mA) had indeed caused a rapid failure. The cause of this premature degradation was a crack in the LED which isolated part of the junction area from the p-contact. The remaining devices driven at the same current level, however, have performed much better. After a relatively fast drop in their output ($10 \div 15\%$ over the first 750 h), their degradation rate slowed (Fig. 11.27).

To speed up the life test, electrical stress under high pulsed-current conditions was applied, which resulted in a degradation of the I-V characteristics with some devices exhibiting a low-resistance ohmic short (40 to 800 Ω). Electron Beam Induced Current (EBIC) imaging pointed to a conductive path extending from the surface (the p-contact metal) to the n-type side of the junction. The high forward current applied to this device has caused metal from the p-contact to migrate across the junction. It was postulated that the columnar growth pattern and the associated defects provide pipes through which metal migrates and causes the device to short, as is evident in the secondary-electron Electron-Beam Induced Current (EBIC) image (Fig. 11.28). One can therefore argue that when and if structural defects are eliminated, this particular degradation path would be eliminated.

Additionally, green LEDs without the deep Zn centers were stressed with pulsed currents of approximately 5 A with a 1 kHz repetition rate and a 10^{-4} % duty cycle yielding an average power dissipation of 25 mW to eliminate heating. Three devices were stressed to failure with a sudden and complete loss of light output. The I-V characteristics were all linear with resistive shorts in the 18 to 140 Ω range. Figure 11.29 exhibits an optical micro-

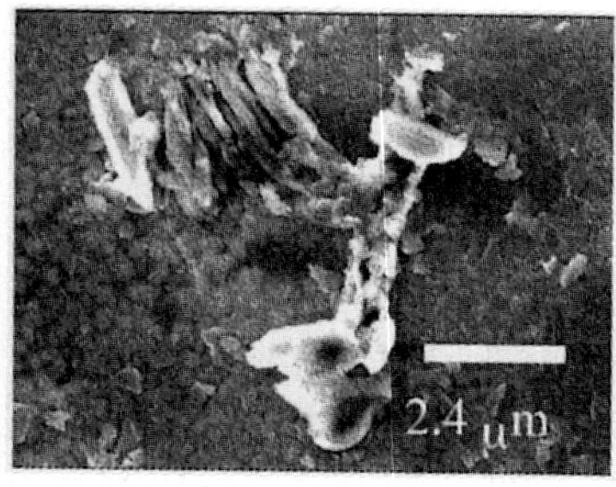

Fig. 11.28. Secondary electron EBIC image of a shorted Nichia LED after the p-contact metal has been removed. Courtesy of M. Osinski, New Mexico State University

Fig. 11.29. An optical micrograph of a failed LED after decapsulation with severe damage to the plastic encapsulation which could not be completely removed due to the damage. Courtesy of M. Osinski, New Mexico State University

graph of the failed LED after decapsulation with severe damage to the plastic encapsulation, which could not be completely removed due to damage. This indicates that most of the damage may be contained in the plastic and not in the LED.

11.8 Luminescence Conversion and White-Light Generation With Nitride LEDs

Availability of violet and blue compact LED emitters has paved the way for alternative approaches to generate blue, green and red primary colors. A blue or a violet LED can be used to pump a medium containing the desired color centers, dyes in organic and phosphors in inorganic materials, to generate the color(s) desired including white. This approach provides an attractive means in that conversion efficiencies of about 90 % are possible in inorganic YAG-based converters without the bounds imposed on the active layer composition. *Baur* et al. [11.32] called this conversion process **LUminescence COnversion** (LUCO) and the resulting LEDs as LUCOLEDs. A schematic cross-sectional diagram of one of such LED is depicted in Fig. 11.30 in which a 450 or a 430 nm InGaN LED is used to pump either an organic or an inorganic medium that contains the color center(s) desired. Figure 11.31 shows the emission from blue, green, red, and white LEDs. The green, red and white colors obtained by pumping, with a blue, a medium containing perylene-based dye molecules as the converter. White-light

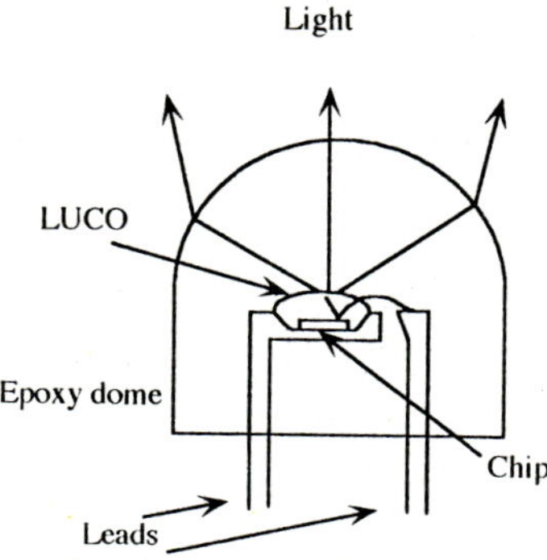

Fig. 11.30. Schematic cross-sectional diagram of a InGaN-based LUminescence COnversion Light Emitting Diode (LUCOLED)

emitting diodes have also been realized when green and red emitting dyes are simultaneously added to the pumped medium.

Referring to Fig. 11.30, the pump LED is mounted in a reflector cup and embedded in an epoxy resin dome by standard LED packing procedures. In Cree LEDs, the 6H-SiC substrate is semiconducting and requires only one bond-wire as indicated in the figure. The LEDs on sapphire, which is insulating, require two leads. In a highly popular approach, the luminescence conversion material, either organic or inorganic, small quantities (approximately $1\,\mu g$) are added to the epoxy resin immediately on top of the semiconductor chip. Alternatively, the entire plastic dome can be filled with highly diluted dyes or phosphors, which results in diffused emission. Similarly, color conversion can also be realized using inorganic phosphors as converters. Color tuning of the YAG:Ce [$Y_3Al_5O_{12}:Ce^{3+}(f')$] phosphor may also be achieved by co-doping with other fluorescent rare-earth or transition-metal ions, such as $Eu^{3+}(4f^6)$ and $Cr^{3+}(3d_3)$. In this case, energy

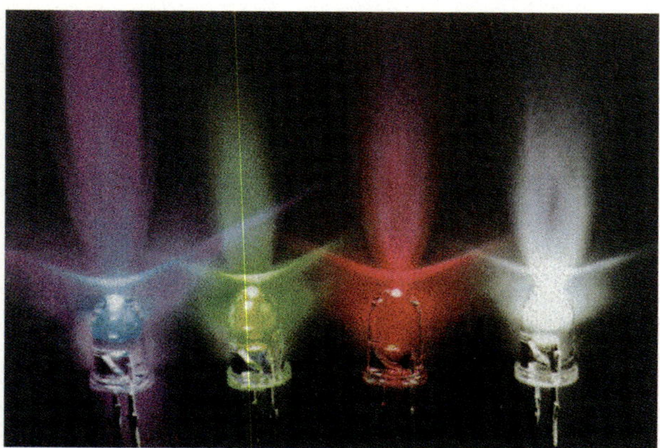

Fig. 11.31. Blue InGaN/GaN-based LED fabricated by Cree, and three LUCOLEDs emitting green, red and white light. Courtesy of J. Schneider, Fraunhofer Institute for Applied Solid-State Physics, Freiburg, Germany

374

transfer from the excited Ce^{3+} levels to the luminescent levels of Eu^{3+} or Cr^{3+} may take place, either by exciton diffusion or by photon reabsorption. The total light output of such a phosphor will then contain an additional red component, and shifts the emission from LUCOLEDs into the "warm-white" spectral range [11.32].

Generation of white light with semiconductor p-n junctions and other compact approaches have always been very attractive. The color-mixing scheme, in which blue, green and red of equal intensities, as perceived by the eye are mixed to produce the white light used in large-area outdoor displays, is not convenient nor appropriate for close-range viewing [11.32]. This requires three LEDs whose current levels are adjusted to produce the white output. A more viable approach is a medium containing phosphors which can be excited with a semiconductor LED. Since the phosphors require a high photon energy for more efficient excitation, a blue or perhaps violet LED can be used to excite a medium containing the proper mixture of phosphors for white-light generation. We should note that red LEDs can also be utilized to excite blue phosphors through two-photon absorption processes but their low efficiency have kept them from the marketplace. Similar to the organic converter case, luminescence conversion employing inorganic phosphors as converters can also be realized. For white-light generation, the phosphor Ce-doped Yttrium Aluminum Garnet (YAG:Ce) $[Y_3Al_5O_{12}:Ce^{3+}(4f')]$ is ideal as it emits yellow light under blue photo-excitation. Consequently, only one converter species is needed for white-light

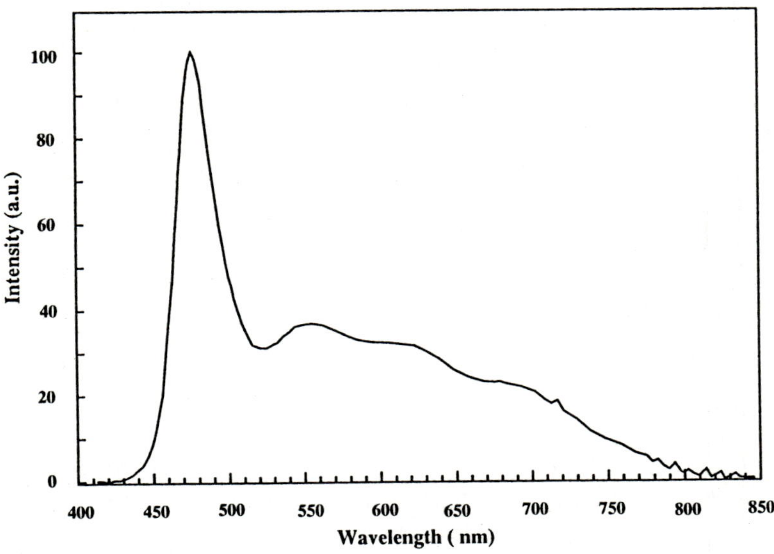

Fig. 11.32. Spectrum of a white Nichia LED lamp which uses a blue InGaN device to pump a YAG containing phosphors. Courtesy of Aixtron, Inc.

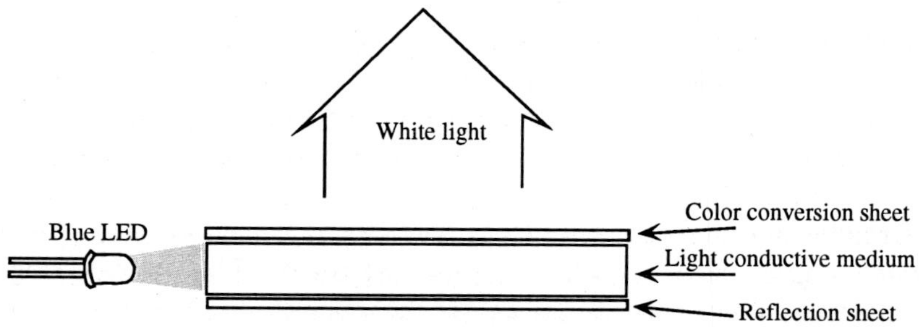

Fig. 11.33. A schematic of the flat-panel scheme of Nichia Chemical utilizing a blue LED for the pump

generation since blue and yellow are complementary colors and add to white light after proper additive mixing.

Nichia Chemical has commercialized a highly efficient white LED lamp which combines a blue GaN LED with a Ce-doped YAG based fluorescent layer to emit white light. Each lamp has a brightness of 5 lumen/W. Two of these devices together are brighter than a typical "dome light" inside a car. Application areas include car dome lights, back-up lights, key-chain lights, etc. As it stands now, the market for white LEDs is larger than that for the other colors. A spectrum of one such a white lamp is depicted in Fig. 11.32.

Many recent applications of LEDs require surface-mounted packages and chip forms. White, blue, and green LEDs are now available in the surface-mount scheme. The surface mount should be popular with manufacturers of color scanners, sensors and indicators which can take advantage of the small footprint and easy assembly offered by the Surface-Mount Technology (SMT). The surface-mount devices exhibit an output power of about 2.0 mW and a luminous intensity of 45 mcd for blue, and 1.2 mW and 150 mcd for green, which is between $50 \div 66\%$ of the perfomance of the domed models. In addition, white flat panels of the size either $40\,\text{mm} \times 136\,\text{mm}$ or $102\,\text{mm} \times 126\,\text{mm}$ are also available for background lighting, etc. A schematic of the flat panel scheme is shown in Fig. 11.33.

11.9 Organic LEDs

Application areas of nitride-based emitters and organic emitters include indoor uses. As such, a succinct description of this emerging technology will be given. Until recently, Light Emitting Polymers (LEPs) were little more than a scientific curiosity. In the wake of rapid scientific progress, particularly in the operation lifetime, a bright future is now seen for organic emitters for indoor displays, background panels, and night lights built around relatively large organic molecules. The large area, the physical flexibility, and the low cost are the attractive features offered by the organic technology. There are also efforts to fabricate transistors based on polymers with the hope of constructing displays having built-in control circuitry in much the same way as liquid-crystal displays. Until recently, the damper was the short longevity and, to some extent, the brightness. Tremendous advances have recently been made on both of these fronts.

In fact, the brightness of some of these emitters is good enough for outdoor applications but the lifetime degrades when they are operated in high-brightness modes. Organic emitters of recent vintage are in some ways similar to the semiconductor varieties taking advantage of multi-layers serving as a hole injector Indium Tin Oxide (ITO), electron injector (contacts such as Mg and Ag:10:1), and a medium for recombination referred to as hydroxyquinoline aluminum (AlQ_3) was demonstrated for the first time by *Tang* and *Van Slyke* [11.33]. Doping AlQ_3 with various substances determines the emission wavelength and along the same lines dopant species can be customized for desired white light [11.34]. This early report demonstrated an efficiency of 1.5 l/W, a brightness of 1000 cd/m^2 coupled with a voltage of 10 V. In the case of white-light generation, the electron-transporting-layer aluminum complex (AlQ_3) emits at 520 nm (green) and Nile red emits in the red at 600 nm. The hole-transporting-layer TriPhenyldiamine Derivative (TPD) emitting around $410 \div 420$ can be utilized for blue. All of these layers together culminate in the generation of the three primary colors which, when their concentrations are adjusted in such a way to generate equal amounts of these the primary colors, as perceived by the eye, white light results. White light with a luminance in the range of 10,000 cd/m^2 is possible. A higher luminance obtained with increased voltage shortens the lifetime by an equivalent amount. Real improvement in brightness must come from new material and efficient structures.

The class of plastic materials for which semiconductor characteristics can be observed are **conjugated polymers** which are the most widely used today. These polymers are known to be conductive polymers but an un-doped state is employed. Extended π-configuration prevents attainment of blue emission which necessitates exploration of non-conjugated polymers, a field which is actively being researched. Additional disadvantages of conjugated polymers are the lifetime and mobility. Mobilities are low due to the largely amorphous nature of conjugated-polymer films. Fortunately, this is not an insurmountable constraint in the formation of diode devices, but solutions such as a more ordered deposition or doping must be applied if transistor applications are to be pursued. Robustness is also of considerable concern. Everyday polymers are not resistant to photo-oxidation to begin with and the situation is exacerbated by the presence of excited states.

Making certain of the presence of excited states is, of course, the key to creating the material's electronic properties in the first place. Storage lifetimes of at least 5 years and operating lifetimes of $>20,000$ hours are typically required. To meet these goals, significant activity is taking place to both develop materials that are more resistant to chemical degradation, such as oxidation, and to improve encapsulation.

12. Semiconductor Lasers

Semiconductor lasers cover the respectable wavelength range of about 0.4 to 11 μm and are very pivotal in many aspects of human life. As *Strite* put it, the Holy Grail of GaN research is the realization of an injection laser which would represent the shortest-wavelength semiconductor laser ever demonstrated [12.1]. Although semiconductor lasers have many applications, for example, in communication as pumping sources, and mundane applications such as pointers, the most salient and imminent application of GaN-based lasers is in Digital Versatile Disks (DVD for short). This is a future version of the compact disk where the spot size and therefore the storage density is diffraction limited [12.2]. The present CD players utilize GaAs infrared lasers produced by Molecular Beam Epitaxy (MBE). The interim approach adopted by the industry relies on red lasers with which pit dimensions of about 0.4 μm can be read. Using a two-layer scheme in a DVD, the density can be increased from today's 1 Gb to about 17 Gb per compact disk [12.3]. The cycle time in the consumer-electronics market is rather short in that even if red-laser-based DVDs are implemented, the blue laser can be introduced some two years after the red lasers. For consumer applications, CW-operation lifetimes on the order of 10,000 hours at 60°C are required. The nitride-based lasers with their inherently short wavelengths, when adopted, offer much increased data storage-capacity possibly in excess of 40 Gb per compact disk. Figure 12.1 presents a photograph of a Nichia InGaN laser emitting near 400 nm, which is intended for such an application.

Since the introduction of CD audio products in 1982, more than 400 million players and six billion discs have been sold. The CD-ROM (CD-Read Only Memory), an extension of the original CD audio format, has proved to be equally successful in personal-computer applications. Ten major electronics companies have agreed on a standard operating-software format and plan to unveil a range of DVD products including DVD movie players and DVD-ROM drives, pending completion of the associated software. The new breed of optical-disk readers will play both existing CDs and DVDs that can store about 14 times more information than current CDs. Moreover, the rate at which the first-generation DVD system plays back data (11 million bits per second) matches that of a fast 9×CD-ROM player. A video made for a DVD player, for example, might not only store an en-

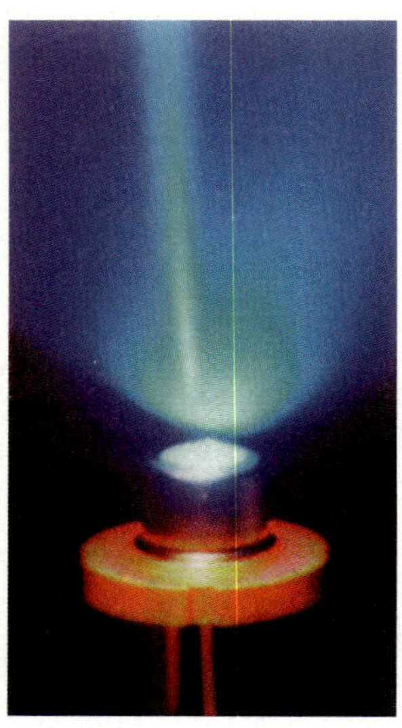

Fig. 12.1. Photographs of a Nichia InGaN laser in action at about 400 nm. Courtesy of S. Nakamura, Nichia Chemical Ltd.

tire movie but it also offers viewers choices between various viewing angles, plots or sound-track languages as well. Within a few years, recordable DVD-RAM (Random-Access Memory) and DVD-R (Recordable) discs and players should reach the marketplace.

The DVD and CD formats share the same basic optical storage technology. Information is represented by microscopic pits, formed on the surface of the plastic disc when the material is injected into a mold. The pitted side of the disc is then coated with a thin layer of aluminum which, in the case of a CD, is followed by a layer of protective lacquer and a label. To read the data, the player shines a small spot of laser light through the disc substrate onto the data layer as the disc rotates. The intensity of the light reflected from the disc's surface varies according to the presence (or absence) of pits along the information track. When a pit lies directly underneath the "read-out" spot, much less light is reflected from the disc than when the spot is over a flat part of the track. A photodetector and other electronics inside the player translate this variation into the 0's and 1's of the digital code representing the stored information. A schematic representation of a CD player is illustrated in Fig. 12.2.

The smallest DVD pits intended for red lasers are only 0.4 μm in diameter whereas the equivalent CD pits are nearly twice as large, or 0.83 μm wide. Also, DVD data tracks are only 0.74 μm apart, whereas the CD data tracks are separated by 1.6 μm. To read the smaller pits, a DVD player's readout beam must reach a finer focus than a CD player by using shorter-wavelength lasers. In addition, DVD players employ a more powerful focusing lens which has a higher numerical aperture than the lens in a CD player. Added density is, in part, due to better Error-Correction and Control (ECC)

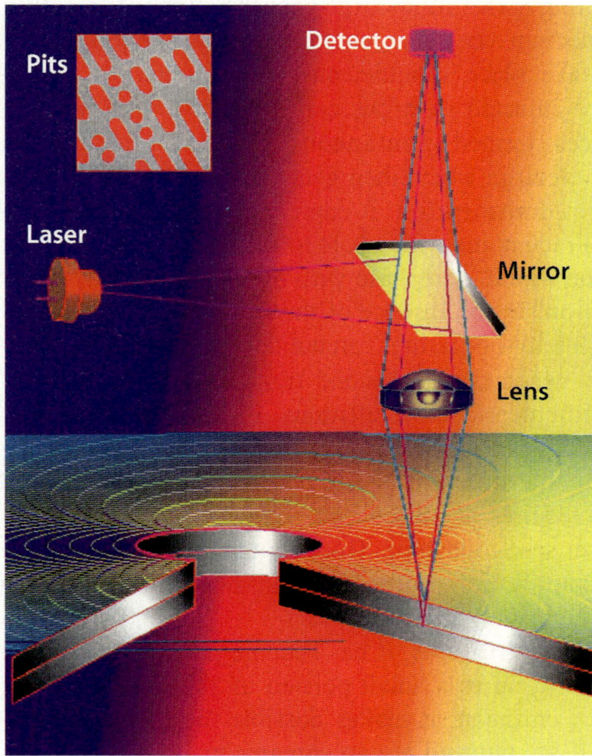

Fig. 12.2. A Schematic representation of how the near-future Digital Versatile Disk (DVD) operates. The diagram at the upper left of the figure shows the pits and their shapes with larger pits causing reduced reflection. The disk itself is of the two-layer type which doubles the capacity. The holographic lens makes it possible to focus on the upper or the lower layer

techniques which require special algorithms that compute additional data bits to be stored along with the user data. These additional bits reduce the fraction of the total disc capacity available.

12.1 A Primer to the Principles of Lasers

Before delving into the details of GaN lasers, a succinct description of the concepts making lasers to operate as they do, is called for. A semiconductor laser is a highly monochoromatic and directional light source which emanates from a leaky optical cavity. For a long time, semiconductors were not considered for lasers as they are absorptive and would attenuate band-edge light emission. It is, of course, now well known that a semiconductor can be made transparent by injecting (generating) large concentrations of electrons and holes in the lasing medium via a p-n junction (optical pumping). In a laser, several processes take place simultaneously with important consequences. For example, the injected electrons can recombine with holes to give off photons in a process termed **spontaneous emission**; a photon can be absorbed by a valence electron which then gets excited into the conduction band in a process termed **absorption**. It contributes to the loss. A conduction-band electron can recombine with a hole to generate a photon in the presence of another photon which is in phase with the initial photon, this is termed **stimulated emission**. In the absence of the last process, the overall loss is positive and amplification, or lasing, can not be obtained. In the presence of the last process, which occurs beyond transparency, gain is obtained and sustains the aforementioned oscillations. In calculating the conditions to yield lasing, all three processes which are strong functions of the conduction- and valence-band structures must be considered. Inefficient photon generation which requires phonon cooperation for momentum conservation precludes indirect semiconductors from being used for lasers. The energy released in the recombination process is given away to the lattice as heat and further reduces the photon density.

With the advent of heterostructures, it has become possible to confine the carriers and the optical field to a small portion of the semiconductor maximizing the stimulated emission; it makes possible the room-temperature and CW operations. If the pumped region contains many non-radiative recombination centers, not only is the photon generation impeded but also the lattice is heated which feeds the non-radiative processes since they generally require phonon-coupled non-radiative recombination. Just as in the case of any semiconductor, GaN-based materials have suffered and still do from such defects. To everyone's delight though, even in the light of large structural defects, the material seems to be more robust against the production of non-radiative recombination centers, even in the presence of a large flux of high-energy photons characteristic of wide-bandgap semiconductors.

12.2 Fundamentals of Semiconductor Lasers

As briefly discussed in the preceding section, absorption and emission processes occur simultaneously in a semiconductor laser. These two processes are sketched in Fig. 12.3 and the present discussion parallels that of [12.4].

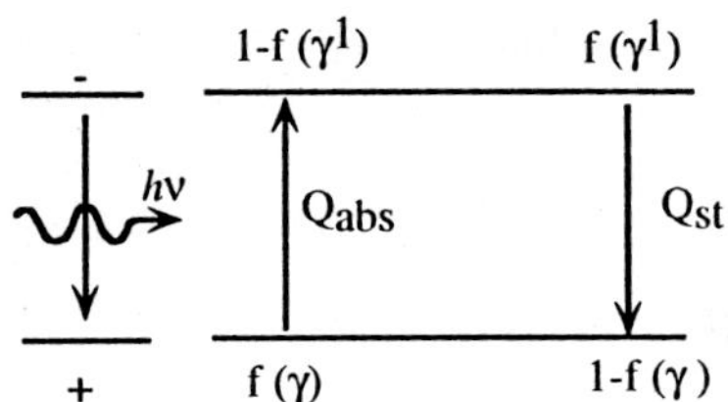

Fig. **12.3**. Schematic representation of upward and downward transition in a two-level system with γ_1 representing the upper and γ the lower state

The electromagnetic power absorbed during the process can be written as

$$Q_{abs} = \frac{1}{V}B(\gamma,\gamma^1)f(\gamma)[1 - f(\gamma^1)]h\nu \tag{12.1}$$

where $B(\gamma,\gamma^1)$ is the transition probability between the γ and γ^1 states, $f(\gamma)$ is the probability of finding an electron in the γ state, $[1 - f(\gamma^1)]$ is the probability of the γ^1 state being unoccupied, V is the sample volume, and $f(\gamma)$, $f(\gamma^1)$ are the electron distribution functions. A similar expression describing the downward stimulated emission, which requires an upper state γ^1 to be occupied and a lower state γ to be empty, is

$$Q_{st} = \frac{1}{V}B(\gamma^1,\gamma)f(\gamma^1)[1 - f(\gamma)]h\nu \ . \tag{12.2}$$

The energy balance is given by the difference between (12.1 and 2) as

$$Q_{tot} = Q_{abs} - Q_{st} = \frac{h\nu}{V}B(\gamma^1,\gamma)[f(\gamma) - f(\gamma^1)] \ . \tag{12.3}$$

The photon-emission rate involves transitions from an energy E to an energy $E-h\nu$ and is related to the product of the density of the occupied states, $N_c(E)f_n(E)$, and the density of empty states, $N_c(E-h\nu)[1 - f_p(E-h\nu)]$, through the relation

$$Q_{st} = B\int N_c(E)F_n(E)N_v(E-h\nu)[1 - f_p(E-h\nu)]|\langle M\rangle|^2 dE \ . \tag{12.4}$$

Here, B is the recombination rate and M is the matrix element which is a function of the valence and conduction bands. The matrix element is a measure of the efficiency of the carrier recombination. The case of lasing involves excitonic transitions, as has been suggested for semiconductors with large exciton binding energies. The matrix element would be different to reflect the exciton participation. The recombination rate can be expressed as

$$B = \frac{4\pi n_r q h\nu}{m^2 \epsilon_0 h^2 c^2} V \qquad (12.5)$$

where V is the volume of the pumped region.

Similarly, for the absorption process which requires occupied states at the energy $E-h\nu$ and empty states at the energy E, one can write

$$Q_{abs} = B \int N_v(E-h\nu)[1 - f_v(E-h\nu)N_c(E)[1 - f_c(E)]|\langle M \rangle|^2 dE . \qquad (12.6)$$

For $Q_{tot} > 0$ absorption prevails and for $Q_{tot} < 0$ the electromagnetic wave is amplified. No net interaction between the electromagnetic wave and the semiconductor takes place for

$$Q_{tot} = 0 , \qquad (12.7)$$

since amplification balances attenuation. This represents the **condition of transparency**. The basic **condition for light amplification** is thus

$$f(\gamma) - f(\gamma^1) < 0 . \qquad (12.8)$$

This means that the probability of finding electrons in the upper level is higher than the probability of finding them in the lower level, thus the term **population inversion** which implies that population of the levels is inverted in comparison to the equilibrium state. In the equilibrium state, the population of the states decreases, as their energy increases according to the Fermi-Dirac statistics. In addition, even in the non-equilibrium case where an overall equilibrium does not exist, carriers in a given band follow the same Fermi-Dirac statistics and the occupation probability is

$$f(E) = \frac{1}{1 + \exp[(E-E_F)/k_B T]} \qquad (12.9)$$

where E is the electron energy, and E_F is the Fermi energy or the Fermi level. E_F is a parameter that determines the filling of the states by electrons.

The population inversion in laser diodes is provided by the injection of non-equilibrium electrons into the conduction band, and holes into the valence band in the active region. Utilizing the occupation probabilities for electrons, f_c, and holes, $1-f_v$, with the quasi-Fermi levels F_n and F_p for electrons and holes in conduction and valence bands, respectively, the condition (12.8) reduces to

$$\frac{1}{1 + \exp[(E_\gamma^1 - F_n)/k_B T]} > \frac{1}{1 + \exp[(E_\gamma - F_p)/k_b T]} \tag{12.10}$$

where E_γ^1 is the energy of the state γ^1 in the conduction band, and E_γ is the energy of the state γ in the valence band. In semiconductor lasers, E_γ^1 and E_γ correspond to the conduction and valence bands E_c and E_v, respectively. As the minimum difference between these states is the bandgap energy E_g, we obtain

$$F_n - F_p > E_g \; . \tag{12.11}$$

This relation was advanced by *Bernard* and *Duraffourg* [12.5]. It also represents the transparency condition meaning that the semiconductor itself no longer absorbs (this is in addition to the waveguide and end losses). This condition simply expresses the fact that the separation of the quasi-Fermi levels must exceed the bandgap energy and that the electron quasi-Fermi level would lie in the conduction band since the density of states in the conduction band is much smaller than that in the valence band. Since carrier injection is through a diffusion process, the concentration of injected carriers in a homojunction laser cannot exceed the carrier concentrations in the n- or p-emitters. Consequently, to satisfy the condition (12.11), the equilibrium Fermi level in the emitter must also be shifted towards the corresponding band, as shown in Fig. 12.4. This figure depicts two cases where in (a) the conduction- and valence-band effective masses are equal, and in (b) the valence-band effective mass is larger than that of the conduction band causing the quasi-Fermi level to enter the conduction band for the lasing condition to be satisfied. Typically the donor and acceptor concentrations in the n-emitter and p-emitter layers are in the range of $(1 \div 5) \cdot 10^{18}$ cm^{-3}. With a heterojunction confinement of the carriers and light, this point is a mute one. The less the volume to be pumped, the smaller the injection current required to reach the transparency condition. Before the advent of heterostructures, the thickness of the pumped region needed to be comparable to the wavelength of the radiation, so the light traveling along this region diffracted severely into absorptive passive regions. This occurs despite the waveguiding effect that is provided by the decreased refractive index of the emitter caused by the high carrier concentration. To maintain a population inversion in the laser diode with an excited-region thickness of

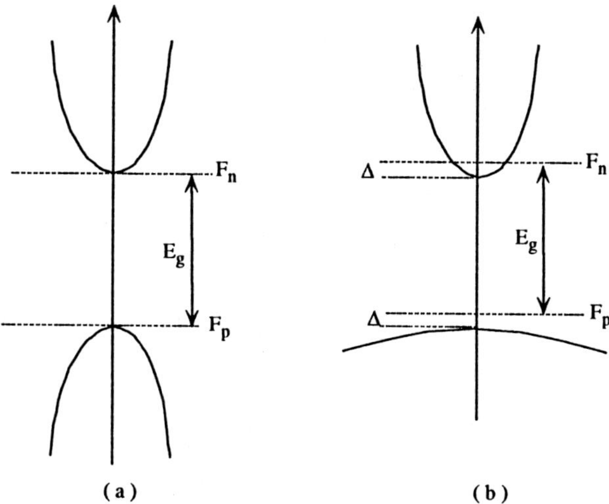

Fig. 12.4a,b. The approximate positions of quasi-Fermi levels in a laser structure under injection of minority carriers and at the threshold of lasing: (**a**) depicts the hypothetical case where the effective masses of electrons and holes are about the same. Strained-layer systems approach this case reasonably closely. (**b**) illustrates the more realistic case of asymmetric electron and hole effective masses. This case where the effective mass of holes is larger than that for electrons, represents all the III-V compound semiconductors used for lasers

$2 \div 3$ μm, current densities of about $20 \div 30$ kA/cm^2 are required. The power to be dissipated by a laser diode at those current densities is so high that p-n junction lasers were able to operate only under pulsed excitation at room temperature, with a very low duty cycle of about 10^{-4} or less, to avoid overheating and catastrophic failure. A similar situation would also occur if the non-radiative processes would be so dominant that the threshold current remains high despite the thin active region employed in the case of a present-day nitride-based laser. Heterojunctions allow the much needed flexibility in the design of laser structures in that it is possible to confine the injected carriers to a very small region while providing a waveguide due to the favorable spatial variation of the refractive index.

Referring to Fig.2.11, the bandgap of the ternary AlGaN alloy increases monotonically with increasing Al content. Thus, a GaN/AlGaN or InGaN/GaN/AlGaN DH laser is essentially a combination of a layer of narrow-bandgap material straddled by two layers of wider-bandgap n- and p-materials. Under forward bias, carriers are injected from wide-bandgap layers (emitters) into the narrow-bandgap (active) layer, where they are confined. The maximum concentration of injected carriers in the active layer does not strictly depend on the equilibrium carrier concentration in the emitters. Even when the concentration of injected carriers exceeds those in the emitters, diffusion back into an emitter is inhibited by heterojunction barriers; it removes the need for very high doping levels in the emitters.

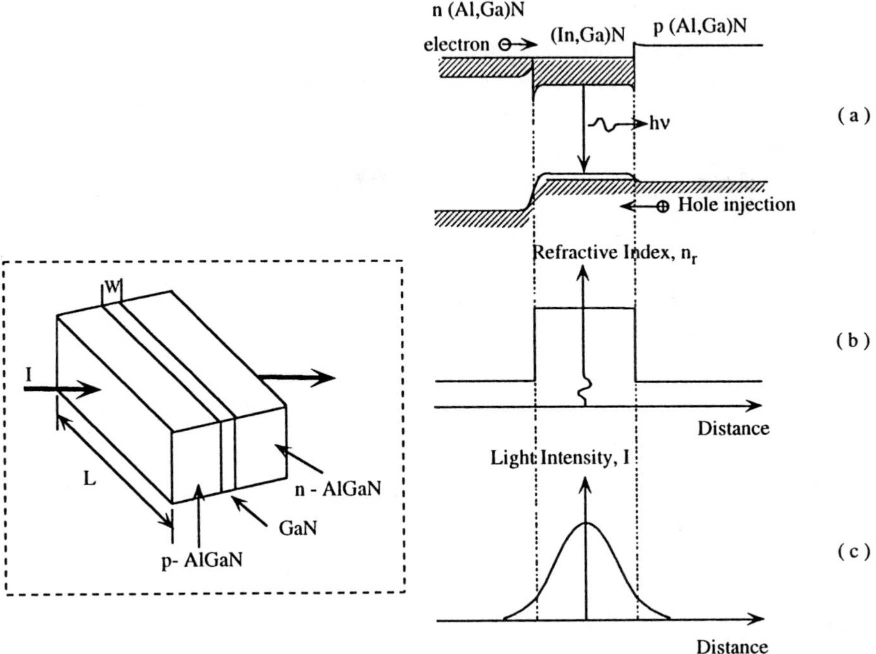

Fig. 12.5a,b. The energy position of the band edges and quasi-Fermi level(s) vs. distance (energy-band diagram) for p-n junction of the injection laser under forward bias (a). The regions occupied by electrons are shown schematically by the shaded areas. (b) and (c) indicate the refractive index and optical field. The inset shows a schematic representation of DH bulk GaN laser with AlGaN cladding layers

In DH lasers, the conductivity type of the active layer also becomes less important. An n-type active layer will be assumed, as in InGaN lasers, and electrons are injected from the p-GaN/n-InGaN heterojunction. These electrons which are confined to the active layer make it negatively charged, and attract additional majority (but non-equilibrium) carrier holes from the p-AlGaN emitter. In general, injection of non-equilibrium minority carriers from the p-n heterojunction rearranges the potential at the opposite n-n (or p-p) heterojunction to facilitate the flow of non-equilibrium majority carriers into the active layer; thereby, they maintain the overall charge neutrality.

The flow and confinement of carriers into the active region is illustrated in Fig. 12.5a. For simplicity, the waveguide is indicated to be of some composition of InGaN, one end point of which is GaN, and the active region to be some composition of InGaN. If AlGaN cladding layers are employed, the active layer can be made of GaN. On the other hand, if the cladding layers are made of GaN, the active layer must be formed of InGaN; this is representative of injection lasers reported in the literature. In this example, the injected electrons are restricted by an energy barrier to the n-(In, Ga)N/p-(Al, Ga)N side. The height of this barrier must be sufficient, as compared to the electron energy in the active layer for carrier confinement. If we neglect the rearrangement of the potential at the p-n heterojunction due to the injection of electrons into the active layer, being on the order of a few kT, one can deduce from Fig. 12.5a that the barrier for electrons is approximately equal to the bandgap difference between the constituents forming the heterojunction. It is given by

$$\Delta E_c^{pn} = \Delta E_c + \Delta E_v = \Delta E_g \; . \tag{12.12}$$

This requirement applies to most of the heterojunctions that are formed by III-V compounds and their alloys. Following the same argument, the barrier

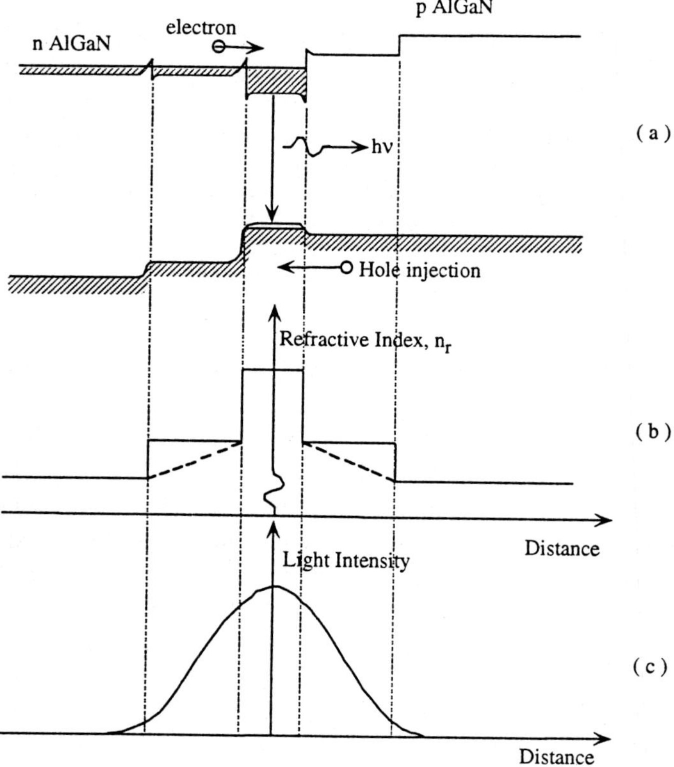

Fig. 12.6a,b. Energy-position diagram along with carrier injection/confinement in an SCH structure. (a) indicates the schematic representation of carrier injection, confinement and photon emission. (b) and (c) depict the refractive-index profile (*solid lines* for SCH and *dashed lines* for graded-index SCH, the latter has not been implemented in the nitride system under the discussion), and the field profile, respectively

388

for holes at the N-n hetero-or homojunction under a high forward bias on the n-(Al, Ga)N/n-(In, Ga)N side is given by

$$\Delta E_v^{nn} = \Delta E_c + \Delta E_v = \Delta E_g .$$
(12.13)

Figures 12.5b and c exhibit the refractive-index and optical-field profiles in a DH laser. The inset of Fig. 12.5 depicts the current flow which is perpendicular to the junction. In DH structures with a small active-layer thickness, the concentration of non-equilibrium carriers in the active layer increases and allows a population inversion to be obtained at lower currents, as the active layer is made thinner and attention is paid to radiation (optical-field) confinement. Otherwise, the radiation leaks out and causes a steep increase in the threshold current. Similar arguments hold for quantum-well lasers of the separate confinement variety, with constant and/or graded refractive index, as depicted in Fig. 12.6. The parts (a), (b), and (c) depict the energy-versus-position diagram, the refractive-index profile, and the optical-field profile, respectively. As in the case of Double Heterojunctions (DHs), (a) indicates a schematic representation of the carrier injection, the confinement and photon emission. Parts (b) and (c) illustrate the refractive-index profile [solid lines for Separate Confinement Heterostructure (SCH)] and dashed lines for graded-index SCH, the latter has not yet been implemented in the nitride system under the discussion), and the field profile, respectively.

12.3 Waveguiding

One of the unique attributes of DH lasers is the built-in optical waveguide without which CW semiconductor lasers cannot exist. Waves suffering total internal reflections are supported by the waveguide and are called the **waveguide modes**. As mentioned earlier, the refractive index of the narrower-bandgap active layer at the operating wavelength is larger than the refractive indices of the wide-bandgap emitters forming the waveguide (Fig. 12.5b and 6b). The transverse field distribution of such a mode, vertical to the growth direction, can be described by a cosine or a sine function inside the layer. It decays exponentially in the outer, low-refractive index media.

12.3.1 Analytical Solution to the Waveguide Problem

Let us consider a three-slab symmetrical waveguide with the core parameters designated with the subscript 1 and cladding-layer parameters with the subscript 2. The problem of waveguiding has been treated in many texts, we follow the treatment of [12.6, 7]. Total internal reflection of an electromagnetic wave or the wave guidance requires the refractive index of the guide to be larger than that for the cladding layers. Specifically, an electromagnetic wave with the incidence angle with respect to the normal equal to or larger than the critical angle ϑ_c

$$\vartheta_c = \sin^{-1}(n_2/n_1) \tag{12.14}$$

will suffer total internal reflection provided that $n_1 > n_2$.[1]

Before we launch into the waveguide expressions, let us define the media parameters. The complex dielectric function ϵ can be related to the refractive index n and the extinction coefficient κ according to

$$\epsilon = \epsilon' + i\epsilon'' = \epsilon_0 (n + i\kappa)^2 . \tag{12.15}$$

Since the power of an electromagnetic field propagating along the z direction is proportional to

$$\exp(-2\kappa k_0 z) \tag{12.16a}$$

where

$$k_0 = 2\pi/\lambda_0 \tag{12.16b}$$

is the free-space wave vector. The absorption coefficient is then given by

$$\alpha = 2\kappa k_0 , \tag{12.17a}$$

$$\kappa = \frac{\epsilon''}{2\epsilon_0 n} . \tag{12.17b}$$

It is clear that the imaginary part of the dielectric constant is responsible for the loss term.

A TE-mode plane wave propagating along the z direction within the guide has even and odd solutions. The analytical treatment here will be limited to the even-mode solution in a lossless core as the waveguide problem

---

[1] The fact that the waveguide width is effectively larger [12.8] due to the Goos-Haenchen shift [12.9] will be ignored in this first-order treatment.

has been covered in countless textbooks. The even-solution field components, in phasor notation, are given by

$$E_y = \hat{\mathbf{y}}\overline{E}\cos(\kappa x)e^{-i\beta z} , \qquad (12.18a)$$

$$H_x = -\hat{\mathbf{x}}\frac{\beta}{\omega\mu_0}\overline{E}\cos(\kappa x)e^{-i\beta z} , \qquad (12.18b)$$

$$H_z = -\hat{\mathbf{z}}\frac{i\kappa}{\omega\mu_0}\overline{E}\sin(\kappa x)e^{-i\beta z} . \qquad (12.18c)$$

Here, overlining means that the field amplitude $\overline{E}$ is used in phasor notation, and the angular frequency ω is related to the frequency ν by $\omega = 2\pi\nu$, $\beta^2 = \omega^2\mu\epsilon$.

The notion of guidance within the guide explicitly implies that the wave should decay outside the waveguide. For this to happen, the exponent in the x-direction should be real. Then

$$E_{y2} = \hat{\mathbf{y}}\overline{E}\cos(\kappa d/2)e^{-\gamma(x-d/2)}e^{-i\beta z} \qquad (12.19)$$

where $\gamma^2 = \beta^2 - n^2 k_0^2$, $k_0 = 2\pi/\lambda_0$ and $\lambda_0 = c/\nu$, with c being the speed of light in vacuum.

For the wave not to propagate, or attenuate outside the core of the waveguide, γ should be positive which leads to the condition that $\beta^2 > n_2^2 k_0^2$ which brings us back to the total internal reflection condition of (12.14).

Similarly, the magnetic-field components outside the core are

$$H_{x2} = -\hat{\mathbf{x}}\frac{\beta}{\omega\mu_0}\overline{E}\cos(\kappa d/2)e^{-\gamma(x-d/2)}e^{-i\beta z} , \qquad (12.20a)$$

$$H_{z2} = -\hat{\mathbf{z}}\frac{i\gamma}{\omega\mu_0}\overline{E}\sin(\kappa d/2)e^{-\gamma(x-d/2)}e^{-i\beta z} . \qquad (12.20b)$$

Applying the boundary conditions that the y-component of the E field and the z-component of the H field must be continuous at $x = d/2$, we get

$$\tan\left(\frac{\kappa d}{2}\right) = \frac{\gamma}{\kappa} = \left(\frac{\beta^2 - n_1^2 k_0^2}{n_1^2 k_0^2 - \beta^2}\right)^{1/2} \quad \text{or} \quad \frac{\kappa d}{2}\tan\left(\frac{\kappa d}{2}\right) = \frac{\gamma d}{2} . \qquad (12.21)$$

Here, β is the unknown, and its solution can be found from the transcendental equation either numerically or graphically.

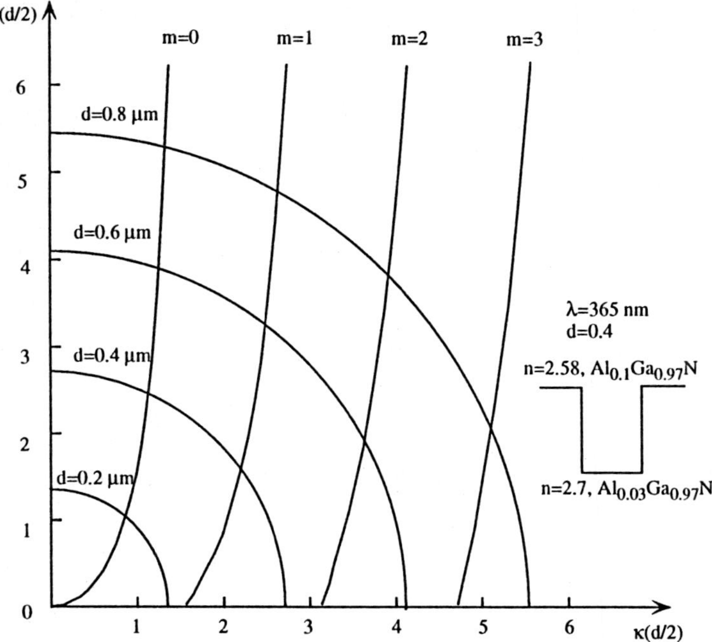

Fig. 12.7. Graphical approach to determine the waveguide modes. The quarter circles represent the solution for (12.21b). The solution for (12.21a) with m as the parameter is also shown. The intersections of (a) and (b) are solutions for modes supported by the waveguide

Recognizing that $\kappa^2 = n_1^2 k_0^2 - \beta^2$ and $\gamma^2 = \beta^2 - n_2^2 k_0^2$, and adding them leads to

$$(n_1^2 - n_2^2)\left[\frac{k_0 d}{2}\right]^2 = \left[\frac{\kappa d}{2}\right]^2 + \left[\frac{\gamma d}{2}\right]^2 . \tag{12.22}$$

The joint solution of (12.21) is that which satisfies the guidance condition. Eq.(12.22) represents a circle with radius $r = (n_2^2 - n_2^2)^{1/2}(k_0 d/2)$. Intersections of the graphical representations of (12.21) and (12.22) are solutions, as illustrated in Fig. 12.7. It is clear that, for small values of d, only the fundamental-mode solution exists. Figure 12.8 depicts the field distribution of the fundamental and second-order modes. It is evident that the fundamental mode is necessary for sufficient confinement. Figure 12.9 displays the refractive indices of GaN, AlGaN and InGaN. The data for GaN have been measured, while those for AlGaN and InGaN have been deduced on grounds that the refractive indices for all three are equal at the respective band edge of each semiconductor. While this deduction is reasonable for AlGaN, the InGaN figures so derived may have more error.

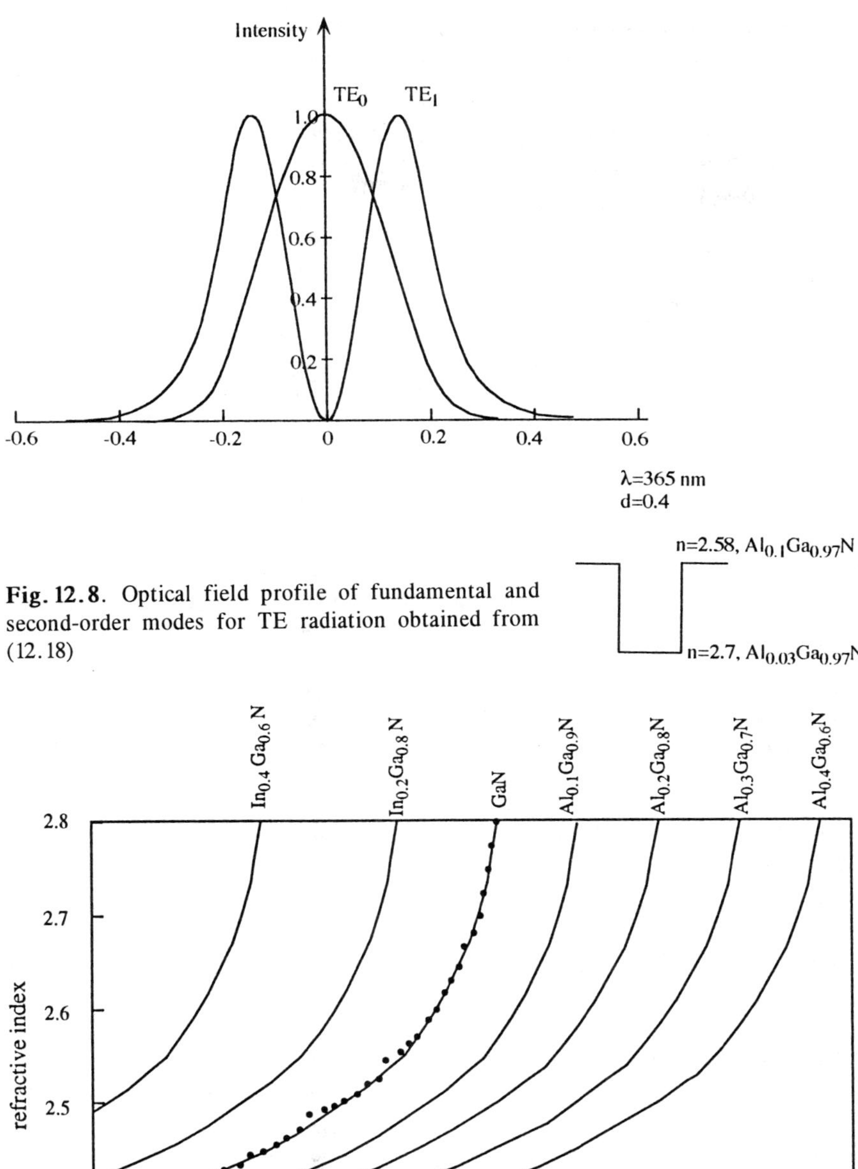

Fig. 12.8. Optical field profile of fundamental and second-order modes for TE radiation obtained from (12.18)

Fig. 12.9. The refractive indices of the AlGaN, GaN and InGaN compounds. The GaN data have been measured whereas the data for the ternaries have been generated from the GaN data with a lateral rigid shift on the assumption that the refractive index is identical at the band edge. This method is reasonable for AlGaN, but its accuracy is questionable for InGaN

12.3.2 Numerical Solution of the Waveguide Problem

Assume for the general case that the number of layers is ℓ. Each layer has the (complex) refractive index n_j and the thickness Δt_j. Since the dimension of the waveguide is much larger in the y-direction than in the x-direction, and z is the propagation direction, the amplitude of the electromagnetic oscillation for a certain mode in the waveguide may be written as

$$E_y(x, y, t) \;=\; E_y(x)\, e^{i(\omega t - \Gamma_j z)} \tag{12.23}$$

where Γ is the complex propagation constant for the mode. For the jth layer, the wave equation describing the optical propagation is given by [12.10]

$$\frac{\partial}{\partial x^2} E_{y,j}(x) - (\Gamma_j^{\,2} - k_0^{\,2} n_j^{\,2}\, E_{y,j}(x) \;=\; 0\,, \tag{12.24}$$

where $k_0 = 2\pi/\lambda$ is the wave number in vacuum, and n_j is the refractive index of the jth layer, which may be complex. The solution of this equation can generally be written as

$$E_{y,j}(x) \;=\; A_j\, e^{\alpha_j (x - t_j)} + B_j\, e^{-\alpha_j (x - t_j)} \tag{12.25}$$

where

$$\alpha_j \;=\; \sqrt{\Gamma_j^{\,2} - k_0^{\,2} n_j^{\,2}}\,. \tag{12.26}$$

Our task now is to find the propagation constant Γ_j. It can be accomplished by using the boundary conditions for each mode. Here, for example, the TE mode requires that $E_y(x)$ and its first derivative be continuous at each interface, which leads to

$$E_{y,j}(t_j) \;=\; E_{y,j+1}(t_j)\,, \tag{12.27a}$$

$$\frac{\partial}{\partial x} E_{y,j}(T_j) \;=\; \frac{\partial}{\partial x} E_{y,j+1}(t_j)\,. \tag{12.27b}$$

From (12.26-27) it can be shown that the amplitude coefficients A_j and B_j for adjacent layers are related as

$$\begin{bmatrix} A_{j+1} \\ B_{j+1} \end{bmatrix} = \begin{bmatrix} (1+\gamma_{j+1})e_j^{\delta} & (1-\gamma_{j+1})e^{-\delta_j} \\ (1-\gamma_{j+1})e_j^{\delta} & (1+\gamma_{j+1})e^{-\delta_j} \end{bmatrix} \begin{bmatrix} A_j \\ B_j \end{bmatrix} = T_j \begin{bmatrix} A_j \\ B_j \end{bmatrix} , \qquad (12.28)$$

where $\gamma_{j+1} = \alpha_j/\alpha_{j+1}$ and $\delta_j = \alpha_j \Delta t_j$. In addition, because the oscillation will be confined within the waveguide, we can assume that

$$E_{y,0}(-\infty) = E_{y,\ell}(+\infty) = 0 . \qquad (12.29)$$

Hence, we can derive a relation between the solutions for the first and the last layer, and obtain

$$\begin{bmatrix} 0 \\ B_\ell \end{bmatrix} = T_{WG} \begin{bmatrix} A_0 \\ 0 \end{bmatrix} = \begin{bmatrix} t_{11} & t_{12} \\ t_{21} & t_{22} \end{bmatrix} \begin{bmatrix} A_0 \\ 0 \end{bmatrix} . \qquad (12.30)$$

where
$$T_{WG} = T_{\ell-1}*T_{\ell-2}*\cdots T_1*T_0 \qquad (12.31)$$

is called the **transmission matrix** of the waveguide. It is easy to see that in order to satisfy the relation, t_{11} must be zero. This fact is taken advantage of to solve the wave equations by numerically searching on the complex surface for the value of Γ_j that gives the least absolute value of t_{11}. After Γ_j has been obtained, (12.25) can be utilized to calculate the electromagnetic wave whose mode squared gives the intensity distribution. The confinement factor, which commonly employs the symbol Γ, is obtained by determining the ratio of the total intensity within the active layers to that within all layers.

Another important parameter in lasers is the **confinement factor** Γ which is defined as the ratio of the optical power overlapping with the active layer to the total optical power in that mode. Mathematically

$$\Gamma = \frac{\displaystyle\int_{-d/2}^{+d/2} E(x)^2 \, dx}{\displaystyle\int_{-\infty}^{+\infty} E(x)^2 \, dx} \qquad (12.32)$$

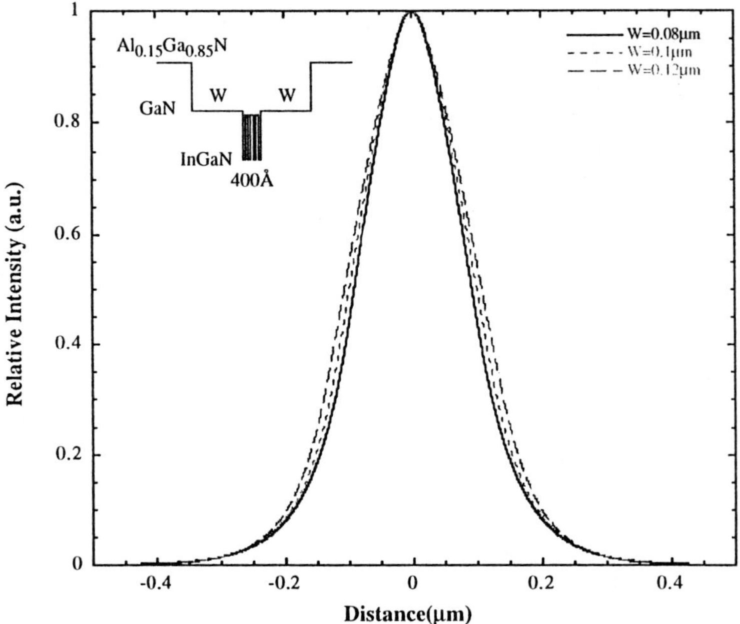

Fig. 12.10. The optical field (E field) distribution for a four-well MQW InGaN active layer, GaN waveguide, and AlGaN cladding layer containing 15% Al which represent a Nichia InGaN laser. The variable parameter is the thickness W of the GaN on each side of the active layer and is half the total thickness of the waveguide

where E(x) is the distribution of the optical-field amplitude in the direction x. The squared exponent is indicative of the need to use optical power.

The method described above has been applied to waveguides employed in lasers that have so far been reported, namely those utilizing InGaN active regions. Potential laser structures relying on GaN active layers and AlGaN waveguide and cladding layers, with different mole fractions of course, have been treated, too. Figure 12.10 exhibits the optical-field (E field) distribution for a MQW InGaN active layer (GaN waveguide) and an AlGaN cladding layer containing 15% Al. The variable parameter is the thickness W of the GaN film on each side of the active layer. The active layer consists of four wells made of InGaN with 15% In and barriers made of InGaN with 5% In having a total MQW-region thickness of 400 Å; it replicates a Nichia laser structure. Here, W is allowed to have the values 0.08, 0.1 and 0.12 μm. Figure 12.11 exhibits the field for the waveguide depicted in Fig. 12.10 where the thickness of the GaN guide on one side is fixed at 0.1 μm while the AlN mole fractions in the cladding layers are allowed to have the values of 0.1, 0.2 and 0.3. Shown in Fig. 12.12 is the field and refractive-index distributions calculated for the same waveguide (Fig. 12.10) and ignoring the 200 Å Al$_{0.2}$Ga$_{0.8}$N employed to keep the InGaN MQW from dissociating during the growth of the top waveguide. To interrogate the

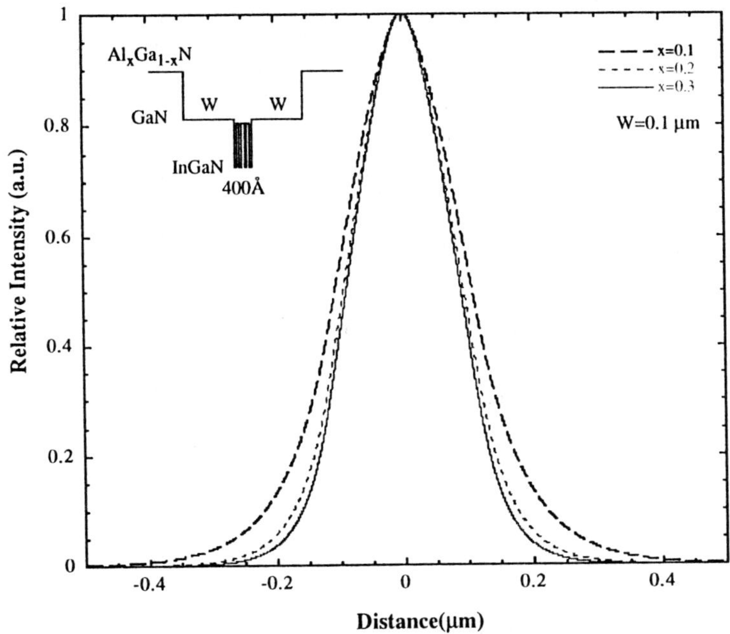

Fig. 12.11. The field for the waveguide depicted in Fig. 12.10 where the thickness of the GaN guide on one side is fixed at 0.1 μm while the AlN mole fraction in the cladding layers are allowed to have the values of 0.1, 0.2 and 0.3.

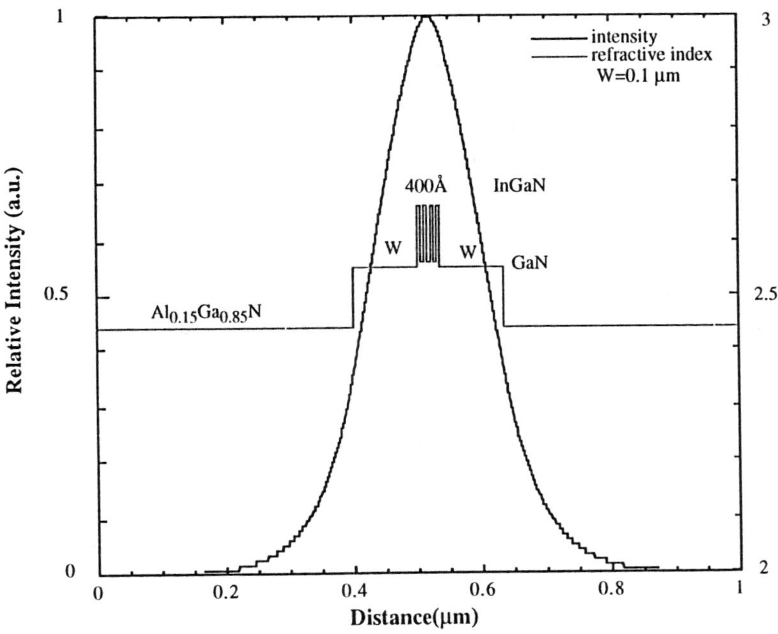

Fig. 12.12. The field and refractive-index distributions calculated for the waveguide as in Fig. 12.10.

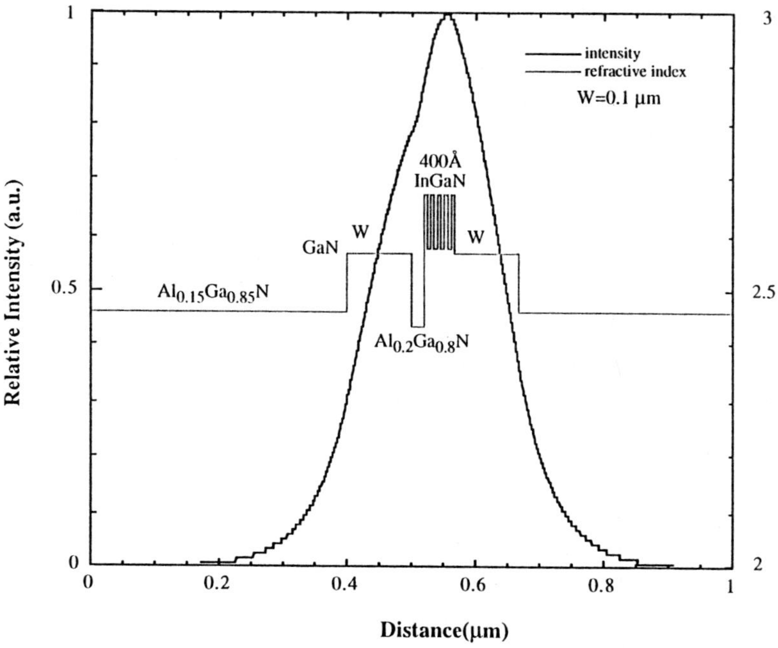

Fig. 12.13. The field and refractive-index distributions taking the 200 Å thick Al$_{0.2}$Ga$_{0.8}$N layer into consideration, which is designed into the Nichia laser to prevent the InGaN MQW region from dissociation during the waveguide growth. This takes place about $100 \div 200\,^\circ$C above the InGaN deposition temperature

effect of the 200 Å Al$_{0.2}$Ga$_{0.8}$N layer designed to prevent the dissociation of the InGaN MQW region during the high-temperature growth, Fig. 12.13 displays the field and refractive index distributions taking this 200 Å thick Al$_{0.2}$Ga$_{0.8}$N layer into consideration. Figure 12.14 depicts the variation of the confinement factor with the half-guide thickness. The parameter is the mole fraction in the cladding layers. The other parameters of the waveguide are as indicated in the inset. Figure 12.15 exhibits the confinement factor for a guide, the characteristics of which are shown in the upper left corner, with varying quantum-well thickness. The expected linear dependence is apparent.

Let us now turn our attention to potential laser structures relying on GaN active layers, and AlGaN waveguide and cladding layers. Figure 12.16 presents the optical-field (E field) distribution for a 50 Å GaN active layer (Al$_{0.03}$Ga$_{0.97}$N waveguide) and Al$_{0.2}$Ga$_{0.8}$N cladding layers. The variable parameter is the thickness W of the GaN film on each side of the active layer, which ranges from 300 to 600 Å. Figure 12.17 plots the field for the waveguide depicted in Fig. 12.16 where the thickness of the GaN guide on one side is fixed at 0.05 μm while the AlN mole fraction in the cladding layers are allowed to have the values 0.1, 0.2 and 0.3. Shown in Fig. 12.18

398

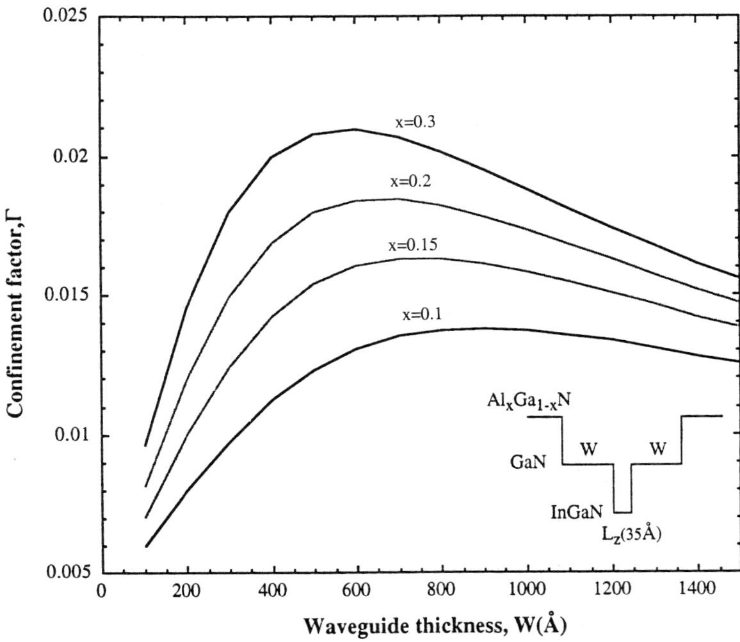

Fig. 12.14. Variation of the confinement factor with the half-guide thickness. The details of the guide are shown on the lower right-hand corner of the figure

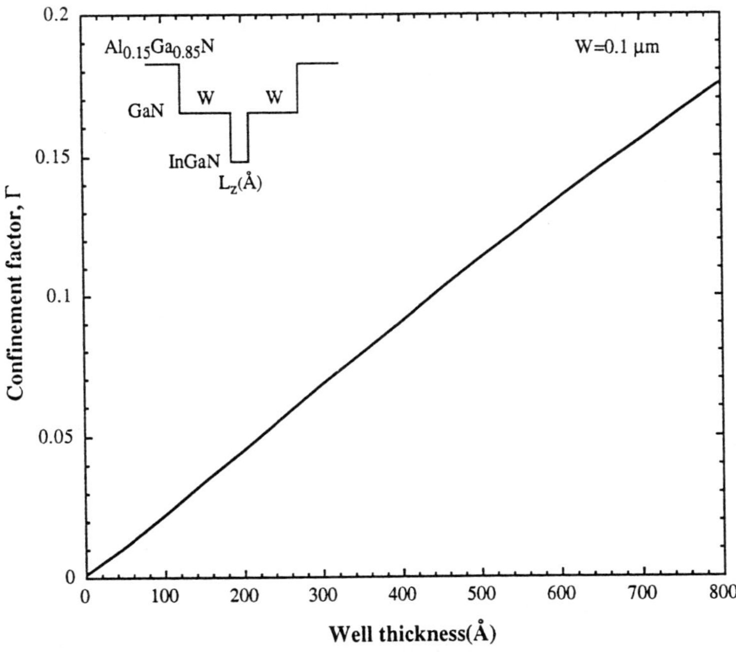

Fig. 12.15. Confinement factor for a guide, the characteristics of which are shown in the upper left corner, with varying quantum-well thickness

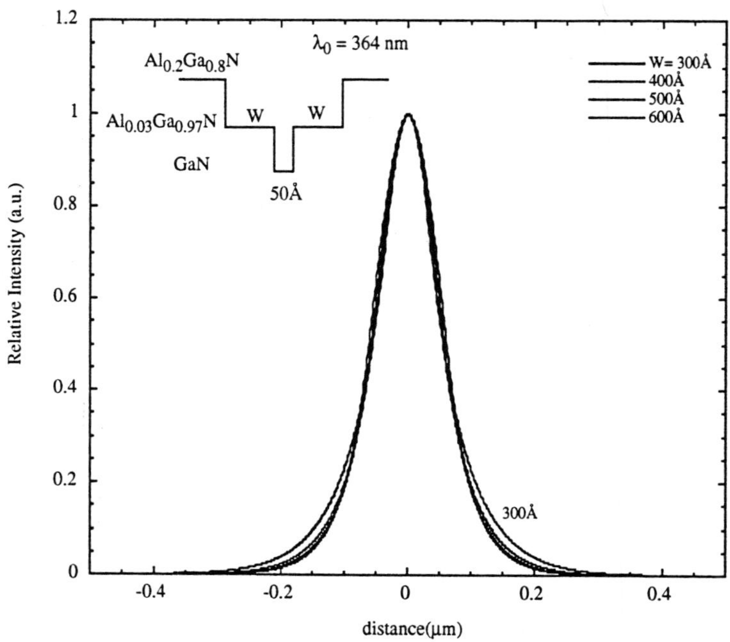

Fig. 12. 16. Optical field (E field) distribution for a 50 Å GaN active layer, $Al_{0.03}Ga_{0.97}N$ waveguide, and $Al_{0.2}Ga_{0.8}N$ cladding layers

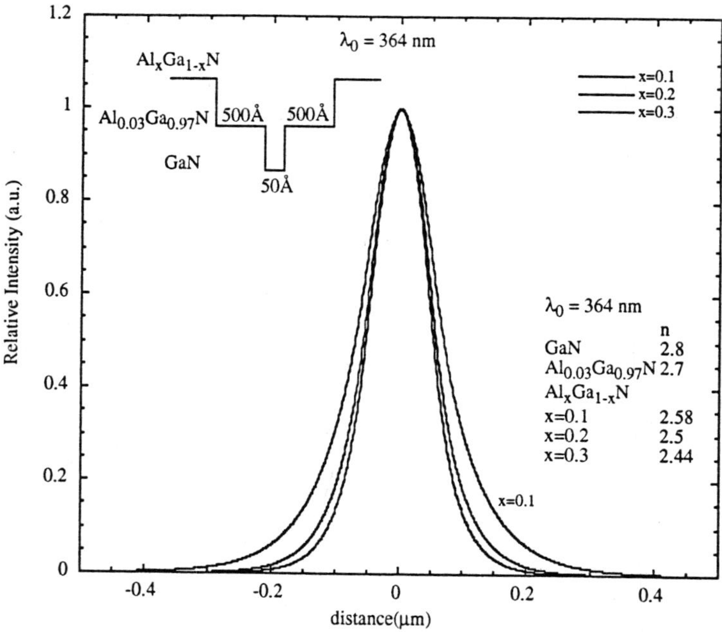

Fig. 12. 17. Field distribution for the waveguide of Fig. 12. 16 where the thickness of the GaN guide on one side is fixed at 0.05 μm while the AlN mole fraction in the cladding layers were allowed to have the values of 0.1, 0.2 and 0.3

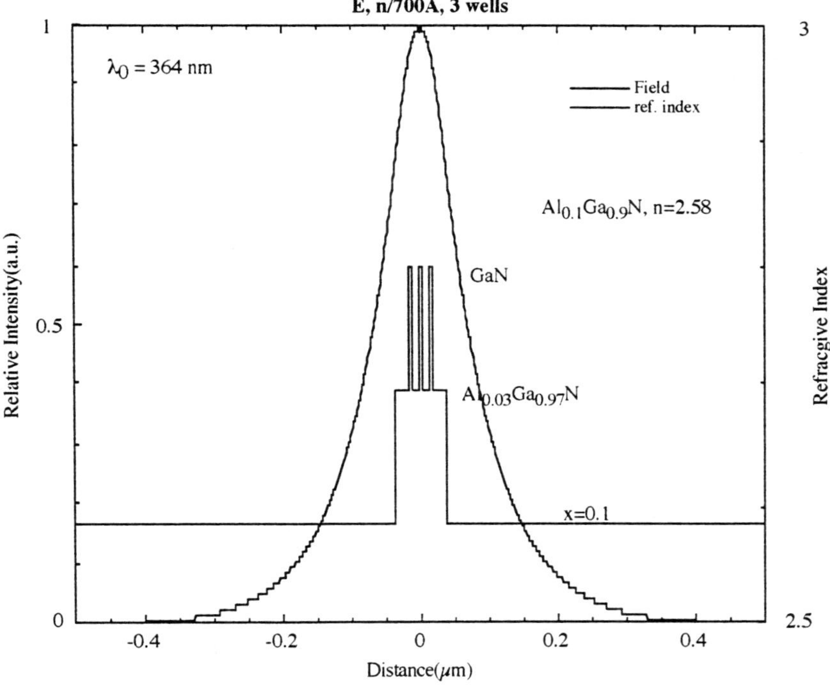

Fig. 12.18. Field and refractive-index distributions calculated for the waveguide with 3-well cladding layers: $Al_{0.03}Ga_{0.97}N$/GaN MQW, 700 Å $Al_{0.03}Ga_{0.97}N$ guide, and $Al_{0.1}Ga_{0.9}N$

is the field and refractive-index distributions calculated for the waveguide with three $Al_{0.03}Ga_{0.97}N$/GaN MQW wells, a 700 Å $Al_{0.03}Ga_{0.97}N$ guide and two $Al_{0.1}Ga_{0.9}N$ cladding layers. Figure 12.19 exhibits the variation of the confinement factor with the half-guide thickness. The parameter is the mole fraction in the cladding layers. The other parameters of the waveguide are as indicated in the inset. Figure 12.20 depicts the confinement factor for a guide, the characteristics of which are given in the inset, with a varying QW thickness for the half-guide thicknesses 400, 500 and 600 Å. The expected linear dependence is apparent.

The method described above treats each layer as being the same, which means that the algorithm requires the same amount of computational time per layer. When the number of QWs in the active layer is very large, this method would become very time consuming, and would not be worthwhile since each QW layer is usually too thin to affect the final results. One way to simplify the calculation involving a large number of QWs is to treat all of them as one thick layer with an effective alloy content x which can be determined by the weighted average of all well and barrier layers. Once a single refractive index (**effective refractive index**) n is assigned to this thick

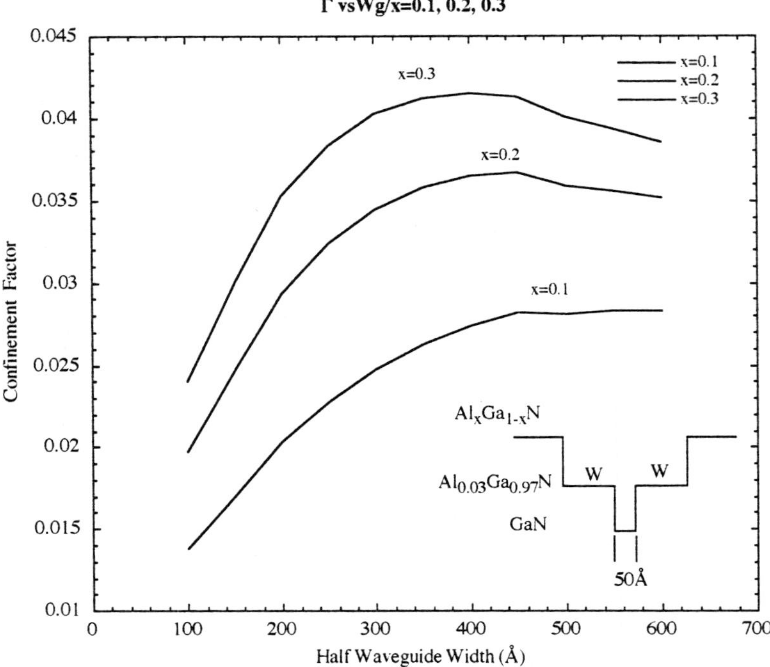

Fig. 12.19. Variation of the confinement factor with the half-guide thickness. The guide details are stated on the bottom right-hand side

layer, the procedure described above can be carried out. Since there is mutual coupling between adjacent QW layers, the effective index depends strongly on the structural parameters of the single layer as well as the value of x. Various methods have been advanced to determine the best value of n based on these parameters [12.11, 12].

It should be pointed out that the lack of experimental data, particularly the refractive indices of $In_xGa_{1-x}N$ and to a lesser extent $Al_xGa_{1-x}N$, necessitated linear interpolation. While the most up-to-date data have been sought, the accuracy of the calculation is certainly limited by the uncertainty in the refractive index, particularly in InGaN, as the mole fractions employed are not accurately known by the practitioners.

12.3.3 Far-Field Pattern

A knowledge of the radiation characteristics, i.e. the far-field pattern, of a semiconductor laser is necessary for a proper collection of the radiated power. It is also a measure of the waveguiding properties which give a real-

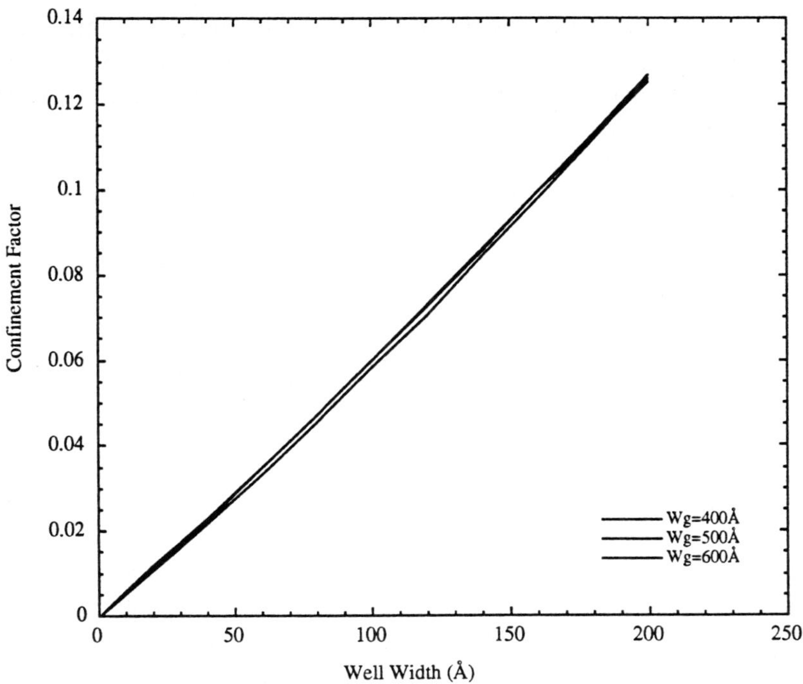

Fig. 12.20. Confinement factor for a guide, the characteristics of which are shown in the inset, with varying quantum-well thickness for half-waveguide thicknesses of 400, 500 and 600 Å

ity check on the calculations of the field distribution and the validity of the parameters such as the refractive indices which are functions of not only the composition but also the doping and the temperature. The radiation pattern emanating from a semiconductor laser is exhibited in Fig. 12.21 where

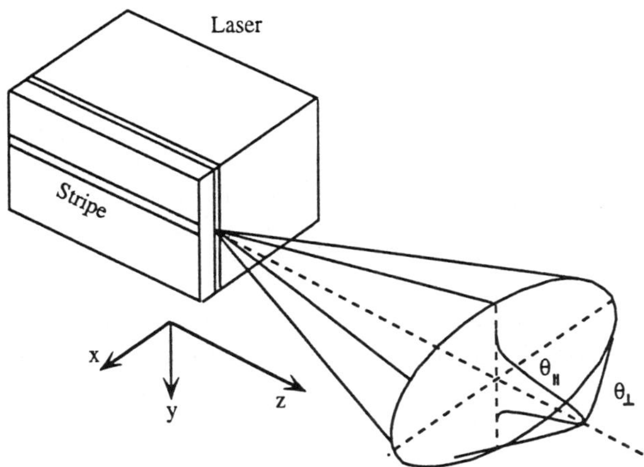

Fig. 12.21. Schematic representation of the far-filed characteristics of an injection laser

$\theta_\perp$ and $\theta_\parallel$ represent the full angles at half power in the directions perpendicular and parallel to the plane of the p-n junction, respectively. Since the dimension perpendicular to the junction plane is small, considerable divergence of the beam occurs in this direction.

To a first approximation, a uniform aperture radiates a single lobe of which the full width at half power, $\theta_\perp$, changes inversely with the thickness W of the emitting region [12.13], as given by

$$\theta_\perp \simeq 1.2\lambda/W . \tag{12.33}$$

Similarly, for a diode with a strong confinement in the lateral direction such as in index-guided lasers of width Z, the beam width in the plane of the junction is, again for a uniform aperture,

$$\theta_\parallel \simeq 1.2\lambda/Z . \tag{12.34}$$

There exist higher-order modes as well, which give rise to complex radiation patterns which can not be represented by (12.32 and 33). The semiconductor is not impartial to which mode propagates, and the mode which has the largest fraction of its intensity distribution overlap the gain region, will have the largest gain. Moreover, the facet reflectivity has an effect, but it is mainly manifested in the polarization. For example, a plane wave propagating in a waveguide whose E field lies in the facet experiences a lower facet loss than other polarizations, culminates in the TE mode that is supported.

The far-field pattern can be obtained by considering the TE waves, as we have done so in conjunction with waveguiding, and solving the wave equation in free space, the details of which can be found in many textbooks including [12.7, 13-16]:

$$\frac{I(\theta_\perp)}{I(0)} = \frac{\cos^2\theta_\perp \left| \int_{-\infty}^{+\infty} E_y(x,0)\exp(j\sin\theta_\perp k_0 x)dx \right|^2}{\left| \int_{-\infty}^{+\infty} E_y(x,0)dx \right|^2} \tag{12.35}$$

where k_0 is the magnitude of the propagation vector in free space, and E_y is the y component of the electric field, it is the only component since the field is a TE mode.

12.4 Loss and Threshold

In addition to losses germane to the semiconductor and the waveguide properties, the end losses must also be compensated for by amplification inside the laser cavity. There are also other losses distributed inside the cavity. For instance, losses associated with transitions of electrons (holes) inside the conduction (valence) band via absorption are called **free-carrier losses**. Other losses which obey the usual absorption law $I(z) = I_0 e^{-\alpha z}$, I_0 being the incident light intensity, α the absorption coefficient, and z the distance, also come into play.

Recognizing that the distributed losses in the active layer and in the cladding layers (α_a and α_c, respectively) are not identical, we can write for the internal loss

$$\alpha_i = \Gamma\alpha_{tot} = \Gamma\alpha_a + (1-\Gamma)\alpha_c , \qquad (12.36)$$

where α_{tot} represents the total loss, and Γ denotes the total confinement factor, as defined by (12.32). Qualitative conditions for the onset of laser action or the threshold condition are obtained when the gain or negative absorption, at the population inversion condition, compensates for all losses sustained during a round trip in the laser cavity. This is given by

$$I_0 R_1 R_2 e^{2(g_{th} - \alpha_i)L} = I_0 \qquad (12.37)$$

the amplitude portion of which leads to

$$g_{th} = \alpha_i + \frac{1}{2L} \ln\left(\frac{1}{R_1 R_2}\right) \qquad (12.38)$$

where g_{th} is the threshold gain, α_i represents the distributed losses, R_1, R_2 are the reflection coefficients of the laser facets (which are generally unequal), and L is the cavity length (i.e., the length of the laser diode). The first term in (12.38) is the internal-loss term while the second term represents the end losses from the facets.

The phase portion of (12.37), i.e., the condition that a wave making a full round trip in the cavity must be in phase for sustained oscillations, leads to

$$\frac{4\pi n_r L}{\lambda_0} = 2m\pi \qquad (12.39)$$

where m is an integer taking the values 1, 2, 3,..., n is the refractive index, and λ_0 and λ_0/n_r are the free-space and inside-the-cavity wavelengths, respectively.

The longitudinal-mode spacing can be obtained from the derivative of the phase condition as

$$dm\lambda_0 + md\lambda_0 = 2Ldn_r . \tag{12.40}$$

By setting $dm = -1$ and substituting m from (12.39), one can find the adjacent mode spacing [12.7, 17]

$$d\lambda_0 = \frac{\lambda_0{}^2}{2n_r L\left[1 - \dfrac{\lambda_0}{n_r}\dfrac{dn_r}{d\lambda_0}\right]} . \tag{12.41}$$

Polynomial fits to the data presented in Fig. 12.9 can ameliorate the calculation of the longitudinal modes determined by (12.41).

In a semiconductor laser, (12.38) must be modified to take the confinement factor into account

$$\Gamma g_{th} = \alpha_i + \frac{1}{2L}\ln\left(\frac{1}{R_1 R_2}\right) \tag{12.42}$$

where $\Gamma g_{th} = g_{nth}$ is the modal gain with g_{nth} being the intrinsic or material gain at threshold (**intrinsic gain**). All the waveguide losses other than the end loss will be lumped into $\alpha_i = \Gamma\alpha_a + (1-\Gamma)\alpha_c$ where α_i is the internal loss.

12.5 Optical Gain

The gain is a figure of merit which indicates how well an electromagnetic field is amplified as it traverses through the semiconductor-laser medium. It is needed to overcome internal losses as well as external losses such as radiation out of the facets which are also referred to as **end losses**. Calculation of the gain is rather complicated and requires a very accurate knowledge of the genesis of the lasing mechanisms as well as the semiconductor band structure which includes effective masses. The gain can be calculated from the spontaneous-emission spectrum or the absorption spectrum if we know the recombination mechanism, i.e., the electron-hole plasma or the excitonic origin. The gain expression is directly related to the band structure, the transition probabilities, and the occupation probabilities. At high injected electron and hole densities, which is the case in semiconductor lasers, Coulomb interactions are screened completely in small-bandgap semiconductors with low exciton binding energies and imply the dissociation of excitons. In such a case, electrons and holes can be treated independently as plasma, except for the many-body effects such as bandgap renormalization. As a first approximation, additional assumptions are made in which the electrons are assumed to interact with the electromagnetic field but not with other electrons and phonons; this is the basis for the so-called **single particle model**.

12.5.1 Gain in Bulk Layers

In the single-particle model, the photon-absorption and emission processes conserve the crystal momentum, which means the k selection rule holds. In this case, the gain in a bulk semiconductor for a given energy $h\nu$ can be expressed as [12.7, 17]

$$g(h\nu) \; = \; \frac{\pi \hbar q^2}{n_r \epsilon_0 m_0{}^2 ch\nu} \int_{E_c}^{\infty} \rho_{cv}(E) |M|^2 [f_c(E) + f_v(E) - 1] dE \; , \quad (12.43)$$

where E is the photon energy, M is the transition matrix element, ρ_{cv} is the joint density of states, and f_c and $1-f_v$ are the occupation probabilities of electrons and holes in the conduction and valence bands, respectively. The symbols c, ϵ_0 and $\hbar$ are universal physics constants, and n_r is the refractive index of the active layer.

In general, however, the single-particle model with k conservation can not be used. This is particularly true in semiconductors with significant band mixing, which is the case in the valence band of GaN and related materials. In such a case, the gain expression (12.43) must be modified as follows [12.18, 19]

$$g'(h\nu) \; = \; \frac{\pi \hbar q^2}{n_r \epsilon_0 m_0{}^2 ch\nu}$$

$$\times \iiint_{E_c}^{\infty} dk_x dk_y \, \rho_{cv}(E) |M|^2 [f_c(E) + f_v(E) - 1] dE \; . \quad (12.44)$$

The integral extends over k_x and k_y as well as over the energy (from the conduction band edge, E_g or E_c, to infinity). The integration over momentum and even energy can be replaced with summations if numerical techniques are employed. Neglecting momentarily the factor in front of the integral, the optical gain is actually the product of three terms.

The first term, $\rho_{cv}(E)$, is the Joint Density Of States (JDOS) function which is determined by the band structure. For bulk semiconductor, applying the parabolic approximation, we have

$$\rho_{cv}(E) \; = \; \frac{1}{2\pi^2}\left[\frac{2m_r}{\hbar^2}\right]^{3/2} (E - E_g)^{1/2} \tag{12.45}$$

where the reduced mass $1/m_r = 1/m_n + 1/m_p$, m_n and m_p being the effective density-of-states masses in the conduction and valence bands, repectively. The effective density-of-states mass in the conduction band is the same as the electron's effective mass. But due to band mixing, the same does not hold for the valence band. For details, see Chap. 3. While the conduction band in nitrides can be assumed parabolic, the valence band is much more complex with heavy- and light-hole band mixing. In this case a numerical integral over the k space is warranted.

Since the electron's effective mass is much smaller than the hole mass, JDOS is dominated by the electron mass in both the Wz and ZB structures, whereas the conduction band is s-like and isotropic. A reduced JDOS implies lower gain. But the same also requires a lower transparency current and thus a lower threshold current; this is the premise of **strained-layer quantum well lasers** in the InGaAs/GaAs system [12.4]. In the ZB case, the Valence Band Maximum (VBM) is degenerate and is lifted in QW structures. This, in turn, leads to a reduced density of states and therefore lower transparency current. On the other hand, VBM in the Wz structure is separated into three two-fold degenerate bands, one Γ_9 and two Γ_7 bands. These bands are not affected much by quantum wells grown in the [0001] direction, as discussed in detail in Chap. 3.

The factor M_{cv} which is the squared Optical Transition Matrix (OTM) element, is given by [12.11]

$$M_{cv} \; = \; \langle \Phi_c | \mathbf{p} | \Phi_v \rangle \; , \tag{12.46}$$

where $\mathbf{p}$ is the momentum operator, and Φ_c and Φ_v are the actual electron and hole wave functions, respectively. Since $\mathbf{p}$ is a vector, we expect M_{cv} also to have three components along the x, y and z-directions, with z being the growth direction. Because the splitting of the p-like states at VBM is determined by the crystal field, the matrix elements are dependent on the symmetry. For the TE mode, the E field is in the plane of the junction and so is the matrix element M_{cv}. For the TM mode, the E field is normal to the plane of the junction and so is the matrix element. The dipole-momentum matrix element for a Wz material in the c-direction is [12.20]

$$|M_{cv}|^2 = \frac{\hbar^2}{2m_0}\left(\frac{m_0}{m_n^{\|}} - \frac{m_0}{m_n^{\perp}}\right)$$

$$= \frac{(E_g + 2\Delta_2)(E_g + \Delta_1 + \Delta_2) - 2(\Delta_3)^2}{E_g + 2\Delta_2} \quad , \qquad (12.47)$$

where $m_n^{\|}$ and $m_n^{\perp}$ represent the c-direction electron effective masses along the k_z direction and $k_{x,y}$ plane, respectively. The Δ_1, Δ_2, Δ_3 parameters represent the diagonal and off-diagonal terms of the 6×6 Hamiltonian that describe the 3 valence bands. The momentum matrix element in the direction perpendicular to the c axis is about the same as that along the c-direction because of a small anisotropy. With the quasi-cubic approximation, that is $\Delta_{cr} = \Delta_1$, $\Delta_{so} = 3\Delta_2 = 3\Delta_3$, the matrix element reduces to

$$|M_{cv}|^2 = \frac{\hbar^2}{2m_0}\left(\frac{m_0}{m_n^*} - 1\right)\frac{E_g(E_g + \Delta_{cr} + \Delta_{so}) + 2\Delta_{cr}\Delta_{so}/3}{E_g + 2\Delta_{so}/3} \quad , \qquad (12.48)$$

where $E_g = 0.25$ Ry,[2] $\Delta_{cr} = 72.9$ meV $= 5.36$ mRy and $\Delta_{so} = 5.17$ meV $= 0.38$ mRy.

The third term in the integrals of (12.43 and 43) is the occupation factor which describes the carrier-density distribution

$$f_c = \frac{1}{1 + \exp[(E_c - F_n)/k_B T]} \quad , \qquad (12.49)$$

$$f_v = \frac{1}{1 + \exp[(E_v - F_p)/k_B T]} \quad , \qquad (12.50)$$

where f_c and f_v are the Fermi-Dirac distribution functions for the conduction and valence bands, respectively.

As in the case of the matrix elements, the Fermi level too is dependent on the density of states in the valence band where the degeneracy is governed by the crystal symmetry as well [12.14].

---

[2] The dimensionless Rydberg (Ry) denotes energy in term of the lowest H energy level (13.6 eV), i.e. 1 mRy = 13.6 meV.

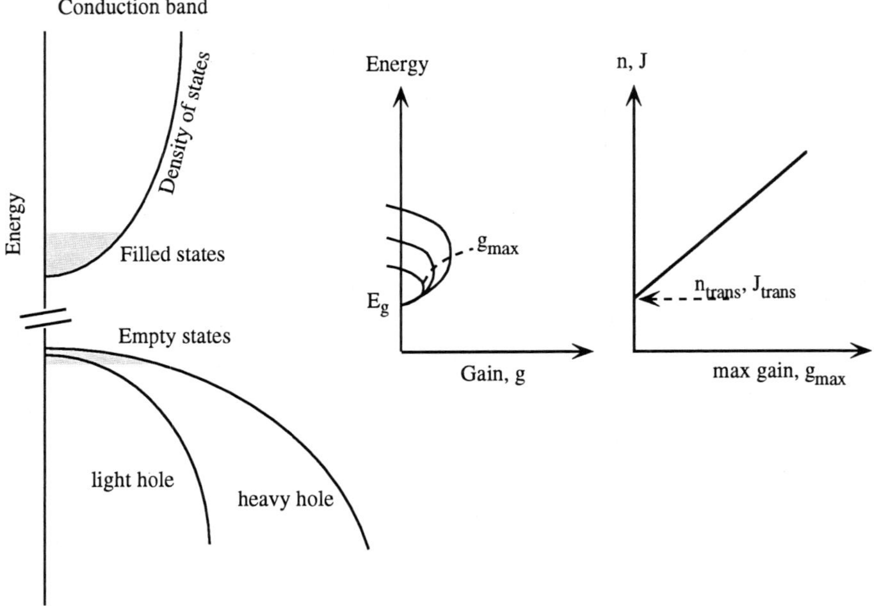

Fig. 12.22. Schematic representation of conduction, heavy-hole and light-hole band densities of states, and gain vs. energy for increasing injection, and the peak gain as a function of injection charge and/or current in WZ GaN. Note the transparency injection charge or current

The joint density of states call for an increase in the gain with increasing energy. On the other hand, increasing energy reduces the occupation factor. These two competing processes result in a maximum in the gain, the **maximum gain** or the **peak gain**, with respect to energy. In addition, the gain increases with the injected-carrier density and thus the injection current. This is schematically depicted in Fig. 12.22.

12.5.2 Gain in Quantum Wells

Due to the quantization in the z-direction (the growth direction), the crystal momentum in this direction takes on discrete quantized values and shifts up the effective bandgap. In the plane of growth, however, the band remains as in the bulk (Fig. 3.6a and b) unless the in-plane effective masses change such as in the case of ZB strained wells [12.4]. Another pertinent change is a staircase DOS function. Consequently, in writing the gain expression for a quantum well, the integral over E in the bulk portion of (12.43) is replaced by a summation over all transitions between the conduction-energy levels and the valence-energy levels [12.22]. In addition, since numerical methods are employed, the integrals over momentum can be replaced by summations

$$g'(E_{eh}) \; = \; \frac{\pi \hbar q^2}{n_r \epsilon_0 m_0{}^2 c E_{eh}}$$

$$\times \iint \sum_{i,j} dk_x dk_y \, \rho_{cv}(E) \, |M_{CV}^{i,j}|^2 \, [f_n(E^i) - f_v(E^j)] , \qquad (12.51)$$

where E_{eh} is the photon energy corresponding to transitions between an electron-energy level and an acceptor-like state or a hole-energy level. As in the case of bulk lasers, the gain expression (12.51) must be modified to include carrier scattering, bandgap renormalization, and many-body effects [12.23, 24]. Assuming that the conduction and valence bands are parabolic, neglecting band mixing of the heavy and light holes in the valence band, and assuming the occupation of only the heavy-hole band, the joint DOS is

$$\rho_{cv}^{i,j}(E) \; = \; \frac{1}{\pi h^2 L_z} m_r \, \hat{u}(E - E_c^i - E_v^j) , \qquad (12.52)$$

where $\hat{u}$ denotes the unity function. The reduced mass here represents the reduced effective DOS mass. In general, the valence-band mass includes the effect of band mixing of the heavy- and light-hole bands, which is also applicable to nitrides. However, if the disparity between the heavy-hole-band mass and light-hole-band mass is very large, and the momentum conserving transitions are the only processes taking place, the DOS hole mass can be equated to the heavy-hole mass.

In an ideal system with no perturbation and at the $k = 0$ (Γ) point, the optical transitions from conduction sub-band states to the valence sub-band states obey the selection rule $\Delta n = 0$. The squared Optical Transition Matrix (OTM) for the TE mode represents mainly the contribution from electrons to heavy-hole transitions, while for the TM mode the contribution comes from electrons to light-hole and split-off hole-band transitions. This is because M_z does not include the heavy-hole wave function, while M_x (or M_y) includes it as well as the light-hole and split-off-hole wave functions. However, a semiconductor laser with a high injection of carriers, particularly a nitride-based one, does not represent this case, and symmetry breaking transitions would be allowed. This necessitates numerical approaches, especially for gain calculations, which include the entire band structure over the momentum and energy spaces. This is also applicable to the calculation of the OTM elements.

The second factor M_{cv} in (12.51) is the squared OTM element [12.19]:

$$M_{cv}^{i,j} \; = \; \langle \Phi_c^i | \mathbf{p} | \Phi_v^j \rangle . \qquad (12.53)$$

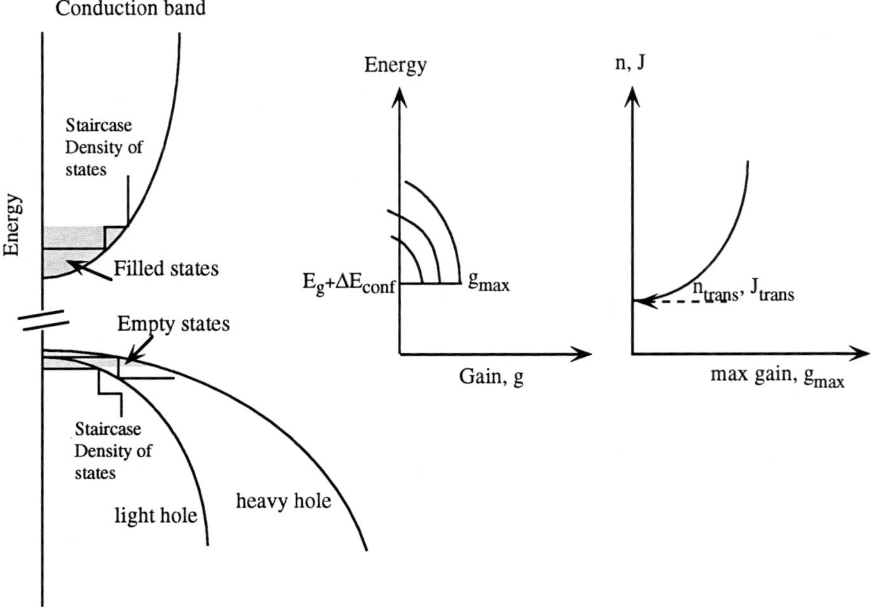

Fig. 12.23. Schematic representation of conduction, heavy-hole and light-hole band densities of states, and gain vs. energy for increasing injection, and the peak gain as a function of injection charge and/or current in a Wz GaN quantum well structure. Note the transparency injection charge or current

The third term in (12.51) is the occupation factor, which describes the carrier density distribution.

$$f_c = \frac{1}{1 + \exp[(E_c^i - F_n)/k_B T]} \, , \tag{12.54}$$

$$f_v = \frac{1}{1 + \exp[(E_v^j - F_p)/k_B T]} \, . \tag{12.55}$$

The terms are the same as those described for the bulk case. As shown schematically in Fig. 12.23, the dependence of the gain on energy in quantum-well lasers differs from that of the bulk (Fig. 12.22) due to the unique staircase-like density of states in which the gain maximum occurs at the confinement energy. This is because the density of states remains constant while the occupation terms decrease exponentially (Fig. 3.6b).

12.6 Coulombic Effects

In the case of wide-bandgap semiconductors such as those employed for GaN-based lasers, the exciton binding energies are comparable to kT at room temperature, and Coulombic effects may not automatically be neglected without further consideration [12.24]. If such is the case, then the parameters used in the gain expression (12.51) must be modified to reflect this. Moreover, broadening effects such as carrier scattering must be considered. The traditional approach is to convolute (12.51) with a spectral line-broadening or line-shape function. One expression which takes into consideration broadening due to the lifetime of the final state [12.19, 23, 24] is

$$g(\hbar\omega) \ = \ \int_{E_c}^{+\infty} g'(E)\, L(\hbar\omega - E)\, dE \tag{12.56}$$

with

$$L(\hbar\omega - E) \ = \ \frac{1}{\pi} \frac{\hbar/\tau_{in}}{(E - \hbar\omega)^2 + (\hbar/\tau_{in})^2} \ ,$$

where τ_{in} and $\hbar\omega$ represent the intra-band relaxation time and the photon energy, respectively. The intra-band relaxation time τ_{in} is not really known for Wz GaN at present. However, gain calculations so far have assumed that this time constant is similar to those of the well characterized ZB semiconductors such as GaAs for which it is on the order of 0.1 ps.

The linewidth-broadening expression can be generalized to take into account any scattering mechanism that causes broadening, in which case we replace $\hbar/\tau_{in}$ with the general broadening factor $\gamma(\hbar\omega - E)$. This factor is energy dependent and must take into account processes in both the conduction and valence bands [12.25].

The resulting gain spectrum converges to the square-root dependence of the absorption coefficient, as described by (10.13), at low carrier densities for the bulk and the parabolic-band cases. With increasing the injected-carrier density, a positive gain peak develops. Keep in mind that the linewidth broadening in GaN-based structures is much larger than that in more mature semiconductor technologies due to large concentrations of defects in relation to other more-developed semiconductor-laser materials, which can not be analysed with (12.56).

The gain in wide-bandgap semiconductors is smaller and, in general, the peak gain is further reduced due to spectral broadening. Both are implicit in many of the gain expressions commonly used. It is thus instructive to consider another form of the gain expression [12.26], namely

$$g(\hbar\omega) \; = \; \frac{c^2 J \eta_{in}}{8\pi q n_r^2 (\hbar\omega)^2 \Delta\nu d} \tag{12.57}$$

which takes into account the line broadening. The terms have their usual meanings: $\hbar\omega$ and $\Delta\nu$, and d represent the emission energy, the spontaneous emission spectral half-width in terms of frequency, and the thickness of the active region, respectively. At first glance, the larger the spontaneous spectral width, the smaller the gain. Likewise, the higher the frequency, as is the case for short-wavelength lasers, the smaller the gain. Generally, any increase in the bandgap is accompanied by a reduced broadening due to the large phonon energies and other factors. However, broadening caused by inhomogeneities prevents narrow spectral widths in GaN and related ternaries. This is also true for excitonic transitions in case they are involved in the lasing action. It is therefore imperative that nitride-based semiconductors be prepared to produce sharp spontaneous linewidths. Reducing the doping level in the semiconductor's active region also reduces the linewidth with the adverse consequence that the low doping levels cause the parasitic resistance to increase. Fortunately, in quantum-well semiconductor lasers, the active region which determines the linewidth, is very thin; even when left undoped, it does not cause the resistance to be unacceptably large.

Coulomb interaction between oppositely charged carriers helps to keep them closer to each other. Since the recombination efficiency increases with a reduced distance between the recombining electrons and holes, an increase in the spontaneous emission rate and consequently in gain would result. The question, of course, is whether the injected carrier concentration in available GaN-based lasers is too large to allow this phenomenon to show up. Specifically, present GaN lasers require injected carrier concentrations in the range of 10^{19} to 10^{20} cm^{-3} for transparency; and as will be discussed in Sect. 12.9, a good deal of broadening due to inhomogeneities occurs in InGaN, which dampens the effect of a Coulombic interaction. Consequently, it is highly unlikely for this process to be of much importance at this stage of development. However, with improved materials quality, the carrier concentration that must be injected for transparency and beyond, should be substantially reduced in which case the Coulombic interaction may indeed play a role [12.27]. In CdZnSe single quantum-well lasers and at an injected carrier concentration of $5.5 \cdot 10^{18}$ cm^{-3}, the Coulombic-interaction term increases the gain by a factor of 2.35 and is accompanied by a blue shift of 6 meV in its spectrum. Carrier scattering alone produces a blue shift in the gain spectrum, as the injected carrier concentration is increased. A similar behavior occurs when the cavity length is enlarged. Shorter cavity lengths increase the end losses and cause the threshold gain to become larger and thus they require higher injected-carrier concentrations. Bandgap renormalization alone causes a red shift with increasing injection, which implies a red shift with reduced cavity lengths. The Coulomb interaction causes a blue shift with higher injected-carrier concentration. All of the three factors involved lead to a blue shift of the gain with an injected carrier concentration and a reduced cavity length. The emission energy in CdZnSe quantum-well lasers reveal a blue shift with decreasing cavity length [12.28]. This has been suggested as an evidence that Coulomb attraction indeed plays a role in these lasers although the experimental emission energies lagged the calculated ones by about 10 meV. This has been attributed to inaccuracies in the structural parameters.

There are other terms which can affect the gain among which is a form of many-body effect, specifically the bandgap renormalization. This is brought about by the large carrier concentrations which screen the repulsive processes between the valence and conduction electrons. It causes a change in the electron distribution which, in turn, results in the reduction of the transition energy (effective bandgap). This decrease is called the **Coulomb-hole contribution to the bandgap renormalization**. At very high injection levels, which may not be sustainable, both the ionic and covalent contributions to the bandgap may change. The Coulomb-hole contribution to the bandgap renormalization is expressed as [12.27-29]

$$\Delta E_{CH} = -2E_R a_B \lambda_s \ln\left(1 + \sqrt{\frac{8\pi n d}{\lambda_s^3 a_B}}\right), \tag{12.58}$$

where d is the thickness of the active layer with the product specifying the areal density of electrons, and E_R, a_B and λ_s represent the Rydberg energy, the Bohr radius for electrons (excitons if excitons dominate the gain process), and the screening length, respectively.

The Bohr radius is given by

$$a_B = \frac{4\pi \hbar^2 \epsilon}{q^2 m} \tag{12.59}$$

where ϵ is the dielectric constant, and m is the effective mass of the electrons (reduced mass of excitons in case the gain is governed by excitons).

The Rydberg energy can be written as

$$E_R = \frac{\hbar^2}{2ma_B} \, .$$

(12.60)

The screening length for quantum wells can be stated by [12.30]

$$\lambda_s^2 = \frac{q^2}{\pi \hbar^2 \epsilon} \sum_j \left[\frac{m_{cj} \, f_c(E_{cj})}{d_{nj}} + \frac{m_{vj} \, f_v(E_{vj})}{d_{vj}} \right]$$

(12.61)

where

$$d_{nj} = j \frac{\pi \hbar}{\sqrt{2m_n E_{nj}}} = j \frac{\pi}{k_{nj\perp}}$$

and

$$d_{vj} = j \frac{\pi \hbar}{\sqrt{2m_v E_{vj}}} = j \frac{\pi}{k_{vj\perp}}$$

(12.62)

where $k_{nj\perp}$ and $k_{vj\perp}$ represent the equivalent out-of-plane wave vector for the j state which would assume values of 1, 2, etc., depending on the number of quantized states that participate. For injection levels in the range between low and high 10^{18} cm^{-3}, the bandgap reduction is roughly between 40 and 60 meV.

Additional contributions to the bandgap renormalization come from the screened-carrier exchange. Both of these contributions have been treated elsewhere in detail [12.27]. Thus, the exchange contribution to the renormalization can be expressed as

$$\Delta E_{sx} = \frac{2E_R}{\lambda_s} \int_0^\infty dk^\| k^\| \frac{1 + \lambda_s a_B k^{\|2}/8\pi nd}{1 + k^\|/\lambda_s + \lambda_s a_B k^{\|2}/8\pi nd}$$
$$\times \left[f_c(E_{cjk^\|}) + f_v(E_{vjk^\|}) \right] \, .$$

(12.63)

For injection levels in the range between low and high values of 10^{18} cm^{-3}, the bandgap reduction is roughly between 20 and 40 meV.

Assuming carrier-injection levels in the low to high 10^{18} cm^{-3}, the total bandgap renormalization due to the Coulomb-hole contribution and the screened-carrier-exchange correction adds up roughly to values between 60 and 100 meV at room temperature. A phenomenological expression devised

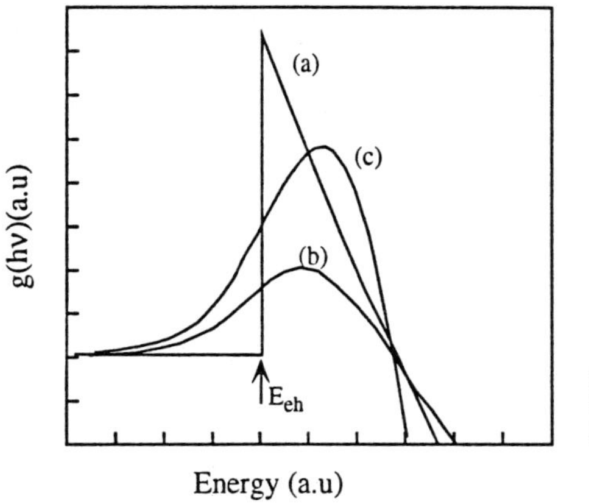

Fig. 12.24. Spectral gain in a quantum-well laser for a given injection level: (a) without the many-body effects, (b) with carrier scattering, and (c) with Coulomb interaction

for GaAs relates the bandgap shrinkage to the sum of the cubic powers of the electron and hole concentrations [12.7].

Figure 12.24 depicts schematically the spectral gain in a quantum-well laser for a given injection level without the many-body effects (a). Carrier scattering causes a broadening of the gain accompanied with a reduction of the peak gain (b). The Coulomb interaction increases the gain due to a reduction of the carrier-to-carrier distance due to the attraction between recombining carriers. This reduces the distance between one another and thereby increases the recombination efficiency (c).

12.7 Gain Calculations for GaN

Despite a relatively short debut, GaN-based lasers and, in particular, their optical gain, have attracted a great deal of theoretical attention due to the all but certain use of these lasers in compact-disk players. In fact, following the successful demonstration of bright LEDs, the theoretical work aimed at determining the valence-band structure and the optical gain in both the zinc-blende and wurtzite phases of GaN has progressed in parallel with efforts to demonstrate the first laser. While the bulk of the initial attempts involved the ZB phase, possibly because of the availability of computer codes developed for other zincblende systems, the attention quickly gave way to the wurtzite phase as this phase appears more and more to be the technologically important one. Consequently, the treatment here will be focused on the Wz phase with some mention of the ZB phase. Due to the non-paraboli-

city of the valence band, the p-like states, strain, and band mixing, gain calculations for GaN require the knowledge of the full bands and numerical
methods applied. Last but not least, the lasing mechanism must be known,
i.e. free electron-hole based or exciton based. Calculation of the nitride
band structure was covered in Chap.3, and the treatment of the gain which
follows builds on our band-structure knowledge.

12.7.1 Optical Gain in Bulk GaN

Several researchers have attempted to calculate the gain in bulk GaN assuming an ideal semiconductor with no inhomogeneities and under the assumption that transitions are governed by electron-hole recombinations. Contributions due to a Coulombic interaction, such as excitons, have also been
considered. The Coulombic interaction serves to keep the carriers of opposite charge closer together, which enhances the recombination efficiency
and, therefore, the gain. Assuming the electron-hole plasma to be the governing process, *Suzuki* and *Uenoyama* [12.22, 31, 32], and *Meney* and
O'Reilly [12.31] calculated the gain in bulk Wz GaN utilizing the band
structure determined by an adaptation of the Full-Potential Linearized Augmented Plane Wave (FPLAPW) [12.34]. The hole masses of the uppermost
two valence bands along the growth direction are very heavy. The mass of
the CH band is light due to a strong coupling with CBM through the $k_z p_z$
perturbation. In the x-direction, in the plane of the layers, CBM is coupled
to the mixed state of the uppermost two valence bands and, thus, the hole
mass of the mixed band is light. The lowest valence band, CH, supports the
optical gain of the TM mode. For this mode to become important, an
extremely large carrier injection is required. This paves the way for the
dominance of the optical gain by the TE mode under normal conditions.

In the ZB case the three valence bands, the uppermost two bands are
degenerate and the hole mass is heavy in all directions. This causes DOS to
be larger than that for the Wz phase as long as the crystal-field splitting in
the latter is larger than the spin-orbit splitting. This results in the gain to be
lower in the ZB phase than in the Wz one. Since the TM mode is supported
by the light-hole and spin-orbit split bands, the TE and TM modes are both
available in the ZB phase. Figure 12.25 shows the optical gain in Wz (for
the TE mode with its E field in the c plane) and in ZB bulk GaN (for the TE
mode, and TM mode with its E field in the c-direction). The topic of gain in
bulk lasers did not get as much attention as in quantum wells which will be
discussed in the next section.

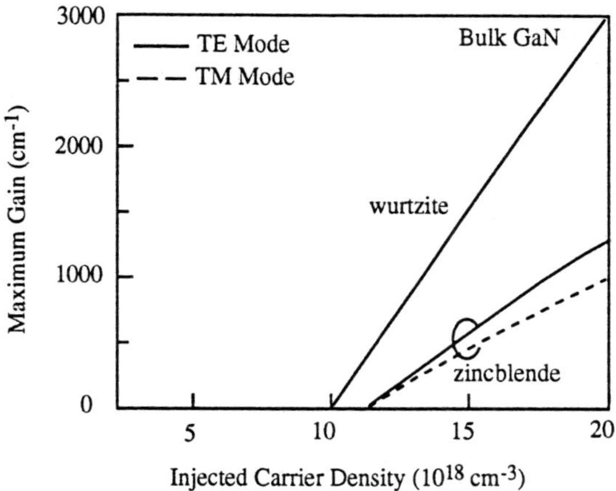

Fig. 12.25. Maximum gain versus injected-carrier density for the TE mode in bulk Wz GaN, and TE (*solid*) and TM (*dashed*) modes in bulk ZB GaN. After [12.31]

12.7.2 Gain in GaN Quantum Wells

Quantum wells at the very least serve to reduce the volume of the semiconductor that must be pumped, leading to very low threshold currents. More importantly, by virtue of thin layers employed in quantum wells, coherently strained systems can be obtained. Strain-induced reduction of the in-plane hole mass in ZB structures leads to a much reduced transparency density and thus a reduced threshold. Consequently, the degrees of freedom in laser design have been enhanced immensely with the advent of quantum wells. The favorable effect of strain germane to ZB systems are not predicted to occur in Wz structures. The heterostructures used in lasers based on wide-bandgap nitrides are not lattice matched; this will eventually require the use of thin layers. In addition, InGaN has not proven to be easily produced in bulk form and necessitates thin layers which are provincially referred to as quantum wells. In what follows, the optical gain in quantum wells, first without strain and later with strain will be discussed.

12.7.3 Gain Calculations in Wz GaN QW Without Strain

As reported by *Suzuki* and *Uenoyama* [12.22], in bulk Wz GaN, the hybridization of the CH band with HH and LH bands[3] is negligible at VBM due to small spin-orbit splitting energies. In Wz GaN QWs though, the splitting be-

---

[3] The shorthands HH, LH, and CH refer to the uppermost, one lower, and lowest subbands of the valence band, respectively.

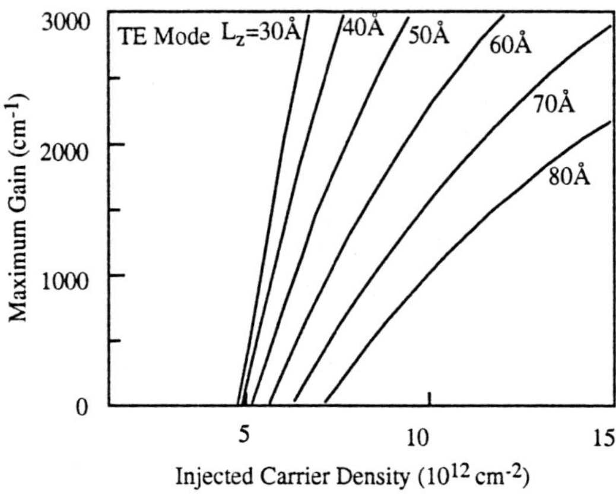

Fig. 12.26. Maximum optical gain in Wz $GaN/Al_{0.2}Ga_{0.8}N$ single QWs with an active-layer width varying from 30 to 80 Å with respect to the injected areal carrier concentration. The initial slope of the maximum gain in the QW structure is somewhat larger than that in the bulk. Due to gain saturation, the slope is reduced with increasing injection of carriers. After [12.32]

tween the CH bands, and HH and LH bands is more pronounced, but the band mixing is similar to that in bulk GaN, which implies that DOS is not substantially reduced and that the TM mode gain is not supported, leaving the TE mode to be dominant. In structures with pronounced quantization, i.e. with well widths comparable to the Bohr radius, the separation of the HH and LH bands is more pronounced, and VBM DOS is somewhat reduced. The resulatnt effect manifests itself as a noticeable gain production at relatively low carrier densities. Results of the simulations by *Suzuki* and *Uenoyama* [12.22] for a series of $GaN/Al_{0.2}Ga_{0.8}N$ quantum wells are shown in Fig. 12.26.

As (12.56) clearly indicates, the intra-band scattering broadens the spectral gain and reduces the maximum gain. In the absence of any experimental value for this parameter, *Suzuki* and *Uenoyama* [12.22] performed gain calculations for a series of relaxation times in the range of 0.01 ps to infinity for a 60 Å active layer. The results are presented in Fig. 12.27 in the form of gain versus the injected sheet carrier concentration.

12.7.4 Gain Calculations in WZ QW With Strain

As discussed in Sect. 3.2, the band structure, and thus the gain, are affected by strain. Calculations performed by *Suzuki* and *Uenoyama* [12.22] indicate that the biaxial strain lowers the injected-carrier density for a given maximum optical gain. As a companion to the strain-free case illustrated in

420

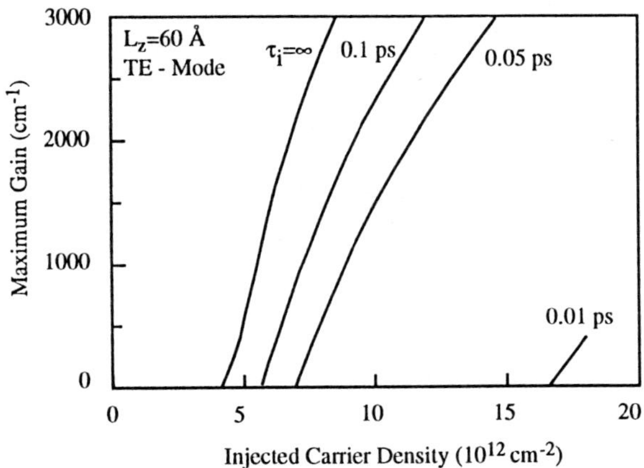

Fig. 12.27. Effect of the intra-band relaxation time τ_i on the gain for TE mode in a GaN/Al$_{0.2}$Ga$_{0.8}$N structure with a 60 Å active layer. After [12.22]

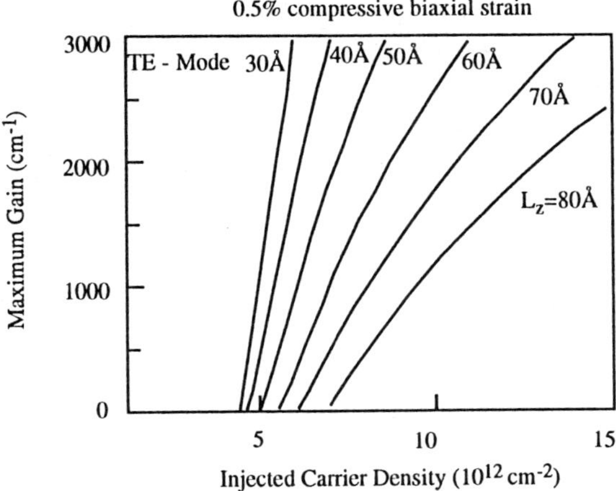

Fig. 12.28. Maximum optical gain in Wz GaN/Al$_{0.2}$Ga$_{0.8}$N single QW with an active-layer width varying from 30 to 80 Å with respect to the injected carrier concentration for 0.5% compressive strain. After [12.32]

Fig. 12.26, the maximum gain for a series of SQW is depicted in Fig. 12.28 for a 0.5% compressive biaxial strain. On the other hand, the tensile strain has an adverse effect on the gain. Figure 12.29 exhibits the effect of biaxial and uniaxial (in the c-plane) compressive and tensile strains. The uniaxial compressive strain is along the y-direction which is perpendicular to the direction of propagation (x-direction). The growth direction is the z-direction or the c-direction. The tensile uniaxial strain is perpendicular to the polarization vector, meaning along the x-direction. Clearly, both biaxial and uniaxial tensile strains reduce the gain while the compressive strain increases it. The effect of a biaxial compressive strain is, however, not suffi-

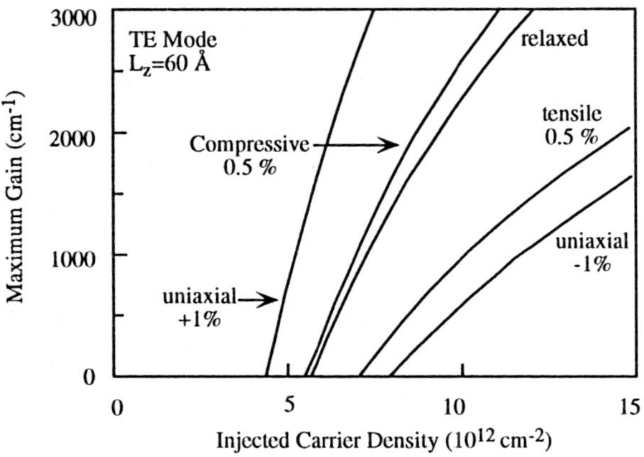

Fig. 12.29. Gain in a 60 Å active-layer GaN/$Al_{0.2}Ga_{0.8}N$ structure without strain, with $\pm 0.5\%$ compressive biaxial strain, and with 1% uniaxial compressive in y-direction, and tensile strain along the x direction. After [12.22]

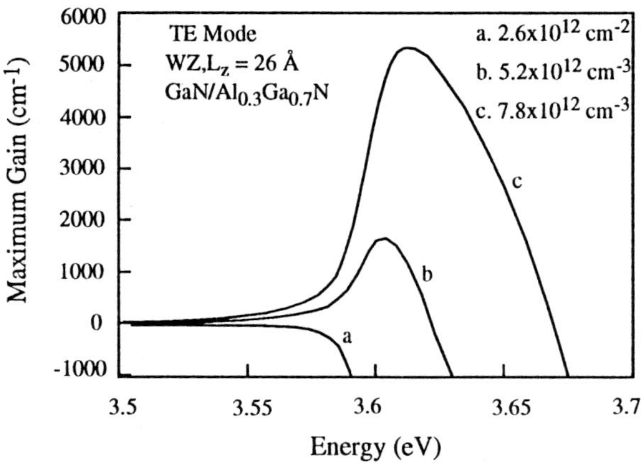

Fig. 12.30. Spectral gain in a 26 Å strained Wz GaN/$Al_{0.3}Ga_{0.7}N$ quantum well structure for injected-carrier volume densities of $3\cdot10^{19}$ cm^3, $2\cdot10^{19}$ cm^{-3}, and $1\cdot10^{19}$ cm^{-3} for TE polarization, as the gain for the TM mode is rather weak. The strain caused by the lattice mismatch between the barrier and the active material has been taken into consideration. After [12.35]. Reprinted with permission from IEEE

ciently large to make a notable difference in the threshold current. Uniaxial strain in the c plane only, if it were possible, would cause a notable increase in the gain due to reduced DOS at VBM through anisotropic splitting in the k_x-k_y plane. The literature is fairly rich in terms of gain calculations for both Wz and ZB quantum-well structures. The gain spectra in a 26 Å strained Wz GaN/$Al_{0.3}Ga_{0.7}N$ quantum-well structure for injected-carrier volume densities of $1\cdot10^{19}$ cm^{-3}, $3\cdot10^{19}$ cm^{-3}, and $1\cdot10^{19}$ cm^{-3}, corresponding to 2.6, 5.2 and $7.8\cdot10^{12}$ cm^{-2}, are plotted in Fig. 12.30 for the

TE mode. The gain maximum in this particular example is higher than that calculated by *Suzuki* and *Uenoyama* [12.22] for comparable injection levels. The discrepancy may have originated from the different band-structure parameters employed. The strain caused by the lattice mismatch between the barrier and the active material has been taken into consideration [12.35].

12.7.5 Gain in ZB QW Structures Without Strain

GaN grown in the [111] direction promotes the Wz polytype. It is therefore natural to consider a ZB quantum well that is grown along the [001] direction with the resultant quantization along the same direction. In attempting to calculate the gain, *Suzuki* and *Uenoyama* [12.22] assumed this orientation for ZB GaN. As indicated on numerous occasions, quantization removes the degeneracy at VBM of a ZB structure (**Γ point**) and separates the HH and LH bands. Band mixing is affected in such a way as to reduce the HH mass, even below that of the LH, causing the DOS at VBM to drop, which would by itself lead to a reduction in the transparency density or the current. However, the relatively small LH and SH separation in GaN leads to increased coupling between the two bands with the adverse effect of an increased mass and a density-of-states causing the transparency current in ZB to be larger than that in a Wz structure. This enhanced coupling also manifests itself by the emergence of a noticeable TM mode. Figure 12.31 displays the maximum gain in a 60 Å active layer ZB GaN/$Al_{0.2}Ga_{0.8}N$ structure where the dominance of the TM mode is clearly seen. Compared to Fig. 12.26, which represents the Wz polytype, the gain is somewhat smaller and suggests the inevitable conclusion that these calculations do not support the rationale for developing a ZB polytype for laser applications. In

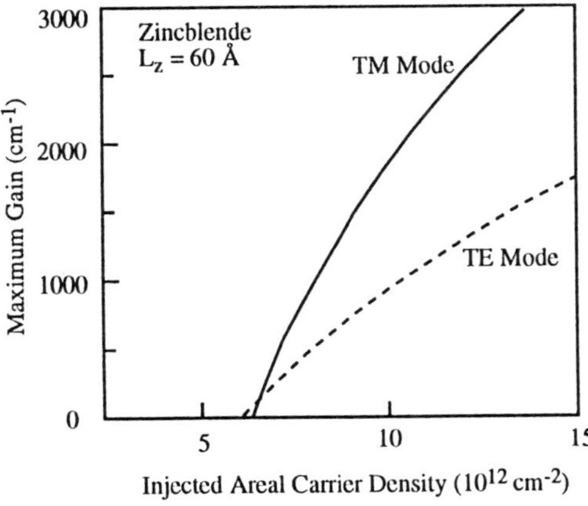

Fig. 12.31. Optical gain in an unstrained 60 Å active-layer ZB GaN/$Al_{0.2}Ga_{0.8}N$ single quantum wells which is loosely referred to as quantum well. The *solid* and *dashed* lines correspond to TM and TE modes, respectively. After [12.22]

fact, unlike a Wz polytype, there has been no report of stimulated emission in ZB GaN.

12.7.6 Gain in ZB QW Structures with Strain

It is a well-known phenomenon that strained quantum wells of the GaAs and InP systems exhibit much reduced DOS in the valence band, a property that has successfully been exploited in the form of strained QW lasers [12.4]. Following the same path, the effect of compressive and tensile biaxial strains on the maximum gain has been considered by many researchers [12.20-21, 31, 35-39]. Shown in Fig. 12.32 are the maximum gain for TE and TM modes versus the injected areal-sheet carrier density in a 60 Å active layer

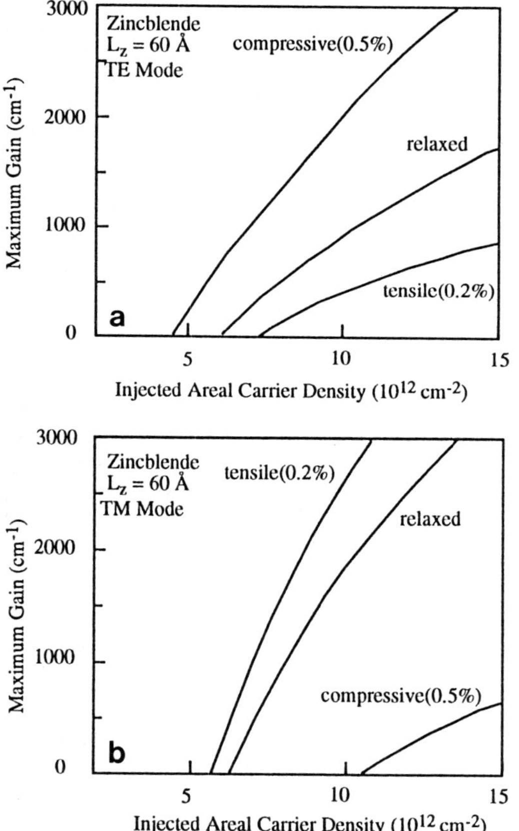

Fig. 12.32. Optical gain in a strained 60 Å active-layer ZB GaN/Al$_{0.2}$Ga$_{0.8}$N single quantum wells with (a) and (b) corresponding to TE and TM modes, respectively. Strain free, 0.5% compressive strain and 0.2% tensile strain cases are shown. After [12.22]

424

ZB $GaN/Al_{0.2}Ga_{0.8}N$ single well structure which is relaxed and either under compressive or tensile strain. The compressive strain strongly reduces the TM mode gain while enhancing the TE mode gain in relation to the relaxed case. On the contrary, a tensile biaxial strain causes the reverse to occur. In short, this is a consequence of compressive strain lifting of the HH and spin-orbit hole band (SH), the two uppermost states (referred to as **X** and **Y** states), well above the LH state (referred to as the Z state) which supports the TE mode. On the other hand, the tensile strain lifts the LH state (the Z state or the p_z state) above the HH and SH states (X and Y states or p_x and p_y states), which proceeds and supports the TM mode at the detriment of the TE mode.

(a) Pathways Through Excitons and Localized States

Excitons

Wide-bandgap semiconductors such as GaN- and ZnSe-based ones have large exciton binding energies of nearly 20 meV. They may not dissociate at the carrier concentration levels that cause the semiconductor to reach transparency. Naturally, the question then arises as to whether or not excitons would be involved in the lasing process taking place in these semiconductors. In a two-dimensional system, the exciton phase-space filling density was given by *Ding* and *Nurmikko* [12.40] as

$$n = 1/\pi a_B{}^2 \tag{12.64}$$

which for GaN with an exciton Bohr radius of about 28 Å, see (12.59), leads to a concentration of $3 \cdot 10^{12}$ cm^{-2}. This is comparable to the transparency carrier concentrations that one predicts for lasers fabricated in ideal GaN. Because of this high dephasing density, one can not automatically rule out excitonic processes in GaN-based lasers. Moreover, the exciton binding energy increases in quantum wells making excitons all the more important in such structures. It should be mentioned that the lack of high-quality heterostructures prevents such lasers from being fabricated at this point in time. It is not too early to explore the exact pathway(s) leading to exciton participation in the lasing process and the extent of it. Calculations have already been performed to determine the gain enhancement due to Coulombic (excitonic in this case) processes which improve the matrix element for inter-band transitions. In other words, the recombination efficiency is enhanced.

    Chow et al. [12.36] estimated the Coulombic enhancement to be about 30 % in ZB GaN-based laser structures. *Uenoyama* [12.41] presented a three-level picture involving excitons where the point has been made that the excitonic enhancement is present but the supply of electrons and holes for the excitonic transitions must involve coupling to the two (conduction and valence) bands. Since experience with GaN-based lasers is very limited, the effort expended in the ZnSe realm, due to its similarities, will be utilized to provide a basis for the discussion of any exciton participation.

While the detailed expressions which account for the exciton participation can be found in [12.23, 36, 41], let us first phenomenologically consider the ideal case in which there exist conduction and valence bands, and a well defined excitonic level at some 20 meV below the conduction band (in the test semiconductor GaN). The excitons are states bound through Coulombic interaction between holes and electrons. For the sake of argument, let us further assume that the band edge of the semiconductor is abrupt and that the exciton-linewidth broadening is negligible. This simplified picture represents an ideal three-band or three-level lasing model in which the carriers in the conduction and valence bands, placed by the injection of carriers and/or absorption of optical radiation (Fig. 12.33), provide fuel for the excitons. If the coupling between the two bands and the exciton state is strong, the supply of electrons and holes for excitonic transitions will be sufficient. In this case, the **quasi-Fermi level separation** needs to be larger than only the bandgap minus the exciton binding energy. This is a condition rendering the semiconductor transparent to the radiation emanating from the excitonic recombination.

As mentioned on several occasions, the excitonic recombination is more efficient than that of electrons and holes in the conduction and valence bands, respectively, due to the Coulombic interaction and the resulting spatial proximity of the carriers with opposite charges. This satisfies the condition that the recombination rate at the excitonic level be larger than the band-to-band recombination, to achieve the required population inversion. In the absence of broadening, this picture illustrates an ideal three-level laser. It is obvious that in this ideal picture, a semiconductor laser based on excitonic recombination would offer a very low threshold current. It is also clear that lasing that would occur at the excitonic level would be red shifted with respect to the band edge by an amount equal to the exciton

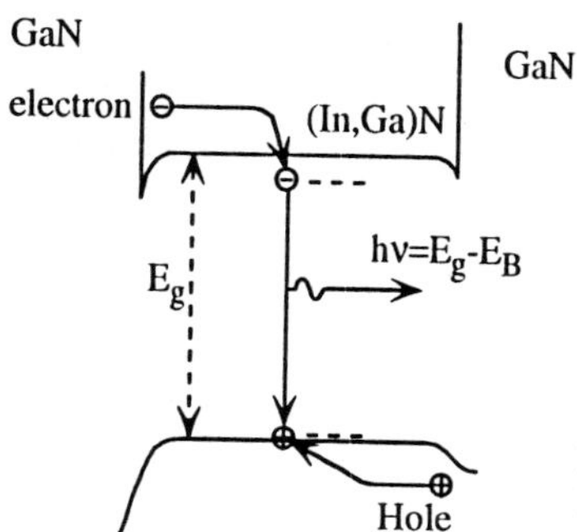

Fig. 12.33. Schematic representation of a three-level laser involving an excitonic state, and conduction and valence bands

binding energy. In reality, however, homogeneous and inhomogeneous broadening coupled with thermal broadening cause dispersion of the excitonic transitions. This is particularly so in an alloy such as InGaN. It simply means that the ideal picture described above, while shedding much needed light, would not alone be sufficient to describe the mechanisms involved. In addition, exciton lasing without a clearly defined third level is unrealistic and is a point that was advanced by *Uenoyama* [12.41] and *Cingolani* [12.42]. Realistically, weakly localized states are much more likely to be the mechanism for lasing in the GaN system, as will be discussed below.

Linewidth broadening in the form

$$\Gamma(T) \; = \; \Gamma_0 + \Gamma_1(T) + \frac{\Gamma_{LO}}{\exp(\hbar\omega_{LO}/kT) - 1} \tag{12.65}$$

has been used to describe the excitonic dispersion in group II-VI semiconductor-based lasers by *Ding* et al. [12.40]. An expression similar to this was employed to describe the energy-band dispersion in GaN by *Petalas* et al. [12.43]. For bandgap broadening, the first term depicts the temperature-independent broadening such as that caused by compositional and/or strain-induced inhomogeneities, impurity and surface scattering of carriers, electron-electron interaction and Auger processes. The second term is the homogeneous-broadening term resulting from carrier scattering due to acoustic phonons. The third term depicts scattering due to LO phonons, a phenomenon which is dominant in polar semiconductors such as GaN at high temperatures. However, due to the large LO-phonon energy of 91 meV in GaN, the impact of the third term is diminished somewhat. As for exciton linewidth broadening, the description above applies with the exception that the term *exciton* must be substituted for carriers and electrons. Depending on the quality, the temperature-independent broadening term in GaN for the bandgap parameter could be as high as 50 meV with the figure expected to be much larger for InGaN and other alloys of nitrides. *Ding* et al. [12.40], through a fitting to the experimentally observed line broadening due to excitons in ZnCdSe-based quantum wells, reported the Γ_{LO} term to be 36 meV. This value was also used to predict a larger broadening. Reduced LO-phonon scattering or exciton dissociation rates by LO phonons lower the broadening. Broadening caused by the aforementioned processes coupled with occupation factors and DOS distributions in the conduction and valence bands may collectively cause the gain spectrum to have its maximum at an energy different from the excitonic state under discussion. In such a case, a blue shift may occur.

Localized States

The net result of the discussion of the preceding subsection is that in GaN and related materials the poor quality of the layers would most likely rule out the three-level laser picture that involves excitons. Further support for this hypothesis is provided by the fact that the injected-carrier concentration for the transparency is more than one order of magnitude larger than concentrations at which the screening processes would rule out Coulombic interaction between carriers. It should be mentioned that the lifetimes used in calculating the transparency current may be overestimated. In short, it is very likely that some other pathway is involved in GaN-based lasers. It is also very probable, and may, in fact, be supported by experiments, that localized states may be involved. In such as case, a localized state below the bandgap, as shown in Fig. 12.34a and b for the case of compositional and/or strain inhomogeneities and Fermi-level fluctuations, the latter may be coupled to the conduction band where the electrons are supplied to the localized states. Compositional and strain inhomogeneities as well as localized defect states all may occur in the same structure. Again, for lasing to occur, the quasi-Fermi-level separation must exceed the energy separation between the localized states and the valence-band maximum, and the recombination rate must be larger than the band-to-band recombination rate. In alloys, the line between the issues of localized states, and compositional fluctuations and other inhomogeneities may be fuzzy. In this case, the effective band edge

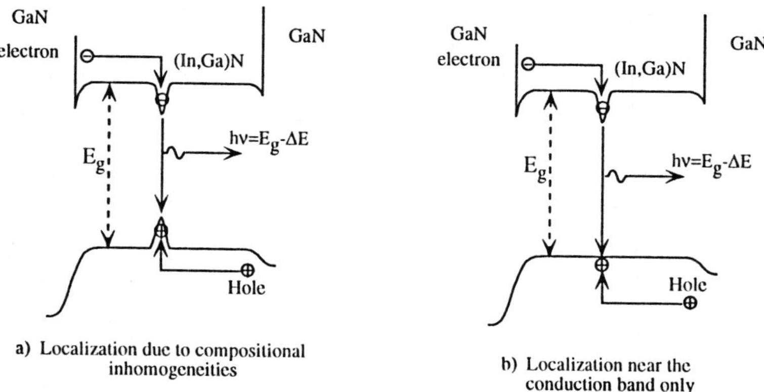

a) Localization due to compositional inhomogeneities

b) Localization near the conduction band only

Fig. 12.34. Schematic representation of a three-level laser involving localized states caused by (**a**) compositional and/or strain inhomogeneities, and (**b**) localized states caused by, for instance, defects. In general though, the localized state can be a deep donor or a deep acceptor-like state, (**a**) exemplifies the case where the localized state is caused by strain and/or compositional inhomogeneities reducing the local effective bandgap. Local tensile strain reduces the effective bandgap and if and when accompanied by a juxtaposition compressive strain, those parts under compressive strain would have an increased bandgap. Compositional and strain inhomogeneities as well as localized defect states may all occur in the same structure

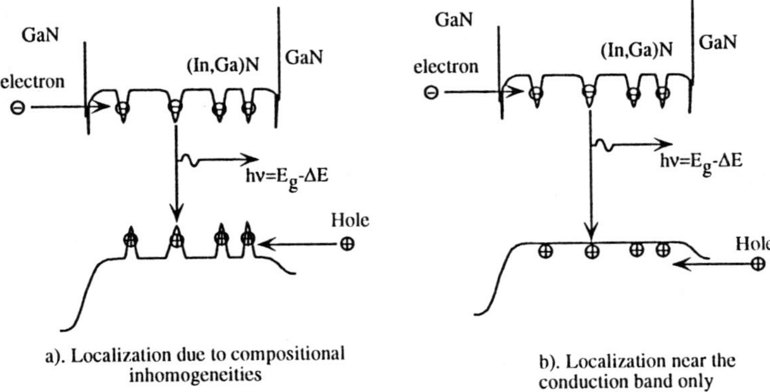

a). Localization due to compositional inhomogeneities

b). Localization near the conduction band only

Fig. 12.35a,b. Schematic interpretation of a case where the localized-state concentration is sufficiently high for the substantial overlapping of those states which cause band tails states. Assuming similar properties in the barrier materials, the carrier transport would be due to defect assisted tunneling

of the conduction band and possibly the valence band, would extend into the gap. The case of a high-density localized state is depicted in Fig. 12.35. If the inohomogeneities are wide spread, band-tail states would result and cause a soft band edge in the absorption spectrum and a red shift (**Stokes shift**) in lasers (Fig. 12.35). If the recombination rate between the localized states is not much larger than that for the band-to-band transitions and/or the coupling between the conduction- and valence-band states is not sufficiently large to not limit the carrier recombination between the localized states, lasing at the localized level(s) will be curtailed. The lasing energy will be shifted towards that of the bandgap if everything else is equal. In the case of substantial overlap between these localized states, band-tail states would appear (Fig. 12.36).

Localized-states arguments have implications on the current conduction as well. If the localized-states density in the barrier layers is also high with band-tail states resulting, it follows that the carrier-transport paths may not be limited to conduction and valence bands. Additional paths through these band-tail states / gap states could take place, which lead to photon assisted tunneling. Preliminary experimental current-voltage characteristics suggest that the current conduction in p-n junctions is through photon-assisted tunneling, as discussed in Chap. 11.

Since research on GaN-based lasers is in an embryonic state and seriously complicated by inhomogeneities in the InGaN active layers employed, the ZnSe lasers, having been developed much farther than their GaN counterparts, will be used as examples to debate the likely mechanisms involved in lasing, and to discount others. *Ding* et al. [12.40], by means of optical pumping experiments in ZnSe-based heterostructures, argued in favor of

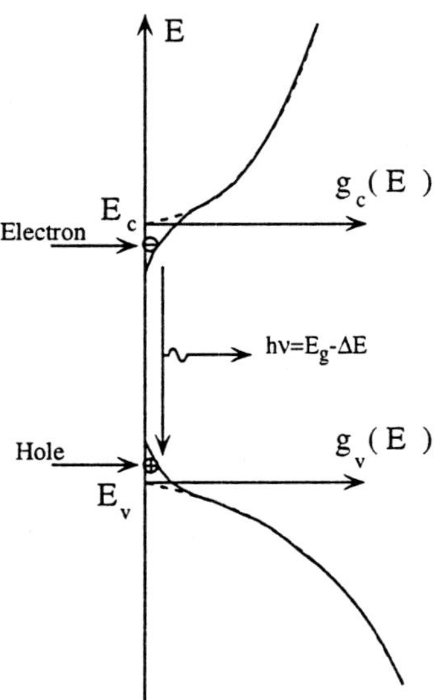

Fig. 12.36. Schematic representation of band-tail states due to large concentrations of localized states

excitons for the gain mechanism. They calculated the gain by employing a phenomenological exciton-driven gain mechanism which is in agreement with experiments conducted at 77 K. On the other hand, *Diessel* et al. [12.44] made a case for the exciton recombination in processes that lead to lasing but not partaking in the actual lasing process due to exciton bleaching. The main controversy appears to have its origins in the observation of excitonic peaks in photoluminescence-excitation experiments with pumping intensities well in excess of the threshold intensity. However, *Diessel* et al. [12.44] argued that this observation is no more than an evidence for carrier generation via excitonic-absorption pathways which are accessible during the leading edge of the excitation pulse. According to *Diessel* et al., pulse probe experiments, where a pulse beam is used to prime the semiconductor while a probe beam gauges the optical activity at an excitation level engendered by the pump beam. It interrogates the excitonic activity more independently of the carrier-density dynamics during the leading edge of the pump beam. Such experiments led to exciton bleaching which justifies the argument against the gain mechanism being solely governed by excitons, and in favor of an electron-hole plasma. This conclusion is consistent with the independent report by *Cingolani* et al. [12.45]. Recently, *Cingolani* et al. [12.46] performed pump and probe experiments in high magnetic fields, which support the exciton bleaching thesis. It has been observed that shallow quantum wells, those with low Cd mole fractions, have reduced the enhancement of the exciton binding energy. They exhibit the typical free-carrier fueled lasing which occurs after excitons have been bleached and the shift of the lasing line with magnetic field follows the Landau shift; this is endemic of free carriers. Wells with high Cd mole fractions whose exciton binding energies are larger than the LO-phonon energy, lase at power densities where the exciton resonances are not bleached out. The lasing-line shift follows the diamagnetic shift which is very small. The set of samples contained well widths of 3 to 20 nm and Cd mole fractions in the range of 10 to 30 %. The results can satisfactorily be described by a self-consistent many-body model based on the mass-action law which includes exciton and free-carrier energy renormalization. The excitons and free carriers have been described by a phase diagram analogous to the dissociation of a diatomic gas. The relative density of the two phases depends on the stability of the excitons. In short, both exciton and free-carrier fueled lasing can be present depending on the conditions, while the primary factor is the strength of the exciton binding energy. However, what still clouds the picture is the localized states which can present signatures similar to excitons. As the Cd mole fraction is increased, so is the likelihood of inhomogeneities and the impact of localized states.

12.7.7 Measurement of Gain in Nitrides

Calculation of the gain is rather convoluted in that it requires an accurate knowledge of key parameters as well as of the critical mechanisms involved. For nitrides, the respective mechanisms and many of the key parameters have yet not been determined experimentally. They must therefore be calculated. It is thus imperative that gain measurements be made to provide the basis for obtaining an insight into the operation of nitride-based injection lasers even before they become widely available and to provide the needed calibration for the calculations. At the onset, it should be stressed that high-quality laser material is required to measure the pertinent parameters with the needed degree of confidence. Ironically, as is always the case, good material is not available in the early phases of development when it is needed the most. Since carrier injection is a major impediment, particularly early on, due to the lack of a good p-n junction and the many leakage pathways, optical-excitation / pumping is generally employed to glean an insight into the processes in new semiconductors, such as nitrides, under high excitation densities.

(a) Gain Measurement via Optical Pumping

The optical pumping experiment provides an excellent environment to investigate the genesis of stimulated emission in semiconductors, particularly if pump and probe experiments are carried out. Lack of metallization, an easy accessibility of the active layer with optical pump and probe beams, the absence of absorbing contacts and the ease with which the sample lattice temperature can be changed, are among the reasons for the attractiveness of the method. Moreover, optical-pumping experiments do not require the formation of good p-n junctions and complicated fabrication procedures, and as such are precursory to the current injection-laser development. One of the methods used for studying the optical gain is the stripe excitation technique which means that the light excitation source is focused on a stripe region. If the sample is excited sufficiently, spontaneously emitted light traveling along the excited stripe gets amplified by the stimulated emission. The intensity of light in such a case is then given according to *Frankowsky* et al. [12.47] as

$$\frac{dI}{dz} = \Gamma g(h\nu)I + \beta'\Gamma r_{sp}(h\nu) \tag{12.66}$$

where $r_{sp}(h\nu)$ and β' are the spontaneous emission rate and spontaneous emission factor, respectively. The other terms have their usual meanings. When solved for a stripe length of L with the boundary condition $I(0) = 0$, we obtain

$$I(L) = \frac{\beta'r_{sp}(h\nu)}{g(h\nu)}(e^{\Gamma gL} - 1) \ . \tag{12.67}$$

The spectrum of the amplified spontaneous emission can be measured for various values of L and used in conjunction with (12.67) to determine the gain. This can be repeated for various wavelengths and pump intensities to obtain the gain spectra as well as the intensity dependence of the gain.

Of course, the nature of the transitions which causes the optical gain must be known before an earnest effort can be made to calculate it. To this extent, experiments and theory go hand in hand to examine the groundwork that has to be laid. Unfortunately, GaN technology is not yet at a point of perfection necessary to perform the definitive and probing experiments in this regard. While we await improved GaN-based heterostructures to become available so that appropriate experiments leading to the genesis of lasing can be performed, a good deal of information can be gleaned from ZnSe-based lasers. Based on what little data are available for GaN-based lasers, it appears that there is a good deal of similarity between ZnSe- and GaN-based lasers. Despite the raging controversy as to the genesis of the lasing mechanism in ZnSe-based lasers, there is a lot to be learned from the results of various investigations. In this vein, the results of an optical-pumping experiment are displayed in Fig. 12.37. The absorption and emission spectra along with the pump-beam spectrum concerning a ZnCdSe active-

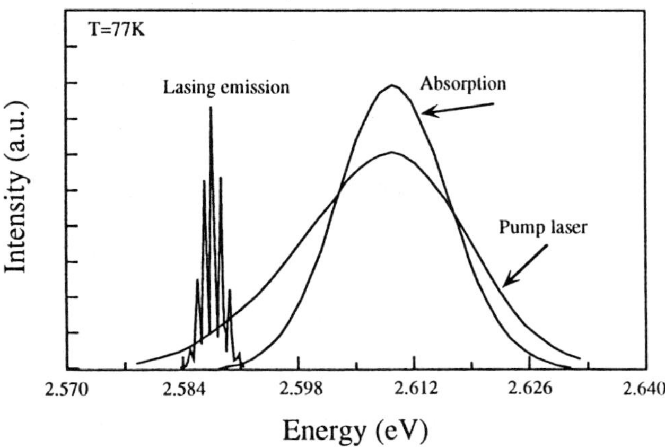

Fig. 12.37. Optical absorption spectrum near the n=1 C to HH exciton resonance for a 6 period and 30 Å ZnCdSe MQW laser structure at 77 K. The pump laser spectrum and the laser emission are also displayed with a red shift of 60 meV. After [12.40]

layer MQW laser structure show a red shift of about 60 meV [12.40]. Though the data are sketchy at this point in time, GaN-based lasers also red shift with respect to the absorption spectrum, as will be discussed in detail in Sect. 12.12.

Optically pumped stimulated emission has been achieved in GaN almost from the inception of epitaxial GaN in the early 1970s [12.48] to the present day with the lattice-temperature rising. The threshold optical power required for stimulated emission has decreased to the point where 40 kW/cm^2 is presently sufficient to achieve stimulated emission at room temperature [12.49].

Frankowsky et al. [12.47] employed optical pumping to investigate the gain in GaN and InGaN heterostructures. GaInN/GaN double heterostructures and quantum wells employed consisted of a 500 nm GaN buffer layer, a 25 Å thick GaInN active layer, and a 150 nm top confinement layer. The GaN/AlGaN double heterostructures had a 500 nm $Al_{0.1}Ga_{0.9}N$ buffer layer, a GaN-active layer, and a 150 nm AlGaN top layer. The modal gain shows strong polarization of the TE mode (electric field in the plane of the sample) being the dominant mode due to the Wz nature of the crystal. This results from symmetry-induced splitting of the valence band and the almost exclusive coupling of the heavy hole to the TE polarized light.

Optical gain as a function of pump power (Fig. 12.38) exhibits a single maximum close to the bandgap of the ternary InGaN that is accompanied with an apparent red shift of the gain maximum with increasing pump power. The shape of the gain spectra is characterized by a sharp reduction at the higher energy side due to population statistics. At the lower-energy side, the gain shows a sub-linear plateau which deserves further attention. This is

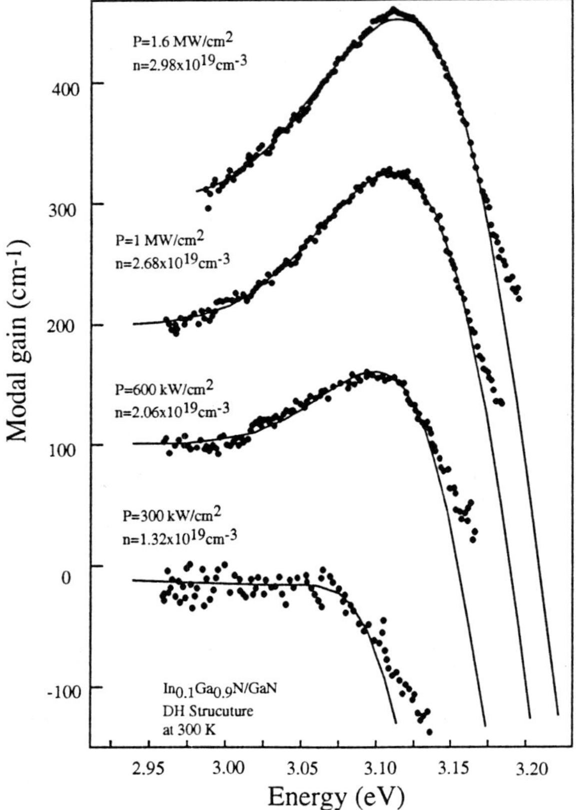

Fig. 12.38. Optical gain measured at optical excitation levels of 300, 1000 and 1600 kW/cm^2 in an InGaN/GaN DH structure. After [12.47]

indicative of the fact that the transparency carrier concentration is very high and unlike that displayed in Fig. 12.37 where the plateau observed in InGaN at the low-energy side is not apparent. Using available and/or estimated effective electron (0.22) and hole (1.64), estimates vary between 0.7 and 2 masses and an inter-band momentum matrix element (9.3eV), the injected-carrier concentrations were estimated, knowing the gain, while letting the confinement factor and the broadening factor be adjustable parameters. The resultant carrier concentrations were deduced to be between 1.3 and $3 \cdot 10^{19}$ cm^{-3} for excitation levels of 300, 1000 and 1600 kW/cm^2. The optical gain spectra for a GaN/AlGaN double-heterostructure sample, the active layer of which is relatively better known, with $L_z = 100$ nm was investigated at three different pump-power levels. In contrast to the GaInN/GaN structure, two distinct gain maxima at ≈ 3.33 eV and about 3.24 eV were observed. Within the range of pump-power levels accessible, only the amplitude of the lower-energy peak was found to depend strongly on the

434

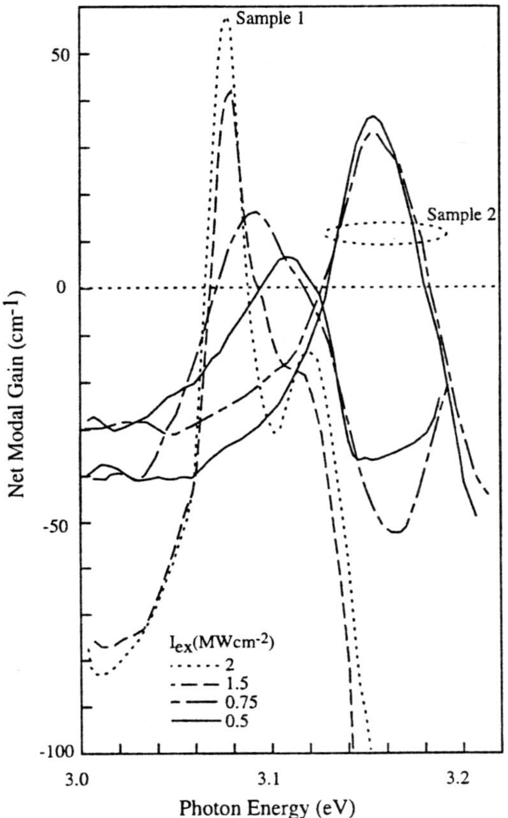

Fig. 12.39. Optical gain measured in two Nichia InGaN/GaN DH laser structures. The origin of the absence of gain spectral broadening with the optical pump intensity is not clear. One structure exhibits a narrower gain spectrum (sample 1) than the other (sample 2). Purportedly, sample 1 has more InGaN compositional fluctuations. Courtesy of S. Nakamura, Nichia Chemical Ltd

pump power. A comparison of the gain and the absorption spectra indicates that the higher-energy gain peak appears within the low-energy tail of the absorption spectrum. A suggestion has been introduced that the lower-energy peak is due to exciton-LO-phonon scattering which is about 90 meV and is close to the LO-phonon energy. The weak structure observed, about 90 meV above the localized-exciton peak, has been interpreted as an anti-Stokes exciton-LO-phonon gain peak. It supports the case of excitonic gain due to localization at least in GaN. The gain spectra have also been measured by the optical-pumping method in representative Nichia InGaN/GaN DH laser structures (Fig. 12.39). The reason for the absence of gain spectral broadening with an increasing optical pump intensity is not yet clear. One of the structures exhibits a narrower gain spectrum (sample 1) than the other (sample 2). Purportedly, sample 1 has more InGaN compositional fluctuations.

To ascertain whether localization prevails, *Cingolani* et al. [12.49] investigated the stimulated-emission peak in a separate GaN/AlGaN confinement MQW heterostructure as a function of magnetic field and tempera-

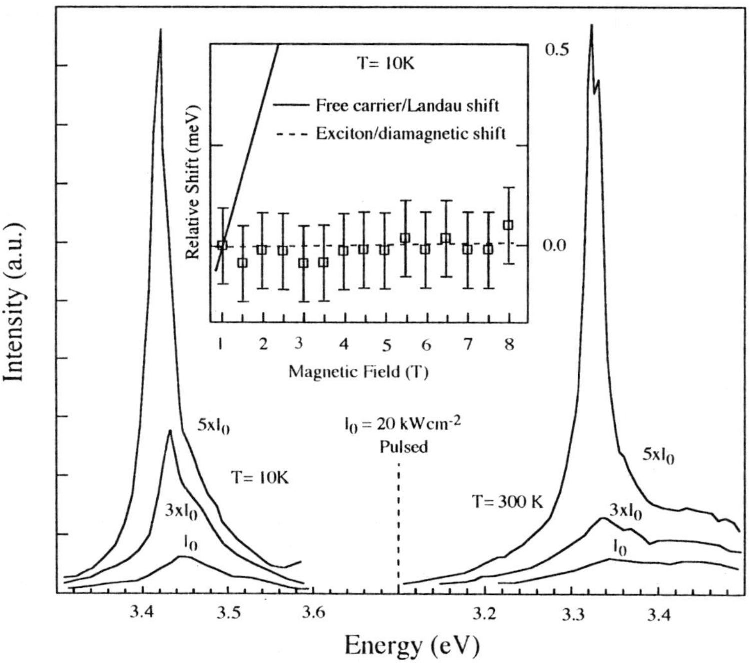

Fig. 12.40. Shift of the stimulated emission in a high magnetic field. The continuous line represents the calculated excitonic diamagnetic shift (very small). The dashed line is the Landau shift expected for free-carrier recombination Inset: stimulated emission band vs. magnetic field from 0 to 6 Tesla. After [12.49]

ture. In parallel, photoluminescence and photoluminescence-excitation measurements were conducted with appropriate line-shape fitting to determine and identify the peaks in the spectra. The peak positions will be contrasted later to the lasing line in an effort to determine if there is a red shift and if so to what extent. As shown in Fig. 12.40, the broad spontaneous-emission band generated by pulsed photo-pumping evolves from the low-energy tail of the CW-PL spectrum, which suggests a dominant localized-exciton recombination. At threshold a sharp spike appears at 3.34 eV (300 K) and grows super-linearly with an increasing photo-generation rate. On the average, the lasing line is red shifted by about 110 meV below the $n = 1$ gap of the QW, which is too large to be accounted for by bandgap renormalization at threshold, and is suggestive of population inversion occurring at the low-energy tail of an inhomogeneously broadened density of states, i.e. **lasing from localized states**. To test this hypothesis further, the sample was subjected to a high magnetic field. The recombination mechanism can be identified by exploiting the difference in the diamagnetic shift of the free-carrier states (linear in a field with about 0.3 meV/T blue shift) and the excitonic states (very small and quadratic in field, with about $0.1\,\mu\text{eV/T}^2$ blue shift). The results of this experiment are depicted in Fig. 12.40, where the negligible shift of the emission line (dots) is displayed in comparison to the calculated exciton diamagnetic shift (continuous line) and the free-carrier shift (dashed line). The data support the hypothesis that lasing occurs through inhomogeneously broadened localized states in the sample under investigation. Their origin can be attributed to strain inhomogeneities and/or defect states.

(b) Gain Measurement via Electrical Injection (Pump) and an Optical Probe

Optical pump-probe experiments used to deduce the gain spectra can be replicated in injection devices as well. In this case, the injected carriers render the semiconductor transparent, or push the active region of the semiconductor to even beyond transparency, while a wavelength-selective probe pulse gauges the gain/loss activity. If the semiconductor provides any gain at all, the probe pulse would be amplified with the degree being dependent on the magnitude of the gain. Complications are that one must be content with getting the probe pulse into the device without any unaccountable absorption external to the gain medium. The advantage is that an injection device which is much closer to an actual laser is used and renders the results much more applicable to injection lasers. Recently, *Kuball* et al. [12.50] applied this method to a light-emitting device with an InGaN active layer. As is commonly done, the absorption spectrum was obtained by using the device as a detector where the dependence of the photo-current on the incident light with varying wavelength would represent the functional form of the absorption spectrum. For the particular InN mole fraction, the absorption peak in a single pass occurred at 3.3 eV, and indicated an absorption coefficient of $(5\pm0.5)\cdot10^4$ cm^{-1} at this wavelength. Though the structure contained five 25 Å quantum wells, no confinement was evident and the structure was characterized by a significant low-energy tail. With the help of absorption as calibration for an absolute measurement, the gain/loss spectrum of the device under forward injection at 233 K was deduced, as shown in Fig. 12.41, along with the superimposed EL spectrum. Figure 12.42 depicts the dependence of the maximum differential gain at the peak of the differential absorption on current density. At low injection levels there is what appears to be a residual plateau as the device reaches transparency. At higher injection levels, the expected linear behavior is observed before saturation sets in, which at this point is most likely due to thermal effects as opposed to intrinsic saturation of the joint density of states. At a current density of $J = 3.5$ kA/cm^2, the single-pass total differential gain is about $6\cdot10^{-3}$ leading to a gain coefficient of about 4800 cm^{-1} (the total active layer thickness of 125 Å multiplied by the gain per unit length yields the total single-pass gain).

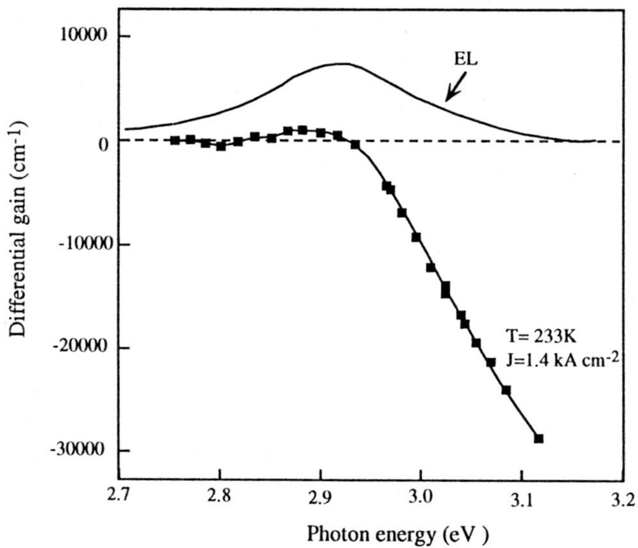

Fig. 12.41. Differential absorption at 233 K at a current density of 1.4 kA/cm^2 with the *solid line* being a guide to the eye (*lower curve*). The gain spectrum at the same device of the same current density (*upper curve*). After [12.50]

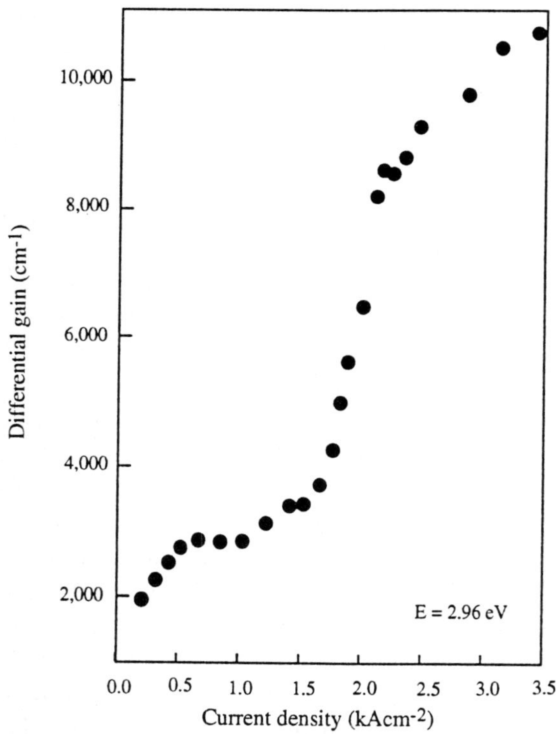

Fig. 12.42. Maximum differential absorption coefficient/gain coefficient as a function of current density. After [12.50]

A point of concern is the large linewidths and the spectral position of the gain relative to the joint density-of-states (absorption) and the EL spectra. The gain bandwidth is estimated at 100 meV versus about 30 meV in ZnCdSe. With increasing injection, the gain emanates in the low-energy tail of the rather graded-absorption edge. This is similar to observations with injection lasers of *Nakamura* et al. [12.50, 51] which are discussed in Sect. 12.12.

12.8 Threshold Current

The gain can be related to the spontaneous emission rate as [12.21, 38]

$$R_{sp}(E_{eh}) = \frac{g'(E_{eh})}{\pi^2 n_r^2 E_{eh}^2 c^2 \hbar^3} \ . \tag{12.68}$$

Once the spontaneous emission rate is calculated versus energy, the radiative current supporting the spontaneous emission rate (12.68) can be expressed as

$$J_{rad} = q \int_{E_{eh}}^{\infty} R_{sp}(E)\,dE \quad [A/cm^3] \ . \tag{12.69}$$

In the case of a two-dimensional (2D) system this expression should be replaced by

$$J_{rad} = qL_z \int_{E_{eh}^{i,j}}^{\infty} R_{sp}(E)\,dE \quad [A/cm^2] \ , \tag{12.70}$$

where L_z is the total quantum-well thickness.

One approach for finding the gain and current in an injection laser is to start with a certain concentration of injected carriers and calculate the quasi-Fermi levels. They pave the way to determine the occupation probabilities. One can then calculate the gain as a function of energy from (12.43, 50) with the prerequisite, of course, that we have an accurate knowledge of the lasing mechanism, the band structure, and the relevant matrix elements associated with the active layer. From the gain expression, the spontaneous emission rate and thus the radiative current can be calculated for that particular gain. The current should be consistent with the injected-carrier concentration assumed at the beginning of this procedure. The radiative current density must be divided by the internal quantum efficiency to get the current density.

12.9 Analysis of Injection Lasers with Simplifying Assumptions

The calculations mentioned above, which were performed for GaAs lasers, together with experimental data culminated in the conclusion that the gain is linearly proportional to the injected-carrier concentration for a range of carrier densities and thus current, which is called the **linear region**. This holds reasonably well for bulk lasers where the density of states, and thus the gain, does not saturate. In quantum-well lasers, the absorption coefficient is a staircase function of energy. Nevertheless, the linear approximation has been shown to explain the observed results away from the extremes even in quantum wells. It is therefore very useful to the understanding and the diagnosis of semiconductor lasers.

The gain of a semiconductor laser in the linear region can be expressed as [12.9, 18]

$$g \; = \; A_0(n - n_{transp}) \; = \; \beta(J - J_{transp}) \; , \tag{12.71}$$

where n is the injected- carrier concentration and n_{transp} is the injected carrier concentration to render the semiconductor transparent, J is the current density, J_{transp} is the current density which causes the semiconductor to be transparent. A_0 and β are the differential gain and the gain coefficient, respectively.

At threshold, the net modal gain can be expressed as

$$g_{th} \; = \; A_0(n_{th} - n_{transp}) \; = \; \beta(J_{th} - J_{transp}) \; . \tag{12.72}$$

The injected-carrier concentration can be written as

$$n \; = \; \frac{g}{A_0} + n_{transp}$$

and at threshold $\tag{12.73}$

$$n_{th} \; = \; \frac{g_{th}}{A_0} + n_{transp} \; .$$

The injected-carrier concentration needed to reach threshold can be calculated by

$$n = \frac{1}{A_0}\left[\frac{\alpha_i}{\Gamma} + \frac{1}{2\Gamma L}\ln\left(\frac{1}{R_1 R_2}\right)\right] + n_{transp} \; . \tag{12.74}$$

The current density is given by

$$J = \frac{qd}{\tau_s}n = \frac{qd}{\tau_s}\left(\frac{g}{A_0} + n_{transp}\right) \tag{12.75}$$

where d represents the thickness of the pumped region, and τ_s the carrier recombination lifetime, which is related to radiative and non-radiative recombination lifetimes through

$$1/\tau_s = 1/\tau_{rad} + 1/\tau_{nonrad} \; . \tag{12.76}$$

At threshold, the current density reduces to

$$J_{th} = \frac{qd}{\tau_s}n_{th}$$

$$= \frac{qd}{\tau_s}\left\{\frac{1}{A_0}\left[\frac{\alpha_i}{\Gamma} + \frac{1}{2\Gamma L}\ln\left(\frac{1}{R_1 R_2}\right)\right] + n_{transp}\right\} \; . \tag{12.77}$$

When the quantum efficiency is less than one, that is $\eta = \tau_s/\tau_r$, the threshold current density can alternatively be written as

$$J_{th} = \frac{qd}{\eta\tau_r}n_{th} = \frac{qd}{\eta\tau_r}\left\{\frac{1}{A_0}\left[\frac{\alpha_i}{\Gamma} + \frac{1}{2\Gamma L}\ln\left(\frac{1}{R_1 R_2}\right)\right] + n_{transp}\right\} \tag{12.78}$$

or

$$J_{th} = \frac{1}{\beta}\left[\frac{\alpha_i}{\Gamma} + \frac{1}{2\Gamma L}\ln\left(\frac{1}{R_1 R_2}\right)\right] + J_{transp} \; . \tag{12.79}$$

Here, the transparency current is that calculated with the injected-carrier concentration and the radiative lifetime. In reality, this figure would increase by the inverse of the efficiency or the factor of radiative lifetime over the carrier lifetime.

From (12.74) one can relate the gain to measurable quantities such as the current density and the differential gain by

$$g = \frac{\tau_s A_0}{qd} J - A_0 n_{transp} \; .$$ (12.80)

The threshold gain can be obtained by replacing the current density with that at threshold.

12.10 Recombination Lifetime

The carrier recombination lifetime is an important parameter in that, if measured accurately, it gives one a qualitative assessment of recombination processes vis-a-vis radiative and non-radiative varieties. Moreover, as indicated in the preceding section, the lifetime is needed in relating the current to important parameters such as the transparency carrier density. The lifetime τ_s can be measured from the delay τ_d of the optical pulse emanating from the laser in response to a current pulse. This is determined from the rate equation which states that the time rate of change of carriers is equal to the rate of carrier injection minus the rate of recombination. Mathematically, we have

$$\frac{dn}{dt} = \frac{kJ}{qd} - R(n) = \frac{kJ}{qd} - \frac{\partial R(n)}{\partial n} n = \frac{kJ}{qd} - \frac{n}{\tau_s(J)}$$ (12.81)

where J, d, n, and τ_s represent the current density (assumed to be uniform across the active layer), the active-layer thickness, the injected-carrier concentration, and the carrier lifetime which is dependent on current. R(n) is the recombination rate, and k is the injection efficiency which ranges between 0 and 1 and accounts for the leakage current and the carrier overflow. The value of kJ relates to the part of the current that really participates in the radiative recombination.

Assuming a current-independent recombination time, the solution of (12.81) with the boundary condition n(0) = 0, meaning that the laser is not pre-biased, can be written as

$$n(t) = \frac{kJ\tau_s}{qd}(1 - e^{-t/\tau_s}) \; .$$ (12.82)

At the onset of laser oscillations, the carrier concentration reaches the threshold value

$$n_{th} = \frac{kJ_{th}\tau_s}{qd} \; .$$

(12.83)

Referring to the time needed to reach threshold and providing that k does not change with the injection current, one obtains

$$\tau_d = \tau_s \ln\left(\frac{J}{J - J_{th}}\right) .$$

(12.84)

In the case when the laser is pre-biased with the current density J_0, which is done to reduce τ_d, the delay between the electrical pulse and the onset of lasing, the delay time becomes

$$\tau_d = \tau_s \ln\left(\frac{J - J_0}{J - J_{th}}\right) .$$

(12.85)

Plotting the delay time versus the logarithm term gives a strait line from which τ_s can be extracted. Once τ_s is determined and knowing the thickness of the pumped region, the injected-carrier concentration can be calculated for a given current density above threshold from (12.75). The lifetime here is assumed to be constant in the range of injection levels which are above threshold, very high.

In general, the recombination lifetime is not independent of current. The current-dependent lifetime τ can be determined from the 3 dB point of the modulation bandwidth through [12.52-54]

$$F(\omega) = 10\log\left(\frac{F_0}{|1 + i\omega\tau|}\right)$$

(12.86)

for a series of injection currents. Here $F(\omega)$ and F_0 depict the frequency response and DC response of the laser power, respectively. The differential lifetime decreases with injection current as the excess carrier concentration increases, especially between a low-injection level and the level of transparency.

Similarly, the injected-carrier concentration can also be deduced from this method. When Auger recombination can be neglected, the lifetime can be expressed as

$$\frac{1}{\tau^2} = a^2 + \frac{4bkJ}{qd} \; .$$

(12.87)

From a plot of τ^{-2} vs. J one obtains a and bk/d. Again, from the dependence of the lifetime vs. injection current, the excess electron concentration can be calculated from

$$n(J) = \frac{k}{qd} \int_0^J \tau(J')\,dJ' \; .$$

(12.88)

Before we analyse lasers and attempt to determine pertinent parameters such as lifetime, efficiency, gain, and loss, a brief graphical description of the laser structure under examination is called for. Figure 12.43 sketches the layered structure of a laser with InGaN MQW active layers, fabricated by *Nakamura* et al. [12.51]. The layer thicknesses and mole fractions are chosen for carrier and light confinement while avoiding strain-induced cracks. When in the AlGaN cladding layers they exceed certain values cracks occur. On the other hand, thin and/or low AlN nitride mole-fraction

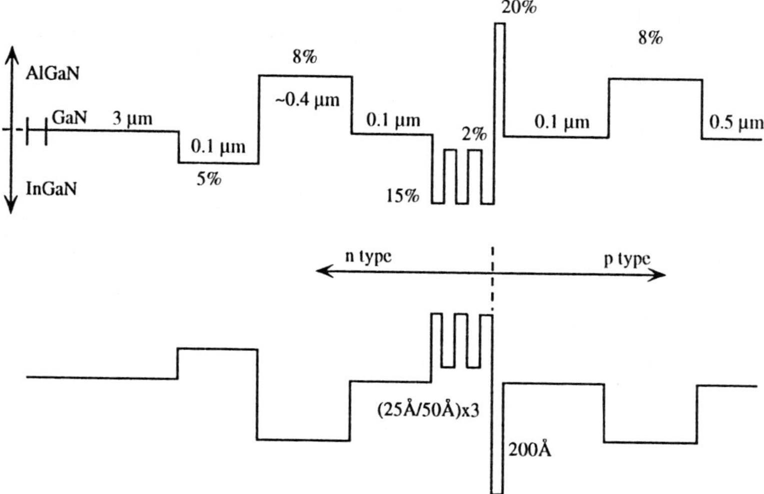

Fig. 12.43. Schematic diagram of the layer structure of an InGaN MQW active-layer Nichia laser structure. Courtesy of S. Nakamura, Nichia Chemical Ltd.

cladding layers are not so effective in guiding the radiation. One is then forced into making an undesirable compromise unless other means are found. As discussed in Sect. 4.15 use of GaN/AlGaN Short-period SuperLattices (SLS) that emulate bulk AlGaN can ameliorate this compromise in the designer's favor in that thicker layers and/or higher mole fractions can be employed if the bulk cladding layers are replaced with SLS layers. This idea has been implemented by researchers at Nichia Chemical (Fig. 12.44). Moreover, modulation doping, where the dopants are placed in the larger-bandgap material, was also employed in cladding layers to reduce the effective binding energy of the dopant impurities and to increase the carrier concentration over that obtainable with bulk layers. In fact, uniformly doped SLS layers would automatically lead to modulation doping and would benefit from carriers transferred from the larger-bandgap material as well as the those given off from the ionized impurities. The p-type layers benefit from this concept to a larger extent because of deep Mg acceptors. The carrier transport through the heterojunction barriers relies on tunneling. The structure in Fig. 12.44 also takes advantage of lateral growth whose details are discussed in Sect. 4.7.3 for defect reduction. Figure 12.45 presents an artistic view of an index-guided ridge laser. Figure 12.46a plots the light and voltage vs. current characteristics of such a Nichia InGaN laser at 20°C with a threshold current of 16 mA and a forward voltage of about 5 V. Figure 12.46b depicts the light and voltage vs. current characteristics of a Nichia ridge laser at 20°C with a threshold current of 50 mA and forward voltage of about 5 V relying on the deposition of an unpatterned c-plane

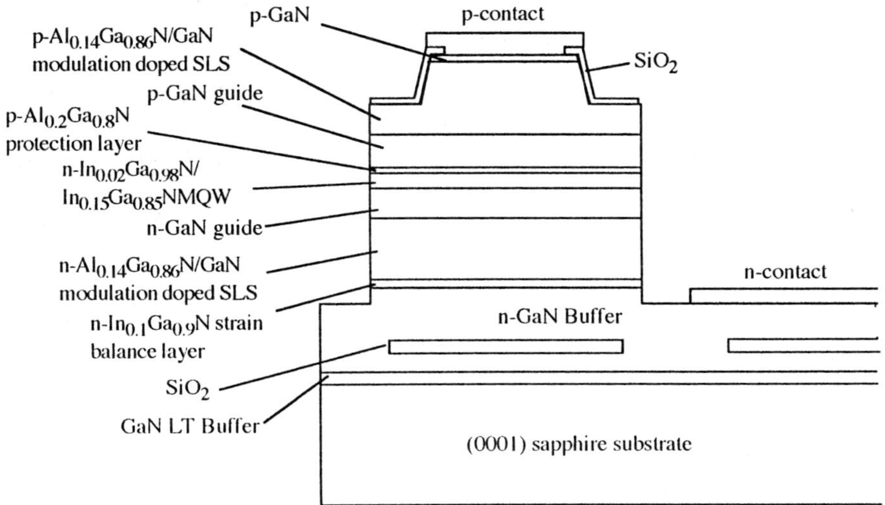

Fig. 12.44. Cross-sectional view of a Nichia InGaN/GaN/AlGaN DH laser structure which take advantage of short-period superlattice cladding layers and lateral growth. Courtesy of S. Nakamura, Nichia Chemical Ltd.

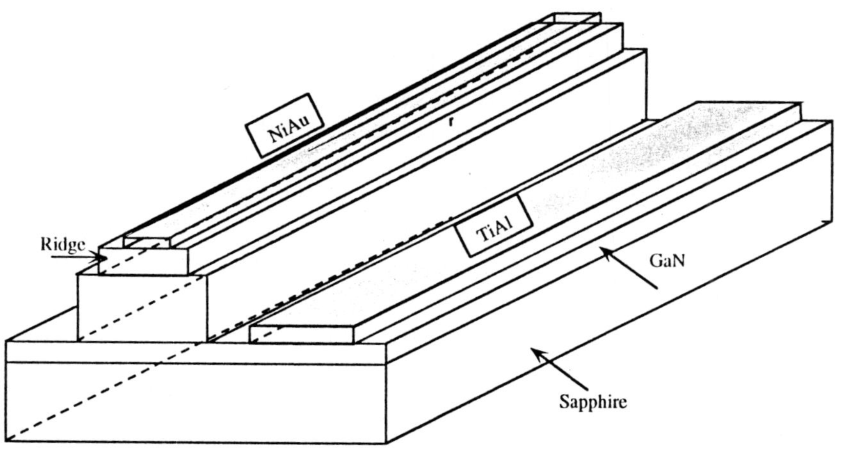

Fig. 12.45. An artistic view of an index-guided DH laser on a sapphire substrate indicating facets and electrode geometries

onto a sapphire substrate. Figure 12.46c displays the light and voltage vs. current characteristics of another Nichia ridge laser at $20°$, $60°$, $80°$, and $100°$C. The laser structure of Fig. 12.46c exhibits a room-temperature threshold current of 75 mA and forward voltage of $4 \div 5$ V.

Returning to the discussion of carrier lifetime, Fig. 12.47 depicts the delay time vs the natural log of the injection-current density divided by the injection current density above the threshold current density, as described by (12.84) [12.51]. For this particular device the threshold current density was 7 kA/cm^2 and the total active layer thickness was 150 Å. A fit to the data of Fig. 12.47 results in a carrier lifetime of 5 ns. It should be mentioned that this figure is unreasonably long and the underlying assumption that the recombination lifetime is current independent in deriving (12.84) may not be applicable. From the knowledge of the threshold current, the total thickness of the entire pumped region, which is 150 Å, and the lifetime just determined, one can calculate the carrier-injection level at threshold, which comes out to be about $1 \cdot 10^{20}$ cm^{-3}; this is extremely high. There are only three data points in Fig. 12.47 which do not allow one to determine if, indeed, the delay at the onset of the lasing oscillation, varies linearly with the parameter on the abscissa. Since the carrier lifetime decreases as the injection level is increased, a method not requiring a constant lifetime such as in the modulation-bandwidth measurements described by (12.86-88) is warranted.

446

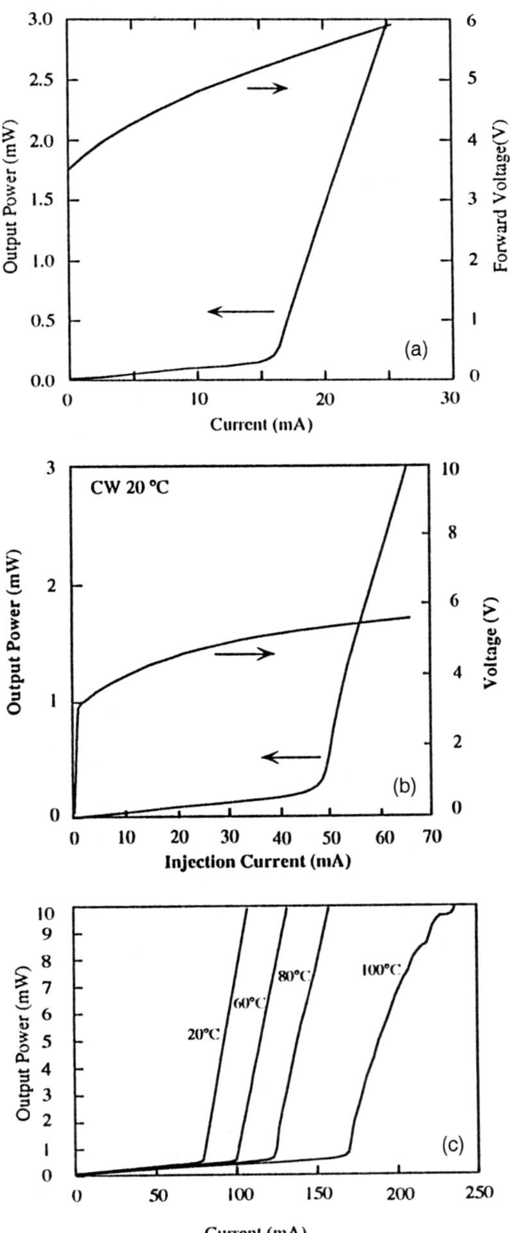

Fig. 12.46. (a) Light and voltage vs. current characteristics of a Nichia ridge laser at 20°C with a threshold current of 16 mA and a forward voltage of about 5 V. Lateral growth over SiO$_2$ stripes and modulation doping have been employed. (b) Light and voltage vs. current characteristics of a Nichia ridge laser at 20°C with a threshold current of 50 mA and a forward voltage of about 5 V relying on direct unpatterned growth on the c-plane of a sapphire substrate. (c) Light vs. current characteristics of a Nichia ridge laser at 20°, 60°, 80°, and 100°C. The laser structure of (b) exhibits a room-temperature threshold current of 75 mA and forward voltage of 4÷5 V. After [12.49]

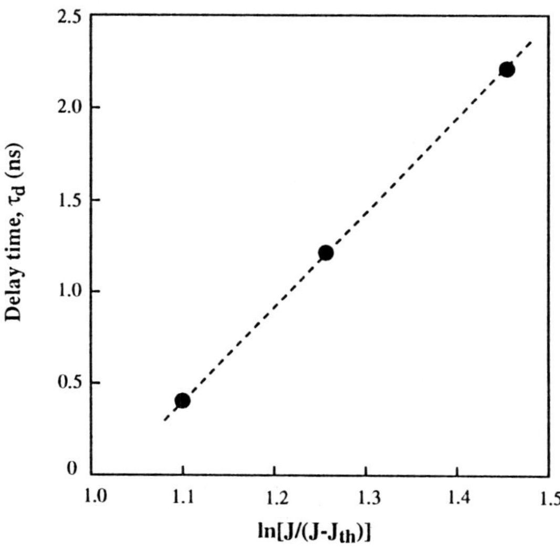

Fig. 12.47. The delay of the laser-oscillation onset with respect to the current pulse as a function of the current drive level in an InGaN MQW Nichia laser. Courtesy of S. Nakamura, Nichia Chemical Ltd.

12.11 Quantum Efficiency

The external differential efficiency is a very useful parameter from which the **internal quantum efficiency** and, in turn, the internal loss can be determined. The differential quantum efficiency is simply the rate of change of the optical power with respect to the injection current. If the power is measured from one facet, the differential figure is doubled to account for two facets providing that the reflectivities of both mirrors are the same. The external efficiency of a laser is simply related to the internal efficiency through

$$\eta_{ext} = \eta_{int}\frac{\text{end losses}}{\text{total loss}} = \eta_{int}\frac{(2L)^{-1}\ln(1/R_1R_2)}{(2L)^{-1}\ln(1/R_1R_2)+\alpha_i} \tag{12.89}$$

which reduces to

$$\frac{1}{\eta_{ext}} = \frac{1}{\eta_{int}} - \frac{1}{\eta_{int}}\frac{\alpha_i}{2L}\ln(R_1R_2) . \tag{12.90}$$

448

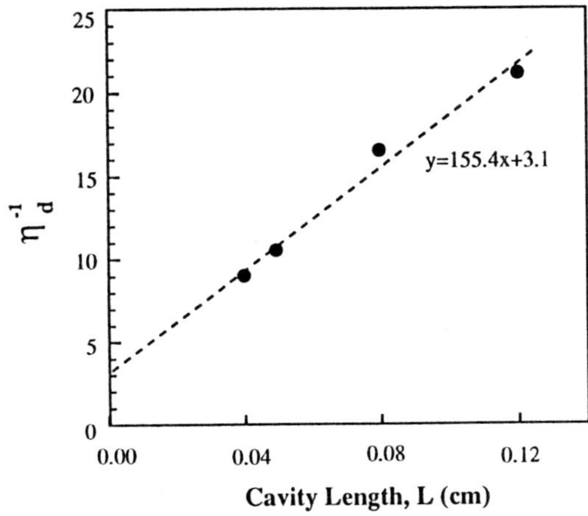

Fig. 12.48. Plot of the inverse of the external quantum efficiency against the cavity length for a series of lasers with all the other parameters remaining the same. Courtesy of S. Nakamura, Nichia Chemical Ltd.

In other words, the **external quantum efficiency** beyond threshold is measured as the rate of change of the optical power converted to current vs. the injection current. One method to obtain the **internal quantum efficiency** is to determine the external quantum efficiency of a series of lasers with varying cavity lengths. Plotting the inverse of the measured external quantum efficiency as a function of the cavity length leads to a straight line. In the limit of an infinitely long cavity, the end loss is zero at which point the internal quantum efficiency is the same as the external efficiency. From the slope of the straight line one can obtain the internal loss. The threshold gain is then equal to the sum of the internal loss just deduced and the end loss $[(2L)^{-1}\ln(1/R_1 R_2)]$ which can be calculated knowing the facet reflectivities and the cavity length. Figure 12.48 depicts the results of such an exercise with a plot of the inverse of the external quantum efficiency versus the cavity length. The data fit to the straight-line expression $155.4x + 3.1$. The inverse of 3.1 is the internal quantum efficiency which is about 33%. Using a reflectivity of 50% for the facets (coated), a confinement factor of 2.5%, and with the aid of (12.42) one finds the gain at threshold to be 3200 cm^{-1}.

12.12 Gain Spectra of InGaN Injection Lasers

The gain spectra can be obtained employing, for example, the method developed by *Hakki* and *Paoli* [12.56]. In this method, the device is pumped to just below threshold, and the gain is calculated from the interference fringes. This can be done as a function of injection current and wavelength, i.e.,

$$R_1 R_2 e^{-2\alpha L} = \frac{\text{power at peak}}{\text{power at valley}} \tag{12.91}$$

where α is the *negative* loss which is equal to the modal gain minus the internal loss. In other words,

$$-\alpha = \Gamma g - \alpha_{\text{tot}} = \frac{1}{2L}\left[\ln\left(\frac{1}{R_1 R_2}\right) + \ln\left(\frac{\text{power at peak}}{\text{power at valley}}\right)\right] . \tag{12.92}$$

Having already found the internal loss, the gain can be calculated from (12.92) as a function of wavelength for a given current. The peak power

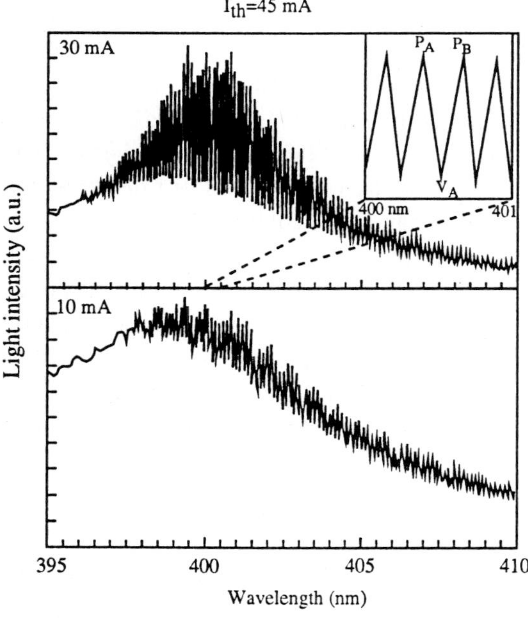

Fig. 12.49. Light intensity of a ridge laser vs. wavelength below the threshold current (45 mA) for current levels of 10 and 30 mA indicating the modes used in calculating the modal gain through the method of *Hakki* and *Paoli* [12.56]. Courtesy of S. Nakamura, Nichia Chemical Ltd.

450

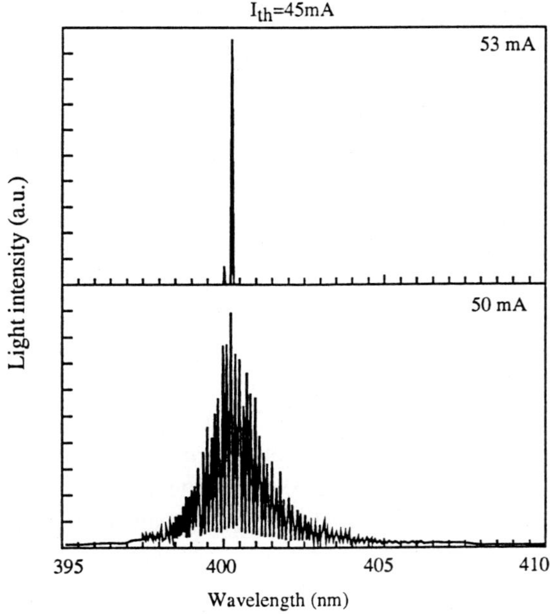

Fig. **12.50**. Light intensity of a ridge laser vs. wavelength above the threshold for current levels of 50 and 53 mA. Courtesy of S. Nakamura, Nichia Chemical Ltd.

and the valley power can be determined at any wavelength. It is customary to average the intensity of the two adjacent peaks for the peak power (Fig. 12.49). The minimum (VA) straddled by the two adjacent maxima (PA and PB) is taken as the power-at-valley. The output power of the ridge laser (Fig. 12.46b) versus wavelength above threshold for current levels of 50 and 53 mA is shown in Fig. 12.50. The modal gain vs. the injection current for

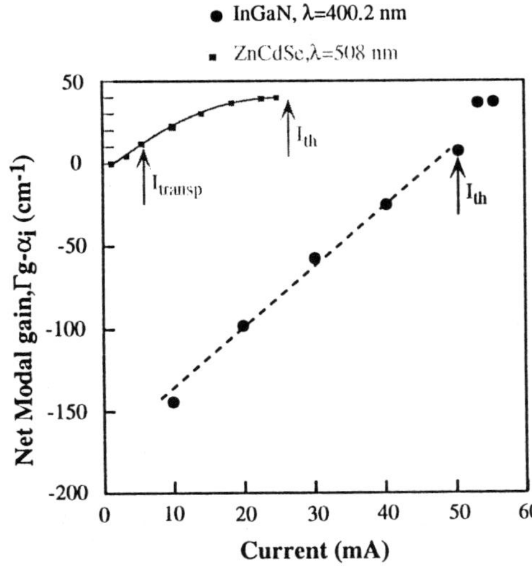

Fig. **12.51**. Net modal gain vs. the injection current for an InGaN laser at a wavelength of 400.2 nm which can be expressed as $0.03J - 180$ [cm^{-1}]. InGaN laser data: Courtesy of S. Nakamura, Nichia Chemical Ltd.

451

an InGaN laser at the wavelength of 400.2 nm is depicted in Fig. 12.51. For the InGaN laser employed in this experiment, the threshold current density is 8.8 kA/ cm^2, the internal loss is 46 cm^{-1}, and the recombination lifetime is 3.5 ns. The methods used in the determination of these parameters are those described in Sects. 12.10 and 11. Superimposed on the figure are measurements of a ZnSe-based laser for comparison [12.57]. The modal gain of the InGaN laser can be represented by a straight line expressed by − 180 + 0.03J [cm^{-1}]. It should be noted that the first term is constant whereas the second term is a function of the injection current. This is in agreement with the form of (12.72) which applies to the linear operating regime. The constant term and the slope can be utilized to determine the differential gain coefficient and the transparency current, which for the device under test are $5.8 \cdot 10^{-17}$ cm^2 and $9.3 \cdot 10^{19}$ cm^{-3}, respectively. Again, the question arises how to find the carrier lifetime, as was alluded to in Sect. 12.10. If the carrier lifetime were about 0.5 ns, the transparency density would reduce to about $1 \cdot 10^{19}$ cm^{-3} which is much closer to the calculations of *Suzuki* and *Uenoyama* [12.22] for GaN. The single-carrier model indicates a transparency carrier density to be around $5 \cdot 10^{18}$ cm^{-3}. When InGaN is employed for the active layers, the carrier-injection density at transparency should even be smaller. In defective materials, it is possible to have transparency currents that are extremely high because of non-radiative recombination and the carrier leakage, but a figure in the 10^{20} cm^{-3} range appears questionably high.

The net modal gain can be deduced through the *Hakki* and *Paoli* method as a function of wavelength for a series on injection currents up to below the threshold. Using data such as those shown in Fig. 12.49, the net modal gain vs. wavelength for the injection currents of 10, 30, 50 and 53 mA has been obtained [Ref. 12.51, Fig. 12.52]. Clearly, the 53 mA is too high as this

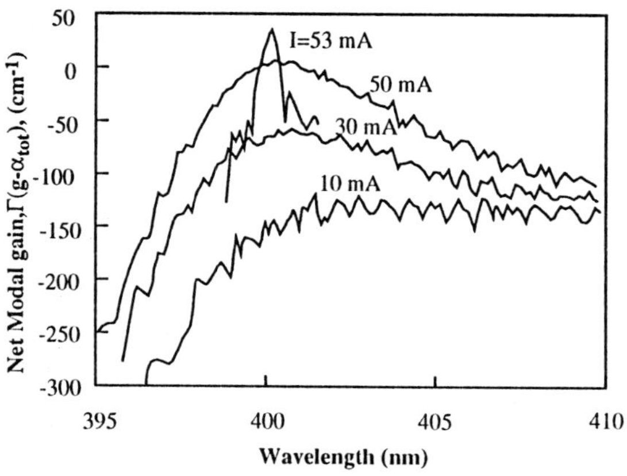

Fig. 12.52. Net modal gain vs. wavelength for an InGaN laser for a series of injection currents of 10, 30, 50 and 53 mA. Courtesy of S. Nakamura, Nichia Chemical Ltd.

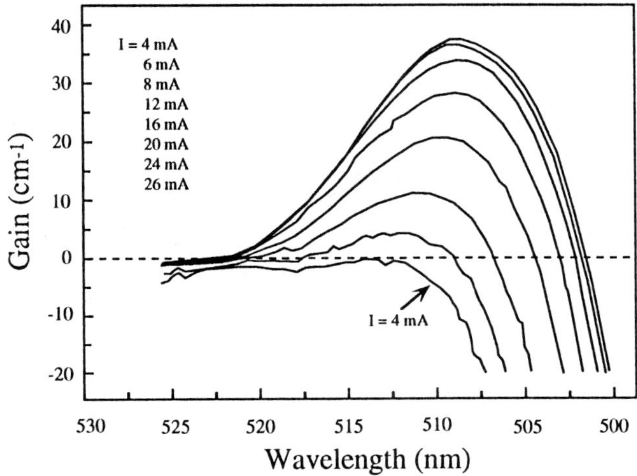

Fig. 12.53. Net modal gain vs. wavelength for a ZnCdSe MQW laser for a series of injection currents of 4, 6, 8, 12, 16, 20, 24, 26 mA. After [12.57]

is past the onset of lasing in which case the gain spectra are skewed in favor of the lasing mode. The transparency current determined at zero net modal gain is about 50 mA. Consistent with the point we made earlier, which has to do with the embryonic stage of the GaN laser research, the well resolved and behaved modal gain spectra of a ZnCdSe MQW laser is depicted in Fig. 12.53 for comparison. The transparency current here is about 4 mA above which a net positive gain exists.

As the spectra displayed in Fig. 12.49 indicate, many modes are supported by a gain vs. energy (wavelength) - cavity combination. Even though only one mode appears when the injection current is at 53 mA, single-mode operation is not maintained, as the current is changed and/or the temperature is controlled. Figure 12.54 illustrates the mode hopping behavior with current. It appears that there is a slight blue shift at each mode as the injec-

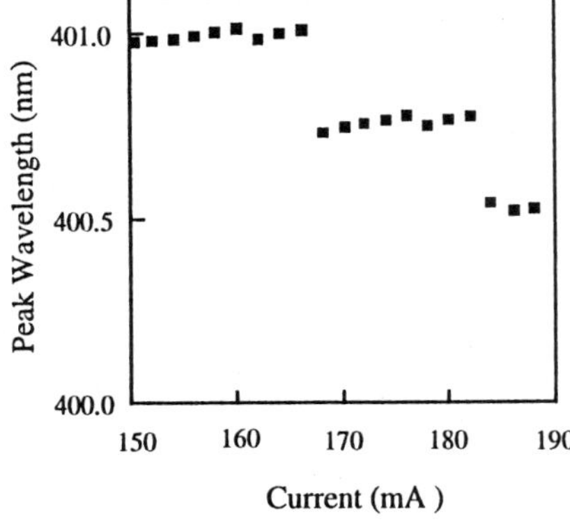

Fig. 12.54. Mode hopping behavior of an InGaN laser with an injection current in the range of 150 to 190 mA. Courtesy of S. Nakamura, Nichia Chemical Ltd.

453

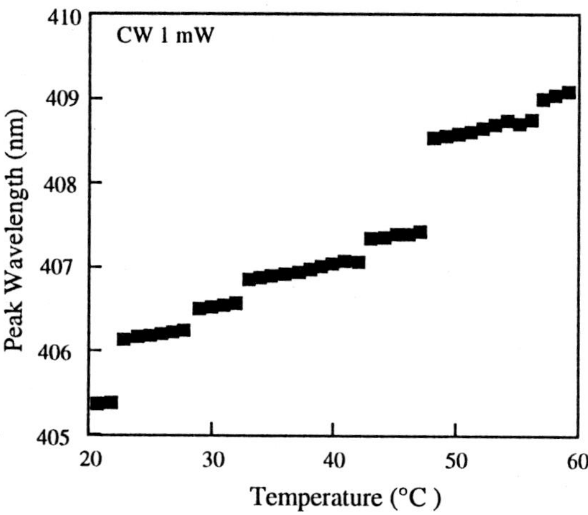

Fig. 12.55. Mode hopping as the case temperature is increased from $20°$ to $60°$ C in an InGaN laser, with the red shift being clearly observed with increasing temperature. Courtesy of S. Nakamura, Nichia Chemical Ltd.

tion current is increased, which appears to be indicative of filling effects. On the other hand, Fig. 12.55 exhibits the mode hopping behavior as the deviece temperature is increased from $20°$ to $60°$C. As expected, a red shift is clearly observed with increasing temperature.

The Nichia lasers were CW-operated at room temperature with reported lifetimes of about 3.000 hours. In life testing, the current drive is automatically adjusted to maintain a per facet power of 1.5 mW. Figure 12.56 displays the evolution of the current through two InGaN lasers up to about 1.000 hours without any marked degradation. These devices were fabricated in layers utilizing the lateral growth concept discussed in Sect. 4.7.3. Operating lifetimes in excess of 10.000 hours or longer has been extrapolated at room temperature under CW operation at 2 mW optical power. As is the case with any commercial laser, the lifetime and far-field pattern issues will much dominate the future discussion as well as the understanding of the governing processes of lasing. In the context of the latter, the absorption, as determined by photo-current experiments (by using the LD device as a detector), PL excitation and stimulated emission spectra of a laser are presented in Fig. 12.57. Clearly, there is substantial (nearly 190 meV) red shift of the laser line with respect to the absorption spectrum, caused by localized states. The absence of localized states in the excitation spectrum may have to do with the small joint density of states. Resonant measurements may have been warranted to observe absorption by localized states. In weakly localized systems, the resultant proximity of electrons and holes would increase the injection efficiency. Strongly localized states, which may not occur here as this case requires strong localization poten-

454

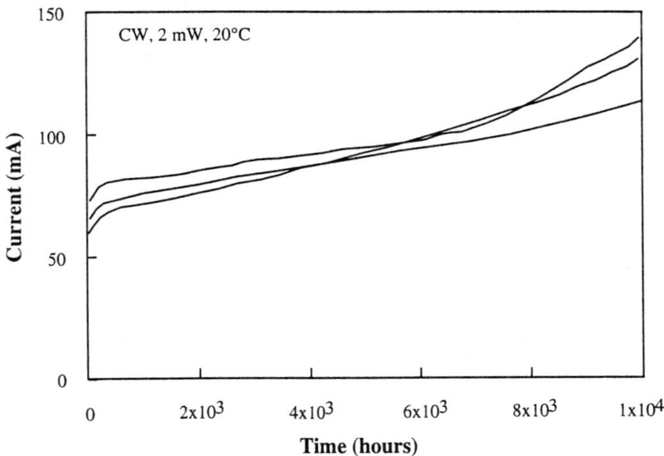

Fig. 12.56. Evolution of the current through a CW Nichia InGaN lasers with an initial threshold currents in the range of $60 \div 70$ mA. The injection current was adjusted to maintain the light output at 2 mW. All devices were still operative after 10,000 hours at room temperature. Courtesy of S. Nakamura, Nichia Chemical Ltd.

tials, and the spread in the reciprocal space would reduce the recombination efficiency and thus increase the carrier lifetime. In contrast to the GaN case, the Stokes shift in ZnCdSe lasers is about 30 meV or less. This red shift is most likely due to the lasing action taking place at localized states.

In parallel developments, InGaN/GaN/AlGaN double heterostructures grown on SiC substrates have been reported to operate first pulsed with a duty cycle of 0.1 % and later CW at room temperature [12.58]. Unlike the laser and LED structures on sapphire substrates, cursory investigations pur-

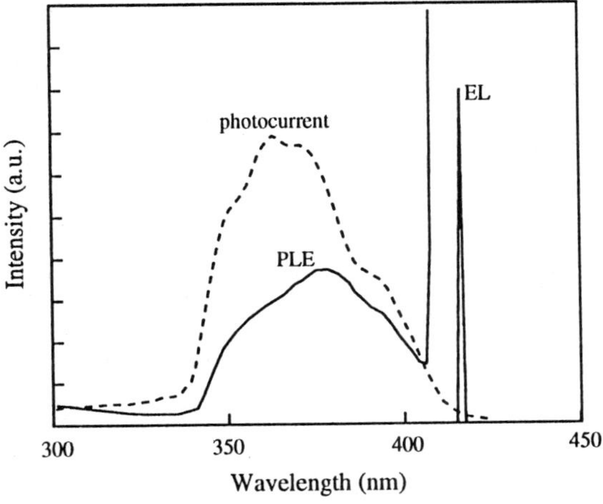

Fig. 12.57. Absorption spectrum as determined by photo-current (using the LD device as a detector) and PL excitation, and the stimulated-emission spectrum of an InGaN laser. Courtesy of S. Nakamura, Nichia Chemical Ltd.

portedly led to the conclusion that compositional inhomogeneities are not present. Moreover, the red shift which is prevalent in lasers on sapphire was presumably absent. The relative ease with which SiC substrates can be cleaved, manifested itself in sharp linewidths on the order of a few Ångstroms and attests to the quality of the cavity. However, as in the case of lasers on sapphire, the mode spacing in the spectral response of the lasers did not correspond to the physical cavity length. This anomaly which is endemic to all nitride-based lasers may have its roots in some internal reflections within the cavity.

12.13 Observations

Although the available data in injection lasers in nitrides are limited, and further confined to InGaN, a picture is beginning to emerge:

- Nominally, there is a substantial red shift in the lasing wavelength as compared to the absorption spectrum, which happens to be true to a lesser degree for just about all semiconductors.
- The I-V characteristic does not show any current transport associated with processes in the conduction and/or valence bands.
- The transparency current or the transparency carrier concentration is too high.
- Lasing action is limited to a narrow wavelength range around 410 nm.

As discussed earlier, the red shift is most likely due to lasing at localized states. Lasing governed by excitonic processes, which would show a red shift of about 20 meV under ideal conditions, is not likely as the transparency carrier concentration is well above the dephasing concentration by an order of magnitude or more; the Coulombic interactions would be screened and the band edge of InGaN is not sufficiently sharp. Localized states provide the three-level picture needed for lasers which would account for the red shift and so would bandgap renormalization. Chances are that both play a role. Of course, the localized state(s) must be coupled to the conduction and/or the valence band with a scattering rate which is faster than the band-to-band recombination rate. On top of these processes, there is the siphoning non-radiative recombination that takes place in parallel. The lack of diffusion current and generation-recombination current (unless it is due to the non-ohmic p-type contact in conjunction with defects in the p-type semiconductor under the said contact that prevents its observation) provides the basis for the suggestion that the current flow may be due to photon-assisted tunneling. The localized-state argument is also supported by compositional and strain inhomogeneities observed in the InGaN material. Lasers with GaN active layers would not suffer from compositional inhomogeneities and may shed the needed light for a better understanding of the processes involved, but do not exist in the injection form. It appears that compositional non-uniformity in InGaN active layers paves the way for, not hinders lasing because of an inherent strain and defect minimization processes that occur as a result of clustering. Outside clusters, the semiconductor may be of smaller InN composition, has a relatively large bandgap, and is transparent to the radiation generated in the regions of large InN mole fractions. If so, and if these regions are the ones that are pumped, local heating that may occur may also be responsible for the red shift. Based on transmission electron microscopy, energy dispersive X-ray micro-analysis, and low-temperature PL at excitation levels near the transparency conditions, the suggestions have been made that these clusters are actually quantum dots [12.59] which localize the excitons and govern lasing. Finally, the columnar growth may be detrimental to coherent-light propagation. This process is particularly troubling, as the shorter the wavelength the more light is scattered.

One point of wonderment and agreement in the maze of GaN-based injection lasers is that GaN again is a paradigm buster in that it has been able to provide the medium for an injection-laser CW operating at room temperature for as long as 3.000 h (extrapolated to 10.000h). No other semiconductor containing as many defects as GaN has been able to accomplish this. On this note, one can not help but assume that the compositional and strain inhomogeneities provide that narrow pathway to CW-lasing oscillations, albeit mainly in one laboratory so far.

12.14 A Succinct Review of the Laser Evolution in Nitrides

The semiconductor laser, though it has important applications, is analogous to a race car. As is the case in a race car, the components of an injection laser are put through the acid test in that high concentrations of injected carriers, high photon densities, and large heat generation exist simultaneously, and will challenge the robustness of the device. Since the first report of optically-pumped stimulated emission in GaN at 2 K [12.50], great strides have been made at an astonishing rate particularly following the paper on p-type GaN by the group of I. Akasaki in 1989 (Chap.5). Utilizing $Al_{0.1}Ga_{0.9}N/GaN$ separate confinement laser structures, the optical pumping required for lasing at room temperature was reduced to well under 100 kW/cm^2 from several MW/cm^2 in the early versions relying on bulk GaN [12.60-61].

Building on a successful commercialization of bright (blue followed by green) LEDs, *Nakamura* et al. [12.62] reported the observation of laser oscillations at room temperature in InGaN quantum wells utilizing GaN waveguides and AlGaN cladding layers. Although the initially reported laser utilized 26 periods of $In_{0.2}Ga_{0.8}N/In_{0.05}Ga_{0.95}N$ MQW structures consisting of 25 Å thick $In_{0.2}Ga_{0.8}N$ well layers and 50 Å thick $In_{0.05}Ga_{0.95}N$ barrier layers, recent structures incorporate MQWs with as little as 7 periods [12.61], which were recently reduced to 3 periods. A 200 Å thick p-type $Al_{0.2}Ga_{0.8}N$:Mg layer was employed to prevent dissociation of the InGaN layers during the growth of the subsequent GaN and AlGaN layers, which require much higher substrate temperatures. The 0.1 μm thick layer of n-type $In_{0.1}Ga_{0.9}N$ was imbedded in the buffer layer to prevent cracking. Since it is difficult to cleave the c-face sapphire substrates, Reactive Ion Etching (RIE) was employed to form the cavity facets. High-reflectivity facet coatings ($60 \div 70\%$) were used to reduce the threshold current. A Ni/Au contact was evaporated onto the entire area of the p-type GaN layer and a Ti/Al contact onto the n-type GaN layer. The earlier versions of injection lasers fabricated by *Nakamura* et al. suffered from large forward voltages, as high as 30 V and which is some 7 times larger than it should be; this increases the power dissipation by a factor of almost 50. Later, that voltage was reduced to about 5 V (Fig. 12.46). The threshold current density of the latest variety of Nichia lasers fabricated in structures grown by the lateral growth technique has been reduced to approximately 1.5 kA/cm^2. *Akasaki* et al. [12.64] also reported lasing action in a single $In_{0.1}Ga_{0.9}N$ well Separate-Confinement Heterostructure (SCH) at 376 nm, the shortest of any injection semiconductor laser (which has not been reproduced). A threshold current density of 2.9 kA/cm^2 under pulsed operation with a duty cycle of 1% and a pulse width of 0.3 μs has been observed at room temperature. At the time of this writing, there were some six laboratories reporting lasing action, by current injection, in InGaN DH structures.

Lasers utilizing films grown on the c-plane of sapphire had etched cavities as the cleavage is not practical because sapphire does not cleave well. This is complicated further in that the GaN colums are rotated with respect to the underlying sapphire, making it impossible to align the cleavage plane to a plane of GaN and sapphire. It is for this reason that *Nakamura* et al. [12.65] explored lasers grown on the a-plane of sapphire. Even though the materials quality does not compare to that on the c-plane, improved cavity formation along the $(1\bar{1}00)$ plane of GaN outweighs the reduced materials quality. Moreover, laser structures on (111) $MgAl_2O_4$ (spinel substrates) which lead to Wz GaN along the c-plane, have also been explored for optical pumping [12.66] and injection-laser experiments [12.67]. In this scheme, spinel cleaves along the {100} planes which are inclined to the surface with cleavage following one of the $(1\bar{1}00)$ planes of GaN about where the epilayers are reached. Even though the facet quality in this scheme was the best among the aforementioned approaches, materials-quality degradation is too severe to pull it ahead of the other approaches. *Nakamura* et al. [12.51] may have found a way to produce etched mirrors on c-plane sapphire substrates, perhaps eliminating the undesired compromise between the best crystalline quality and the best mirrors.

References

Chapter 1

1.1 S.N. Mohammad, A. Salvador, H. Morkoç: Proc. IEEE **83**, 1306 (1995)

1.2 S.N. Mohammad, H. Morkoç: *Progr. Quantum Electron.* **20**, Nos. 5 and 6, 361 (Pergamon, Oxford 1996)

1.3 M. Razheghi, A. Rogalski: J. Appl. Phys. **79**, 7433 (1996)

1.4 G.Y. Xu, A. Salvador, W. Kim, Z. Fan, C. Lu, H. Tang, H. Morkoç, G. Smith, M. Estes, B. Goldenberg, W. Yang, S. Krishnankutty: Appl. Phys. Lett. **71**, 2154 (1997)

1.5 H. Morkoç: Beyond SiC! III-V nitride based heterostructures and devices, in *SiC Materials and Devices*, Semiconductors and Semimetals, Vol. 52 (Academic, San Diego, CA 1998)

1.6 W.C. Johnson, J.B. Parson, M.C. Crew: J. Phys. Chem. **36**, 2561 (1932)

1.7 H.P. Maruska, J.J. Tietjen: Appl. Phys. Lett. **15**, 327 (1969)

1.8 E. Tiede, M. Thimann, K. Sensse: Chem. Berichte **61**, 1568 (1928)

1.9 S.V. Biryukov, Yu.V. Gulyaev, V.V. Krylov, V.P. Plessky: *Surface Acoustic Waves in Inhomogeneous Media*, Springer Ser. Wave Phenom., Vol. 20 (Springer, Berlin, Heidelberg 1995)

1.10 R. Juza, H. Hahn: Z. Anorgan. Allgem. Chem. **239**, 282, (1938)

1.11 J.I. Pankove, E.A. Miller, J.E. Berkeyheiser: RCA Rev. **32**, 383 (1971)

1.12 H. Amano, N. Sawaki, I. Akasaki, Y. Toyoda: Appl. Phy. Lett. **48**, 353 (1986).

1.13 H. Amano, M. Kito, K. Hiramatsu, I. Akasaki: Jpn. J. Appl. Phys. **28**, L2112 (1989)

1.14 S. Nakamura, T. Mukai, M. Senoh: Jpn. J. Appl. Phys. **30**, L1998 (1991)

1.15 S. Nakamura, T. Mukai, M. Senoh: Appl. Phys. Lett. **64**, 1687 (1994)

1.16 S. Nakamura, M. Senoh, S. Nagahama, N. Iwasa, T. Yamada, T. Matsushita, H. Kiyoku, Y. Sugimoto: Jpn. J. Appl. Phys. **35**, L74 (1996)

Chapter 2

2.1 W.C. Johnson, J.B. Parson, M.C. Crew: J. Phys. Chem. **36**, 2561 (1932)

2.2 H.P. Maruska, J.J. Tietjen: Appl. Phys. Lett. **15**, 327 (1969)

2.3 J.I. Pankove: J. Electrochem. Soc. **119**, 1110 (1972)

2.4 T.L. Chu: J. Electrochem. Soc. **118**, 1200 (1971)

2.5 E. Lakshmi: Thin Solid Films **83**, L137 (1981)

2.6 Y. Morimoto: J. Electrochem. Soc. **121**, 1383 (1974)

2.7 A. Shintani, S. Minagawa: J. Electrochem. Soc. **123**, 706 (1976)

2.8 K. Itoh, H. Amano, K. Hiramatsu, I. Akasaki: Jpn. J. Appl. Phys. **30** 1604 (1991)

2.9 K. Ito, K. Hiramatsu, H. Amano, I. Akasaki: J. Cryst. Growth **104**, 533 (1990)

2.10   S.N. Mohammad, A. Salvador, H. Morkoç: Proc. IEEE **83**, 1306 (1995)

2.11   Y. Morimoto: J. Electrochem. Soc. **121**, 1383 (1974)

2.12   M. Furtado, G. Jacob: J. Cryst. Growth **64**, 257 (1983)

2.13   S.P. Gordienko, G.V. Samsonov, V.V. Fesenko: Sov. J. Phys. Chem. **38**, 1620 (1964)

2.14   Z.A. Munir, A.W. Searcy: J. Chem. Phys. **42**, 4223 (1965)

2.15   R. Groh, G. Gerey, L. Bartha, J.I. Pankove: Phys. Status Solidi A **26**, 353 (1974)

2.16   Z.A. Munir, A.W. Searcy: J. Chem. Phys. **42**, 4223 (1965)

2.17   R.A. Logan, C.D. Thurmond: J. Electrochem. Soc. **119**, 1727 (1972)

2.18   C.D. Thurmond, R.A. Logan: J. Electrochem. Soc. **119**, 622 (1972)

2.19   S.N. Mohammad, H. Morkoç: Progr. Quantum Electron. **20**, Nos.5,6, 361 (Pergamon, Oxford 1996)

2.20   I. Akasaki, H. Amano: In *Properties of Group III Nitrides*, ed. by J.H. Edgar, EMIS Data Reviews Series (IEE, London 1994) p.222

2.21   H.P. Maruska, L.J. Anderson,, D.A. Stevenson: J. Electrochem. Soc. **121**, 1202 (1974)

2.22   O. Lagerstedt, B, Monemar: Phys. Rev. B **19**, 3064 (1979

2.23   H.P. Maruska, L.J. Anderson, D.A. Stevenson: J. Electrochem. Soc. **121**, 1202 (1974)

2.24   A.U. Sheleg, V.A. Savastenko: Vestsi Akad. Nauk, Ser. Fiz.-Mat. Nauk (SSSR) (1977) p.126

2.25   E.K. Sichel, J.I. Pankove: J. Phys. Chem. Solids **38**, 330 (1977)

2.26   G.A. Slack: J. Phys. Chem. Solids **34**, 321 (1973)

2.27   I. Basin, O. Knacke, O. Kubaschewski: *Thermochemical Properties of Inorganic Substances* (Springer, Berlin, Heidelberg 1977)

2.28   D. Elwell, M.M. Elwell: J. Cryst. Growth **17**, 53 (1988)

2.29   J. Karpinski, S. Porowski: J. Cryst. Growth **66**, 11 (1984)

2.30   S. Porowski, I. Grzegory: In *Properties of Group III Nitrides*, ed. by J.H. Edgar, IEE EMIS Datarev. Series, No.11 (Inspec, London 1994) ps.71,76,83

2.31   V.A. Savastenko, A.U. Sheleg: Phys. Status Solidi A **48**, K135 (1978)

2.32   T. Detchprohm, K. Hiramatsu, K. Itoh, I. Akasaki: Jpn. J. Appl. Phys. **31**, L1454 (1992)

2.33   I.F. Chetverikova, M.V. Chukichev, L.N. Rastorguev: Inorg. Mater. **22**, 53 (1986)

2.34   M.E. Sherwin, T.J. Drummond: J. Appl. Phys. **69**, 8423 (1991)

2.35   K. Miwa, A. Fukumoto: Phys. Rev. B **48**, 7897 (1993)

2.36   Y.N. Xu, W.Y. Ching: Phys. Rev. B **48**, 4335 (1993)

2.37   P.E. Van Camp, V.E. Van Doren, J.T. Devreese: Solid State Commun. **81**, 23 (1992)

2.38   D. Gerlich, S.L. Dole, G.A. Slack: J. Phys. Chem. Solids **47**, 437 (1986)

2.39   P. Perlin, C.J. Carillon, J.P. Itie, A.S. Miguel, I. Grzegory, A. Polian: Phys. Rev. B **45**, 83 (1992)

2.40   D.D. Manchon Jr., A. S. Barker Jr., P.J. Dean, R.B. Zetterstrom: Solid State Commun. **8**, 1227 (1970)

2.41   A. Cingolani, M. Ferrara, M. Lugara, G. Scamarcio: Solid State Commun. **58**, 823 (1986)

2.42   S. Miyoshi, K. Onabe, N. Ohkouchi, H. Yaguchi, R. Ito, S. Fukutsu, Y. Shraki: J. Cryst. Growth **124**, 439 (1992)

2.43   A.A. Oliner (ed.): *Acoustic Surface Waves*, Topics Appl. Phys., Vol.24 (Springer, Berlin, Heidelberg 1978)

2.44   G.A. Slack: J. Phys. Chem. Solids **34**, 321 (1973)

2.45   G.A. Slack, T.F. McNelly: J. Cryst. Growth **42**, 560 (1977) (AlN from pellet drop, sublimation, and final growth)

2.46   G.A. Slack, R.A. Tanzilli, R.O. Pohl, J. W. Vandersande: J. Phys. Chem. Solids **48**, 641 (1987)

2.47   G.A. Slack, S.F. Bartram: J. Appl. Phys. **46**, 89 (1975)

2.48   W.M. Yim, R.J. Paff: J. Appl. Phys. **45**, 1456 (1974)

2.49   G.A. Slack, T.F. McNelly: J. Cryst. Growth **34**, 263 (1976)

2.50   W.J. Meng: In *Properties of Group III Nitrides*, ed. by J.H. Edgar, IEE EMIS Datarev. Series, No.11 (Inspec, London 1994) p.22

2.51   C.F. Cline, J.S. Kalm: J. Electrochem. Soc. **110**, 773 (1963)

2.52   K.M. Taylor, C. Lenie: J Electrochem. Soc. **107**, 308 (1960)

2.53   K. Kawabe, R.H. Tredgold, Y. Inuishi: Elect. Eng. Jpn. **87**, 62 (1967)

2.54   J. Edwards, K. Kawabe, G. Stevens, R.H. Tredgold: Solid State Commun. **3**, 99 (1965)

2.55   G.A. Cox, D.O. Cummins, K. Kawabe, R.H. Tredgold: J. Phys. Chem. Solids **28**, 543 (1967)

2.56   W.M. Yim, E.J. Stofko, P.J. Zanzucchi, J.I. Pankove, M. Ettenberg, S.L. Gilbert: J. Appl. Phys. **44**, 292 (1973)

2.57   S. Yoshida, S. Misawa, Y. Fujii, S. Takada, H. Hayakawa, S. Gonda, A. Itoh: J. Vac. Sci. Technol. **16**, 990 (1979)

2.58   T.L. Chu, D.W. Ing, A. J. Noreika: Solid State Electron. **10**, 1023 (1967)

2.59   R.F. Rutz: Appl. Phys. Lett. **28**, 379 (1976)

2.60   R.F. Rutz, E.P. Harrison, J.J. Cuome: IBM J. Res. Sev. **17**, 61 (1973)

2.61   J. Edwards, K. Kawabe, G. Stevens, R.H. Tredgold: Solid State Commun. **3**, 99 (1965)

2.62   K. Kawabe, R.H. Tredgold, Y.Inuishi: Elect. Eng. Jpn. **87**, 62 (1967)

2.63   J.H. Harris, R.A. Youngman: In *Properties of Group III Nitrides*, ed. by J. H. Edgar, IEE EMIS Datarev. Series, No.11 (Inspec, London 1994) p.203

2.64   S. Pacesova, L. Jastrabik: Czech. J. Phys. B **29**, 913 (1979)

2.65   R.A. Youngman, J.H. Harris: J. Am. Ceram. Soc. **73**, 3238 (1990)

2.66   J.H. Harris, R.A. Youngman, R.G. Teller: J. Mater. Res. **5**, 1763 (1990)

2.67   J. Pastrnak, L. Souckova: Phys. Status Solidi **9**, K71 (1963)

2.68   J. Pastrnak, L. Roskovcova: Phys. Status Solidi **11**, K73 (1965)

2.69   P.B. Perry, R.F. Rutz: Appl. Phys. Lett. **33**, 319 (1978)

2.70   F. Karel, J. Pastrnak, J. Hejduk, V. Losik: Phys. Status Solidi **15**, 693 (1966)

2.71   F. Karel, J. Pastrnak: Czech. J. Phys. B **19**, 78 (1969)

2.72   F. Karel, J. Pastrnak: Czech. J. Phys. B **20**, 46 (1970)

2.73   F. Karel, J. Mares: Czech. J. Phys. B **22**, 847 (1972)

2.74   F. Karel, J. Mares: Czech. J. Phys. B **23**, 652 (1973)

2.75   T.L. Tansley, C.P. Foley: J. Appl. Phys. **59**, 3241 (1986)

2.76   H. Hahn, R. Juza: Z. Anorg. Allg. Chem. **244**, 111 (1940)

2.77   T.L. Tansley: In *Properties of Group III Nitrides*, ed. by J.H. Edgar, IEE EMIS Datarev. Series, No.11 (Inspec, London 1994) p.35

2.78   P.E. van Camp, V.E. van Doren, J.T. Devreese: Phys. Rev. B **41**, 1598 (1990)

2.79   K. Kubota, Y. Kobayashi, K. Fujimoto: J. Appl. Phys. **66**, 2984 (1989)

2.80 A.V. Sheleg, V.A. Savastenko: Vestsi Akad. Nauk USSR Ser. Fiz. Mat. Nauk **3**, 126 (1976)

2.81 T.L. Tansley, C.P. Foley: *Proc. 3rd Int'l Conf. on Semi-insulating III-V Materials* (Warm Springs, OR 1984), ed. by J.S. Blakemore (Shiva, London 1985)

2.82 W.R. Bryden, S.A. Ecelberger, M.E. Hawley,, T.J. Kistenmacher: In *Diamond, SiC, Nitride Wide Bandgap Semiconductors*, ed. by C.H. Carter Jr., G. Gildenblat, S. Nakamura, R.J. Nemanich, MRS Proc. **339**, 497 (Mater. Res. Soc., Pittsburgh, PA 1994)

2.83 A. Wakahara, T. Tsuchiya, A. Yoshida: J. Cryst. Growth **99**, 385 (1990)

2.84 K. Kubota, Y. Kobayashi, K. Fujimoto: J. Appl. Phys. **66**, 2984-2988 (1989)

2.85 A.F. Wright, J.S. Nelson: Phys. Rev. B **51**, 7866 (1995)

2.86 T.L. Tansley, C.P. Foley: Electron. Lett. **20**, 1066 (1984)

2.87 V.A. Tyagai, O.V. Snitko, A.M. Evstigneev, A.N. Krasiko: Phys. Status Solidi B **103**, 589 (1981)

2.88 K. Osamura, S. Naka, Y. Murakami: J. Appl. Phys. **46**, 3432 (1975)

2.89 T. Matsuoka: In *Properties of Group III Nitrides*, ed. by J.H. Edgar, IEE EMIS Datarev. Series, No. 11 (Inspec, London 1994) pp. 231-238

2.90 S. Yamasaki, S. Asami, N. Shibata, M. Koike, K. Manabe, T. Tanaka, H. Amano, I. Akasaki: Appl. Phys. Lett. **66**, 1112 (1995)

2.91 H. Amano, T. Takeuchi, S. Sota, H. Sakai, I. Akasaki: ???book title ?????, ed. by F.A. Ponce, T.D. Moustakas, I. Akasaki, B. Menemar. MRS Proc. **449**, 1143 (1997)

2.92 S. Yoshida, S. Misawa, S. Gonda: J. Appl. Phys. **53**, 6844 (1982)

2.93 D.K. Wickenden, C.B. Bargeron, W.A. Bryden, J. Miragliova, T.J. Kistenmacher: Appl. Phys. Lett. **65**, 2024 (1994)

2.94 Y. Koide, H. Itoh, M.R.H. Khan, K. Hiramatsu, N. Sawaki, I. Akasaki, J. Appl. Phys. **61**, 4540 (1987)

2.95 M.A. Khan, J.M. Van Hove, J.N. Kuznia, D.T. Olson: Appl. Phys. Lett. **58**, 2408 (1991)

2.96 T. Tanaka, A. Watanabe, H. Amano, Y. Kobayashi, I. Akasaki, S. Yamazaki, M. Koike: Appl. Phys. Lett. **65**, 593 (1994)

2.97 S. Nakamura, T. Mukai: J. Vac. Sci. Technol. A **13**, 6844 (1995)

2.98 T. Nagatomo, T. Kuboyama, H. Minamino, O. Omoto: Jpn. J. Appl. Phys. **28**, L1334 (1989)

2.99 N. Yoshimoto, T. Matsuoka, A. Katsui: Appl. Phys. Lett. **59**, 2251 (1991)

2.100 T. Matsuoka, N. Yoshimoto, T. Sasaki, A. Katsui: J. Electron. Mater. **21**, 157-163 (1992)

2.101 S. Nakamura, T. Mukai: Jpn. J. Appl. Phys. **31**, L1457 (1992)

2.102 S. Nakamura, T. Mukai, M. Seno: Jpn. J. Appl. Phys. **31**, L16 (1993)

2.103 S. Nakamura, N. Iwasa, S. Nagahama: Jpn. J. Appl. Phys. **32**, L338-341 (1993)

2.104 W.R. Bryden, T.J. Kistenmacher: In *Properties of Group III Nitrides*, ed. by J.H. Edgar, IEE EMIS Datarev. Series, No. 11 (Inspec, London 1994) pp. 117-118

2.105 K. Starosta: Phys. Status Solidi A **68**, K55-K57 (1981)

2.106 K. Kubota, Y. Kobayashi, K. Fujimoto: J. Appl. Phys. **66**, 2984 (1989)

2.107 T.J. Kistenmacher, S.A. Ecelberger, W.A. Bryden: J. Appl. Phys. **74**, 1684 (1993)

2.108 H. Morkoç, S. Strite, G. B. Gao, M.E. Lin, B. Sverdlov, M. Burns: J. Appl. Phys. Rev. **76**, 1363 (1994)

2.109 *Thermal Expansion*, ed. by Y.S. Touloukina, R.K. Kirby, R.E. Taylor, T.Y.R. Lee, Thermal Properties of Matter, Vol. 13 (Plenum, New York 1977)

2.110 S. Nakamura, M. Senoh, S. Nagahama, N. Iwasa, T. Yamada, T. Matsushita, H. Kiyoku, Y. Sugimoto: Jpn. J. Appl. Phys. **35**, L217 (1996)

2.111 S. Nakamura, M. Senoh, S. Nagahama, N. Iwasa, T. Yamada, T. Matsushita, H. Kiyoku, Y. Sugimoto: Appl. Phys. Lett. **68**, 2105 (1996)

2.112 A.S. Zubrilov, V. I. Nikolaev, D.V. Tsevetkov, V. A. Dmitirev, K. G. Irvine, J.A. Edmond, C.H. Carter: Appl. Phys. Lett. **67**, 533 (1995)

2.113 J.-J. Song: Priv. communication (1996)

2.114 M. Leszczynski, I. Grzgory, M. Bokowski: J. Crystal Growth **126**, 601 (1993) (GaN grown at high temperature and pressure)

2.115 G. Slack, McNelly: J. Crystal Growth **42**, 560 (1977)

2.116 A.U. Sheleg, V.A. Savastenko: Izv. Akad. Nauk SSSR, Neorg. Mat. **15**, 1598 (1979) and reproduced in M. Suzuki, T. Uenoyama: Jpn. J. Appl. Phys. **36**, 1420 (1996)

2.117 M. Suzuki, T. Uenoyama: Electronic and optical properties of GaN based quantum wells, in *Group III Nitride Semiconductor Compounds: Physics and Applications*, ed. by B. Gil (Oxford Univ. Press, London 1988)

2.118 S. Kamiyama, K. Ohnaka, M. Suzuki, T. Uenoyama: Jpn. J. Appl. Phys. **34**, L821 (1995)

2.119 K. Tsubouchi, N. Mikoshiba: IEEE Trans. SU-**32**, 634 (1985)

2.120 S. Loughin, R.H. French: In *Properties of Group III Nitrides*, ed. by J.H. Edgar, EMIS Datareview Ser. (IEE, London 1994) p. 175

2.121 L.E. McNeil, M. Grimsditch, R.H. French: J. Am. Ceram. Soc. **76**, 1132 (1993)

2.122 P. Perlin, A. Polian, T. Suski: Phys. Rev. B **47**, 2874 (1993)

2.123 Landolt-Börnstein: *Numerical and Functional Relationsips in Science and Technology*, Vol. 17, *Semiconductors* (Springer, Berlin, Heidelberg 1982)

2.124 ZnO data from W. Harsch of Eagle Picher, Joplin, MO, USA (1997)

Chapter 3

3.1 W.R.L. Lambrecht, B. Segall, J. Rife, W.R. Hunter, D.K. Wickenden: Phys. Rev. B **51**, 13516 (1995)

3.2 W.R.L. Lambrecht, B. Segall: In *Optical Properties of III-V Nitrides*, ed. by J. I. Pankove, T.D. Moustakas (Academic, San Diego, CA 1997)

3.3 E. Wimmer, H. Krakauer, M. Weinert, A. Freeman: Phys. Rev. B **24**, 864 (1981)

3.4 H.L. Skriver: *The LMTO Method*, Springer Ser. Dolid-State Sci., Vol. 41 (Springer, Berlin, Heidelberg 1984)

3.5 H. Morkoç, S. Strite, G.B. Gao, M.E. Lin, B. Sverdlov, M. Burns: J. Appl. Phys. Rev. **76**, 1363-1398 (1994)

3.6 W.R.L. Lambrecht, B. Segall: Band structure of the group-III nitrides, in *Gallium Nitride*, ed. by J.I. Pankove, T.D. Moustakas. *Semiconductors and Semimetalls* **50**, 369-402 (Academic, San Diego, CA 1998)

3.7 M. Suzuki, T. Uenoyama, A. Yanase: Phys. Rev. B **52**, 8132 (1995)

3.8 S.L. Chuang, C.S. Chang: Phys. Rev. B **54**, 54 (1996)

3.9 M. Suzuki, T. Uenoyama: Electronic and optical properties of GaN based quantum wells, in *Group III Nitride Semiconductor Compounds, Physics and Applications*, ed. by B. Gil (Clarendon, Oxford 1998)

3.10 M. Suzuki, T. Uenoyama: Jpn. J. Appl. Phys **35**, 1420 (1996)

3.11 S. Kamiyama, K. Ohnaka, M. Suzuki, T. Uenoyama: Jpn. J. Appl. Phys. **34**, L821 (1995)

3.12 D.C. Reynolds, D.C. Look, W. Kim, Ö. Aktas, A. Botchkarev, A. Salvador, H. Morkoç, D.N. Talwar: J. Appl. Phys. **80**, 594-596 (1996)

3.13 M. Suzuki, T. Uenoyama: Phys. Rev. B **52**, 8132 (1996)

3.14 Yu.M. Sirenko, J.B. Jeon, K.W. Kim, M.A. Littlejohn, M.A. Stroscio: Phys. Rev. B **53**, 1997 (1996)

3.15 B.K. Vainshtein: *Fundamentals of Crystals*, 2nd edn., Modern Crystallography, Vol.1 (Springer, Berlin, Heidelberg 1994)

3.16 B.K. Vainshtein, V.M. Fridkin, V.L. Indenbom: *Structure of Crystals*, 2nd edn., Modern Crystallography, Vol.2 (Springer, Berlin, Heidelberg 1995)

3.17 G. Pikus: Sov. Phys. – JETP **14**, 1075 (1962)

3.18 G.L. Bir, G.E. Pikus: *Symmetry and Strain-Induced Effects in Semiconductors* (Wiley, New York 1974)

3.19 B. Monemar, J.P. Bergman, I.A. Buyanova: Optical characterization of GaN and related material, in *GaN and Related Material*, ed. by S.J. Pearton (Gordon and Breach, New York 1997) pp.85-140

3.20 W. Ludwig, C. Falter: *Symmetries in Physics. Group Theory Applied to Physical Problems*, 2nd edn., Springer Ser. Solid-State Sci., Vol.64 (Springer, Berlin, Heidelberg 1996)

3.21 T. Inui, Y. Tanabe, Y. Onodera: *Group Theory and Its Applications in Physics*, 2nd edn., Springer Ser. Solid-State Sci., Vol.78 (Springer, Berlin, Heidelberg 1996)

3.22 H. Morkoç, B. Sverdlov, G.B. Gao: Proc. IEEE **81**, 492 (1993)

3.23 A.U. Sheleg, V.A. Savastenko: Izv. Akad.Nauk SSSR, Neorg. Mater. **15**, 1598 (1979)

3.24 W.R.L. Lambrecht, B. Segall: General remarks and notations on the band structure of pure group III nitrides, in *Group III Nitrides*, ed. by J.H. Edgar, EMIS Datarev. Series, No.11 (IEE, London 1995) p.125

3.25 K. Kawabe, R.H. Tredgold, Y. Inuishi: Elect. Eng. Jpn. **87**, 62 (1967)

3.26 M. Balkanski, J. de Cloizeaux: J. Phys. Radium **21**, 825 (1960)

3.27 J.J. Hopfield, D.G. Thomas: Phys. Rev. **132**, 563 (1963)

3.28 J.J. Hopfield: J. Phys. Chem. Solids **15**, 97 (1960)

3.29 G. Bastard: Phys. Rev. B **25**, 7584 (1982)

3.30 J.Y. Marzin: Strained superlattices, in *Heterojunction and Semiconductor Superlattices*, ed. by A. Allan, G. Bastard, N. Boccara, M. Lannoo, M. Voos (Springer, Berlin, Heidelberg 1986) pp.161-176

3.31 F.H. Pollak, M. Cardona: Phys. Rev. B **172**, 816 (1968)

3.32 B. Gil, O. Briot, R.L. Aulombard: Phys. Rev. B **52**, R17028 (1995)

3.33 B. Gil, F. Hamdani, H. Morkoç, Phys Rev. B **54**, 7678 (1996)

3.34 D. Ahn: J. Appl. Phys. **76**, 8206 (1994)

3.35 C. Weisbusch, B. Vinter: *Quantum Semiconductor Structures: Fundamentals and Applications* (Academic, San Diego, CA 1991)

3.36 T. Uenoyama, M. Suzuki: Appl. Phys. Lett. **67**, 2527 (1995)

3.37 For a review of electronic states in semiconductor quantum wells, see G. Bastard, J.A. Brum: IEEE J. QE-22, 1625 (1986)

3.38 J.B. Jeon, B.C. Lee, Yu.M. Sirenko, K.W. Kim, M.A. Littlejohn: J. Appl. Phys. **82**, 386 (1997)

3.39   E.L. Ivchenko, G.E. Pikus: *Superlattices and other Heterostructures*, 2nd edn., Springer Ser. Solid-State Sci., Vol. 110 (Springer, Berlin, Heidelberg 1997)
3.40   P. Bigenwald, P. Christol, L. Konczewicz, P. Testud, B. Gil: Presented at E-MRS'97, Strassburg (France)
3.41   P. Bigenwald, B. Gil: Solid State Commun. **91**, 33 (1994)
3.42   R.D. King-Smith, D. Vanderbilt: Phys. Rev. B **47**, 1651 (1990)
3.43   R. Resta: Rev. Mod. Phys. **66**, 899 (1994)
3.44   F. Bernardini, V. Fiorentini: Phys. Rev. B **57**, 1-4 (15 April 1998)
3.45   F. Bernardini, V. Fiorentini, D. Vanderbilt: Phys. Rev. B **56**, R10024 (1997)
3.46   A.D. Bykhovski, V.V. Kaminski, M.S. Shur, Q.C. Chen, M.A. Khan: Appl. Phys. Lett. **69**, 3254 (1996)
3.47   A. Bykhovski, B. Gelmont, M. Shur: J. Appl. Phys. **74**, 6734 (1993)
3.48   A. Bykhovski, B. Gelmont, M. Shur: Appl. Phys. Lett. **63**, 2243 (1993)
3.49   K. Kim, W.R.L. Lambrecht, B. Segall: Phys. Rev. B **56**, 7018 (1997)
3.50   J.G. Gualtieri, J.A. Kosinski, A. Ballato: IEEE Trans. UFFC-**41**, 53 (1994)
3.51   D. Bykhovski, B.L. Gelmont, M.S. Shur: J. Appl. Phys. **81**, 6332 (1997)
3.52   G.D. O'Clock, M.T. Duffy: Appl. Phys. Let. **23**, 55 (1973)
3.53   M.A. Littlejohn, J.R. Hauser, T.H. Glisson: Appl. Phys. Lett. **26**, 625 (1975)
3.54   A. Bykhovski, B. Gelmont, M. Shur: J. Appl. Phys. **77**, 1616 (1995)
3.55   M. Shur, B. Gelmont, A. Khan: J. Electron. Mater. **25**, 777 (1996)
3.56   D.L. Smith, C. Mailhiot: J. Appl. Phys. **63**, 2717 (1988)
3.57   J. Wang, J.B. Jeon, Yu.M. Sirenko, K.W. Kim: IEEE Photon. Lett. **9**, 728 (1997)

Chapter 4

4.1   E. Tiede, M. Thimann, K. Sensse: Chem. Berichte **61**, 1568 (1928)
4.2   W.C. Johnson, J.B. Parson, M.C. Crew: J. Phys. Chem. **36**, 2561 (1932)
4.3   R. Juza, H. Hahn: Z. Anorgan. Allgem. Chem. **239**, 282 (1938)
4.4   H.P. Maruska, J.J. Tietjen: Appl. Phys. Lett. **15**, 327 (1969)
4.5   H. Amano, N. Sawaki, I. Akasaki, Y. Toyoda: Appl. Phys. Lett. **48**, 353 (1986)
4.6   S. Yoshida, S. Misawa, S. Gonda: J. Vac. Sci. Technol. B **1**, 250 (1983)
4.7   I. Akasaki, H. Amano, M. Kito, K. Hiramatsu: J. Luminescence **48/49**, 666 (1991)
4.8   S. Nakamura, T. Mukai: Jpn. J. Appl. Phys. **31**, L1457-L1459 (1992)
4.9   K. Osamura, K. Nakajima, Y. Murakami, P.H. Shingu, A. Ohtsuki: Solid State Commun. **11**, 617 (1972)
4.10   G.A. Slack, T.F. McNelly: J. Cryst. Growth **34**, 263 (1976)
4.11   J. Karpinski, J. Jun, S. Porowski: J. Crys. Growth **66**, 1 (1984)
4.12   S. Porowski, I. Grzegory: Phase diagram of InN, in *Properties of Group III Nitrides*, ed. by J.H. Edgar (INSPEC, London 1994) pp. 71-88
4.13   C.D. Thurmond, R.A. Logan: J. Electrochem. Soc. **119**, 622 (1972)
4.14   T. Sasaki, T. Matsuoka: J. Appl. Phys. **77**, 192 (1995)
4.15   R. Madar, G. Jacob, J. Hallais, F. Fruchart: J. Cryst. Growth **31**, 197 (1975)
4.16   Landolt, Börnstein: *Numerical Data, Fundamental Relationships in Science, Technology*, Vol. 17, *Semiconductors* (Springer, Berlin, Heidelberg 1984)
4.17   *Binary Alloys, Phase Diagrams*, ed. by T.B. Massalski, H. Okamoto, P.R. Subrananinn, L. Kacprak (ASM Int'l, Materials Park, OH 1990) p. 176
4.18   J.D. Latwa: Metal. Progr. **82**, 139 (1962)
4.19   B.E. Wayne: Ceramics **15**, 48 (1964)

4.20   J.A. Van Vechten: Phys. Rev. B **7**, 1479 (1973)

4.21   G.A. Slack, T.F. McNelly: J. Cryst. Growth **42**, 560 (1977)

4.22   M. Leszczynski, I. Grzegory, M. Bockowski: J. Cryst. Growth **126**, 601 (1993)
S. Porowski: Acta Phys. Polon. A **87**, 295 (1995)

4.23   S. Porowski, I. Grzegory, J. Jun: In *High Pressure Chemical Synthesis*, ed. by J. Jurczak, B. Baranowski (Elsevier, Amsterdam 1989) p.21

4.24   I. Grzegory, J. Jun, M. Bockowski, S. Krukowski, M. Wroblewski, B. Lucznik, S. Porowski: J. Phys. Chem. Solids **56**, 639 (1995)

4.25   S. Strite, H. Morkoç: J. Vac. Sci. Technol. B **10**, 1237 (1992)

4.26   Z. Yang, F. Guarin, I.W. Tao, W.I. Wang: J. Vac. Sci. Technol. B **13**, 789 (1995)

4.27   T.P. Pearsall: III-Vs Rev. **9**, 38 (1996)

4.28   M.E. Lin, S. Strite, A. Agarwal, A. Salvador, G.L. Zhou, N. Teraguchi, A. Rockett, H. Morkoç: Appl. Phys. Lett. **62**, 702 (1993)

4.29   P. Vermaut, P. Ruterana, G. Nouet, A. Salvador, H. Morkoç: HREM, CBED studies of polarity of nitride layers with prismatic defects grown over SiC. *III-V Nitrides*, ed. by F.A. Ponce, T.D. Moustakas, I. Akasaki, B.A. Monemar, MRS Proc. **449**, 317 (Mater. Res. Soc., Pittsburgh, PA 1997)

4.30   W. Kim, M. Yeadon, A.E. Botchkarev, S.N. Mohammad, J.M. Gibson, H. Morkoç: J. Vac. Sci. Techn. B **15**, 921 (1997)

4.31   K. Uchida, A. Watanabe, F. Yano, M. Koguchi, T. Tanaka, S. Minagawa: J. Appl. Phys. **79**, 3487 (1996)

4.32   T. Lei, K.F. Ludwig Jr., T. Moustakas: J. Appl. Phys. **74**, 4430 (1993)

4.33   T.L. Tansley, E.M. Goldys, M. Godlewski, B. Zhou, H.Y. Zuo: The contribution of defects to the electrical, optical properties of GaN, in *GaN, Related Materials*, ed. by S. Pearton (Gordon & Breach, Amsterdam 1997) pp.233-295

4.34   K. Doverspike, L.B. Rowland, D.K. Gaskill, J.A. Freitas Jr.: J. Electron. Mater. **24**, 269 (1995)

4.35   H.M. Manasevit, F.M. Erdmann, W.I. Simpson: J. Electrochem. Soc. **118**, 1864 (1971)

4.36   S. Nakamura, Y. Harada, M. Seno: Appl. Phys. Lett. **58**, 2021 (1991)

4.37   K. Hiramatsu, S. Itoh, H. Amano, I. Akasaki, N. Kuwano, T. Shiraishi, K. Oh: J. Cryst. Growth **115** 628 (1991)

4.38   S. Nakamura: Jpn. J. Appl. Phys. **30**. L1705 (1991)

4.39   S.D. Hersee, J.C. Ramer, K.J. Malloy: MRS Bulletin **22**, 45 (July 1997)

4.40   S.N. Mohammad, H. Morkoç: Progr. Quantum Electron. **20**, 361 (1996)

4.41   S. Fujieda, M. Mizuta, Y. Matsumoto: Jpn. J. Appl. Phys. **26**, 2067 (1987)

4.42   H. Okumura, S. Misawa, S. Yoshida: Appl. Phys. Lett. **59**, 1058 (1991)

4.43   J.E. Andrews, M.A. Littlejohn: J. Electrochem. Soc. **122**, 1273 (1975)

4.44   L. Reimer: *Transmission Electron Microscopy*, 4th edn., Springer Ser. Opt. Sci., Vol.36 (Springer, Berlin, Heidelberg 1997)

4.45   A. Kuramata, K. Horino, K. Domen, K. Shinohora, T. Tanahashi: Appl. Phys. Lett. **67**, 2521 (1995)

4.46   S. Nakamura, M. Senoh, S. Nagahama, N. Iwasa, T. Yamada, T. Matsushita, H. Kiyoku, Y. Sugimoto: Jpn. J. Appl. Phys. **35**, L217 (1996)

4.47   C.I. Sun, J.W. Yang, Q. Chen, M.A. Khan, T. George, P. Chang-Chien, S. Mahajan: Appl. Phys. Lett. **68**, 1129 (1996)

4.48   M.A. Herman, H. Sitter: *Molecular Beam Epitaxy*, 2nd edn., Springer Ser. Mater. Sci., Vol.7 (Springer, Berlin, Heidelberg 1996)

4.49   M.B. Panish, H. Temkin: *Gas Source Molecular Beam Epitaxy*, Springer Ser. Mater. Sci., Vol.26 (Springer, Berlin, Heidelberg 1993)

4.50   S.T. Strite, H. Morkoç: Energetic particle assisted MBE, in *Handbook of Thin Film Process Technology*, ed. by S.I. Shah, D.A. Glocker (IoP, Bristol 1995) pp. 1-25

4.51   *The Technology, Physics of Molecular Beam Epitaxy*, ed. by E.H.C. Parker (Plenum, New York 1985)

4.52   H. Lueth: *Surfaces abd Interfaces of Solid Materials*, 3rd edn. (Springer, Berlin, Heidelberg 1995)

4.53   G. Popovici, S.N. Mohammad, H. Morkoç: In *Physics, Applications of Group III Nitride Semiconductor Compounds*, ed. by B. Gil (Oxford Univ. Press, London 1998)

4.54   H. Morkoç, S. Strite, G. B. Gao, M.E. Lin, B. Sverdlov, M. Burns: J. Appl. Phys. (Rev.) **76**, 1363 (1994)

4.55   K.W. Boer: *Survey of Semiconductor Physics* (Van Nostrand Reinhold, New York 1990) Vol.1

4.56   S. Yoshida, S. Misawa, A. Itoh: Appl. Phys. Lett. **26**, 461 (1975)

4.57   S. Winsztal, B. Wauk, H. Majewska-Minor, T. Niemyski: Thin Solid Films **32**, 251 (1976)

4.58   W. Kim, O. Aktas, A.E. Botchkarev, A. Salvador, S.N. Mohammad, H. Morkoç: J. Appl. Phys. **79**, 7657 (1994)

4.59   R.C. Powell, N.-E. Lee, Y.-W. Kim, J.E. Greene: J. Appl. Phys. **73**, 1891 (1993)

4.60   K.R. Evans, T. Lei, R. Kaspi, C.R. Jones: Presented at Topical Workshop on III-V Nitrides, Nagoya, Japan (September 1995)

4.61   R.P. Burns, K.A. Gabriel, D.E. Pierce: J. Am. Ceram. Soc. **76**, 273 (1993)

4.62   J.R. Jenny, R. Kaspi, C.R. Jones, K.R. Evans: J. Cryst. Growth **175/176**, 89 (1997)

4.63   Q. Zhu, A. Botchkarev, W. Kim, Ö. Aktas, B.N.Sverdlov, H. Morkoç: Appl. Phys. Lett. **68**, 1141 (1996)

4.64   A. Botchkarov, A. Salvador, R. Sverdlov, J. Myoung, H. Morkoç: J. Appl. Phys. **77**, 4455 (1995)

4.65   Z.Z. Bandic, R.J. Hauenstein, M.L. O'Steen, T.C. McGill: Appl. Phys. Lett. **68**, 1510 (1996)

4.66   M.V. Averyanova, S. Yu. Karpov, Yu. N. Makarov, I.N. Przhevalskii, M. S. Ramm, R. A. Talalaev: MRS Internet J. Nitride Semicond. Res. **1**, No.31 (1996)

4.67   O. Aktas, Z. Fan, A. Botchkarov, S.N. Mohammad, M. Roth, T. Jenkins, L. Kehias, H. Morkoç: IEEE Electron Dev. Lett. **18**, 293 (1997)

4.68   S.N. Mohammad, A. Salvador, H. Morkoç: Proc. IEEE **83**, 1306 (1995)

4.69   H. Morkoç: Beyond SiC! III-V nitride based heterostructures, devices, in *SiC Materials, Devices*, ed. by Y. S. Park (Academic, San Diego 1998), Vol.52, pp.307-394

4.70   M. Smith, J. Y. Lin, H. X. Jiang, A. Salvador, A. Botchkarev, W. Kim, H. Morkoç: Appl. Phys. Lett. **69**, 2453 (1996)

4.71   R. Singh, D. Doppalaudi, T.D. Moustakas: Appl. Phys. Lett. **69**, 2388 (1996)

4.72   T.J. Schmidt, X.H. Yang, W. Shan, J.J. Song, A. Salvador, W. Kim, Ö. Aktas, A. Botchkarev, H. Morkoç: Appl. Phys. Letts. **68**, 1820 (1996)

4.73   B. Sverdlov, G.A. Martin, H. Morkoc, D.J. Smith: Appl. Phys. Lett. **67**, 2063 (1995)

4.74   T. Matsuoka, N. Yoshimoto, T. Sasaki, A. Katsui: J. Electron. Mater. **21**, 157 (1992)

4.75   F. Hamdani, A. Botchkarev, W. Kim, A. Salvador, H. Morkoç, M. Yeadon, J. M. Gibson, S. C. T. Tsen, D. J. Smith, D. C. Reynolds, D. C. Look, K. Evans, C. W. Litton, W. C. Mitchel, P. Hemenger: Appl. Phys. Lett. **70**, 467 (1997)

4.76   F. Hamdani, M. Yeadon, D.J. Smith, H. Tang, W. Kim, A. Salvador, A. Botchkarev, J.M. Gibson, A.Y. Polydkov, M. Skowronski, H. Morkoç: J. Appl. Phys. **83**, 983 (1998)

4.77   F.A. Ponce, D.P. Bour, W.T. Young, M. Sounders, J.W. Steeds: Appl. Phys. Lett. **69**, 337 (1996)

4.78   A. Gassmann, T. Suski, N. Newmann, C. Kiselowski, E. Jones, E.R. Weber, Z. Liliental-Weber, M. D. Rubin, H. I. Helava, I. Grezegory, M. Bockovski, J. Jun, S. Porowski: J. Appl. Phys. **80**, 2195 (1996)

4.79   M.A.L. Johnson, S. Fujita, W.H. Rowland Jr., W.C. Hughes, Y.W. He, N.A. El-Masry, J.W. Cook Jr., J.F. Schetzina, J. Ren, J.A. Edmond: J. Electron. Mater. **25**, 793 (1996)

4.80   H. Amano, M. Kito, K. Hiramatsu, I. Akasaki: Jpn. J. Appl. Phys. **28**, L2112 (1989)

4.81   S. Nakamura, T. Mukai, M. Senoh: Jpn. J. Appl. Phys. **30**, L1998 (1991)

4.82   T. Tanaka, A. Watanabe, H. Amano, Y. Kobayashi, I. Akasaki, S. Yamazaki, M. Koike: Appl. Phys. Lett. **65**, 593 (1994)

4.83   S. Yamasaki, S. Asami, N. Shibata, M. Koike, K. Manabe, T. Tanaka, H. Amano, I. Akasaki: Appl. Phys. Lett. **66**, 1112 (1995)

4.84   H. Morkoç, B. Sverdlov, G. B. Gao: Proc. IEEE **81**, 492 (1993)

4.85   A. Bykhovski, B. Gelmont, M. Shur: J. Appl. Phys. **81**, 6332 (1997)

4.86   A. D. Bykhovski, B.L. Gelmont, M.S. Shur: J. Appl. Phys. **78**, 3691 (1995)

4.87   J.P. Hirth, J. Lothe: *Theory of Dislocations* (Wiley, New York 1982) pp.231–278

Chapter 5

5.1   K. Hiramatsu, H. Amano, I. Akasaki, H. Kato, N. Koide, K. Manabe: J. Cryst. Growth **107**, 509 (1991)

5.2   Z. Sitar, M.J. Paisley, B. Yan, R.F. Davis: MRS Proc. **162**, 537 (1990)

5.3   F.A. Ponce, J.S. Major Jr., W.E. Plano, D.F. Welch: Appl. Phys. Lett. **65**, 2302 (1994)

5.4   F.A. Ponce: MRS Bulletin **22** (2), 51 (1997)

5.5   F.A. Ponce, D.P. Bour: Nature **386**, 351 (1997)

5.6   R.F. Davis, Z. Sitar, B.E. Williams, H.S. Kong, H.J. Kim, J.W. Palmour, J.A. Edmond, J. Ryu, J.T. Glass, C.H. Carter Jr.: Mater. Sci. Eng. B **1**, 77 (1988)

5.7   W. Seifert, A. Tempel: Phys. Status Solidi A **23**, K39 (1974)

5.8   R.C. Powell: Heteroepitaxial wurtzite and zincblende structure GaN grown by molecular beam epitaxy: Growth kinetics, microstructure, and properties. Ph.D. Thesis, University of Illinois at Urbana-Champaign (1992)

5.9   T. Lei, T.D. Moustakas: MRS Symp. **242**, 433 (1992)

5.10   S. Strite, B. Sariel, D.J. Smith, H. Chen, H. Morkoç: 7th Int'l MBE Conf., Schwäbisch Gmünd, Germany (1992). J. Cryst. Growth **127**, 204 (1993)

5.11   M. Lannoo, J. Bourgoin: *Point Defects in Semiconductors I Theoretical Aspects*, Springer Ser. Solid-State Sci., Vol.22 (Springer, Berlin, Heidelberg 1981)

5.12   J. Bourgoin, M. Lannoo: *Point Defects in Semiconductors II Experimental Aspects*, Springer Ser. Solid-State Sci., Vol.35 (Springer, Berlin, Heidelberg 1983)

5.13   H.P. Maruska, J.J. Tietjen: Appl. Phys. Lett. **15**, 327 (1969)

5.14   J. Neugebauer, C.G. Van de Walle: Appl. Phys. Lett. **69**, 503 (1996)

5.15   J. Neugebauer, C.G. Van de Walle: Phys. Rev. B **50**, 8067 (1994)

5.16   J. Neugebauer, C.G. Van de Walle: *Proc. Int'l Conf. on the Physics of Semiconductors*, ICPS-22 (World Scientific, Singapore 1995) p.2327

5.17   P. Perlin, T. Suzuki, H. Teisseyre, M. Leszczynski, I. Gregory, J. Jun, S. Porowski, P. Boguslawski, J. Bernholc, J.C. Chervin, A. Polian, T.D. Moustakas: Phys. Rev. Lett. **75**, 296 (1995)

5.18   C. Wetzel, W. Walukiewicz, E.E. Haller, J. Ager III: Phys. Rev. B **53**, 1322 (1996)

5.19   C. Wetzel, T. Suchi, J.W. Ager III, E.R. Weber, E.E. Haller, S. Fischer. B.K. Meyer, R.J. Molnax, P. Berlin: Phys. Rev. Lett. **78**, 3923 (1997)

5.20   D.W. Jenkins, J.D. Dow: Phys. Rev. B **39**, 3317 (1989)

5.21   D.W. Jenkins, J.D. Dow, M.-H. Tsai: J. Appl. Phys. **72**, 4130 (1992)

5.22   T.L. Tansley, C.P. Foley: J. Appl. Phys. **59**, 3241 (1986)

5.23   J. Neugebauer, C.G. Van de Walle: *Festkörperprobleme, Advances in Solid State Physics* **35**, 25 (Vieweg, Braunscheweig 1996)

5.24   J. Neugebauer, C.G. Van de Walle: In *Diamond, SiC and Nitride Wide Bandgap Semiconductors*, ed. by C.H. Carter Jr., G. Gildenblatt, S. Nakamura, R.J. Nemanich. MRS Proc. **339**, 687 (1994)

5.25   P. Boguslawski, E.L. Briggs, J. Bernholc: Phys. Rev. B **51**, 17255 (1995)

5.26   W. Kim, A.E. Botchkarev, A. Salvador, G. Popovici, H. Tang, H. Morkoç: J. Appl. Phys. **82**, 219 (1997)

5.27   J. Neugebauer, C.G. Van de Walle: Appl. Phys. Lett. **68**, 1829 (1996)

5.28   M.S. Brandt, N.M. Johnson, R.J. Molnar, R. Singh, T.D. Moustakas: Appl. Phys. Lett. **64**, 2264 (1994)

5.29   W. Götz, N.M. Johnson, J. Walker, D.P. Bour, H. Amano, I. Akasaki: Appl. Phys. Lett. **67**, 2666 (1995)

5.30   D.C. Look, D.C. Reynolds, J.W. Hernshy, J.R. Sizelove, R.L. Jones, R.J. Molnar: Phys. Rev. Lett. **79**, 2273 (1997)

5.31   T.L. Tansley, R.J. Egan: Physica B **185**, 190 (1993)

5.32   T.L. Tansley, E. Goldys, M. Godlewski, B. Zhou, H.Y. Zhou: In *Optical Properties of GaN and Related Materials*, ed. by S.J. Pearton (Gordon and Breach, New York 1997) p.233

5.33   D.C. Reynolds, D.C. Look, B. Jogai, J. Van Nostrand, R.L. Jones, J. Jenny: Solid State Commun. **106**, 701-704 (1998)

5.34   N. Koide, H. Kato, M. Sassa, S. Yamasaki, K. Manabe, M. Hashimoto, H. Amano, K. Hiramatsu, I. Akasaki: J. Cryst. Growth **115**, 639 (1991)

5.35   W. Götz, N.M. Johnson, C. Chen, H. Liu, C. Kuo, W. Imler: Appl. Phys. Lett. **68**, 3144 (1996)

5.36   S. Nakamura, T. Mukai, M. Seno: Jpn. J. Appl. Phys. **31**, 195 (1992)

5.37   B. Goldenberg, J.D. Zook, J. Van Vechten: Bull. Am. Phys. Soc. **38**, 446 (1993)

5.38   H. Morkoç, S. Strite, G.B. Gao, M.E. Lin, B. Sverdlov, M. Burns: J. Appl. Phys. Rev. **76**, 1363 (1994)

5.39   S.N. Mohammad, A. Salvador, H. Morkoç: Proc. IEEE **83**, 1306 (1995)

5.40   S.N. Mohammad, H. Morkoç: Progress and prospects of group III-V nitride semiconductors. *Progress in Quantum Electronics* **20**, 361 (Pergamon, Oxford 1996)

5.41   C.-C. Yi, B.W. Wessels: Appl. Phys. Lett. **69**, 3026 (1996)

5.42   H. Amano, M. Kito, K. Hiramatsu, I. Akasaki: Jpn. J. Appl. Phys. **28**, L2112 (1989)

5.43   S. Nakamura, T. Mukai, M. Senoh: Jpn. J. Appl. Phys. **30**, L1998 (1991)

5.44   I. Akasaki, H. Amare, M. Kitoh, K. Hiramatsu, Z. Akasaki: J. Electro. Chem. Soc. **137**, 1639 (1989)

5.45   V. Fiorentini, M. Methfessel, M. Scheffler: Phys. Rev. B **47**, 13353 (1993)

5.46   S.-H. Wei, A. Zunger: Phys. Rev. B **37**, 8958 (1988)

5.47   J.A. Van Vechten, J.D. Zook, R.D. Hornig, B. Goldenberg: Jpn. J. Appl. Phys. **31**, 3662 (1992)

5.48   J. Neugebauer, C.G. Van de Walle: Phys. Rev. Lett. **75**, 4452 (1995)

5.49   Y. Okamoto, M. Saito, A. Oshiyama: Jpn. J. Appl. Phys. **35**, L807 (1996)

5.50   W. Kim, A. Salvador, A.E. Botchkarev, Ö. Aktas, S.N. Mohammad, H. Morkoç: Appl. Phys. Lett. **69**, 559 (1996)

5.51   W. Götz, N.M. Johnson, J. Walker, D.P. Bour, R.A. Street: Appl. Phys. Lett. **68**, 667 (1996)

5.52   J. Chevallier, B. Clerjaud, B. Pajot: In *Hydrogen in Semiconductors*, ed. by J.I. Pankove, N.M. Johnson. Semiconductors and Semimetals, Vol. 34 (Academic, New York 1991)

5.53   J.A. Van Vechten: Private commun. (1997)

5.54   C. Yuan, T. Salagaj, A. Guray, P. Zawadzki, C.S. Chern, W. Kroll, R.A. Stall, Y. Li, M. Schurman, C.-Y. Hwang, W.E. Mayo Y. Lu, S.J. Pearton, S. Krishnankutty, R.M. Kolbas: J. Electrochem. Soc. **142**, L163 (1995)

5.55   S. Nakamura, N. Iwasa, M. Senoh, T. Mukai: Jpn. J. Appl. Phys. **31**, 1258 (1992)

5.56   J. Neugebauer, C.G. Van de Walle: MRS Proc. **395**, 645 (1996),

5.57   W. Götz, N.M. Johnson, D.P. Bour: Appl. Phys. Lett. **68**, 3470 (1996)

5.58   J.C. Zolper, M.H. Crawford, A.J. Howard, J. Ramer, S.D. Hersee: Appl. Phys. Lett. **68**, 200 (1996)

5.59   J.I. Pankove, M.T. Duffy, E.A. Miller, J.E. Berkeyheiser: J. Luminescence **8**, 89 (1973)

5.60   M. Ilegems, R. Dingle: J. Appl. Phys. **44**, 2434 (1973)
       J.I. Pankove, J.A. Hutchby: J. Appl. Phys. **47**, 5387 (1976)

5.61   F. Bernardini, V. Fiorentini: Appl. Phys. Lett. **70**, 2990 (1997)

5.62   A. Salvador, W. Kim, Ö. Aktas, A. Botchkarev, Z. Fan, H. Morko: Appl. Phys. Lett. **69**, 2692 (1996)

5.63   O. Brandt, H. Yang, H. Kostial, K. Ploog: Appl. Phys. Lett. **69**, 2702 (1996)

5.64   J.I. Pankove, J.A. Hutchby: J. Appl. Phys. **47**, 5387 (1976)

5.65   E. Ejder, H.G. Grimmeiss: J. Appl. Phys. **5**, 275 (1974)

5.66   P. Boguslawski, E.L. Brigs, J. Bernholc: Appl. Phys. Lett. **69**, 233 (1996)

5.67   M. Sato: Appl. Phys. Lett. **68**, 935 (1996)

5.68   C.R. Abernathy, J.D. MacKenzie, S.J. Pearton, W.S. Hobson: Appl. Phys. Lett. **66**, 1969 (1995)

5.69   S. Fisher, C. Wetzel, E.E. Haller, B.K. Meyer: Appl. Phys. Lett. **67**, 1298 (1995)

5.70   J.I. Pankove, E.A. Miller, J.E. Berkeyheiser: RCA Rev. **32**, 383 (1971)

5.71   H. Amano, S. Sawaki, I. Akasaki, Y. Toyoda: J. Cryst. Growth **93**, 79 (1988)

5.72   F. Bernardini, V. Fiorentini, R.M. Nieminen: *Proc. 23rd Int'l Conf. on the Physics of Semiconductors*, Berlin 1996 (World Scientific, Singapore 1996) p.497

5.73   B. Monemar, H.P. Gislason, O. Lagertedt: J. Appl. Phys. **51**, 625 (1980)

5.74   S. Strite: Jpn. J. Appl. Phys. **33**, L699 (1994)

5.75   J.W. Lee, S.J. Pearton, J.C. Zolper, R.A. Stall: Appl. Phys. Lett. **68**, 2102 (1996)

5.76   J.D. MacKenzie, C.R. Abernathy, S.J. Pearton, U. Hommerich, X. Wu, R.N. Schwartz, R.G. Wilson, J.M. Zavada: Appl. Phys. Lett. **69**, 2083 (1996)

5.77   M. Rubin, N. Newman, J.C. Chan, T.C. Fu, J.T. Ross: Appl. Phys. Lett. **64**, 64 (1994)

5.78   S.J. Pearton, C.B. Vartuli, J.C. Zolper, C. Yuan, R.A. Stall: Appl. Phys. Lett. **67**, 1435 (1995)

5.79   R.G. Wilson, S.J. Pearson, C.R. Abernathy, J.M. Zavada: Appl. Phys. Lett. **66**, 2238 (1995)

5.80   R.G. Wilson, C.B. Vartuli, C.R. Abernathy, S.J. Pearton, J.M. Zavada: Solid State Electron. **38**, 11329 (1995)

5.81   J.C. Zolper, R.G. Wilson, S.J. Pearton, R.A. Stall: Appl. Phys. Lett. **68**, 1945 (1996)

5.82   S.C. Binari, L.B. Rowland, W. Kruppa, G. Kelner, K. Doverspike, D.K. Gaskill: Electron. Lett. **30**, 1248 (1994)

5.83   S.J. Pearton, C.R. Abernathy, P.W. Wisk, W.S. Hobson, F. Ren: Appl. Phys. Lett. **63**, 2238 (1993)

5.84   J.C. Zolper, S.J. Pearton, C.R. Abernathy, C.B. Vartuli: Appl. Phys. Lett. **66**, 3042 (1995)

5.85   D.V. Lang: Space-charge spectroscopy in semiconductor, in *Thermally Stimulated Relaxation in Solids*, Topics Appl. Phys., Vo.37 (Springer, Berlin, Heidelberg 1979)

5.86   L.C. Kimerling: In *Defects in Semiconductors*, ed. by S. Narayan, T.Y. Tan. Mater. Res. Soc. Proc., Vol.2 (North-Holland, New York 1981)

5.87   W. Götz, N.M. Johnson, R.A. Street, H. Amano, I. Akasaki: Appl. Phys. Lett. **66**, 1340 (1995)

5.88   P. Hacke, T. Detchprohm, K. Hiramatsu, N. Sasaki: J. Appl. Phys. **76**, 304 (1994)

5.89   W. Götz, N.M. Johnson, H. Amano, I. Akasaki: Appl. Phys. Lett. **65**, 463 (1994)

5.90   W.I. Lee, T.C. Huang, J.D. Guo, M.S. Feng: Appl. Phys. Lett. **67**, 1721 (1995)

5.91   C.D. Wang, Q.Z. Liu, D. Qiao, L.S. Yu, S. Lau, E.T. Yu, W. Kim, A. Botchkarev, H. Morkoç: Appl. Phys. Lett. **72**, 1211 (1998)

5.92   Z.-Q. Fang, D.C. Look, W. Kim, Z. Fan, A.E. Botchkarev, H. Morkoç: Appl. Phys. Lett. **72**, 2277 (1998)

5.93   V. Fiorentini, F. Bernardini, A. Bosin, D. Vanderbilt: *Proc. 23rd Int'l Conf. on the Physics of Semiconductors*, Berlin 1996 (World Scientific, Singapore 1996) p.497

Chapter 6

6.1   S.M. Sze: *Physics of Semiconductor Devices* (Wiley, New York 1981) Chap.5

6.2   E.H. Rhoderick, E.H. William: *Metal Semiconductor Contacts* (Clarendon, Oxford 1988)

6.3   F.A. Padovani, R. Stratton: Solid-State Electron. **9**, 695 (1966)

6.4   K. Suzue, S.N. Mohammad, Z.F. Fan, W. Kim, Ö. Aktas, A.E. Botchkarev, H. Morkoç: J. Appl. Phys. **80**, 4467 (1996)

6.5   I. Suemune: Appl. Phys. Lett. **63**, 2612 (1993)

6.6   J. Bardeen, Phys. Rev. **71**, 717 (1947)

6.7   A.M. Cowley, S.M. Sze: J. Appl. Phys. **96**, 3212 (1965)

6.8   W. Shockley: Rep, AL-TDR-64-207, Air Force Avionics Laboratory, Wright-Patterson AFB, OH (1964)

6.9   H. Murrmann, D. Widmann: Solid-State Electron. **12**, 879 (1969)

6.10   G.K. Reeves, H.B. Harbison: IEEE EDL-**3**, 111 (1982)

6.11   K.K. Shih, J.M. Blum: Solid-State Electron. **15**, 1177 (1971)
6.12   A. Scorzoni, M. Finetti: Materials Sci. Rpt. **3**, 79 (1988)
6.13   T.C. Shen, G.B. Gao, H. Morkoç; J. Vac. Sci. Technol. B **10**, 2113 (1992)
6.14   J.S. Foresi, T.D. Moustakas: Appl. Phys. Lett. **62**, 2859 (1993)
6.15   M.E. Lin, Z. Ma, F.Y. Huang, Z. Fan, L. H. Allen, H. Morkoç: Appl. Phys. Lett. **64**, 1003 (1994)
6.16   S. Nakamura, T. Mukai, M. Senoh: Jpn. J. Appl. Phys. **30**, L1998-2000 (1991)
6.17   S. Nakamura, M. Senoh, T. Mukai: Appl. Phys. Lett. **62**, 2390 (1993)
6.18   C.K. Peng, J. Chen, J. Chyi, H. Morkoç: Appl. Phys. Lett. **64**, 429 (1988)
6.19   Y. Wu, W. Jiang, B. Keller, S. Keller, D. Kapolnek, S. Denbaars, U. Mishra: Solid State Electron. **41**, 165 (1997)
6.20   M.E. Lin, F.Y. Huang, H. Morkoç: Appl. Phys. Lett. **64**, 2557 (1994)
6.21   Z. Fan, S. N. Mohammad, W. Kim, Ö Aktas, A.E. Botchkarev, H. Morkoç: Appl. Phys. Lett. **68**, 1672 (1996)
6.22   K.J. Duxstad: Metal contacts to ZnSe and GaN. PhD Thesis, Dept. of Mater. Sci., University of California at Berkely (1997)
6.23   S. Ruvimov, Z. Liliental-Weber, J. Washburn, K.J. Duxstad, E.E. Haller, S.N. Mohammad, Z. Fan, H. Morkoç: Appl. Phys. Lett. **69**, 1556 (1996)

Chapter 7

7.1   W. Shockley: *Electrons and Holes in Semiconductors* (Van Nostrand, Princeton, NJ 1950)
7.2   W. Götz, N.M. Johnson, C. Chen, H. Liu, C. Kuo, W. Imler: Appl. Phys. Lett. **68**, 3144 (1996)
7.3   D.C. Look, J.R. Sizelove, S. Keller, Y.F. Wu, U.K. Mishra, S.P. Den Baars: Solid State Commun. **102**, 297 (1997)

Chapter 8

8.1   D.L. Rode: Low field electron transport. *Semiconductors and Semimetals* **10**, 1 (Academic, New York 1975)
8.2   D.A. Anderson, N. Apsley: The Hall effect in III-V semiconductors. Semicond. Sci. Technol. **1**, 187 (1986)
8.3   B.R. Nag: *Electron Transport Compound Semiconductors*, Springer Ser. Solid-State Sci., Vol. 11 (Springer, Berlin, Heidelberg 1980)
8.4   D.C. Look: *Electrical Characterization of GaAs Materials, Devices* (Wiley, New York 1989)
8.5   H. Brooks: Adv. Electron. Electron. Phys. **7**, 85 (1955)
8.6   B. Gelmont, B. Lund, K.-S. Kim, G.U. Jensen, M.S. Shur, T.A. Fjeldly: J. Appl. Phys. **71**, 4977 (1992)
8.7   B. Gelmont, M. Shur, M. Stroscio: J. Appl. Phys. **77**, 657 (1995)
8.8   H. Ehrenreich: J. Phys. Chem. Solids **8**, 130 (1959)
8.9   J.D. Wiley: Mobility of holes in semiconductors. *Semiconductors and Semimetals* **10**, 134 (Academic, New York 1975)
8.10   C. Hammar, B. Magnusson: Phys. Scripta **6**, 206 (1972)

8.11 W. Shockley: *Electrons, Holes in Semiconductors* (Van Nostrand, Princeton, NJ 1950)

8.12 D.C. Look, D.K. Lorance, J.R. Sizelove, C.E. Stutz, K.R. Evans, D.W. Whitson: J. Appl. Phys. **71**, 260 (1992)

8.13 V.W.L. Chin, B. Zhou, T.L. Tansley, X. Li: J. Appl. Phys. **77**, 6064 (1995)

8.14 J.C. Phillips: Rev. Mod. Phys. **42**, 317 (1970)

8.15 M.E. Lin, B.N. Sverdlov, S. Strite, H. Morkoç, A.E. Drakin: Electron. Lett. **29**, 1759 (1993)

8.16 Y.J. Wang, R. Kaplan, H.K. Ng, K. Doverspike, D.K. Gaskill, T. Ikedo, H. Amano, I. Akasaki: J. Appl. Phys. **79**, 8007 (1996)

8.17 D.L. Rode: Low field electron transport, in *Semiconductors and Semimetals* **10**, 1 (Academic, New York 1975)

8.18 D.K. Schroeder: *Semiconductor Material and Device Characterization* (Wiley, New York 1990)

8.19 A.D. Bykhovski, V.V. Kaminski, M.S. Shur, Q.C. Chen, M.A. Khan: Appl. Phys. Lett. **68**, 818 (1996)

8.20 W. Shan, T. Schmidt, T.X.H. Yang, J.J. Song, B. Goldenberg: J. Appl. Phys. **79**, 3691 (1996)

8.21 Perlin, E. Litwin-Staszewska, B. Suchanek, W. Knap, J. Camassel, T. Suski, R. Piotrzkowski, I. Grzegory, S. Porowski, E. Kaminska, J.C. Chervin: Appl. Phys. Lett. **68**, 1114 (1996)

8.22 T.L. Tansley: Crystal structure, mechanical properties, thermal properties and refractice index of InN, in *Group III-V Nitrides*, ed. by J.H. Edgar, EMIS Datarev. Series, Vol. 11 (Inspec, London 1994)

8.23 V.W.L. Chin, T.L. Tansley, T. Ostockton: J. Appl. Phys. **75**, 7365 (1995)

8.24 T.L. Tansley, E.M. Goldys, M. Godlewski, B. Zhou, H.Y. Zuo: The contribution of defects to the electrical and optical properties of GaN, in *GaN and Related Materials*, ed. by S. Pearton (Gordon and Breach, Amsterdam 1997)

8.25 S.T. Strite, H. Morkoç: GaN, AlN, and InN: A review. J. Vacuum Sci. Techn. B **10**, 1237-1266 (1992)

8.26 S.M. Sze: *Physics of Semiconductor Devices*, 2nd edn. (Wiley Interscience, New York 1981)

8.27 D.C. Look, J.R. Sizelove, S. Keller, Y.F. Wu, U.K. Mishra, S.P. DenBaars: Solid State Commun. **102**, 297 (1997)

8.28 D.L. Rode: Phys. Status Solidi B **55**, 687 (1973)

8.29 J. Kolnik, I.H. Oguzman, K.F. Brennan, R. Wang, P.P. Ruden, Y. Wang: J. Appl. Phys. **78**, 1033, 1995

8.30 M.A. Littlejohn, J.R. Hauser, T.H. Glisson: Appl. Phys. Lett. **26**, 625-627 (1975)

8.31 M.S. Shur, B. Gelmont, C. Saavedra-Munoz, G. Kelner: Proc. 5th Conf. SiC Related Materials. IoP Conf. Ser. **137**, 155 (IoP, Bristol 1994)

8.32 V.W.L. Chin, T.L. Tansley, T. Ostockton: J. Appl. Phys. **75**, 7365 (1994)

8.33 M. Ilegems, H.C. Montgomery: J. Phys. Chem. Solids **34**, 885 (1973)

8.34 M.A. Khan, J.N. Kuznia, J.M. Van Hove, D.T. Olson, S. Krishnakutty, R.M. Kolbas: Appl. Phys. Lett. **58**, 526 (1991)

8.35 I. Akasaki, H. Amano, Y. Koide, K. Hiramatsu, N. Sawaki: J. Cryst. Growth **98**, 209 (1990)

8.36 S. Nakamura, T. Mukhai, M. Senoh: J. Appl. Phys. **71**, 5543 (1992)

8.37 H. Amano, H. Amano, N. Sawaki, I. Akasaki: Appl. Phys. Lett. **48**, 353 (1986)

8.38   K. Hiramatsu, S. Itoh, H. Amano, I. Akasaki, N. Kuwano, T. Shiraish, K. Oki: J. Crystal Growth **115**, 628 (1991)

8.39   S. Nakamura, T. Mukhai, M. Senoh: J. Appl. Phys. **71**, 5543-5549 (1992)
S. Nakamura: Jpn J. Appl. Phys. **30**, L1705-1707 (1991)

8.40   D.K. Gaskill, L.B. Rowland, K. Doverspike: Electrical properties of AlN, GaN, AlGaN, in *Properties of Group III Nitrides*, ed. by J. H. Edgar, Inspec Ser., Vol. 11 (Inst. Electr. Eng., London, UK 1994) pp. 101-116

8.41   D.L. Rode, D.K. Gaskill: Appl. Phys. Lett. **66**, 1972 (1995)

8.42   D.C. Look, D.C. Reynolds, J.R. Sizelove, R.L. Jones, R.J. Molnar: Phys. Rev. Lett. **79**, 2273 (1997)

8.43   D.C. Look, R.J. Molnar: Appl. Phys. Lett. **70**, 3377 (1997)

8.44   H. Morkoç, H. Ü nlü, G. Ji: *Fundamentals, Technology of MODFETs* (Wiley, New York 1991) Vol. 2

8.45   J. Redwing, M.A. Tischler, J.S. Flynn, S. Elhamri, M. Ahoujja, R.S. Newrock, W.C. Mitchel: Appl. Phys. Lett. **69**, 963 (1996)

8.46   T. Tanaka, A. Watanabe, H. Amano, Y. Kobayashi, I. Akasaki, S. Yamazaki, M. Koike: Appl. Phys. Lett. **65**, 593 (1994)

8.47   S. Yamasaki, S. Asami, N. Shibata, M. Koike, K. Manabe, T. Tanaka, H. Amano, I. Akasaki: Appl. Phys. Lett. **66**, 1112-1114 (1995)

8.48   S. Mohammad. A. Salvador, H. Morkoç: Proc. IEEE **83**, 1306-1355 (1995)

8.49   T.L. Tansley, C.P. Foley: *Proc. 3rd Int'l Conf. on Semiinsulating III-V Materials* (Warm Springs, OR 1984), ed. by J. S. Blakemore (Shiva, London 1985) pp. 497-500

8.50   W.R. Bryden, S.A. Ecelberger, M.E. Hawley, T.J. Kistenmacher: ECR-assisted reactive magnetron sputtering of InN, in *Diamond, SiC, Nitride Wide Bandgap Semiconductors*, ed. by C. H. Carter Jr., G. Gildenblat, S. Nakamura, R. J. Nemanich (MRS, Pittsburgh, PA 1994)

8.51   T.L. Tansley, C.P. Foley: Electron. Lett. **20**, 1066 (1984)

8.52   M. Suzuki, T. Uenoyama: In *Group III Nitride Semiconductor Compounds: Physics and Applications*, ed. by B. Gil (Clarendon, Oxford 1998)

8.53   J. Edwards, K. Kawabe, G. Stevens, R.H. Tredgold: Solid State Commun. **3**, 99 (1965)

8.54   K. Kawabe, R.H. Tredgold, Y. Inuishi: Electr. Eng. Jpn. **87**, 62 (1967)

8.55   D.C. Look, D.C. Reynolds, W. Kim, O. Aktas, A, Botchkarev, A. Salvador, H. Morkoç: J. Appl. Phys. **80**, 2960-2962 (1996)

8.56   W. Kim, A.E. Botchkarev, A. Salvador, G. Popovici, H. Tang, H. Morkoç: J. Appl. Phys. **82**, 219 (1997)

8.57   N.F. Mott, W.D. Twose: Adv. Phys. **10**, 107 (1961)

8.58   T. Matsubara, Y. Toyozawa: Prog. Theoret. Phys. (Kyoto) **26**, 739 (1961)

8.59   J.S. Im, A. Moritz, F. Steuber, V. Härle, F. Scholz, A. Hangleiter: Appl. Phys. Lett. **70**, 631 (1997)

Chapter 9

9.1   A.G. Milnes, D.L. Feucht: *Heterojunctions and Metal Semiconductor Junctions* (Academic, New York 1972)

9.2   E.T. Yu, J.O. McCaldin, T.C. McGill: *Solid State Physics* **46**, 1-146 (Academic, San Diego 1992)

9.3   R.L. Anderson: Solid State Electron. **5**, 341 (1962)

9.4   H.L. Skriver: *The LMTO Method*, Springer Ser. Solid-State Sci., Vol.41 (Springer, Berlin, Heidelberg 1984)

9.5   G.A. Martin, S. Strite, A. Botchkarev, A. Agarwal, A. Rockett, H. Morkoç, W.R.L. Lambrecht, B. Segall: Appl. Phys. Lett. **65**, 610 (1994)

9.6   G.A. Martin, A. Botchkarev, A. Agarwal, A. Rockett, H. Morkoç: Appl. Phys. Lett. **68**, 2541 (1996)

9.7   G.A. Martin: Semiconductor electronic band alignment at heterojunctions of wurtzite AlN, GaN, and InN. PhD. Thesis, Department of Physics, University of Illinois (1996)

9.8   J.P. Waldrop, R.W. Grant: Appl. Phys. Lett. **68**, 2879 (1996)

9.9   J. Baur, K. Maier, M. Kunzer, U. Kaufmann, J. Schneider: Appl. Phys. Lett. **65**, 2211 (1994)

9.10 E.A. Albanesi, W.R.L. Lambrecht, B.E. Segall: *Diamond, SiC and Nitride Wide Bandgap Semiconductors*, ed. by C.H. Carter Jr., C. Gildenblat, S. Nakamura, R.J. Nemanich. MRS. Proc. **339**, 607 (Mater. Res. Soc., Pittsburgh, PA 1994)

9.11 D.L. Smith: Solid State Commun. **57**, 919 (1986)

9.12 S.T. Strite, H. Morkoç: J. Vac. Sci. Technol. B **10**, 1237-1266 (1992)

9.13 A. Bykhovski, B. Gelmont, M. Shur: J. Appl. Phys. **74**, 6734 (1993)

9.14 F. Bernardini, V. Fiorentini, D. Vanderbilt: *III-V Nitrides*, ed. by F.A. Ponce, T.D. Moustakas, I. Akasaki, B.A. Monamar. MRS Proc. **449**, 923 (Mater. Res. Soc., Pittsburgh, PA 1997)

9.15 R.N. Hall: Phys. Rev. **87**, 387 (1952)

9.16 W. Shockley, W.T. Read: Phys. Rev. **87**, 835 (1952)

9.17 C.T. Sah, R.N. Noyes, W. Shockley: Proc. IRE **45**, 1228 (1957)

9.18 A.S. Grove: *Physics and Technology of Semiconductor Devices* (Wiley, New York 1967)

9.19 S.M. Sze: J. Appl. Phys. **38**, 2951 ((1967)

9.20 V.A. Dmitriev: MRS Internet J. **1** (1996) Article 29

9.21 H.C. Casey Jr., J. Muth. S. Krishnankutty, J.M. Zavada: Appl. Phys. Lett. **68**, 2867 (1996)

9.22 P. Perlin, M. Osinski, P. Eliseev, V.A. Smaglev, J. Miu, M. Banas, P. Sartori: Appl. Phys. Lett. **89**, 1680 (1996)

9.23 K.G. Zolina, V.E. Kudryashov, A.N. Turkin, A.E. Yunovich, S. Nakamura: MRS Internet J. **1** (1996) Article 11

9.24 Landolt-Börnstein: *Numerical Data and Functional Relationships in Science and Technology*, Vol.17, *Semiconductors* (Springer, Berlin, Heidelberg 1982)

9.25 V.W.L. Chin, T.L. Tansley, T. Osotchan: J. Appl. Phys. **75**, 7365 (1994)

9.26 K. Kim, W.R.L. Lambrecht, B. Segall: In *Gallium Nitide and Related Materials*, ed. by F.A. Ponce, R.D. Dupuis, S. Nakamura, J.A. Edmond. MRS Proc. **395**, 399 (Mater. Res. Soc., Pittsburgh, PA 1995); values agree well with experimental data from [9.24] and from J. H. Edgar: In *Properties of Group III Nitrides*, ed. by J.H. Edgar, EMIS Data Reviews Series (IEE, London 1994)

Chapter 10

10.1 A. Einstein: Physik Z. **18**, 121 (1917)

10.2 C. Kittel: *Elementary Statistical Physics* (Wiley, New York 1958) p.175

10.3   S. Wang: *Fundamentals of Semiconductor Theory and Device Physics* (Prentice Hall, Englewood Cliffs, NJ 1989)

10.4   P.Y. Yu, M. Cardona: *Fundamentals of Semiconductors* (Springer, Berlin, Heidelberg 1995)

10.5   H.C. Casey, M.B. Panish: *Heterostructure Lasers* (Academic, New York 1978)

10.6   J.I. Pankove: *Optical Processes in Semiconductors* (Prentice Hall, Englewood Cliffs, NJ 1971)

10.7   B. Monemar, J. P. Bergman, I.A. Buyanova: Optical characterization of GaN and related material, in *GaN and Related Material*, ed. by S.J. Pearton (Gordon and Breach, Amsterdam 1997) p.85

10.8   D.C. Reynolds, D.C. Look, W. Kim, Ö. Aktas, A. Botchkarev, A. Salvador, H. Morkoç, D.N. Talwar: J. Appl. Phys. **80**, 594 (1996)

10.9   M. Smith, G.D. Chen, J.I. Li, J.Y. Lin, H. Jiang, A. Salvador, B.N. Sverdlov, A. Botchkarev, H. Morkoç: Appl. Phys. Lett. **67**, 3387 (1995)

10.10 W. Shan, B.D. Little, A.J. Fischer, J.J. Song, B. Goldenberg, W.G. Perry, M.D. Bremser, R.F. Davis: Phys. Rev. B **54**, 16369 (1996)

10.11 R.J. Elliott, Phys. Rev. **108**, 1384, (1957)

10.12 J. J. Song, W. Shan, T. Schmit, X. H. Yang, A. Fischer, S. J. Hwang, B. Taheri, B. Goldenberg, R. Horning, A. Salvador, W. Kim, Ö. Aktas, A. Botchkarev, H. Morkoç, SPIE Proc. **2693**, 86 (1996)

10.13 M. Voos, R. F. Leheny, J. Shah: In *Handbook on Semi-Conductors: Optical Properties of Solids*, ed. by M. Balkanski (North Holland, Amsterdam 1986) Vol.2, p.340

10.14 B. Gil, S. Clur, O. Briot: Solid State Commun. **104**, 267 (1997)

10.15 B. Gil, O. Briot, R.L. Aulombard: Phys. Rev. B **52**, 17028 (1995)

10.16 K.P. Korona, A. Wysmolek, K. Pakula, R. Stepniewski, J.M. Baranowski, I. Grzegory, B. Lucznik, M. Wroblewski, S. Porowski: Appl. Phys. Lett. **69**, 788 (1996)

10.17 J.M. Baranowski, S. Porowski: *Proc. 23rd Int'l Conf. on Physics of Semiconductors*, Berlin 1996 (World Scientific, Singopore 1996) p.497

10.18 J.M. Baranowski, Z. Liliental-Weber, K. Korona, K. Pakula, R. Stepniewski, A. Wysmolek, I. Grzegory, G. Nowak, S. Porowski, B. Monemar, P. Bergman. *III-V Nitrides*, ed. by F.A. Ponce, T.D. Moustakas, I. Akasaki, B.A. Monomar. MRS Proc. **449**, 393 (Mater. Res. Soc., Pittsburgh, PA 1996)

10.19 W. Shan, R.J. Hauenstein, A.J. Fischer, J.J. Song, W.G. Perry, M.D. Bremser, R.F. Davis, B. Goldenberg: Phys. Rev. B **54**, 13460 (1996)

10.20 B. Monemar: Optical properties of GaN, in *Semiconductors and Semimetals* **50**, 305 (Academic, San Diego 1997)

10.21 B. Gil, F. Hamdani, H. Morkoç: Phys. Rev. B **54**, 7678 (1996)

10.22 G. Pikus: Sov. Phys. -JETP **14**, 1075 (1962)

10.23 E.L. Ivchenko, G.E. Pikus: *Superlattices and Other Heterostructures*, 2nd edn., Springer Ser. Solid-State Sci., Vol.110 (Springer, Berlin, Heidelberg 1997)

10.24 A. Gavini and M. Cardona, Phys. Rev. B **1**, 672 (1970)

10.25 G.L. Bir, G.E. Pikus: *Symmetry and Strain-Induced Effects in Semiconductors* (Wiley, New York 1974)

10.26 A. Polian, M. Grimsditch, I. Grzegory: J. Appl. Phys. **79**, 3343 (1996)

10.27 Y.P. Varshni: Physica **34**, 149 (1967)

10.28 B.K. Meyer: In *Free and Bound Excitons in GaN Epitaxial Films*, ed. by F.A. Ponce, T.D. Moustakas, I. Akasaki, B.A. Monemar. MRS Proc. **449**, 497 (1997)

10.29 D.C. Reynolds, D.C. Look, R. Jogai, V.M. Phanse, R.P. Vaudo: Solid State Commun. **103**, 533 (1997)

10.30 J.R. Haynes: Phys. Rev. Lett. **4**, 351 (1960)

10.31 C. Merz, M. Kunzer, U. Kaufmann, H. Amano, I. Akasaki: Semicond. Sci. Technol. **11**, 712 (1996)

10.32 G.W. Hooft, W.A.J.A. van der Poel, L.W. Molenkamp: Phys. Rev. B **35**, 8281 (1987)

10.33 W. Shan, X.C. Xie, J.J. Song, B. Goldenberg: Appl. Phys. Lett. **67**, 2512 (1997)

10.34 G.W. Hooft, W.A.J.A. Van der Poel, L.W. Molenkamp, C.T. Foxon: In *Excitons in Confined Systems*, ed. by R. Del Sole, A. D'Andrea, A. Lapiccirella. Springer Proc. Phys. **25**, 58 (Springer, Berlin, Heidelberg 1988)

10.35 D.G. Thomas, M. Gershenzon, F.A. Trumbore: Phys. Rev. **133**, A269 (1964)

10.36 D.C. Reynolds, K.K. Bajaj, C.W. Litton, D.E.B. Smith, P.W. Yu, W.T. Masselink, R. Fischer, H. Morkoç: Solid State Commun. **52**, 685 (1984)

10.37 M.A.L. Johnson, Z. Yu, C. Boney, W.C. Hughes, J.W. Cook Jr., J.F. Schetzina, H. Zao, B.J. Skromme, J.A. Edmond. MRS Proc. **449**, 271 (1997)

10.38 R. Dingle. M. Ilegems: Solid State Commun. **9**, 175 (1971)

10.39 S.N. Mohammad, H. Morkoç: Progr. Quant. Electron. **20**, 361-525 (1996)

10.40 P. Boguslawski, E.L. Briggs, J. Bernholc: Phys. Rev. B **51**, 17255 (1995)

10.41 D.W. Jenkins, J.D. Dow: Phys. Rev. B **39**, 3317 (1989); ibid. **39**, 3317 (1989)

10.42 J. Neugebauer, C.G. Van de Walle: Phys. Rev. Lett. **75**, 4452 (1995); Appl. Phys. Lett. **68**, 1829 (1996)

10.43 T.L. Tansley, R.J. Eagan: Phys. Rev. B **45**, 10942 (1992)

10.44 P. Boguslawski, E.L. Briggs, J. Bernholc: Phys. Rev. B **51**, 17255 (1995)

10.45 T.L. Tansley, C.P. Foley: J. Appl. Phys. **59**, 3241 (1986)

10.46 S.Y. Ren, J.D. Dow, J. Shen: Phys. Rev B **38**, 10677 (1988)

10.47 E.R. Glaser, T.A. Kennedy, K. Doverspike, L.B. Rowland, D.K. Gaskill, J.A. Freitas, M.A. Khan, D.T. Olson, J.N. Kuznia, D.K. Wickenden: Phys. Rev. B **51**, 13326 (1995)

10.48 S.N. Mohammad, A.E. Botchkarev, A. Salvador, W. Kim, O. Aktas, H. Morkoç: Philos. Mag. B **76**, 131 (1997)

10.49 T.A. Kennedy, E.R. Glaser, J.A. Freitas, W.E. Carlos, M.A. Khan, D.K. Wickenden: J. Electron. Mater. **24**, 219 (1995)

10.50 D.M. Hoffmann, D. Kovalev, G. Steude, B.K. Meyer, A. Hoffman, L. Eckey, R. Heitz, T. Detchprom, H. Amano, I. Akasaki: Phys. Rev. B **52**, 16702 (1995)

10.51 D.C. Reynolds, D.C. Look, B. Jogai, H. Morkoç: Solid State Commun. **101**, 643 (1997)

10.52 M. Smith, G.D. Chen, J.Y. Lin, H.X. Jiang, A. Salvador, W. Kim, Ö. Aktas, A. Botchkarev, H. Morkoç, B. Goldenberg: Appl. Phys. Lett. **68**, 1883, (1996)

10.53 M. Smith, J.Y. Lin, H.X. Jiang, A. Salvador, A. Botchkarev, H. Morkoç: Appl. Phys. Lett. **69**, 2453 (1996)

10.54 K.C. Zeng, J.Y. Lin, H.X. Jiang, A. Salvador, G. Popovici, H. Tang, W. Kim, H. Morkoç: Appl. Phys. Lett. **71**, 1368 (1997)

10.55 S. Ruvimov, Z. Liliental-Weber, T. Suski, J.W. Ager III, J. Wasburn, J. Krueger, C. Kisielowski, E.R. Weber, H. Amano, I. Akasaki: Appl. Phys. Lett. **67**, 990 (1996)

10.56 S. Nakamura, M. Senoh, S. Nagahama, N. Iwasa, T. Yamada, T. Matsushita, Y. Sugimoto, H. Kiyoko: Appl. Phys. Lett. **70**, 1417 (1997)

10.57 D.L. Smith: Solid State Commun. **57**, 919 (1986)

10.58 D.L. Smith, C. Mailhiot: J. Appl. Phys. **63**, 2717 (1988)

10.59 A. Bykhovski, B. Gelmont, M. Shur: J. Appl. Phys. **74**, 6734 (1993)

10.60 Y. Koike, S. Yamasaki, S. Nagai, N. Koide, S. Asami, H. Amano, I. Akasaki: Appl. Phys. Lett. **68**, 1403 (1996)

10.61 S. Nakamura, T. Mukai, M. Senoh, S.I. Nagahama, N. Iwasa: J. Appl. Phys. **74**, 3911 (1993)

10.62 H. Amano, T. Takeuchi, S. Sota, H. Sakai, I. Akasaki: In *Free and Bound Excitons in GaN Epitaxial Films*, ed. by F.A. Ponce, T.D. Moustakas, I. Akasaki, B. Monomar. MRS Proc. **449**, 1143 (1997)

Chapter 11

11.1 D.C. Reynolds, C.W. Litton, R.A. Collins: Phys. Status Solidi **9**, 645 (1965)

11.2 T.P. Lee, A.G. Dentai: IEEE J. QE-**14**, 150 (1978)

11.3 S. Wang: *Fundamental of Semiconductor Theory and Device Physics* (Prentice Hall, Englewoods Cliffs, NJ 1989)

11.4 D. Evans, M. Hodapp, H. Sorensen, K. Jamison, B. Krause: *Optoelectronics/Fiber-Optics Applications Manual*, 2nd edn. (McGraw-Hill, New York 1981) Chap.2

11.5 See, for example, A.A. Berg, P.J. Dean: *Light-Emitting Diodes* (Clarendon, Oxford 1976) pp. 11-35

11.6 H. Morkoç, S.N. Mohammad: Sci. Magazine **267**, 51 (1995)

11.7 D.L. MacAdam: *Color Measurement*, 2nd edn., Springer Ser. Opt. Sci., Vol.27 (Springer, Berlin, Heidelberg 1985)

11.8 M.G. Craford: IEEE Circuits & Devices Mag. **8**, 24 (1992)

11.9 R.M. Bhargava: Optoelectronics - Devices and Technol. **7**, 19 (1992)

11.10 J.I. Pankove, E.I. Miller, J.E. Berkeyheiser: RCA Rev. **32**, 383 (1971)

11.11 H. Amano, M. Kito, K. Hiramatsu, I. Akasaki: Jpn. J. Appl. Phys. **28**, L2112 (1989)

11.12 S. Nakamura, T. Mukai, M. Senoh: Jpn. J. Appl. Phys. **30**, L1998 (1991)

11.13 H. Amano, M. Kito, K. Hiramatsu, I. Akasaki: Jpn. J. Appl. Phys. **28**, L2112 (1989)

11.14 I. Akasaki, H. Amano: J. Electrochem. Soc. **141**, 2266 (1994)

11.15 H. Amano, K. Hiramatsu, I. Akasaki: Jpn. J. Appl. Phys. **27**, L1384 (1988)

11.16 S. Nakamura, M. Senoh, N. Isawa, S. Nagahama: Jpn. J. Appl. Phys. **34**, L797 (1995)

11.17 J.I. Pankove, J.E. Berkeyheiser, H.P. Maruska, J. Wittke: Solid State Commun. **8**, 1051 (1970)

11.18 P. Perlin, V. Iota, B.A. Weinstein, P. Wisniewski, T. Suski, M. Osinski, P. Eliseev: Appl. Phys. Lett. **70**, 2993 (1997)

11.19 P. Perlin, M. Osinski, P.G. Eliseev: *III-V Nitrides*, ed. by F.A. Ponce, T.D. Moustakas, I. Akasaki, B.A. Monomar. MRS Proc. **449**, 1173 (Mater. Res. Soc., Pittsburgh, PA 1997)

11.20 S.D. Lester, F.A. Ponce, M.G. Craford, D.A. Steigerwald: Appl. Phys. Lett. **66**, 1249 (1995)

11.21 K. Osamura, S. Naka, Y. Murakami: J. Appl. Phys. **46**, 3422 (1975)

11.22 S. Nakamura, M. Senoh, S. Nagahama, N. Iwasa, T. Yamada, T. Matsushita, H. Kiyoku, Y. Sugimoto: Jpn. J. Appl. Phys. **35**, L74 (1996)

11.23 K.G. Zolina, V.E. Kudryashov, A.N. Turkin, A.E. Yunovich, S. Nakamura: MRS Internet J. Nitride Semicond. Res. **1**, 11 (1996)

11.24 S. Chichibu, T. Azuhata, T. Sota, S. Nakamura: Appl. Phys. Letts. **69**, 4188 (1996)

11.25 Y. Narukawa, Y. Kawakami, M. Funata, S. Fujita, S. Fujita, S. Nakamura: Appl. Phys. Lett. **70**, 981 (1997)

11.26 P. Perlin, M. Osinski, P. Eliseev, V. A. Smaglev, J. Miu, M. Banas, P. Sarton: Appl. Phys. Lett. **89**, 1680 (1996)

11.27 H.C. Casey Jr., J. Muth, S. Krishnankutty, J.M. Zavada: Appl. Phys. Lett. **68**, 2867 (1996)

11.28 V.A. Dimitriev, K.G. Irvine, J.A. Edmond, C.H. Carter Jr., N.I. Kuznetsov, A.S. Zubrilov, E.V. Kalinina, D.V. Tsvetkov: *Gallium Nitride and Related Materials*, ed. by F.A. Ponce, R.D. Dupuis, S. Nakamura, J.A. Edmond. MRS Proc. **395**, 295 (Mater. Res. Soc., Pittsburgh, PA 1996)

11.29 M. Osinski, J. Zeller, P-C Chiu, S. Phillips: Appl. Phys. Lett. **69**, 898 (1996)

11.30 D. Barton, J. Zeller, B. Phillips, P. Chiu, S. Askar, D. Lee, M. Osinski, K. Malloy: Proc. 33rd Annual IEEE Int'l on Reliability. Phys. Symp. Proc. **33**, 191-199 (IEEE, New York 1995)

11.31 D. Barton, M. Osinski: Priv. commun. (1997)

11.32 J. Baur, P. Schlotter, J. Schneider: *Festkörperprobleme/Advances in Solid State Physics* **37**, 67 (Vieweg, Braunschweig 1997)

11.33 C.W. Tang, S.A. Van Slyke: Appl. Phys. Lett. **51**, 913 (1987)

11.34 J. Kido, M. Kimura, K. Nagai: Sci. Magazine **267**, 1332 (1995)

Chapter 12

12.1 S. Strite: Priv. commun. (1991)

12.2 P. Asthana: IEEE Spectrum **31**, 60 (1994)

12.3 A. Bell: Sci. Am. **275** 42-46 (July 1996)

12.4 H. Morkoç, B. Sverdlov, G.B. Gao: Proc. IEEE **81**, 492-556 (1993)

12.5 M.G.A. Bernard, G. Duraffourg: Phys. Status Solidi **1**, 699 (1961)

12.6 D. Marcus: *Light Transmission Optics* (Van Nostrand-Reinhold, New York 1972)

12.7 H.C. Casey Jr., M.B. Panish: *Heterostructure Lasers*, Pt.A: *Fundamentals Principles* (Academic, New York 1978)

12.8 H. Kogelnik: In *Guided-Wave Optoelectronics*, 2nd edn., ed. by T. Tamir, Springer Ser. Electron. Photon., Vol.26 (Springer, Berlin, Heidelberg 1990)

12.9 H.K.V. Lotsch: Optik **32**, 116, 189, 299, 553 (1970/71)

12.10 K.H. Schlereth, M. Tacke: IEEE J. QE-**26**, 627-630 (1990)

12.11 I. Skinner, R. Shail, B.L. Weiss: IEEE J. QE-**25**, 6-11 (1989)

12.12 G.M. Alman, L.A. Melter, M. Dutta: IEEE J. QE-**28**, 650-657 (1992)

12.13 H. Kressel, J. Butler: Heterojunction laser diode. *Semiconductors and Semimetals* **14**, 65-194 (Academic, New York 1979)

12.14 H. Kressel, J. Butler: *Semiconductor Lasers Heterojunction LEDs* (Academic, New York 1977)

12.15 H. Kressel (ed.): *Semiconductor Devices for Optical Communication*, 2nd edn., Topics Appl. Phys., Vol.39 (Springer, Berlin, Heidelberg 1982)

12.16 M.I. Nathan, A.B. Fowler, G. Burns: Phys. Rev. Lett. **11**, 152 (1963)

12.17 C. Weisbuch, B. Vinter: *Quantum Semiconductor Structures* (Academic, New York 1991)

12.18 T. Honda, A. Katsube, T. Sakaguchi, F. Koyama, K. Iga: Jpn. J. Appl. Phys. **34** Pt. 1, 3527 (1995)

12.19 W.J. Fan, M.F. Li, T.C. Chong, J.B. Xia.: Solid State Commun. **98**, 737 (1996)

12.20 M. Suzuki, T. Uenoyama: Jpn. J. Appl. Phys. **35**, 543 (1996)

12.21 E.A. Albanesi, W.R.L. Lambrecht, B. Segall: J. Vac. Sci. Technol. B **12**, 2470 (1994)

12.22 M. Suzuki, T. Uenoyama: Electronic and optical properties of GaN based quantum wells, in *Group III Nitride Semiconductor Compounds: Physics and Applications*, ed. by B. Gil (Clarendon, London 1998)

12.23 A. Grindt, S.W. Koch, W.W. Chow: Appl. Phys. A **66**, 1 (1998)

12.24 W.W. Chow, A.F. Wright, A. Grind, F. Jahnke, S.W. Koch: Appl. Phys. Lett. **71**, 2608 (1997)

12.25 C.F. Hsu, P.S. Zory, C.H. Wu, M.A. Emanuel: IEEE J. Selected Topics in Quantum Electronics on Semiconductor Lasers **13**, 158 (1997)

12.26 J.I. Pankove: *Optical Processes in Semiconductors* (Prentice Hall, Englewood Cliffs, NJ 1971)

12.27 W.W. Chow, S.W. Koch, M. Sargent III: *Semiconductor Laser Physics* (Springer, Berlin, Heidelberg 1994)

12.28 C.F. Hsu, P.S. Zory, P. Rees, M. Haase: Coulomb enhancement in CdZnSe single quantum well lasers. SPIE Proc. **3001**, 271 (1997)

12.29 H. Haug, S.W. Koch: *Quantum Theory of Optical and Electronic Properties of Semiconductors* (World Scientific, Singapore 1990)

12.30 M. Asada: In *Quantum Well Lasers*, ed. by P.S. Zory (Academic, San Diego 1993) Chap. 2

12.31 M. Suzuki, T. Uenoyama: Optical gain and crystal symmetry in III-V nitride lasers. Appl. Phys. Lett. **69**, 3378 (1996)

12.32 M. Suzuki, T. Uenoyama: Jpn. J. Appl. Phys. **35**, 1420 (1996)

12.33 A.T. Meney, E.P. O'Reilly: Theory of optical gain in ideal GaN heterostructure lasers. Appl. Phys. Lett. **67**, 3013 (1995)

12.34 B. Wimmer, H. Krakauer, M. Weinert, A.J. Freeman: Phys. Rev. B **24**, 864 (1981)

12.35 S.L. Chuang: IEEE J. QE-**32**, 1791 (1996)

12.36 W.W. Chow, A.F. Wright, J.S. Nelson: Appl. Phys. Lett. **68**, 296 (1996)

12.37 P. Rees, S. Cooper, P.M. Smowton, P. Blood, J. Hegarty: IEEE Photon. Techn. Lett. **8**, 197 (1996)

12.38 W.J. Fan, M.F. Li, T.C. Chong, J.B. Xia: J. Appl. Phys. **80**, 3471 (1996)

12.39 D. Ahn: J. Appl. Phys. **76**, 8206 (1994)

12.40 J. Ding, M. Hagerott, T, Ishihara, H. Jeon, A.V. Nurmikko: Phys. Rev. B **47**, 10528 (1993)

12.41 T. Uenoyama: Phys. Rev. B **51**, 10228 (1995)

12.42 R. Cingolani: Priv. commun. (1997)

12.43 J. Petalas, S. Logothetidis, S. Boultadakis, M. Alouani, J.M. Wills: Phys. Rev. B **52**, 8082 (1995)

12.44 A. Diessel, W. Ebeling, J. Gutowski, B, Jobst, K. Schöll, D. Hommel, K. Henneberger: Phys. Rev. B **52**, 4736 (1995)

12.45 R. Cingolani, R. Rinaldi, L. Calcagnile, P. Prete, P. Sciacovelli, L. Tapfer, L. Vanzetti, G. Mula, F. Bassani, L. Sorba, A. Franciosi: Phys. Rev. B **49**, 16769 (1994)

12.46 R. Cingoloni, L. Calcagnile, G. Coli, D. Greco, R. Rinaldi, M. Lomascolo, M. DiDio, A. Franciosi, L. Vancetti, G.C. LaRocca, D. Camp: J. Opt. Soc. Am. B **13**, 1268 (1996)

12.47 G. Frankowsky, F. Steuber, V. Härle, F. Scholz, A. Hangleiter: Appl. Phys. Lett. **68**, 46 (1996)

12.48 R. Dingle, K.L. Shaklee, R.F. Leheny, R.B. Zetterstrom: Appl. Phys. Lett. **19**, 5 (1971)

12.49 R. Cingolani, G. Coli, R. Rinaldi, L. Calcagnila, H. Tang, A. Botchkarev, W. Kim, H. Morkoç: Phys. Rev. B **56**, 1491 (1997)

12.50 M. Kuball, E.-S. Jeon, Y.-Song, A.V. Nurmikko, P. Kozodoy, A, Abare, S.Keller, L.A. Coldren, U.K. Mishra, S.P. DenBaars, D.A. Steigerwald: Appl. Phys. Lett. **70**, 2580 (1997)

12.51 S. Nakamura, M. Seroh, N. Nagahama, N. Zwara, T.H. Umenoto, S. Sano, K. Chocho: Jpn. J. Appl. Phys. **38**, L1568-1571 (1997)

12.52 J.W. Bae, G. Shtengel, D, Kuksenkov, H. Temkin: Appl. Phys. Lett. **66**, 2031 (1995)

12.53 R. Olshansky, P. Hill, V. Lanzisera, W. Powazinik: IEEE J. QE-**23**, 1410 (1987)

12.54 R. Olshansky, C.B. Su, J. Manning, W. Powazinik: IEEE J. QE-**20**, 838 (1984)

12.55 R. Nagarajan, T. Kamiya, A. Krobe: IEEE J. QE-**25**, 1410 (1989)

12.56 B.W. Hakki, T.L. Paoli: J. Appl. Phys. **46**, 1299 (1975): and J. Appl. Phys. **44**, 4113 (1973)

12.57 V. Kozlov, A. Salokatve, A, Nurmikko, D.C. Grillo, Li He, J. Han, Y. Fan, M. Ringle, R.L. Gunshor: Appl. Phys. Lett. **65**, 1863 (1994)

12.58 G.E. Bulman, K. Doverspike, S.T. Sheppard, T.W. Weeks, M. Leonard, H. S. Kong, H. Dieringer, C. Carter, J. Edmond, J.D. Brown, J.T. Swindell, J.F. Schetzina: Electron. Lett. **33**, 1556 (1997)

12.59 Y. Narukawa, Y. Kawakami, M. Funato, S. Fujita, S. Fujita, S. Nakamura: Appl. Phys. Lett. **70**, 981 (1996)

12.60 H. Amano, N. Watanabe, M. Koike, I. Akasaki: Jpn. J. Appl. Phys. **32**, L1000 (1993)

12.61 T.J. Schmidt, X.H. Yang, W. Shan, J.J. Song, A. Salvador, W. Kim, Ö. Aktas, A. Botchkarev, H. Morkoç: Appl. Phys. Lett. **68**, 1820 (1996)

12.62 S. Nakamura, M. Senoh, S. Nagahama, N. Iwasa, T. Yamada, T. Matsushita, H. Kiyoku, Y. Sugimoto: Jpn. J. Appl. Phys. **35**, L74 (1996)

12.63 S. Nakamura, M. Senoh, S. Nagahama, N. Iwasa, T. Yamada, T. Matsushita, Y. Sugimoto, H. Kiyoku: Appl. Phys. Lett. **69**, 1568 (1996)

12.64 I. Akasaki, S. Sota, H. Sakai, T. Tanaka, H. Amano: Electron. Lett. **32**, 1105 (1996)

12.65 S. Nakamura, M. Senoh, S. Nagahama, N. Iwasa, T. Yamada, T. Matsushita, H. Kiyoku, Y. Sugimoto: Jpn. J. Appl. Phys. **35**, L217 (1996)

12.66 A. Kuramata, K. Horino, K. Domen, K. Shinohora, T. Tanahashi: Appl. Phys. Lett. **67**, 2521 (1995)

12.67 S. Nakamura, M. Senoh, S. Nagahama, N. Iwasa, T. Yamada, T. Matsushita, H. Kiyoku, Y. Sugimoto: Appl. Phys. Lett. **68**, 2105 (1996)

Springer Series in
MATERIALS SCIENCE

Editors: R. Hull · R. M. Osgood, Jr. · H. Sakaki · A. Zunger

* The 2nd edition is available as a textbook with the title: *Laser Processing and Chemistry*

Printing: Saladruck, Berlin
Binding: H. Stürtz AG, Würzburg